KAUFFMAN GOLD MEDAL

The Kauffman Gold Medal is awarded to a person who . . . has made a significant contribution toward the most outstanding achievement in the science of geophysical exploration during the previous year, or recognized as such during the previous year A fund has been established by Virgil Kauffman, former president of Aero Service Corporation, to encourage improvement in the science of geophysical exploration.

— SEG Yearbook

The Kauffman Medal was awarded in 1969 for the predecessor of this dictionary.

DEDICATION

This work is dedicated to the memory of Neal Smith (1912-1971), to whom we are all indebted professionally and to whom many of us are indebted personally as well. Neal Smith was a thoroughly professional geophysicist of the greatest integrity, with a high sense of responsibility to his company, to his profession, and to nature.

Instrumental in the creation of this work are some strong beliefs which Neal held and which I share: that each of us owes to our profession the obligation to share his knowledge with others, that each owes to him-self continual learning, and that each has an obligation to go beyond the mere requirements of his job. Neal was instrumental in offering my original Glossary to SEG so that our whole profession might benefit instead of just our own company geophysicists; he encouraged me to keep it current and to expand it into related fields; and he encouraged the assumption of other responsibilities with the SEG, especially (in my own case) along the educational line.

Encyclopedic Dictionary

of

Exploration Geophysics

Robert E. Sheriff
Geophysical Division
Chevron Oil Company

Society of Exploration Geophysicists
Tulsa, Oklahoma

Published by
The Society of Exploration Geophysicists
P.O. Box 3098, Tulsa, Oklahoma 74101

Second Printing 1974

Third Printing 1976

Fourth Printing 1980

Fifth Printing 1980

Sixth Printing 1981

Seventh Printing 1981

Eighth Printing 1982

Printed in the United States of America

Library of Congress Catalog Card No. 72-94565

ISBN 0-931830-03-6

Table of Contents

LIST OF TABLES

PREFACE

The "Glossary of Terms Used in Geophysical Exploration" which was published in Geophysics in February, 1968 occasioned many expressions of appreciation for my attempt to help geophysical understanding and communication. Improvements and additions were suggested, including expansion into related fields. These suggestions resulted in the publication in April, 1969 of an "Addendum" (with expanded coverage of mining geophysics) and the publication in December, 1970 of the "Glossary of Terms Used in Well Logging."

The Glossary developed greater use as a reference than was anticipated. One well-received feature of the original glossary was the "extended entry," wherein a definition was expanded to clarify implications and relate it to other concepts. The expanded entries provided convenient access to basic equations and facts which are only used occasionally. Mr. Milton Dobrin observed that this work has developed into more than a mere glossary and suggested the name, "Encyclopedic Dictionary of Exploration Geophysics."

The following conventions have been used:

(1) Entries begin with the essence of a definition, though occasionally a discussion is given first to provide a frame of reference in which the concept will have more meaning. Restrictions on meanings sometimes are contained in a discussion which follows a definition rather than incorporated into the definition itself. Terms indicated as being synonomous are often used interchangeably even though they may not be identical in all respects. This is especially true of well logs which are based on nearly the same principles but differ in details.

(2) Boldface type in the definition refers the reader to another entry which supplements the meaning. Cross-references are shown only where needed to complete the meaning. One wishing to explore a subject in detail may wish to look up other entries in addition to those in boldface type.

(3) The numbering within an entry indicates different meanings, though the sequence does not indicate preference as to usage. Where the different meanings are contradictory (as they are in some instances) this is pointed out, and, in some cases, the "preferred" usage is indicated and an alternative word suggested to avoid ambiguity. Letters subdivide an entry without implying differences in meaning. Only the specialized meanings used in geophysics are included.

(4) Trade names are included where they are in general use. Where such entries are used for a class of devices, the entry begins

with a lower-case letter even though the special name for the device may begin with a capital. Trademark ownership is shown where I am aware of same. However, neither inclusion nor exclusion should imply judgments about the merits of any devices. Additional trade names and company names are in appendices.

(5) References cited suggest a place to begin looking for further information about an entry. I usually cite only a single reference, which is (hopefully) readily available, to provide the starting point for further looking. Thus, the citation of a reference does not imply the original source nor that it is the most complete or current reference.

(6) Figures have been kept simple to illustrate the terminology and the important features of concepts without attempting to make them realistic or to illustrate completely all the features of devices.

The Encyclopedic Dictionary of Exploration Geophysics was compiled for "practical" geophysicists rather than for researchers or other specialists. The novice surveyor can find the reason for surveying for a "walkaway." Common geologic terms are included because the end objective of most geophysical work is a geologic picture, though geologic time has been relegated to an appendix. Some "popular" information like sea states and earthquake intensity scales is included. The man who's only a seismologist may find that his specialized vocabulary is also used in electrical exploration, and, hopefully, he'll realize that geophysics includes many disciplines. I hope that browsing in this work may show the broad scope of geophysics.

While I believe quantities ought to be expressed in the metric mks units which surely will be used more in the future, most geophysicists use a hybrid of English units and cgs units. Therefore, terms are expressed in both systems because readers need both to relate terms in their own reference frames and also to think of the metric future.

The eclectic nature of this work inevitably involves inconsistencies in deciding which entries and how much information to include. The most probable reason why a particular entry may be missing is that I didn't happen to think of it. While I would have preferred to have a more balanced and complete work, this dictionary seems open-ended as if it could never be complete. To have continued to refine it would only have delayed its availability. Therefore, despite faults, we have proceeded to press. It will be appreciated if readers who find errors or omissions or with other suggestions will inform me.

I do not claim to have invented the terms included here but merely to give the meanings actually in geophysical usage. Numerous references have been consulted to assure reasonable conformity and sometimes this results in repeating someone else's phraseology

Among the works I have used extensively are the following:

The Random House Dictionary of the English Language, Stein and Urdang, editors, 1966, Random House, New York.

1961 IRE Dictionary

Encyclopedia of Oceanography, Fairbridge, editor, 1966, Reinhold, New York.

Cooke and Mahkus, *Electronics and Nucleonics Dictionary,* 1960, McGraw-Hill Book Co. Inc., New York.

International Tectonic Dictionary, Dennis, editor, 1967, AAPG, Tulsa.

Automatic Data Processing Glossary, U.S. Bureau of the Budget; reprinted by Datamation Magazine, New York.

Glossary of Geology and Related Sciences, 1960, American Geological Institute, Washington.

The standard textbooks of Geophysics (which are mostly very out-of-date):

Anstey, Evenden, and Stone, *Seismic Prospecting Instruments,* Vols. 1 and 2, 1970-71, Gebrüder Borntraeger, Stuttgart.

Dix, *Seismic Prospecting for Oil,* 1952, Harper Bros., New York.

Dobrin, *Introduction to Geophysical Prospecting,* 1960, McGraw-Hill Book Co., Inc., New York.

Eve and Keys, *Applied Geophysics,* 1956, University Press, Cambridge.

Grant and West, *Interpretation Theory in Applied Geophysics,* 1965, McGraw-Hill Book Co., Inc., New York.

Heiland, *Geophysical Exploration,* 1968, Hafner Publishing Co., New York.

Howell, *Introduction to Geophysics,* 1959, McGraw-Hill Book Co., Inc., New York.

Jakosky, *Exploration Geophysics,* 1961, Trija, Newport Beach.

Nettleton, *Geophysical Prospecting for Oil,* 1940, McGraw-Hill Book Co., Inc., New York.

Richter, *Elementary Seismology,* 1958, Freeman, Cooper & Co., San Francisco.

White, *Seismic Waves,* 1965, McGraw-Hill Book Co., Inc., New York.

The publications of the Society of Exploration Geophysicsts and the European Association of Exploration Geophysicists:

Geophysics, the bimonthly journal of the SEG.

Geophysical Prospecting, the quarterly journal of the EAEG.

Mining Geophysics, Vols. 1 and 2, 1967, SEG, Tulsa.

Nettleton, *Elementary Gravity and Magnetics for Geologists and Seismologists*, 1971, SEG, Tulsa.

Seismic Filtering, Van Nostrand, editor, 1971, SEG, Tulsa.

Seismic Refraction Prospecting, Musgrave, editor, 1967, SEG, Tulsa.

Many people helped in this compilation, including many whom I have never met and some whose names I do not even know. I thank all for their help even though I do not cite them specifically. I wish to express appreciation for the following specific contributions:

Mr. Milton Dobrin, Chairman of the former SEG Glossary Committee, and Messrs. John W. Allingham, C. M. Clark, C. H. Dix, L. W. Gardner, George V Keller, James A. Long, G. S. MacKenzie, F. A. Van Melle, Matthew Slavin III, and Isidore Zietz, the reviewers of the previous glossary.

Mr. Harold Swanson, who helped with the mining terms in the April 1969 Addendum.

Messrs. G. W. Burns, E. E. Finklea, Noel Frost, D. W. Hilchie, John I. Myung, R. A. Robertson, F. Segesman, Terry Walker, and J. E. Walstrom, who helped with the *Glossary of Terms Used in Well Logging*.

Mr. John S. Sumner who allowed me to extract from his *Glossary of Resistivity and Induced Polarization*.

Mr. T. R. LaFehr, editor of the present manuscript, and his reviewers: Messrs. R. J. Bean, Halbert H. Bybee, R. S. Carmichael, Milton Dobrin, Elmer Eisner, James N. Galbraith, Lloyd P. Geldart, John Haun, H. Robert McCormack, A. W. Musgrave, L. L. Nettleton, George R. Pickett, and Sulhi H. Yungul. Also Jerry W. Henry of the SEG staff.

The Standard Oil Company of California and especially the Chevron Oil Company, Geophysical Division, for permission to publish.

I thank Mrs. Nina Everhart who typed this manuscript several times, whose contribution went beyond just doing her job well; Mrs. Pat Gage and Mrs. Nora Potter who drafted the illustrations; and my daughter and my wife who read for spelling, grammatical, and other errors.

ROBERT E. SHERIFF
Houston, 1972

A

ABC method: A weathering computation method based on refracted arrivals whereby the effect of irregular weathering thickness is determined from shots above the base of the weathering. Shots from *A* and *C* (Figure 1) are recorded at *C, A*, and at *B* (between *A* and *C*). The surface-to-surface times are t_{AC}, t_{AB}, and t_{BC}, given by adding the uphole times to the recorded first-break times. The weathering time below *B* is:

$$t_w = (t_{AB} + t_{BC} - t_{AC})/2.$$

This value is sometimes multiplied by the "K-factor" to give the vertical weathering time:

$$t_v = Kt_w = t_w V_1 (V_1^2 - V_0^2)^{-1/2}$$

AB electrodes: Current electrodes. In resistivity surveying and resistivity well logging, current is passed between the *A* and the *B* electrodes while voltage is measured between the *M* and *N* electrodes.

abnormal events: Coherent events which are not reflections. Refractions, reflected refractions, diffractions, surface waves, and sometimes multiple reflections are included (though there's nothing "abnormal" about any of these).

abnormally high pressure: Pressure of formation fluids which appreciably exceeds the hydrostatic head. Hydrostatic pressure is about 0.465 psi/ft (1.05×10^4 N/m^3, equivalent to 9.2 lb/gallon mud); geostatic or lithostatic pressure (that which would exist if the weight of the section were carried by the fluid) is about 1 psi/ft (2.26×10^4 N/m^3 or 20 lb/gallon mud). Pressures greater than about 0.7 psi/ft are considered "abnormal" and are associated with low seismic velocity and low density.

AB rectangular array: Gradient **array.**

absorption: A process whereby some of the energy of a seismic wave is converted into heat while passing through a medium. Absorption for seismic waves is perhaps of the order of 1/2 db/cycle. Sometimes claimed to be proportional to the frequency squared rather than linear with frequency. See **Q.**

ac-bias recording: A modulation system in which the modulating message biases a constant-amplitude carrier. See Figure 122 (**modulation**). Used in magnetic tape recording to avoid hysteresis distortion. The half-wave rectified portion is sometimes used as the information carrier.

acceleration of gravity: See **gravitational constant.**

accelerometer: 1. A transducer whose output is proportional to acceleration. Accelerometers are used with the shipboard gravimeter and in inertial navigation systems. 2. A seismometer or **geophone** whose output is proportional to acceleration. For example, a moving coil geophone with a response proportional to frequency (as may be the case below the natural frequency) operates as an accelerometer.

acceptance criteria: Criteria which data must satsify in order not to be rejected.

acceptor: p-type semiconductor.

access time: The time it takes a computer to locate data or an instruction in its memory and to transfer it to its arithmetic unit where computations are to be performed, or the time to transfer it back to the proper location in the memory.

ac coupling: See **coupling.**

accumulate error: To have a systematic cause for errors, so that errors in successive observations tend to be of the same sign. Many small errors with little tendency to cancel may add up to a large cumulative error.

accuracy: 1. The degree of freedom from error; the total error compared to the true value. Compare **precision, uncertainty,** and **sensitivity.** 2. The capability of an instrument to follow a true value. Inaccuracy is the departure from the true value due to any instrument error such as lack of repeatability, drift, temperature effects, or other causes.

aclinic line: Magnetic equator.

acoustic: Sonic; pertaining to sound. Usually refers to compressional P waves, sometimes restricted to P-waves in fluids (liquids and gases), sometimes generalized to include other elastic wave types.

acoustic coupler: A device for connecting a computer terminal (such as a teletype unit) to a telephone.

acoustic impedance: Seismic velocity multiplied by density. **Reflection coefficient** depends on changes in

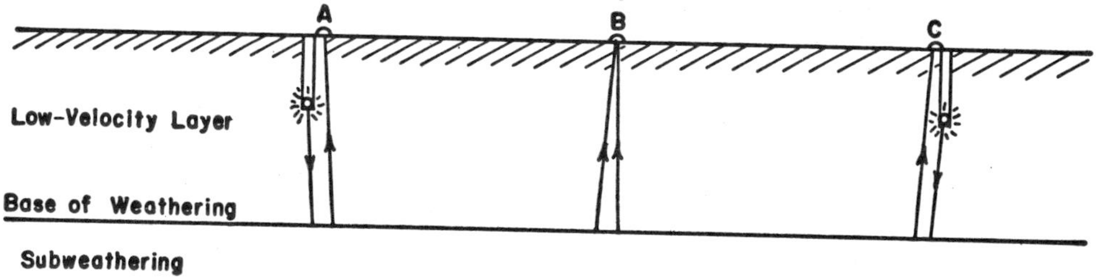

Fig. 1. **ABC weathering method.** Weathering time below *B* is $(T_{AB} + T_{BC} - T_{AC})/2$, where T_{AB} = surface-to-surface time, from *A* to *B*, etc.

acoustic impedance.

acoustic log: 1. A generic term for well logs which display any of several aspects of **acoustic-wave** propagation. In some acoustic logs (**sonic log,** continuous-velocity log), the traveltime of the compressional wave between two points is measured. In others (**amplitude log**), the amplitude of part of the wave train is measured. Other acoustic logs (**character log, three-D log,** VDL-log, microseismogram log, signature log) display part of the wave train in wiggle or variable-density form. Still others (**cement-bond log, fracture log**) are characterized by the objective of the measurements rather than their form. **Borehole televiewer** is also an acoustic log. 2. Specifically, a **sonic log.**

acoustic wave: 1. Sonic wave. An elastic wave train, sometimes restricted to propagation through a fluid. 2. The wave train generated and detected by a sonic-logging sonde (see **acoustic log**). The wave train (Figure 2) is a composite of various modes of energy transfer. The first arrival usually results from compressional (P- or longitudinal) waves traveling in the formation; the inverse of its velocity is measured by the **sonic log.** A second arrival is sometimes identified as shear (S-) wave travel in the formation; it represents a pseudo-Rayleigh wave which travels at approximately the velocity of S-waves. Compressional waves traveling through the mud usually have relatively high frequency content; they are sometimes called *fluid waves.* One or more modes of high-amplitude, low-frequency tube waves (sometimes called *Stoneley waves*) are usually very distinct arrivals. 3. More generally, an elastic wave or **seismic wave.**

acoustilog: Acoustic-velocity log or **sonic log.**

activation logging: A well-logging technique in which the formation is irradiated with neutrons that transmute some nuclei into radioisotopes whose radiation is measured after a short time interval.

active: 1. A system or circuit which includes an energy source. Often pertains to electronic elements such as amplifiers and filters. Compare **passive.** 2. A **positioning** system which involves transmission from the mobile station which is to be located.

activity (electrochemical): The relative tendency of a substance to enter into a reaction; the equivalent concentration of dissolved salts. When the shales adjacent to a reservoir act as perfect cationic membranes and the permeable bed is clean, the **electrochemical SP** can be found from the activities of the formation water and mud filtrate.

A/D: Analog to digital.

adder: A device whose output represents the sum of its inputs. See **half adder.**

address: A label that identifies a specific location in a computer memory where certain information is stored.

adiabatic: Involving no gain or loss of heat.

adjoint: The adjoint or adjugate of a **matrix** is the inverse matrix times the determinant.

admittance: 1. The reciprocal of **impedance;** the complex ratio of current to voltage in a linear circuit. Unit of measure is the mho. 2. The admittance S of a section of horizontal layers of thickness h_i and resistivity ρ_i overlying a basement of very high resistivity is:

$$S = \Sigma (h_i/\rho_i).$$

See also **S rule.** 3. The reciprocal of **acoustic impedance.** 4. Admittance is used in mechanical situations where an analogy is made to an electrical circuit. Thus one might refer to the "admittance of the ground" when discussing the transfer of energy from a source into seismic wave energy.

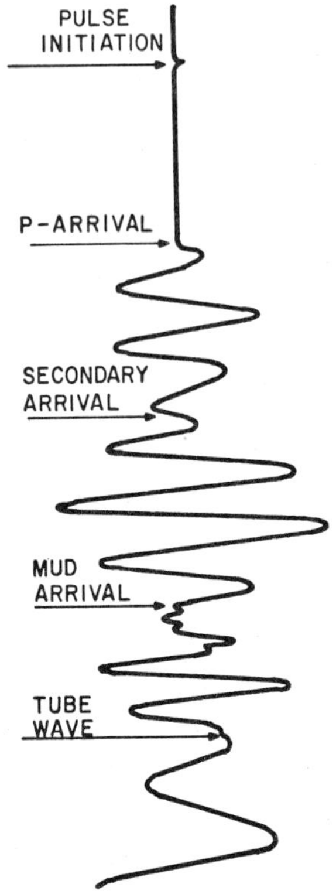

Fig. 2. **Acoustic wavetrain** in a borehole (idealized).

adsorption: An electrostatic chemical process in which a thin layer of molecules becomes fixed to the outer surface of a solid. See **fixed layer.**

AEM: Airborne electromagnetics.

aeolotropy: Anisotropy.

aerated layer: Weathering or near-surface low-velocity layer.

aeromagnetic: Magnetic measurements made from an aircraft.

Afmag method: Audio-frequency magnetic technique; the use of natural electromagnetic noise in the audio-frequency range to study lateral changes in earth resistivity. The quantities measured are the azimuth and inclination or dip of the major axes of the integration ellipsoid of polarization. See **polarization ellipse.** Used in mineral prospecting and mapping faults and shear zones, especially in rugged terrain and heavy vegetation. See **sferics** and Ward et al, 1968, Afmag Interpretation, Geophysics, v. 33, p. 621-644.

AGC: Automatic gain control.

AGC time constant: The time required for the voltage of a system under **automatic gain control** to return to 63 percent (or 1 − 1/e) of its final steady-state value after a sudden change in input level. If the input change is an increase, the constant is called *attack time;* if a decrease, *release time* (which may be different from attack time). AGC time "constants" sometimes depend on factors such as signal level and magnitude of the change. AGC characteristics are usually specified in db/sec. See Figure 16 **(automatic gain control).**

airborne magnetometer: Aerial magnetometer; device used to measure variations in the earth's magnetic field while being transported by an aircraft. Five types are in use: the **flux-gate magnetometer** which measures variations in the total magnetic intensity; the **nuclear-precession magnetometer** and **optically pumped magnetometers,** both of which measure the absolute total magnetic intensity; the **vector magnetometer** which measures total intensity, inclination, and declination; and the cathode-ray magnetometer (Elliot Aspect) which measures total intensity. The first three are the most common.

air drill: A drill which removes the cuttings by circulating air.

air gun: 1. A marine seismic source (Figure 3) which injects a bubble of highly compressed air into the water. Oscillations of the bubble as it alternately expands and contracts generate a sonic wave whose frequency depends on the amount of air in the bubble, its pressure, and the water depth (or water pressure). Arrays of guns of different sizes are sometimes used so that a broader frequency spectrum will be generated. Different from gas gun. See also **waveshape kit. 2.** Air guns have also been adapted for lowering in a borehole or pushed down into the marsh. The adaptations prevent mud, sand, etc., from entering and fouling the air gun. **3.** Air guns are also used in small tanks of water which are set on the surface of the ground as a seismic source on land.

air shooting: A method of generating seismic energy in the earth by detonating charges in the air. Charges are usually placed on poles so that the explosive shock wavefront is distributed over more of the ground surface than if the charge were laid on the surface, although sometimes the latter procedure is also called air shooting. See **Poulter method** and Poulter, 1950, The Poulter Seismic Method of Geophysical Exploration: Geophysics, v. 15, p. 181-207.

air wave: Energy from the shot which travels in the air at the velocity of sound: $1051\,(1 + 0.00218\,F)^{1/2} \approx 1051 + 1.1\,F$ ft/sec, where F = Fahrenheit temperature, or $331.51\,(1 + 0.00366\,C)^{1/2} \approx 331.5 + 0.607\,C$ m/sec, where C = Centigrade temperature.

Airy Hypothesis: A model of compensation for isostasy in which topographically elevated areas are compensated by underlying "roots" of crustal material penetrating into the higher density subcrust. See Figure 104 **(isostasy).**

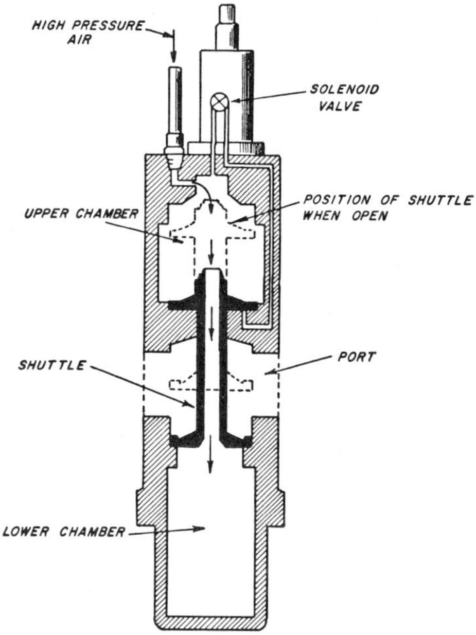

Fig. 3. **Air-gun.** High-pressure air flows continuously into the upper chamber and through the shuttle into the lower chamber. Opening the solenoid valve puts high-pressure air under the upper shuttle seat causing the shuttle to move upward, opening the lower chamber and allowing its air to flow out through ports to form a bubble of high-pressure air in the water. The size of a gun is the size of its lower chamber. (Courtesy Bolt Associates)

Airy phase: A build-up in amplitude of energy traveling by normal-mode propagation in a liquid resting on a solid substratum. Airy waves are associated with a minimum in the group velocity versus frequency curve. The Airy phase is characterized by an abrupt termination of the minimum group velocity.

alert: The time when a navigation satellite should pass within range so that a location fix can be obtained. See **satellite navigation.**

algorithm: A step-by-step procedure for carrying out a numerical or algebraic operation. Compare **heuristic.**

alias: An ambiguity in the frequency represented by sampled data. Where there are fewer than two samples per cycle, an input signal at one frequency yields the same sample values as (and hence appears to be) another frequency at the output of the system. Half of the frequency of sampling is called the "folding" or "Nyquist" frequency f_N; a frequency larger than this,

$f_N + Y$, appears to be the smaller frequency, $f_N - Y$. The two frequencies, $f_N + Y$ and $f_N - Y$ are "aliases" of each other. Thus, for a signal sampled every 4 msec, or 250 times per second, the Nyquist frequency is 125 hz; if, for example, 50 hz is within the passband, then 200 hz will also be passed if an alias filter has not been used and will appear in the output to have a 50 hz frequency (Figure 4). To avoid aliasing, frequencies above the Nyquist frequency must be removed by an alias filter before the sampling. The passbands obtained by folding about the Nyquist frequency are called *alias bands, side lobes,* and *secondary lobes.* Aliasing is an inherent property of all sampling systems. It applies to digital seismic recording, to the sampling which is done by the separate elements of geophone and shotpoint arrays, and to sampling such as done in gravity surveys where the potential field is determined only at discrete stations, etc.

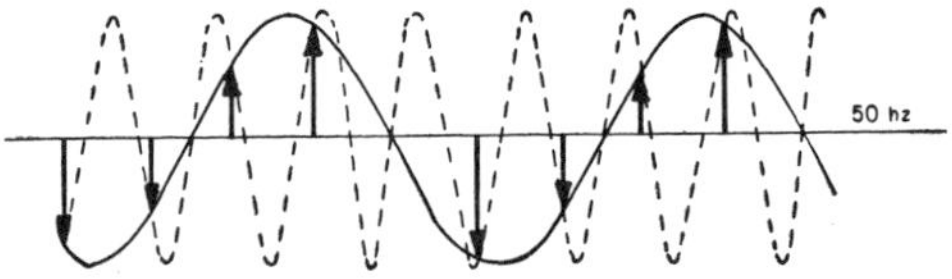

Fig. 4. **Aliasing** of 200 hz as 50 hz. Both 50-hz and 200-hz waves give the same values when sampled at 250 hz (4 msec sampling).

alias filter: A filter used before sampling to remove undesired frequencies which the sampling process would otherwise **alias.** An alias filter should have linear phase response and nearly flat amplitude response over the signal passband and should roll off rapidly above this band to provide high attenuation above the Nyquist frequency; see Figure 75 **(filter)**. Also called *anti-alias filter.*

alidade: A ruler equipped with sights which allows aligning a straight edge in the direction of an object. See Figure 5. A telescope equipped with vertical circle and stadia cross-hairs is mounted on a flat straight-edge base. The alidade is placed on a **plane table** and the object to be located is sighted by looking down into the alidade eyepiece. A line drawn on the plane-table paper along the straight edge indicates the direction to the object.

alkali-vapor magnetometer: See **optically pumped magnetometer.**

alpha: The ratio of **pseudostatic SP** to static SP **(SSP).**

alphanumeric: Containing characters representing letters, numerals, and other characters.

alternating-field demagnetization: A method for determining the stable component of remanent magnetization by partial demagnetization and removal of components with low coercive force. The specimen is placed in a space with nulled field (such as produced with Helmholz coils) and then subjected to an alternating magnetic field which is then reduced gradually to zero by decreasing the current of the field coil or by pulling the specimen from the coil. Also called *ac demagnetization.*

alternator: A rotating electromechanical device for supplying alternating current.

altitude: Angular distance above the horizon.

AM: Amplitude modulation; modulation in which the am-

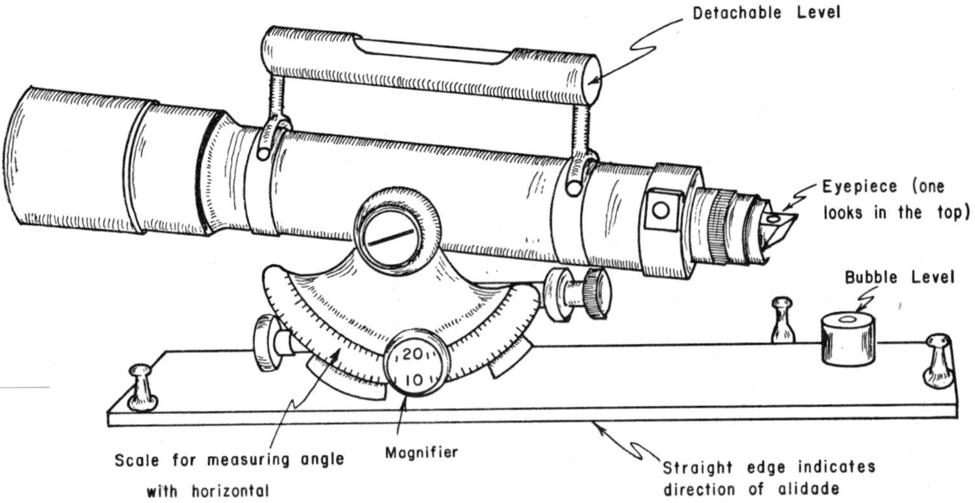

Fig. 5. **Alidade.** The essence of an alidade is simply a telescope aligned with a straight edge.

plitude of the carrier wave varies according to the information being carried. See Figure 122 (**modulation**).

ammeter: An instrument used to measure electrical current.

ammonium nitrate: A commercial fertilizer which becomes an explosive when mixed with (usually) diesel fuel. It is cheaper and safer than most explosives. After bulk ammonium nitrate is mixed with the oil and confined by tamping, it is detonated by the explosion of a primer. It is water-soluble and will not detonate if wet. Ammonium nitrate is also an ingredient of some packaged explosives.

Ampere's law: The magnetic field contribution resulting from a current I through a length of wire Δl at a point is given by:

$$\Delta \mathbf{B} = 10^{-7} \frac{I \sin \theta \, \Delta l}{r^2}$$

See Figure 6. $\Delta\mathbf{B}$ is in webers/m^2 when I is in amperes and r and Δl are in meters; θ is the angle which r makes with Δl; μ is the magnetic permeability. Also called the Biot-Savart law.

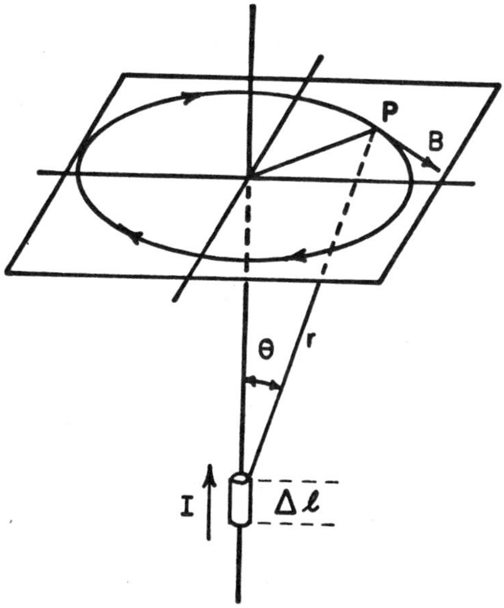

Fig. 6. **Ampere's law.**

amplifier: A device which increases a signal amplitude. The output of a linear amplifier is the input multiplied by a constant factor called the *amplifier gain*. An amplifier is sometimes symbolized by a triangle in a circuit diagram. **a.** Seismic amplifiers may have a gain of 10^6. Schematic drawings of analog and digital seismic amplifiers are shown in Figure 7. See also Evenden, Stone, and Anstey, 1971, Seismic Prospecting Instruments, v. 1, chap. 4, and v. 2, chap. 6 and 8: Berlin, Gebruder Borntraeger. **b.** Amplifiers are sometimes classed as voltage, power, or current amplifiers, depending on which is amplified. A typical amplifier characteristic is shown schematically in Figure 8. Amplifiers are "class A" if they involve only a linear portion of the curve, as "class AB" if the output is over most but not all of the input cycle, as "class B" if they output only during half of an input cycle, or as "class C" if they output during only a very small portion of a cycle.

amplitude log: A borehole log of the amplitude of a portion of the **acoustic wave** used in acoustic logging See **cement-bond log** and **fracture log.**

amplitude modulation: Variations in the amplitude of a high-frequency carrier wave according to low-frequency information; AM. See Figure 122 (**modulation**).

amplitude spectrum: Amplitude-versus-frequency relationship such as is computed in a Fourier analysis. See **Fourier transform.**

analog: 1. A continuous physical variable (such as voltage or rotation) which bears a direct relationship (usually linear) to another variable (such as motion of the earth because of seismic waves) so that one is proportional to the other. 2. Continuous, as opposed to discrete or digital.

analog computer: See **computer.**

analog-digital converter: Device for converting analog signals into digital form.

analog modeling: A method of studying effects of subsurface bodies by comparison with the response of models. Induced-polarization and resistivity surveys, for example, are simulated by measurements in an electrolytic tank using conductive or polarizable shapes to represent subsurface bodies.

analog tape formats: Analog information is written on magnetic tape in several forms. The magnetization (less a constant bias) may be proportional to the input in "direct" or "bias" recording. Information may be carried by frequency variations of a carrier wave in "FM' recording or by the width of square-wave pulses in "pulse-width" recording. See Figure 122 (**modulation**).

analog-to-digital: **Digitizing**; the conversion of analog data to digital form.

anchor: See **charge anchor.**

andesite line: A closed line roughly ringing the Pacific Ocean within which only basic rocks occur (andesite is a basic rock). Surface waves are markedly attenuated at the line.

AND gate: A circuit with multiple inputs which functions only when signal is present at all inputs. If the inputs are A and B, AND is signified as (A · B), (AB), (A ∩ B), (A × B), or the "intersection of A and B." The AND truth table is:

A	B	A·B
0	0	0
1	0	0
0	1	0
1	1	1

See Figure 88 (**gate**).

AN/FO: A mixture of **ammonium nitrate** and fuel oil used as an explosive.

angle of incidence: The acute angle which a raypath

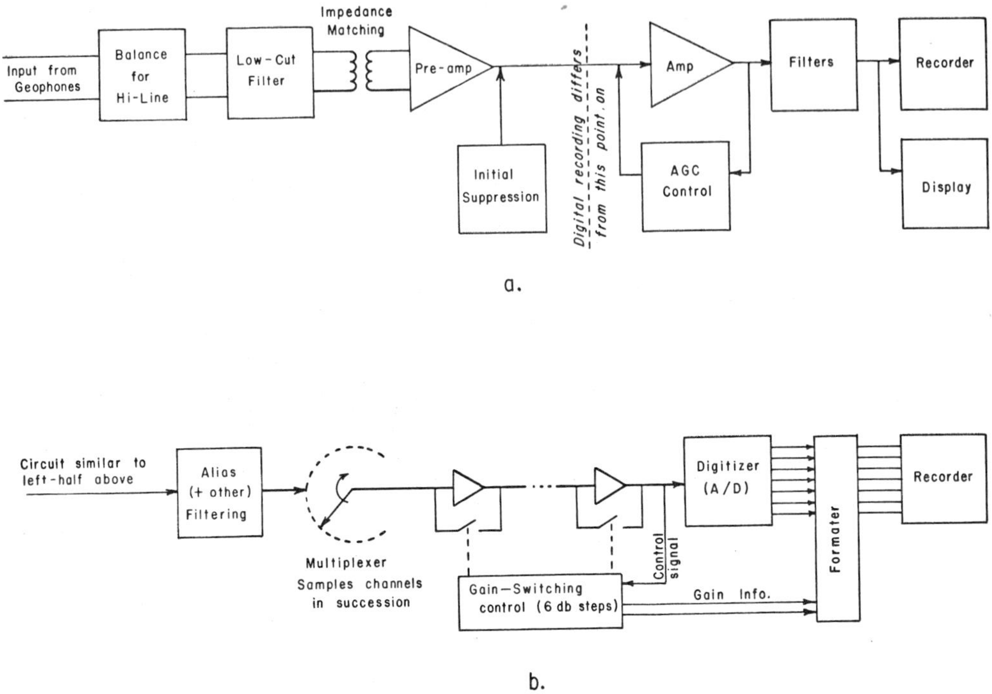

Fig. 7. Seismic **amplifier.** (a) Schematic of seismic analog amplifier. "Amp" is an amplification stage and usually there are several such stages. The AGC control produces negative feedback. Filters may be located at various positions. Usually there are 24 parallel amplifier channels. (b) Schematic of seismic digital amplifier. A monitor record may be made in parallel or by playback of the digital recording. Only one channel is required after the multiplexer. The gain control sometimes is common to all channels and located before the multiplexer.

makes with the normal to an interface. This is the same angle as an approaching wavefront makes with an interface in an isotropic medium. The angle between the raypath and the normal is the angle of incidence in the anisotropic case, the raypath not necessarily being perpendicular to the wavefront in this case. The angle of incidence may be complex for electromagnetic waves.

angles (surveying): The direction of a survey leg with respect to the preceding leg of the survey traverse. Several measuring conventions are used (Figure 9). *Azimuth angles* are measured to the right (clockwise) with respect to north (either true north or magnetic north), occasionally with respect to south. The first leg of the traverse is usually specified by azimuth or compass direction. *Interior angles* are the angles lying inside a closed traverse. *Angles right* are measured clockwise after backsighting on the previous station. A *deflection angle* is the angle between the forward extension of the previous leg and the line ahead.

angular frequency: Repetition rate measured in radians/sec. Where f = frequency in hz, the angular frequency ω is $\omega = 2\pi f$.

anhysteretic: The magnetic state of a sample which has been subjected to a constant magnetic field while a supplemental decaying alternating field has been progressively reduced to zero. This procedure removes isothermal remanent magnetization.

anion: A negatively charged ion.

anisotropy: Variation of a physical property depending on the direction in which it is measured. Geophysical usage of anisotropy is sometimes qualified as *apparent anisotropy* or *effective anisotropy* to distinguish from the point property such as possessed by crystals. **a.** The general elasticity tensor relating stress and strain in anisotropic media contains 21 independent constants. If properties are the same in two directions (*transversely isotropic* medium), these reduce to five independent constants. Isotropic media possess only two independent elastic constants. Anisotropic media have two modes of S and one mode of P body-wave propagation, wavefronts are not necessarily orthogonal to the

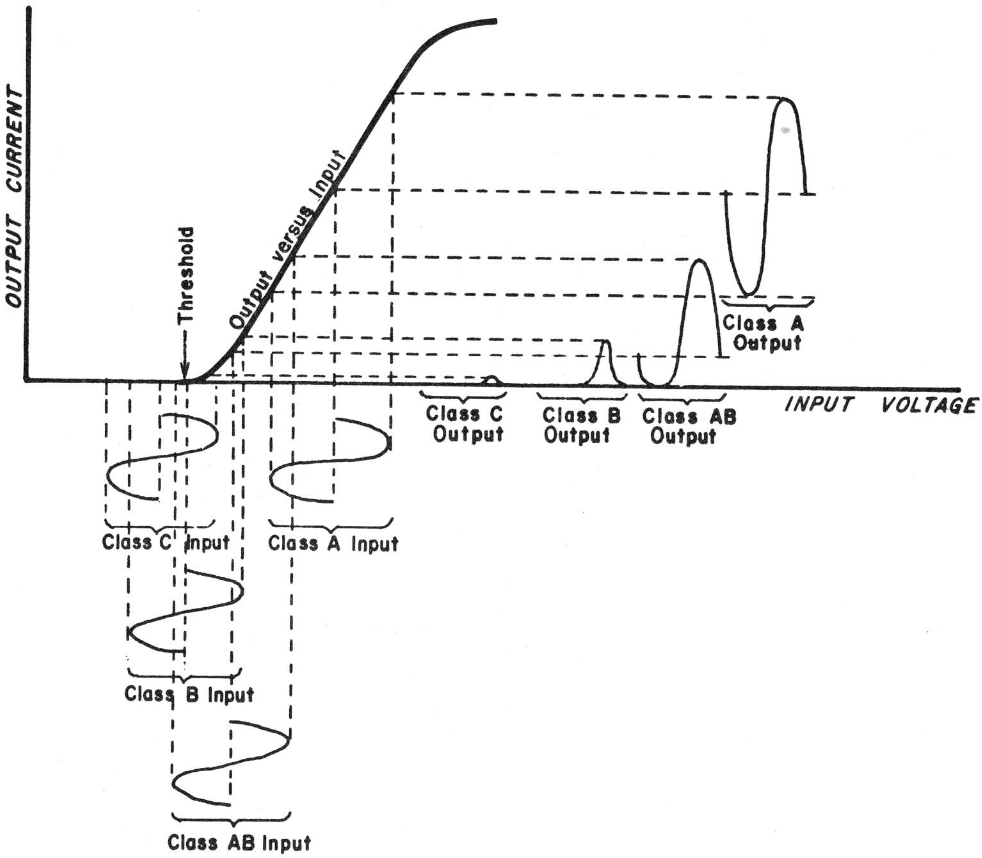

Fig. 8. **Amplifier** "classes" are based on the portion of the input-output curve which is used.

directions of wave propagation, and Snell's law requires modification. *Aeolotropy* is also used for anisotropy. **b.** In seismic usage, a difference between velocity parallel to the bedding plane and the velocity perpendicular to the bedding for a lithologic unit. Velocity along the bedding (as measured by refraction, for example) is typically 10-15 percent higher than velocity measured perpendicularly (as in a well). **c.** Anisotropy is sometimes (improperly) used to denote a difference between the velocity parallel and perpendicular to the bedding for an entire layered sequence. In such sequences, the velocity parallel to the bedding appears greater because the higher-velocity members carry the first energy whereas, in measurements perpendicular to the bedding, all members contribute in proportion to their thickness. **d.** *Microscopic anisotropy* is the variation of a property measured perpendicular to the bedding from that measured parallel to the bedding because plate-like mineral grains and interstices tend to orient themselves parallel to the bedding. **e.** *Macroscopic anisotropy* is the variation due to thin beds whose properties differ appreciably from that of the remainder of the volume over which a measurement is made. **f.** Anisotropy of induced polarization in rocks is less than anisotropy of resistivity. In foliated rocks the true resistivity parallel to foliation is less than that perpendicular to foliation. **g.** The *resistivity anisotropy coefficient* is the square root of the ratio of the resistivity measured perpendicular to the bedding to that measured parallel to the bedding. It usually has a value between 1 and 2.

anisotropy paradox: See **paradox of anisotropy.**

annulus: 1. The space between a drill pipe and the formations through which the returning drilling fluid (mud) returns to the surface. **2.** The space between tubing and casing or between casing and formation. **3.** A low-resistance ring about a borehole sometimes produced by invasion of mud filtrate into hydrocarbon-bearing beds. Because of their greater mobility, hydrocarbons may be displaced farther beyond the invaded

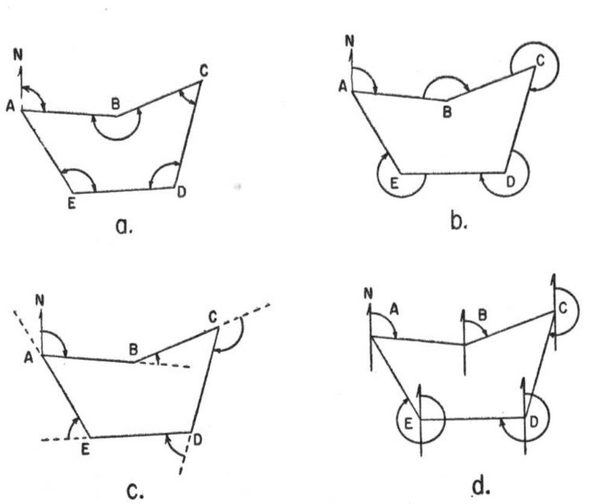

Fig. 9. **Angle** measuring conventions. (a) Specifying by interior angles. (b) Specifying by angles right. (c) Specifying by deflection angles. (d) Specifying by azimuth angles. The first leg of a loop in all conventions is specified by azimuth.

zone than conductive formation water. See Figure 104 **(invaded zone)**.

anode: An electrode (or portion thereof) at which oxidation occurs and electrons are produced (that is, are given up to the electrode). The positive terminal of an electrolytic cell or the negative terminal of a battery.

anomaly: **1.** A deviation from uniformity in physical properties. Especially a deviation from uniformity of exploration interest; for example, a traveltime anomaly, Bouguer anomaly, free-air anomaly. **2.** A portion of a geophysical survey, such as magnetic or gravitational, which is different in appearance from the survey in general. **3.** In seismic usage, generally synonymous with structure. Occasionally used for unexplained seismic events. **4.** A feature which may be associated with petroleum accumulation or mineral deposits. **5.** An induced-polarization anomaly is usually positive and greater than background (or the normal effect) to be economically interesting. An interesting resistivity anomaly is generally less than background.

antialias filter: **Alias filter.**

anticipation function: A function which collapses a wave train into an impulse at the head end of the train. Involved with **recursive filters.**

anticline: A fold or arch away from which the strata dip in opposite directions.

antiferromagnetism: Property of certain magnetic atoms which makes sublattices take on antiparallel ordering of spins (i.e., oriented 180 degrees to each other), such that no net magnetization is observed. Compare **ferromagnetism** and **ferrimagnetism.**

antiparallel: Two vectors which point 180 degrees to each other.

antiroots: Shallow high-density mantle material beneath thin portions of the relatively light crust. Characterized by shallow Moho. The effect is to bring topographically low areas (like ocean basins) into isostatic equilibrium. See Figure 104 **(isostasy).**

antisymmetric: The property of a function which makes it change sign when its argument changes sign:

$$A(x) = -A(-x).$$

Also called *odd function.*

antithetic: **1.** A system of normal faults with rotation of the blocks so that the net slip on each fault is greater than it would have been without the rotation. See Figure 10. **2.** Minor normal faults with throw in opposite direction to the major fault with which they are associated.

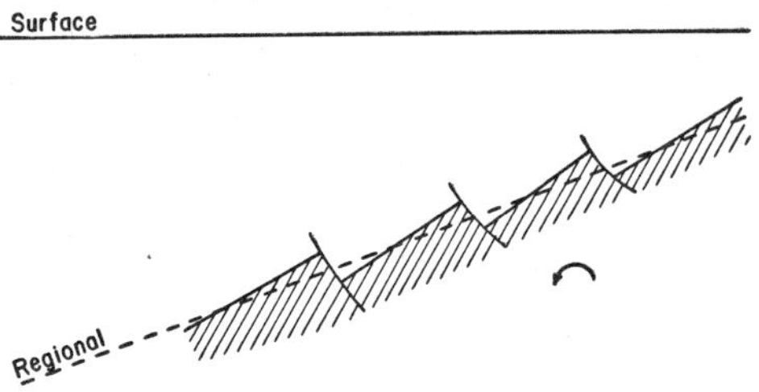

Fig. 10. **Antithetic faulting.** The fault blocks have rotated so as to increase the throw on the faults.

aperture: Gate or **window.**

aperture time: **1.** A designation of the location of the time interval which contains the data utilized in a calculation, usually indicated as starting and finishing time for the aperture (e.g., 1.5 to 2.5 sec). **2.** Sometimes, the time about which the aperture is centered (usually given along with the aperture width).

aperture width: The *effective aperture width* is the width of a boxcar with the same peak height and the same energy.

API unit: **1.** A unit of counting rate for the **gamma-ray log.** The difference between the high and low radioactivity sections in the API calibration pit is defined as 200 API units. **2.** A unit of counting rate for the **neutron log.** The reading in the Indiana limestone portion of the API neutron log calibration pit (which has 19 percent porosity and is saturated with fresh water) is defined as 1000 API units.

API well number: A unique number assigned by the American Petroleum Institute to each well drilled in the United States. A 12-digit decimal number which is broken down as follows:

Digits 1, 2: Code for state; numbers 1 to 49 are alphabetical for the states including the District of Columbia except for Alaska and Hawaii, which are 50 and 51.
Digits 3-5: Code for county, parish, or offshore.
Digits 6-10: Code for the specific well.
Digits 11-12: Code for sidetrack, etc.

See API Bulletins D12 and D12A.

aplanatic surface: The tentatively postulated surface between two media which is the locus of traveltime for wave energy reflected or refracted at the surface. The traveltime between two points via any point on the aplanatic surface is constant. Used, for example, in defining a salt-sediment interface (Figure 11). Each combination of shotpoint and geophone position defines one aplanatic surface. If the velocity and other assumptions are correct, the salt-sediment interface is the common tangent to all the aplanatic surfaces. See Gardner, 1949, Seismograph Determination of Salt Dome Boundaries Using Well Detector Deep on Dome Flank: Geophysics, v. 14, p. 29-38, and Musgrave et al, 1967, Outlining of Salt and Shale Masses by Refraction Methods, *in* Seismic Refraction Prospecting: SEG, Tulsa p. 426-457. See also **bathtub chart.**

apodizing function: A weighting function used in truncating which reduces discontinuity effects. See **window functions.**

apogee: The point on the orbit of an earth satellite which is farthest from the center of the earth; see Figure 69 **(ellipticity)**. The shortest distance is perigee.

apparent: 1. The value indicated by a measurement, as in *apparent velocity*. **2.** The value of a property assuming the ground to be homogeneous, isotropic, and semiinfinite, as distinct from the "true" value. The subscript a is frequently used to indicate that a quantity is apparent, as with $(PFE)_a$, $(MF)_a$, etc.

apparent resistivity: 1. The ground resistivity calculated from measurements and a geometric factor derived for the case where the ground is homogeneous and isotropic. Apparent resistivity ρ_a is an Ohm's law ratio of measured voltage V to applied current I, with a geometric constant K which depends on the electrode array:

$$\rho_a = KV/I.$$

Usually has units of ohm-meters. See **resistivity. 2.** With the electromagnetic methods, quantities such as the moduli of the electric and magnetic field intensities **(E** and **H)** are measured for a certain frequency. If the subsurface were homogeneous and isotropic, these quantities would yield the true resistivity via a certain equation. However, use of the same equation for a heterogeneous subsurface yields a fictitious intermediary parameter called *apparent resistivity* which provides a convenient way of displaying the data. It does not represent the average resistivity and it can be lower than the lowest or higher than the highest resistivity within the subsurface to which it pertains. **3.** The resistivity recorded by an electrical log which differs from the true resistivity of the formation because of the presence of mud column, invaded zone, influence of adjacent beds, etc.

apparent-resistivity curve: A graph of **apparent resistivity** against electrode separation or frequency. Apparent-resistivity curves plotted on logarithmic paper often are compared with **type curves** (normalized theoretical curves) for the purpose of interpreting the resistivity, thickness, and depth of subsurface layers.

apparent velocity: 1. The velocity which a wavefront appears to have along a line of geophones. If the wavefront makes the angle θ with the spread and the true velocity of the wavefront if V, then the apparent velocity is $V/\cos\theta$. See Figure 192 **(wave)**. **2.** The inverse of the slope of a refraction **time-distance curve.**

apparent wavelength: The distance between corresponding points on successive cycles of a wave train as seen by a geophone spread. Differs from actual wavelength if the wave train direction makes an angle with the spread. See Figure 192 **(wave)**.

Aquapulse: Sleeve exploder. Western Geophysical trademark.

Aquaseis: A marine seismic energy source in which an explosive cord is towed and detonated. Imperial Chemical Industries trademark.

Arcer: A high-powered **sparker.** Alpine Geophysical system

Archie's formulas: Empirical relationships between the formation resistivity factor F (sometimes F_R), porosity ϕ, water saturation S_W, and resistivities in clean granular rocks:

$$F = R_o/R_W = a\,\phi^{-m},$$

$$R_o/R_t = S_W^n,$$

where m = porosity exponent or *cementation factor*, which varies between 1.3 and 3,
a = proportionality constant, varying from 0.6 to 1.5,
R_0 = resistivity of the formation when 100 percent saturated with formation water,
R_W = resistivity of the formation water,
R_t = true resistivity of the formation,
n = "saturation exponent"; often $n = 2$.

Archie's law assumes that $m = 2$ and $a = 1$. The **Humble formula** assumes that $m = 2.15$ and $a = 0.65$.

arc shooting: 1. Fan shooting. 2. A method of refraction exploration with geophones spread on the arc of a circle with either the source or the point of incidence of energy on the refractor at the center of the circle.

ARD: Approximate reverberation deconvolution; a digital **deconvolution** program which involves **whitening** the frequency-amplitude spectrum on a minimum-phase basis. A Geophysical Service Inc. tradename.

argument: 1. Angle of a complex number. **2.** An independent variable; the variable which determines the value of a function; e.g., in sin ϕ, ϕ is the argument.

Aries: The *first point of Aries* or the *vernal equinox* is the reference point on the celestial sphere from which right ascension and celestial longitude are measured. It is the intersection point of the celestial equator and the ecliptic where the sun is located on the vernal equinox, March 21.

arm: 1. A bow spring or lever connected to a logging

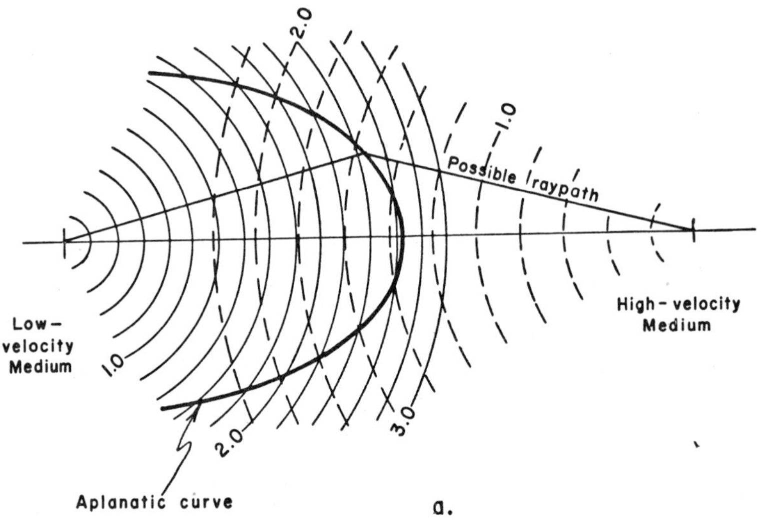

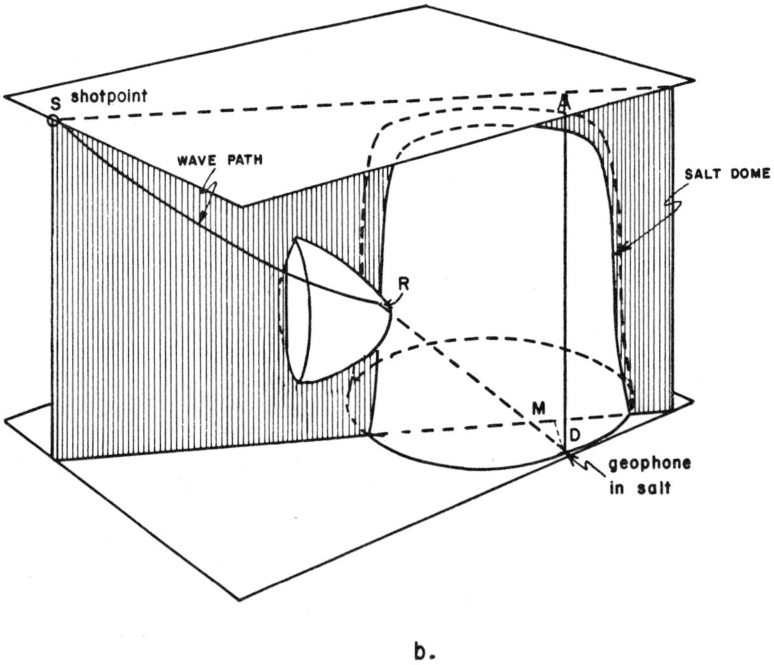

Fig. 11. **Aplanatic Surface.** (a) Aplanatic surface can be found by the intersections of wavefronts about shotpoint and geophone. Shown is the aplanatic curve in the plane containing both shotpoint and geophone for a traveltime of 4.0 sec. (b) Use of aplanatic surface in three-dimensional problem of locating salt dome flank by shooting into a geophone in a borehole in the salt. (From Gardner, Geophysics: v. 14, no. 1, p. 30-31.)

sonde which presses against the borehole wall to centralize the tool, to push the tool to the opposite side of the borehole, or to hold a sensor pad to the borehole wall. **2.** To prepare a blaster or a perforating gun for firing.

arpent: Old French land measure, approximately one acre.

array: **1.** A group of geophones connected to a single recording channel or a group of shotpoints to be fired simultaneously; sometimes called a *pattern* or *patch.* **2.** The arrangement or pattern of a group of geophones or shotpoints (Figure 12). Discrimination against certain events (on the basis of their stepout or apparent wavelength) can be effected by array design. See **directivity.** **3.** A **matrix.** **4.** The arrangement of antennas in electromagnetic prospecting. While there is no generally accepted terminology, Parasnis has suggested the following for moving source methods: Let the traverse direction be the *x*-axis and let the *z*-axis lie in the vertical plane through the *x*-axis; the *z*-axis may not be vertical if the traverse is in a borehole. Let *a, b,* and *c* be the direction cosines of the transmitter coil axis; *d, e,* and *f* those of the line *L* joining transmitter and receiver coils; and *g, h,* and *i* those of the receiver coil axis. Then

$$T(a, b, c)\ L(d, e, f)\ R(g, h, i)\ \tau$$

describes the configuration completely; τ is the transmitter-receiver distance. (If the traverse direction is vertical, the *z*-axis is arbitrarily made north-south.) See Parasnis, 1970, An Elegant, Universal Nomenclature for Electromagnetic Moving Source-Receiver Dipole Configurations: Geophysical Prospecting, v. 18, p. 88-102. Fixed source methods are described in terms of the type of source (long wire, large loop, small loop, etc.), orientation of the source, orientation of the receiver, and the relationship of the traverses to the source. **5.** The arrangement of electrodes in resistivity prospecting, also called "configuration." Several array types are shown in Figures 13 and 14. Resistivity array types include: **a.** *Azimuthal array*—See **azimuthal survey.** **b.** *Dipole-dipole array* or *axial-dipole array*—An array in which one dipole (a connected pair of electrodes) sends current into the ground and the other dipole serves as the potential-measuring pair. The separation between pairs is often comparable to or only a few times greater than the spacing within each pair, so the electrode pairs are not ideal dipoles. In deep resistivity sounding, the separation is larger. The dipole pairs are usually collinear (in line) but other orientations are used (Figure 13). Resistivity and IP data from this array often are displayed as on Figure 14. **(psuedosection).** **c.** *Gradient array* or *AB rectangular array*—An arrangement in which a pair of potential electrodes measure the voltage between points of a rectangular grid between two distant, fixed-current electrodes. A variation of the Schlumberger array. **d.** *Pole-dipole array*—A voltage-measuring pair of grounded potential electrodes is separated successively from one current electrode (pole) while traversing a survey line. The second current electrode (the "infinite electrode") is distant from the first, so far away that its location has negligible effect on the measurements. Data can be plotted below the mid-point between the current and the near potential electrode on a pseudosection. Called a "three array" if the electrodes are equally spaced. **e.** *Pole-pole array* or *two array*—Two electrodes (poles), a current and a potential, are traversed or successively expanded on a survey line. The other current and potential electrodes are located so far away that their location has negligible effect on the measurements. Data are plotted either at the potential electrode or halfway between the two poles. **f.** *Radial array*—See **azimuthal survey.** **g.** *Schlumberger array*—The inner voltage-measuring pair of potential electrodes is closer together than the outer current-electrode pair, by a factor of about 6. **h.** *Wenner array*—Four equally spaced in-line electrodes. Either the electrodes are moved along a traverse or their interval is successively expanded. The usual or α *configuration* has the center two electrodes as the potential electrodes; the β *configuration* has the first two electrodes as potential electrodes; and the γ *configuration* alternates current and potential electrodes.

array factor: See **geometric factor.**

array processor: A subcomputer controlled by another computer which carries out special functions such as matrix manipulations more efficiently than can be done with the general-purpose computer.

array station: Earthquake-detection station which uses an array of seismometers. Twenty seismometers spread over about 20 km may be employed so that spatial filtering can be used to improve the detectability of weak signals. The **LASA** array (Figure 109) uses 525 seismometers distributed over 200 km.

arrival: An event; a lineup of coherent energy signifying the arrival of a new wavetrain.

arrival time: **1.** The time from shot or other energy release to the time which is picked for an event, making allowance for static and dynamic corrections. **2.** The time of an event without corrections.

arrow plot: A **tadpole plot** display of dipmeter or drift data. See Figure 53 **(dipmeter).**

artificial magnetic anomalies: Local magnetic fields caused by man-made features such as transmission and telegraph lines, electric railways, steel drill casing, pipe lines, tanks, etc.

ascension: See **right ascension.**

aseismic: Free of natural earthquakes.

assemble: To prepare a machine-language program from a symbolic-language program; to make an object program from a source program.

astatic: Having a negative restoring force which aids a deflecting force, thereby rendering the instrument more sensitive and/or less stable. The idea of the astatic balance is illustrated in Figure 15. Used in **gravimeters.**

asthenosphere: A yielding zone in the earth's mantle; the region below the upper mantle thought to be involved in convection currents and in isostatic compen-

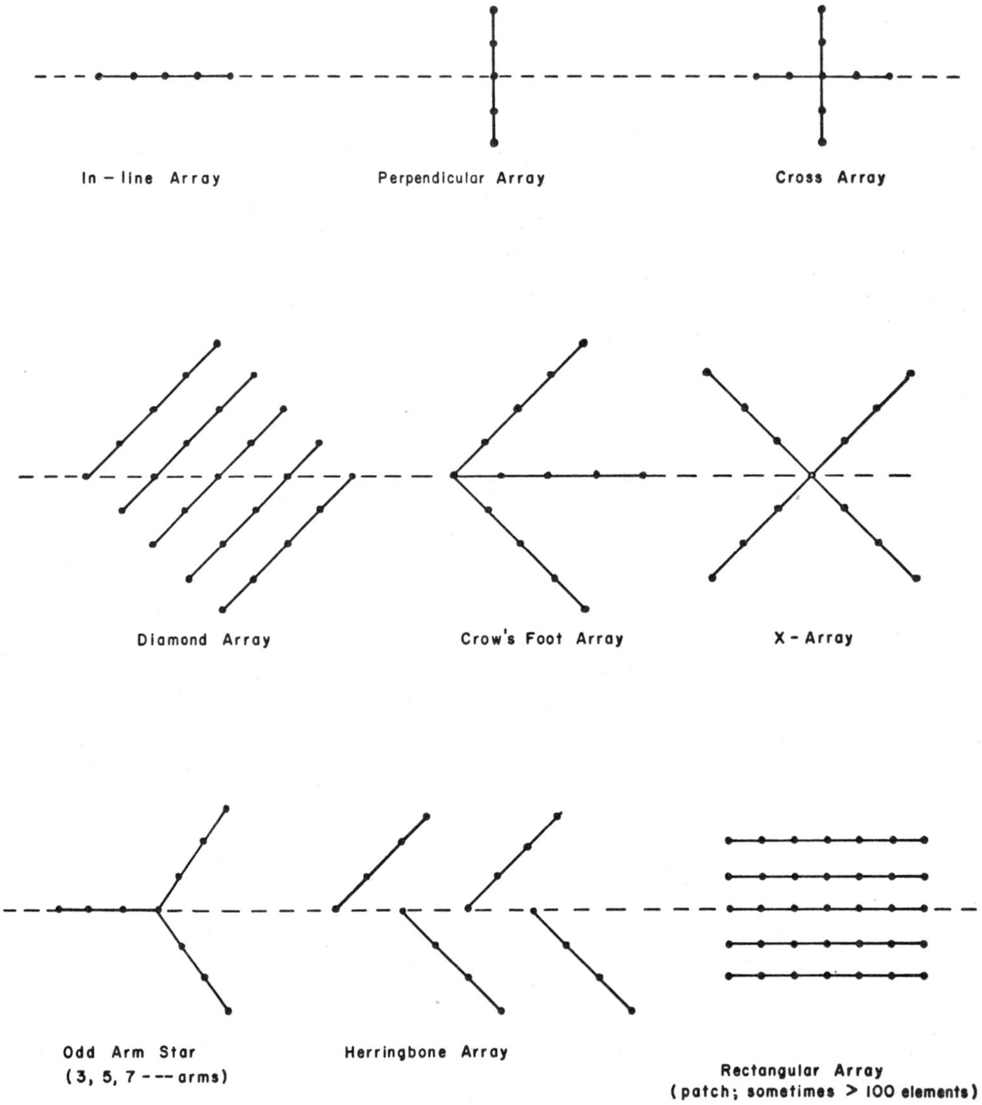

Fig. 12. **Arrays** used for geophone or shot groups.

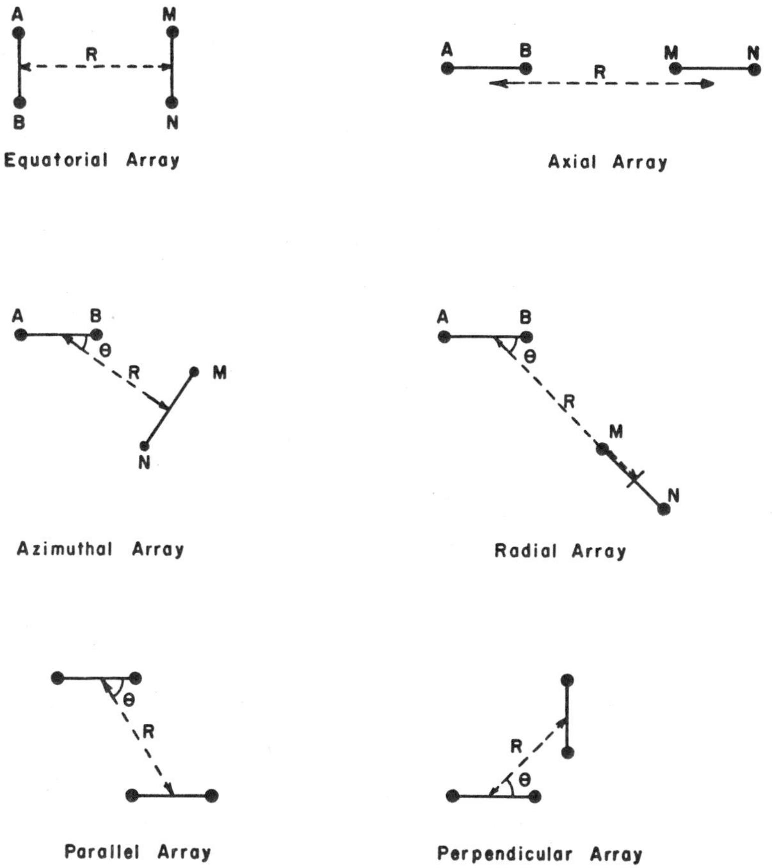

Fig. 13. Dipole-dipole **arrays** used in resistivity and induced-polarization exploration.

sation. The asthenosphere lies between the lithosphere and the mesosphere. It is approximately 200 km thick, has high attenuation of seismic energy (low *Q*), and little strength. See Figure 63 **(earth layering)**.

astronomic latitude: Latitude measured with respect to the vertical and the stars. Differs from "geodetic latitude" by a few seconds of arc where the mass distribution distorts the geoid equipotential surface, as near the roots of mountain ranges. See Figure 89 **(geodetic latitude)**.

attack time: See **AGC time constant**.

attenuation: 1. A reduction in amplitude or energy, such as might be produced by passage through a filter. 2. A reduction in the amplitude of seismic waves, such as produced by **divergence,** reflection and scattering, and absorption. 3. That portion of the decrease in seismic or sonar signal strength with distance not dependent on geometrical spreading. This decrease depends on the physical characteristics of the transmitting media, involving reflection, scattering, and absorption. 4. If the amplitude of a plane wave is reduced by the factor $e^{-\alpha(f)x}$ in traveling a distance of x meters , then the attenuation factor is $\alpha(f)$. Often thought to be linear with frequency, sometimes thought to be quadratic with frequency. 5. Regarding the rapid attenuation of electromagnetic waves in the earth, see **skin depth**.

attenuator: 1. An adjustable passive device for reducing the amplitude of a signal without introducing distortion. 2. A calibrated resistive circuit used on a voltage-measuring instrument so that voltages can be measured in a dynamic range of several decade intervals of voltage. Sometimes called a *multiplier* or *voltage divider.*

attitude: The relation of a feature to horizontal. The strike and dip for a bed or other planar feature, the trend direction and plunge for a linear feature such as an anticline.

A-type section: A three-layer model in which the resis-

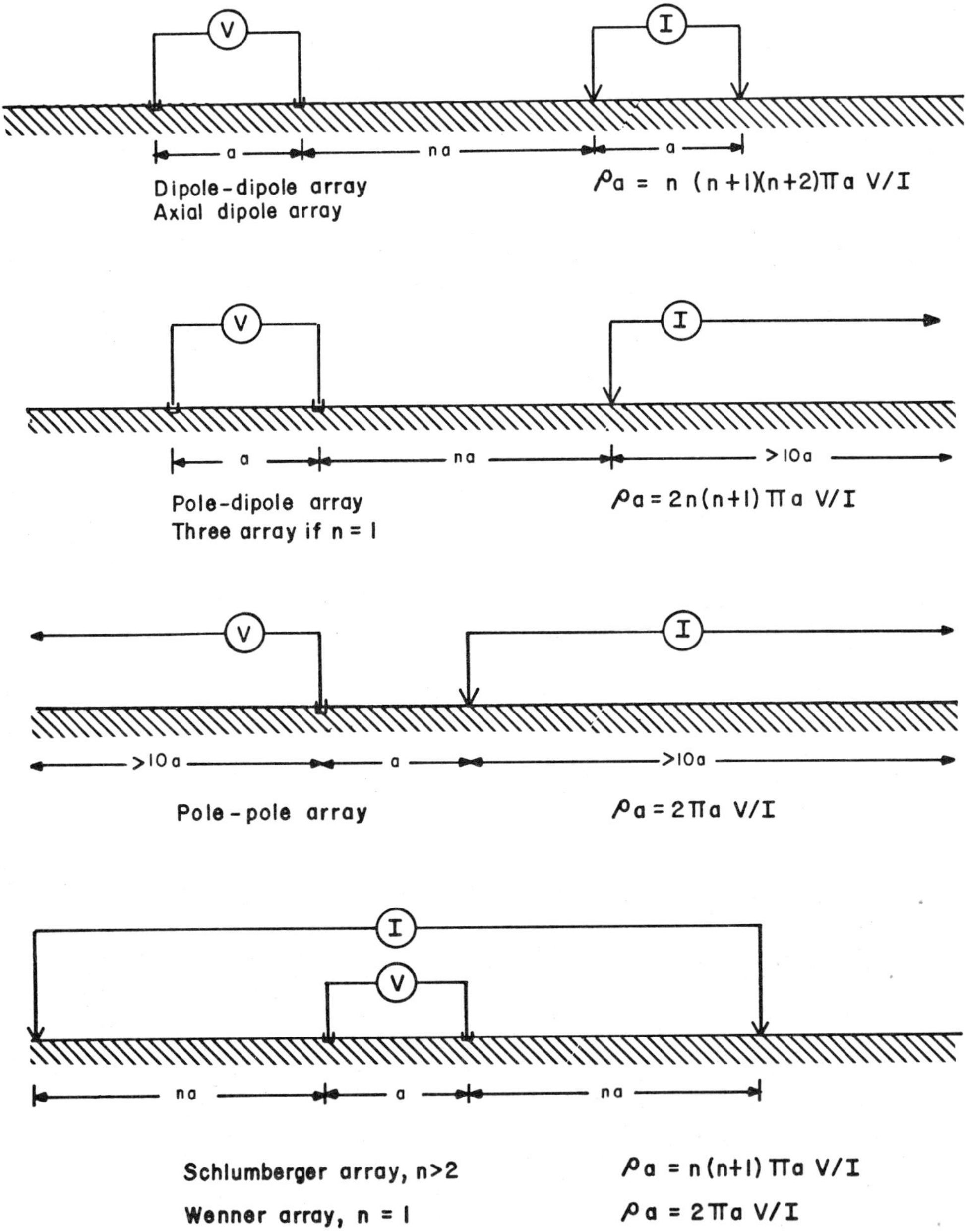

Fig. 14. **Arrays** used in resistivity surveying. ρ_a is the apparent resistivity equation for each array.

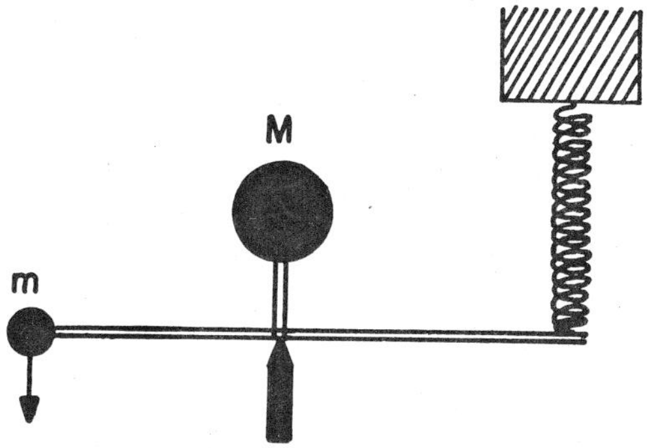

Fig. 15. **Astatic balance** principle. In this example, the large weight *M* exerts no net force if the system is balanced, but throws the system farther out of balance once unbalance occurs.

tivities increase with depth. See Figure 180 (**three-layer resistivity**).

audio-frequency magnetic method: Afmag method.

auger: A drilling tool designed so that the cuttings are carried to the top of the hole continuously during the drilling operation by helical grooves on a rotating drill pipe. In the *wet auger,* fluid is injected at the bit to assist in the removal of cuttings.

autoconvolution: Convolution of a function with itself. See **retrocorrelation.**

autocorrelation: A special case of crosscorrelation, a waveform correlated with itself. The normalized autocorrelation function $\phi_{11}(\tau)$ for a continuous stationary waveform is defined as:

$$\phi_{11}(\tau) = \lim_{T\to\infty} \frac{\int_{-T/2}^{T/2} F(t)\, F(t+\tau)\, dt}{\int_{-T/2}^{T/2} F^2(t)\, dt}$$

where *F(t)* represents a waveform (or seismic trace) and τ is the time shift or lag. Where the integrals are taken only between certain limits, the *apparent autocorrelation function* results, but the distinction "apparent" is usually dropped. For equally sampled (digital) data this becomes:

$$\phi_{11}(\tau) = \lim_{N\to\infty} \frac{\sum_{k=-N}^{N} F_k\, F_{k+\tau}}{\sum_{k=-N}^{N} F_k^2}$$

With long waveforms the sums are usually limited to a relatively short interval (as from T_1 to T_2) called the gate or **window.** The denominators in the preceding equations are the normalizing factors and sometimes are not included. The autocorrelation function is a measure of the statistical dependence of the waveform at a later time τ on the present value, or the extent to which future values can be predicted from past values. The autocorrelation function contains all the amplitude-frequency information in the original waveform but none of the phase information. An autocorrelation function is symmetrical about zero shift. Autocorrelation is also equivalent to passing a waveform through its **matched filter.** See Anstey, 1964, Correlation Techniques Review: Geophysical Prospecting, v. 12, p. 355-382.

autocorrelogram: A display of half (positive time shifts only) of the autocorrelation function of seismic traces, usually in record-section format.

autocovariance: Similar to an autocorrelation except that the mean value $\bar{F}$ is subtracted before the integration, and normalization is not done:

$$\int [F(t) - \bar{F}][F(t+\tau) - \bar{F}]\, dt.$$

Hence, for functions which have a zero mean, autocovariance is the same as an autocorrelation function which is not normalized.

automatic gain control: AGC, a system in which the output amplitude is used for automatic control of the gain of an amplifier. Seismic amplifiers usually have individual AGC for each channel, although multichannel control is sometimes used. See Figure 16 and **gain control.**

a.
b.

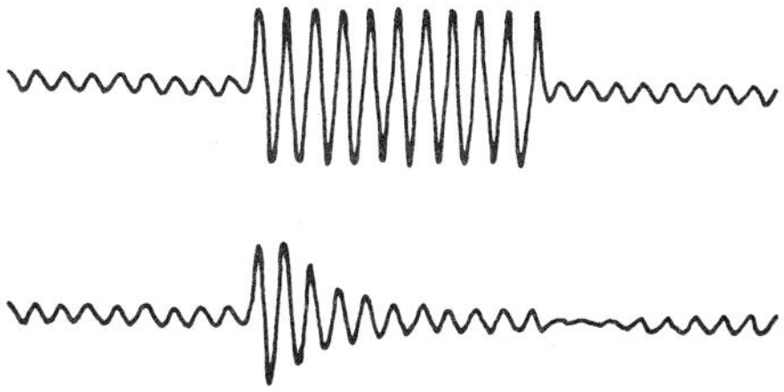

Fig. 16. **Automatic-gain-control** action. (a) Input. (b) Output. The AGC restores the output to the same level after an increase or decrease in input amplitude.

autoregressive series: A time series generated from another time series as the solution of a linear-difference equation. Usually previous values of the output enter into the determination of a current value.

AVC: Automatic volume control or **automatic gain control.**

average: The arithmetic mean over some ensemble, population, etc.

average velocity: The distance traversed by a seismic wavelet divided by the time required, both often corrected to a reference datum plane. For reflections, often refers to a ray reflected at normal incidence. See **velocity**.

axial surface: The surface about which a **fold** is more or less symmetrical. Sometimes called the *axial plane,* though not geometrically "plane." Also called *ab plane.* See Figures 20 **(b-axis)** and 81 **(fold)**.

axis: See **b-axis**.

azimuth: The horizontal angle usually specified clockwise from true north.

azimuthal projection: A map projection in which a spherical surface is developed on a tangent plane. Azimuths to any point from the point of tangency are correctly represented.

azimuthal survey: **1.** A survey in which current electrodes at the ground surface on specific azimuths from a drill hole are used with one or both potential electrodes in the hole. The electrodes in the borehole may be successively raised to develop a log with the objective of determining the azimuthal direction toward better mineralization. **2.** A survey method in which the potential electrodes are moved along radii about a drill hole which contains a fixed current electrode. The second current electrode ("infinite electrode") is a great distance away. Also called *radial survey.* **3.** Azimuthal and radial "arrays" usually refer to dipole-dipole surveys where all electrodes are on the surface in contrast to an azimuthal "survey" involving a drill hole.

azimuth bar: An established survey point near (200 to 2000 ft) a triangulation station, used to aid in orienting a transit.

B

background: Average **noise** level, whether systematic or random, upon which a desired signal (such as a reflection) is superimposed. Usually refers to the total noise, independent of the presence of the signal.

background polarization: The relatively weak IP response exhibited by many unmineralized rocks, particularly those containing abundant clay minerals or layered or fibrous minerals. Also can be due to weak, broad-scale, pervasive mineralization which is not of economic interest. Sometimes known as *normal effect.*

backsight: 1. A sight on a previously established survey point with the objective of determining the position and elevation of the survey instrument. The closing sight of a traverse or level-line loop is not considered a backsight. 2. In plane-table traversing, orientation of the table by aligning the alidade on an established mapped point.

back-to-back: Processes which follow each other sequentially without any judgment being exercised in intermediate stages.

Backus filter: An inverse filter which removes the effects of reverberation involving a simple water bottom. The filter's *z*-transform expression is

$$1 + 2kz^q + k^2 z^{2q},$$

where *k* is the water-bottom reflection coefficient and qt_s is the two-way traveltime through the water layer, if t_s is the sample interval. See Backus, 1959, Water Reverberations: Their Nature and Elimination: Geophysics, v. 24, p. 233-261.

balanced input: A symmetrical input circuit having equal impedance from both input terminals to ground.

band: 1. A range of frequencies between definite limits, such as that passed (band-pass) or rejected (band-reject) by a filter. 2. The names given to the frequency ranges of the electromagnetic spectrum, as shown in Figure 68 (**electromagnetic spectrum**). 3. A **track** on magnetic tape.

band-limited function: A function whose Fourier transform vanishes (or is very small) outside some finite range of frequencies

band-pass: The range of frequencies which a system passes with little attenuation, while substantially attenuating frequencies outside these limits. See **filter.**

band-reject filter: A filter which attenuates a range of frequencies. May be considered the inverse of a band-pass filter.

bandwidth: 1. The range of frequencies over which a given device is designed to operate within specified limits. 2. The difference between half-power points, i.e., the frequencies at which the power drops to half the peak power (or 3 db). 3. The *effective bandwidth* is

$$P(f)\,df/P_{\mathrm{MAX}},$$

where *P(F)* is the power at the frequency *f* and P_{Max} is the maximum power. It is the width of a boxcar with the same total power and the same peak power.

bar: 10^5 N/m^2 (newtons per square meter). A unit of pressure, approximately one atmosphere or 14.5 psi.

barite: Barytes or barium sulphate, used to make drilling mud heavier.

barium titanate: A ceramic having piezoelectric properties. Used in **transducers** such as **hydrophones.**

barker word: The distinctive word in the message from a navigation satellite which indicates the beginning of the two-minute transmission cycle.

Barkhausen noise: Noise introduced by the discreteness of magnetic structure so that magnetization occurs as a series of small steps rather than continuously.

barn: 10^{-28} m^2; a unit for measuring **capture cross-section.**

Barthelmes method: A refraction interpretation method involving continuous profiling. See Figure 17. Differences Δt in the delay time from adjacent profiles give the dip on the refractor (ΔZ = difference in depth). For the two layer case,

$$\Delta Z = \frac{V_1 V_2}{2(V_2^2 - V_1^2)^{1/2}} \Delta t,$$

where a refractor of velocity V_2 is covered by overburden of velocity V_1. The depth *Z* to the refractor has to be obtained by assuming the velocity distribution above the refractor, tying the refractor to subsurface control, etc. The relief on the refractor is displaced by the displacement distance ΔS, with respect to the surface observation point. For the two layer case, ΔS is given by:

$$\Delta S = V_1 Z/(V_2^2 - V_1^2)^{1/2}.$$

See Barthelmes, 1946, Continuous Profiling and Refraction Shooting: Geophysics, v. 11, p. 24-42.

barytes: Barite.

base: 1. The reference integer in a number system. Also called *radix.* 2. The transistor element that corresponds to the grid of a vacuum tube. 3. Base station.

base line: 1. A line used as a reference for measurements. See **shale base line.** 2. The line between two radio-positioning base stations which form a pair.

base-line extension: The straight-line extension of a base line beyond the base station.

base-line shift: A change in the location of the shale base line on the **SP** curve. A shift may occur when waters of different salinities are separated by shale beds which do not act as perfect cationic membranes, when the formation water salinity changes within a permeable bed, or when the resistivity of the mud changes with depth in the borehole.

basement: 1. *Geologic basement* is the surface beneath which sedimentary rocks are not found; the igneous, metamorphic, granitized, or highly folded rock underlying sedimentary rocks. 2. Petroleum *economic basement* is the surface below which there is no current exploration interest, even though some sedimentary units may lie deeper. 3. *Magnetic basement* is the

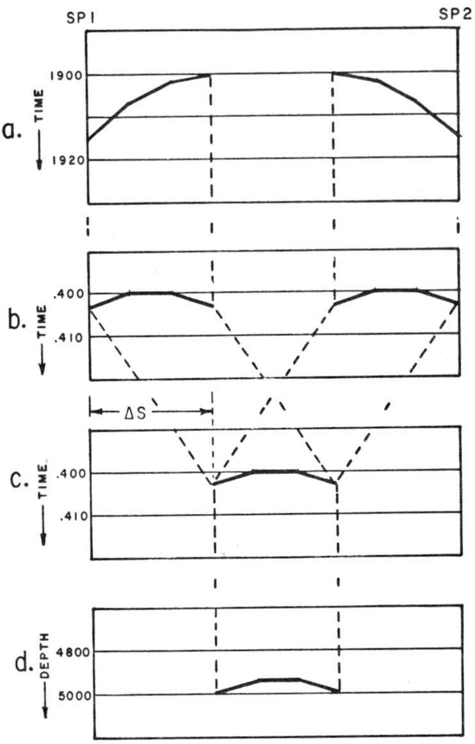

Fig. 17. **Barthelmes refraction method.** (a) Time-distance curves from reversed profiles. (b) Reduced time-distance (delay-time) curves. (c) Displaced delay-time curves. (d) Conversion to depth yields refractor surface. (From Barthelmes, Geophysics, v. 11, no. 1, p.32.)

upper surface of igneous or metamorphic rocks which are magnetized much more than sedimentary rocks. **4.** *Electrical basement* is the surface below which resistivity is very high and hence variations below this surface do not affect electrical-survey results significantly. The different types of basement may or may not coincide.

base of weathering: The boundary between the surface layer of low seismic velocity and an underlying layer of appreciably higher velocity. May or may not correspond to the *geologic weathering* or to the water table. The boundary is involved in deriving time corrections for seismic records. See also **weathering.**

base station: 1. A reference station that is used to establish additional stations in relation thereto. Quantities under investigation have values at the base station that are known (or assumed to be known) accurately. Data from a base station may be used to normalize data from other stations as in the **telluric-current method. 2.** Accurately located fixed station for radiopositioning.

batch processing: A procedure in which work for computer processing that is similar is accumulated and submitted together in order to increase efficiency.

bathtub chart: A chart of the loci of reflection times for an offset geophone; see Figure 18. The loci are also called *aplanatic surfaces.* A reflector tangent anywhere to one of the curves satisfies the reflection time appropriate to that curve. In the limit of zero offset, a bathtub chart is simply a wavefront chart.

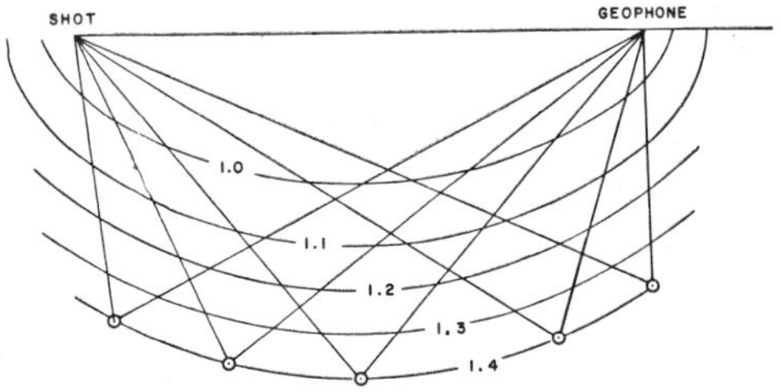

Fig. 18. **Bathtub chart** shows locus of equal traveltime for reflections to offset geophone.

baud: A unit of measure of the ability of a channel to transmit information; one bit per second.

Baumgarte ray-stretching method: 1. A graphical reflection interpretation method in which the positions of successive layers are constructed as surfaces tangent to fictitious wavefronts which are projected backward from the observing stations. **2.** A graphical refraction interpretation method. The construction is shown in the upper diagram of Figure 19 for the two-layer case. A reversed time-distance plot is shown at the top. Let the reversed traveltime (which is the same in either direction) be T_1. The points B and B' are on the first refractor's time-distance line (or extrapolations of it) at $T_1/2$; their abscissas are D and D'. The distances AC and $A'C'$ are $T_1V_1/2$. Right triangles ACD' and $A'C'D$ intersect on the refractor surface. Other lines such as E and E' which are parallel to and equidistant from CD' and $C'D$ also intersect on the refractor. Solution for a third layer is shown in the lower diagram. Let the reversed traveltime on the next refractor be T_2. The points F and F' are on this refractor's time-distance line (or extrapolations) at $T_2/2$, and their abscissas are G and G'. The distances AH and $A'H'$ are $T_2V_1/2$; AH and $A'H'$ intersect the first refractor at I and I'. The distance U is $[T_2/2 - (AI/V_1)]V_2$ and $I'J'$ is $[(T_2/2) - (A'I'/V_1)]V_2$. The intersection point K is on the third layer. Other lines parallel to and equidistant from JK and $J'K$ also intersect on this refractor and so can be used to define it. This method can be extended to additional layers and also to cases where the different refractors strike in different directions. See Baumgarte, 1955, Konstruktive Darstellung von Seismischen Horizonten: Geophysical Prospecting, v. 3, p. 126-162.

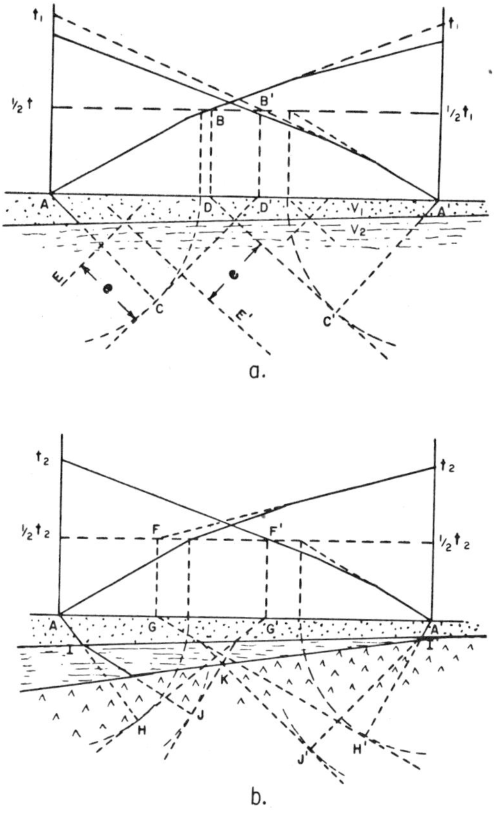

Fig. 19. **Baumgarte refraction method.** Construction of the refracting interface for (a) first refractor and (b) second refractor. (From Baumgarte, Geophysical Prospecting, v. 3, no. 2, p. 149.)

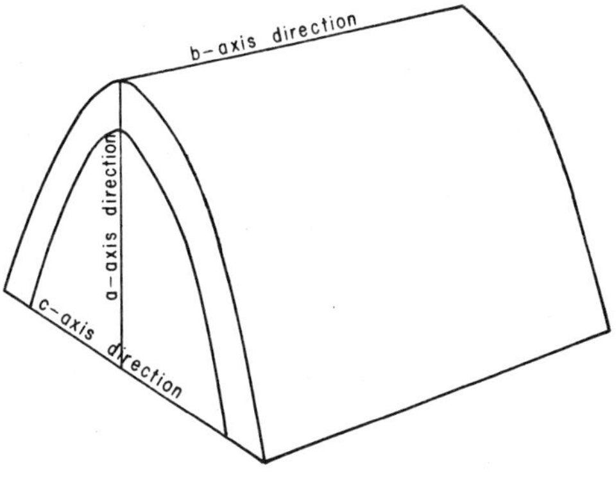

Fig. 20. **B-axis** direction for a fold.

b-axis: Beta axis, the longitudinal axis of a **fold** structure. Defined on a Schmidt net by the intersection of great circles which represent foliation surfaces. See Figure 20.

bay: A transient magnetic disturbance having a period of about an hour or so and the appearance of a bay along the seacoast on an otherwise relatively undisturbed magnetic record. Other transient magnetic disturbances include micropulsations (small rapid variations), giant pulsations, and magnetic storms. The onset of bays is usually accompanied by a micropulsation burst which accounts for some of the telluric noise observed in IP measurements.

bay cable: A marine seismic cable which is laid in place on the water bottom, as opposed to a drag cable and streamer which are towed into place.

BCD: Binary-coded decimal.

beacon: 1. A fixed navigation aid. Beacons may be either passive or active **(transponders). 2.** A radar reflector (such as a corner reflector mounted on a buoy) used as a navigation or positioning aid or to locate the tail of a seismic streamer.

Beaman: A unit of measure of the quantity $50(1 + \sin 2\alpha)$ used in calculating vertical displacement of a stadia rod with respect to the transit, where α is the angle which the line of sight makes with the horizontal. See also **stadia reduction tables.**

beam pointing: Velocity filtering.

beam steering: A method for emphasizing energy from a particular direction by delaying successive channels so that events of a certain dip moveout (or apparent velocity) occur at the same time, and then summing them. This procedure can be repeated for a succession of different dip moveouts to steer for other dips. See also **sonogram** and **Rieberize.**

bearing: The horizontal direction of one point with respect to another, usually measured as a clockwise angle. *True bearing* is the same as *azimuth* with respect to true north.

Beaufort number: A numerical scale indicating wind speed. See Table 1.

bed of nails: A two-dimensional **comb** or the two-dimensional sampling function. An impulse is located at the intersection of all integral coordinate values.

bell-shaped distribution: Normal or **Gaussian distribution.**

benchmark: 1. An established location which can be used as a survey reference. **2.** A test which can be used to evaluate performance.

bender: A type of piezoelectric transducer used in **hydrophones.** Two thin plates of piezoelectric material with metallic film on opposite surfaces are bonded onto a brass block (Figure 21) so that only the ends of the plates are supported. Pressure bends the plates, producing a voltage across each plate.

Benioff zone: Subduction zone; a dipping zone containing earthquake epicenters. Where plates of the earth's

Table 1. Beaufort wind scale and Douglas sea-state scale

Beaufort wind scale	Wind speed Knots	km/hr	Weather bureau terms	Observed effects	Sea description	Wave heights ft	Douglas sea-state
0	<1	<1	Calm	Sea like a mirror	Calm	0	0
1	1-3	1-5	Light	Ripples; smoke drifts	Smooth	<1	<1
2	4-6	6-11	Light breeze	Small wavelets; wind felt	Slight	1-3	2
3	7-10	12-19	Gentle	Waves begin to break; leaves in constant motion	Moderate	3-5	3
4	11-16	20-28	Moderate	Numerous whitecaps; dust and leaves blow	Rough	5-8	4
5	17-21	29-38	Fresh	Some spray; small trees sway			
6	22-27	39-49	Strong wind	Large waves; white foam crests; Large branches in motion			
7	28-33	50-61	Stiff wind	White foam blown downwind			
8	34-40	62-74	Stormy wind		Very rough	8-12	5
9	41-47	75-88	Strong gale		High	12-20	6
10	48-55	89-102	Whole gale		Very high	20-40	7
11	56-63	103-117			Mountainous	40	8
12	>64	>118	Hurricane		Confused		9

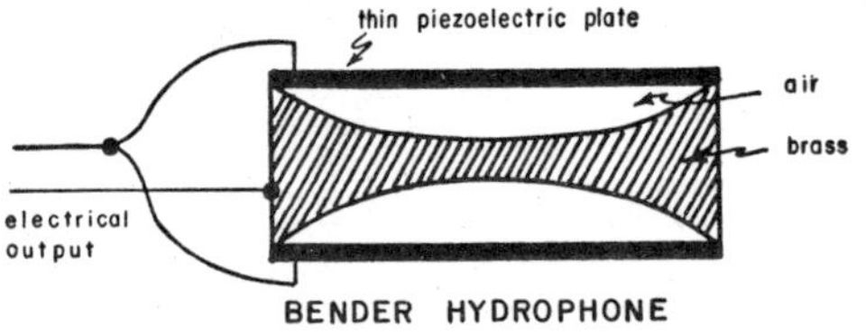

Fig. 21. Bender hydrophone.

lithosphere converge so that one moves downward beneath another, earthquakes resulting from their movement have their epicenters displaced away from the trench associated with the plate convergence, along the surface of plate junction. See Figure 136 (**plate tectonics**).

bentonite: A highly plastic, colloidal clay which increases its volume upon addition of water. Used in drilling mud. Largely made up of the mineral montmorillonite.

beta curve: A type of curve used in interpreting pulse IP data, also called *pulse curve.* Beta, the ratio of observed apparent chargeability to the true chargeability of the lower medium, a bilogarithmic weighting function developed from IP theory, is proportional to IP phase angle. A beta curve for a simple single horizontal layer where only the lower material is polarizable shows the **resistivity contrast factor** as a funtion of the ratio of array interval to depth (alpha) and beta. See Seigel, Chap. 2, in Wait, 1959, Over-voltage Research and Geophysical Applications: New York, Pergamon Press.

beta diagram: A *cyclographic diagram;* a **stereographic projection** in which the direction of a plane is represented by its great circle intersection.

BHC: Borehole-compensated sonic log; see **compensated log.** BHC is a Schlumberger trademark, BHC Acoustilog a Dresser Atlas tradename.

BHT: Bottom hole temperature.

BHTV: Borehole televiewer.

bias: 1. The amount by which the average of a set of values departs from a reference value. 2. Superposing an additional magnetic field upon the magnetic field associated with the signal during magnetic tape recording. Magnetic biasing is used to obtain a linear relationship between the amplitude of the signal and the remanent flux density in the recording medium. See **ac-bias recording.** 3. A voltage which is maintained at a point in a circuit so that the device will operate with desired characteristics. 4. A diode is said to have *forward bias* when the voltage across it is such that current flows through it, and *back bias* when the opposite polarity is maintained so that no current flows.

bimodal: A set of observations which includes members of two populations that have different distributions.

binary: Composed of only two elements. A number system in which two digits, 0 and 1, are used, the position of the digits representing powers of two; e.g., 11011 represent $2^4 + 2^3 + 2^1 + 2^0 = 16 + 8 + 2 + 1$ or the decimal number 27. See Table 3 (**Number system**). Negative numbers may be represented by a minus sign or by codes such as the **one's-complement** or **two's-complement** codes.

binary-coded decimal: BCD; a number system code in which decimal digits are represented by four binary digits; see Table 3 (**number system**); e.g., in the BCD number 0010 0111, the 0010 represents $2^1 \times 10 = 20$ and the 0111 represents $2^2 + 2^1 + 2^0 = 7$, giving the decimal number 27. The **excess-three code** (BCD×S3) is the BCD number plus 0011 (three).

binary digit: Bit; the marks 0 or 1, as used in the binary system.

binary gain: A **gain-control** system in which amplification is changed only in discrete steps by factors of 2. The action of binary-gain control in a seismic amplifier is shown in Figure 22. The times at which the gain steps occur is recorded so that the amplitude can be recovered later.

binate: Decimate, retaining every other sample of sampled data.

Biot-Savart law: Ampere's law.

bipole: A **dipole** electrode arrangement in which the two electrodes of the dipole are an appreciable distance apart.

biquinary: A number-system code in which a decimal digit is represented by a digit pair, $5A + B$, where A

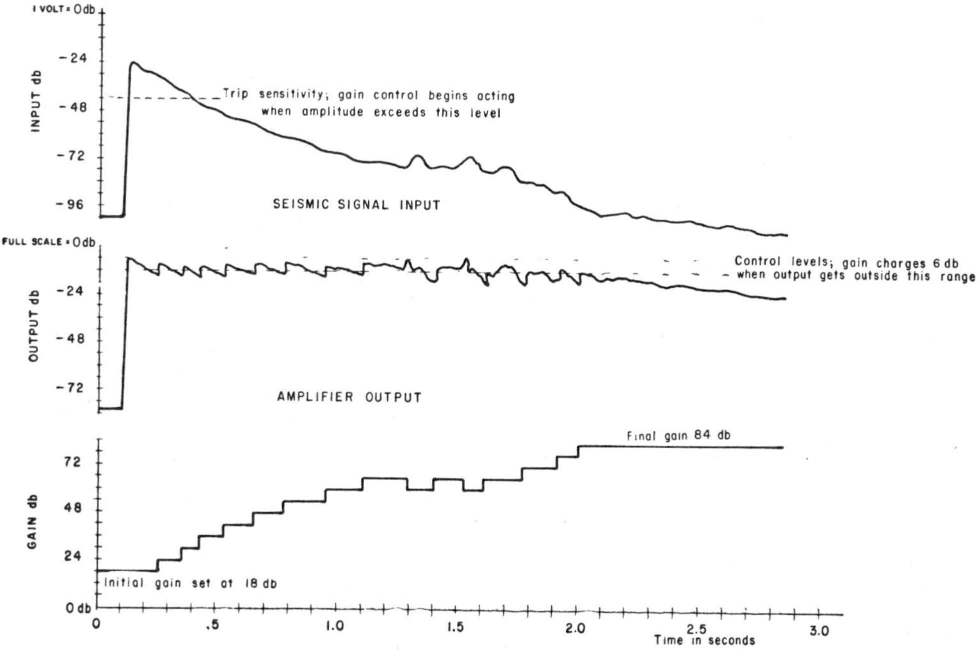

Fig. 22. **Binary-gain** action in seismic amplifier. (Courtesy Seismic Data Service.)

is 0 or 1 and *B* is 0, 1, 2, 3, or 4. Thus the decimal 7 is represented as 12 (1 × 5 + 2). See Table 3 (**Number system**).

bird: 1. A sensor (such as a magnetometer) suspended from an aircraft by a cable to make geophysical measurements. Sensors can be mounted on beams extending from the aircraft tail *(tail stingers)* and in other ways. 2. A **depth controller.**

bird-dog: The representative of the company which hires a geophysical crew, responsible for the work of the crew. Usually a staff or liaison position not carrying line responsibility.

bit: 1. A binary digit, the smallest unit of information; the necessary and sufficient information to distinguish between two choices. A bit may represent zero or one, yes or no, on or off, etc. 2. A magnetized spot on a digital magnetic tape conveying a binary digit. 3. The element on the end of a drill pipe which actually does the cutting (Figure 23). The teeth on *drag bits* tear into soft formations like sand and clay as the drill stem is rotated. The teeth on *rock bits* or *roller bits* are on wheels which turn as the drill stem is rotated, putting pressure on the rock; used to drill hard rock formations.

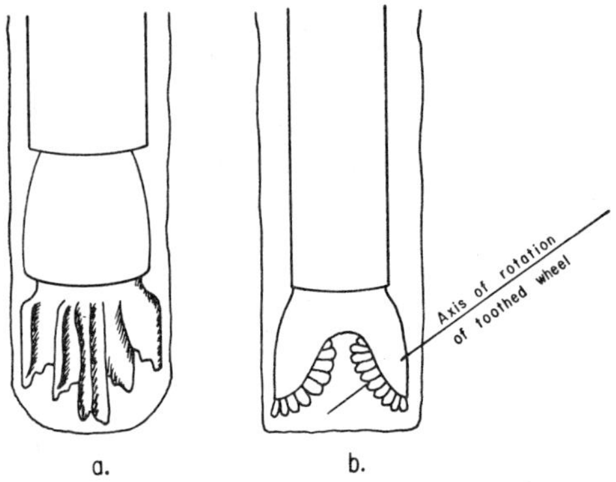

Fig. 23. Drill **bits.** (a) Drag bit or fishtail bit. (b) Rock bit or roller bit.

bit density: Packing density; the number of bits per unit length of magnetic tape, measured in bits per inch per track (BPI). Thus a 9-track tape at 800 BPI actually contains 7200 bits of information in one inch of tape.

bit shift: Multiplying a binary number by a power of two by shifting the decimal place.

black box: A unit or device whose basic function is specified but whose method of operation is not specified.

blank: To set equal to zero; to **mute.**

blaster: Device used (a) to detonate an explosive charge by sending an electric current through a blasting cap and (b) to transmit the time-break to the recording unit. Usually also includes (c) a current-limited ohmmeter for checking the blasting circuit, (d) phone or radio for communication with the recording unit, and (e) connections for the uphole geophone.

blind hole: A borehole characterized by lost circulation of the drilling fluid.

blind zone: 1. A layer which cannot be detected by refraction methods, also called **hidden layer.** The blind zone (a) may have a velocity lower than that of shallower refractors, in which case it will lead to an overestimate of the depth of deeper refractors, or it (b) may have a velocity intermediate between those of layers above and below but not have sufficient velocity difference or thickness to produce first-arrivals; in this case it is apt to cause an underestimate of the depth of deeper refractors. 2. A zone from which reflections do not occur; a *shadow zone.* 3. A zone from which no drill cuttings are returned to the surface. 4. A portion of a formation in which a logging tool response is too low. A blind zone occurs because of the finite size or configuration of the logging tool. For the lateral curve a blind zone (abnormally low reading) is recorded when a bed which is highly resistive compared to the overlying and underlying formations is present between current and measuring electrodes. 5. A layer which cannot be detected by electrical methods because its resistivity is not sufficiently different from the resistivity of other layers or because it is too thin.

block: In digital recording, a group of words or files considered as a unit.

block diagram: 1. Diagram showing the functions of a system or devices and how they are interrelated without showing construction details. 2. Diagram showing the component operations of a computer program.

Blondeau method: A method of determining vertical time to a predetermined depth based on first-break data and the assumption that the instantaneous velocity is proportional to a power of the depth. The modified Blondeau method assumes that the constant of proportionality and the exponent can vary with depth. See Musgrave and Bratton, 1967, Practical Application of Blondeau Weathering Solution, *in* Seismic Refraction Prospecting: Tulsa, SEG, p. 231-246.

blowup: 1. To become unstable; to fail to converge. 2. To enlarge, such as by photographic means.

body waves: P- and S-waves which travel through the body of a medium, as opposed to surface waves. See **P-wave** and **S-wave.**

bomb: 1. Explosive charge used as a marine seismic energy source. 2. To fail to execute on the computer, as with a data processing operation.

Boolean algebra: An algebra of elementary logical properties of statements; a system involving yes or no decisions used in computer design and programming. The rules of Boolean algebra are often shown in truth tables where 0 = no and 1 = yes:

A	B	$\bar{A}$	AND
0	0	1	0
1	0	0	0
0	1	1	0
1	1	0	1

OR	EXCEPT	NOR	NAND
0	0	1	1
1	1	0	1
1	1	0	1
1	0	0	0

Symbolic representations of "A and B" are A.B, AB, A×B, and A B: representations of "A or B" are A+B and AUB. Inversion is often shown by a superscript line: $\overline{A}$ = not A. The laws of Boolean algebra include:

$$A.0 = 0 \qquad A + 0 = A$$

$$A.1 = A \qquad A + 1 = 1$$

$$A.A = A \qquad A + A = 0$$

$$A.\overline{A} = 0 \qquad A + \overline{A} = 1$$

$$\left.\begin{aligned}\overline{A.B} &= \overline{A} + \overline{B}\\ \overline{A}.\overline{B} &= \overline{A + B}\end{aligned}\right\}\text{deMorgan's theorems}$$

EXCEPT (exclusive OR) can be shown as $A.\overline{B} + \overline{A}B$, NAND by $\overline{A.B}$, and NOR by $\overline{A + B}$. See also Figure 88 (gate).

Boomer: 1. A marine seismic-energy source in which capacitors are charged to high voltage and then discharged through a transducer in the water. The transducer consists of a flat coil with a spring-loaded aluminum plate (or plates). Eddy currents force the plates to separate sharply, producing a low pressure region between the plates into which the water rushes, generating a pressure wave by the implosion. EG & G Trademark. **2.** A very strong, usually low-frequency reflection event identified with a distinctive massive reflector.

booster: An intermediate explosive which has to be detonated by another explosive (often a blasting cap), the purpose of which is to detonate the main explosive charge.

boot: Protective shield placed around hydrophone, cable connector, sleeve-exploder unit, etc.

bootstrapping: 1. A system bring itself into a desired state by its own actions. A bootstrap program might be used to instruct the computer to load an operating program, for example. **2.** A feedback technique used to increase the input impedance of an amplifier.

borehole effect: A distortion of a well log because of the size and influence of the borehole or (sometimes) the invaded zone.

borehole gravimeter: A remote reading gravimeter which can be lowered through a borehole and held at constant depth while a gravity value is measured. Differences between the gravity readings at different depths give the mean density ρ between the depths:

$$\begin{aligned}\rho &= 3.686 - 128.5\ \Delta g/\Delta h_1\\ &= 3.686 - 39.18\ \Delta g/\Delta h_2,\end{aligned}$$

where Δg is the gravity difference in mgal, Δh_1 is the depth difference in meters, and Δh_2 is the depth difference in feet.

borehole log: Well log.

borehole televiewer: Seisviewer; a well log wherein a pulsed, narrow acoustic (sonar) beam scans the borehole wall in a tight helix as the tool moves up the borehole. A display of the amplitude of the reflected wave on a cathode ray tube (television screen) is photographed yielding a picture of the borehole wall which reveals fractures, vugs, etc; see Figure 24. BHTV is a Mobil Oil Trademark. See Zemanek et al, 1970, Formation Evaluation by Inspection with the Borehole Televiewer: Geophysics, v. 35, p. 254-269.

bottom lock: Situation where **doppler-sonar** measurements are based on reflections from the sea bottom (the normal operational mode in less than 400-1000 ft of water). As opposed to the *water-scatter* mode which occurs in deeper water.

Bouger anomaly: 1. The residual value obtained after latitude correction, elevation correction (including both free-air and Bouguer corrections), and terrain correction have been applied to gravity data. Pronounced bōō-gāy, sometimes bōō-zhāy. **2.** Sometimes, a departure from smoothness in the contours on a map showing Bouguer gravity values (an anomaly in Bouguer anomaly values).

Bouguer correction: A correction to gravity data because of the attraction of the rock between the station and the elevation of the datum (often sea level) or, in the case of stations below the datum elevation, for the rock that is missing between the station and datum. The Bouguer correction is 0.4185 ρh mgal, where ρ is the specific gravity of the intervening rock and h is the difference between the station and datum elevations in meters (or 0.01276 ρh mgal if h is in feet). In surface ship gravity data, the Bouguer correction replaces the sea water with rock and ρ in the preceding expression is the difference in specific gravity of the rock and that of sea water (usually taken as 1.03).

boundary condition: A constraint that a function must satisfy along certain boundaries, such as that it must equal certain values (often zero).

boundary-value problem: A differential equation together with boundary conditions.

boundary wave: A mode of wave propagation along the interface between media of different properties. Called **surface waves** where the boundary is a free surface such as the earth-to-air interface.

bound layer: Fixed layer.

bound water: Water absorbed in or chemically combined with shales, gypsum, or other material and which is not free to flow under natural conditions.

bow tie: The appearance of a **buried focus** on a seismic record section.

boxcar: A rectangular function. The unit boxcar [symbolically II*(t)* or rect*(t)*] has unit width and unit height. See **window.**

$$\text{II}(t) = \begin{cases}1 & -\tfrac{1}{2} < t < \tfrac{1}{2}\\ \tfrac{1}{2} & t = \pm\tfrac{1}{2}\\ 0 & t < -\tfrac{1}{2} \text{ or } t > \tfrac{1}{2}\end{cases}$$

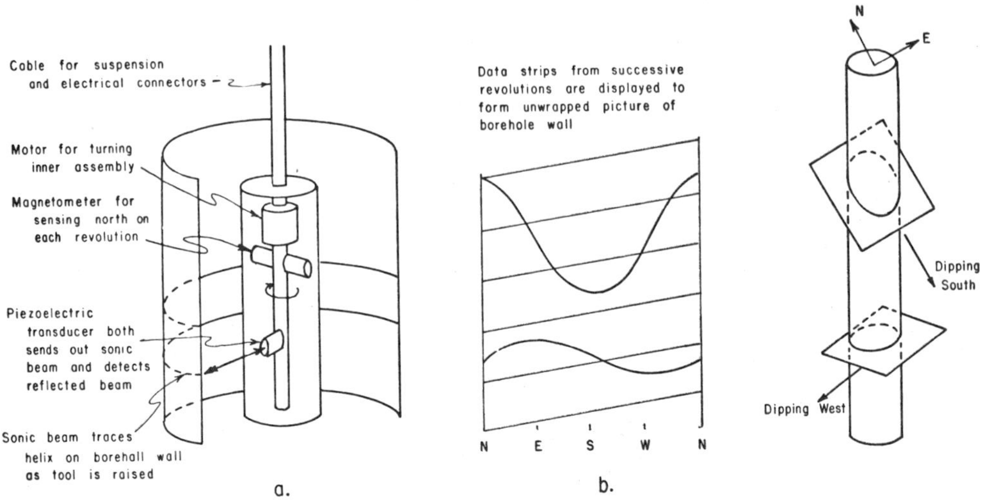

Fig. 24. Borehole televiewer. (a) Schematic of the sonde in the borehole. (b) Schematic appearance of two plane fractures, one with steep south dip and one with gentle west dip. (From Zemanch et al, Geophysics; v. 35. no. 2, p. 255 and 257.)

The Fourier transform of a boxcar is a sinc function:

$$\Pi(t) \rightarrow \mathrm{sinc}(f) = \frac{\sin \pi f}{\pi f} = \frac{\sin(\pi/t)}{(\pi/t)} = \frac{\sin(\omega/2)}{(\omega/2)},$$

where $f = 1/t$ = frequency and ω = angular frequency. For a boxcar of width T, i.e.,

$$\Pi(t/T) = \begin{cases} 0 & t < -T/2 \text{ or } t > T/2 \\ \tfrac{1}{2} & t = \pm T/2 \\ 1 & -T/2 < t < T/2 \end{cases},$$

the transform is

$$\Pi(t/T) \rightarrow T \,\mathrm{sinc}\,(Tf) = \frac{\sin(\pi T f)}{\pi T f}.$$

BPI: Bits per inch; refers to the spacing of bits along a single track on magnetic tape in the direction of motion.

brachistochrone: 1. Least-time path. Any raypath (in the usual sense) is a brachistochrone. **2.** Table of reflection time versus depth.

branches: 1. More than one reflecting event may be observed at a given location from a given reflector where it is concavely curved or discontinuous; the two or more events are called branches. See **buried focus. 2.** The sets of values for a multivalued function. **3.** More than one refraction event may be observed at a given point because of the configuration of the refractor.

branch point: 1. A point where a decision between alternatives must be made. A singularity of a function of a complex variable. **2.** The point of a Riemann surface at which two or more branches of a multivalued analytic function come together.

breadboard: An electrical circuit connected temporarily, such as to test out a new circuit before the design is finalized.

break: Onset of an event, especially the **first break.** A burst of energy indicating the arrival of new energy. See **time break** and **uphole time.**

bridge: 1. An obstruction in a drill hole above the bottom of the hole, usually formed by caving, blocking the passage of drilling tools or the explosive charge. A bridge plug may also be set deliberately. **2.** An electrical network usually having four or more arms connected in a diamond arrangement and containing one branch (the bridge) which connects two points of equal potential when the circuit is properly balanced; see Figure 25. Used for measuring electrical impedance. **3.** A jumper or wire used to short-circuit around an electrical circuit.

bridge rectifier: A full-wave rectifier circuit in which there are four arms, each containing a rectifier. See Figure 25.

bridle: 1. To connect in parallel a group of amplifiers to a common input. **2.** A seismic record produced with the amplifiers bridled. **3.** An arrangement for towing a seismic streamer. **4.** The insulation-covered lower portion of the cable to which the logging tool is connected.

broadside: 1. A reflection shooting arrangement in which the shotpoint is appreciably (more than say two hundred ft) outside the line of the spread. Also called L spread or T spread, depending on whether the shotpoint is opposite one end or the center of the spread. See Figure 171 **(spread). 2.** A refraction shooting tech-

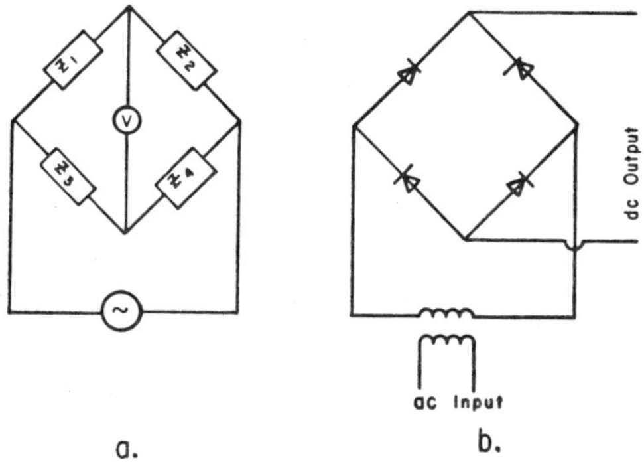

Fig. 25. (a) **Bridge** circuit. When the bridge is balanced, no voltage is measured by *V* and $Z_1/Z_2 = Z_3/Z_4$. Used to measure one unknown *Z* in terms of the other 3 known *Z*s. (b) Bridge rectifier for full-wave rectification.

nique in which the spread is perpendicular to the line connecting it with the shotpoint. The shot-to-spread distance is usually kept nearly constant. **3.** Electromagnetic-surveying procedure in which the transmitter coil is moved along one line while the receiver coil is moved along a parallel line. Compare **inline.**

broomstick charge: A **directional charge** consisting mainly of a long helical coil of Primacord. The pitch of the helix is such that the speed of the explosion front along the axis of the helix is approximately equal to the seismic velocity in the surrounding medium. Used to build up the sharpness of the downward traveling wave to help reduce ghosting.

Brunton: A small pocket compass with sights and a mirror reflector, used for measuring angles as well as for determining directions. See Figure 26. Various arrangements of the sights and mirror permit one to sight on objects and simultaneously read the scale. Often the Brunton is used merely as a protractor.

bubble: An oscillating bubble of gases, such as formed by an explosion in water, which generates **bubble pulses.** Most high-energy marine seismic sources generate bubbles; the bubble oscillation frequency is given by **the Rayleigh-Willis relation.** Various arrangements such as **Flexotir, wave-shape kits,** and **Maxipulse** are used to attenuate bubble effects or correct for them. Bubble oscillations can be prevented if the bubble is generated close enough to the surface that the gases vent before the bubble collapses. This occurs if an explosive of *w* pounds of dynamite is fired shallower than *d* ft, where $d = 3.8\ w^{1/3}$.

bubble pulses: 1. Pulses set up by oscillation of a bubble of high-pressure gas in the water. See Figure 27. **2.** Repetitions on a seismic record of the first arrivals and all other shot-generated events because of oscillations of the bubble. Bubbles are seen occasionally in marsh or land shooting.

bucking electrodes: Guard electrodes.

buffer: 1. An intermediate storage device which accommodates differences between the rate at which information is fed into a computer and the rate at which the computer can receive it, or which performs the same function for information at the computer output or between component parts of a computer. **2.** An isolating circuit to prevent a reaction of the circuit which is driven from affecting the circuit which is doing the driving. Thus a buffer amplifier may follow a critical stage to isolate it from load impedance variations in subsequent stages.

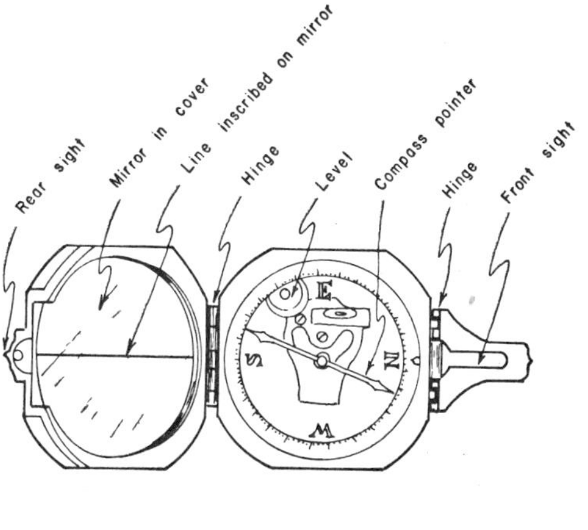

Fig. 26. **Brunton compass.**

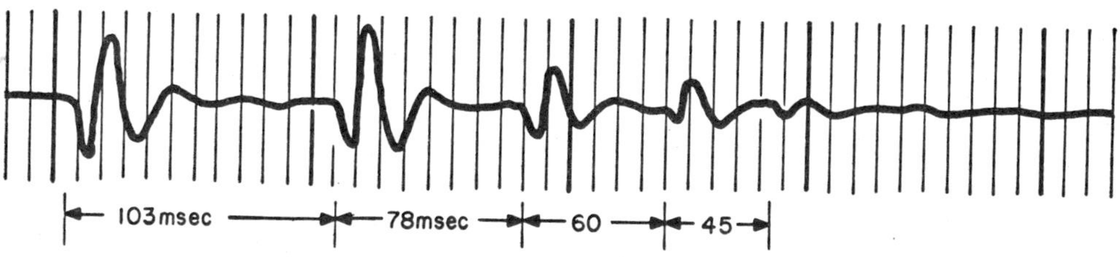

Fig. 27. **Bubble pulses** from small underwater explosion. The time between successive implosions decreases as energy is dissipated. (Courtesy Chevron Oil Company.)

bug: Error or malfunction in equipment, computer program, etc.

bulk modulus: See **elastic constants.**

bullet: **1**. A device for obtaining **sidewall cores. 2.** A device for **perforating.**

buried focus effect: A situation where the concave upward curvature of a reflector is large enough that the energy focuses before it reaches the recording plane. A buried focus situation causes several branches (usually three) of a reflection to be observable from the same surface location (i.e., the law of reflection is satisfied for several points on the reflector from the same surface location); see Figure 28. The portion which passes through the focus is called the *reverse branch*. For zero offset and constant velocity, a buried focus occurs if the center of curvature lies beneath the recording plane. Less curvature is required to produce the buried-focus effect for offset traces than for traces at normal incidence. Hence buried-focus effects are more likely on long offset traces as well as more likely deeper in the section. Velocity gradients and curvature of isovelocity surfaces (which are apt to occur in structural areas) also affect buried-focus effects. Curvature of the reflector out of the plane of the seismic line can also cause multiple branches. There is a quarter-wave phase shift in the reverse branch.

burst out: A large sudden increase in amplitude which threatens to exceed the linear capabilities of a system.

bus: Common connector for electrical power or computer data.

bust: A failure to tie a survey loop within acceptable standards.

butterfly filter: A velocity filter that rejects a prescribed band of apparent velocities or moveouts. As opposed to a *Pie-slice* filter which passes a band of moveouts. Texas Instruments trademark.

Butterworth filter: A type of frequency filter characterized by a very flat passband, often used as an alias filter. Sometimes called a *maximally flat filter.*

button: The small circular electrode on a microresistivity sonde.

byte: A portion of a digital tape word consisting of several bits (as those across the width of a tape) which are manipulated as a unit. See also **character**.

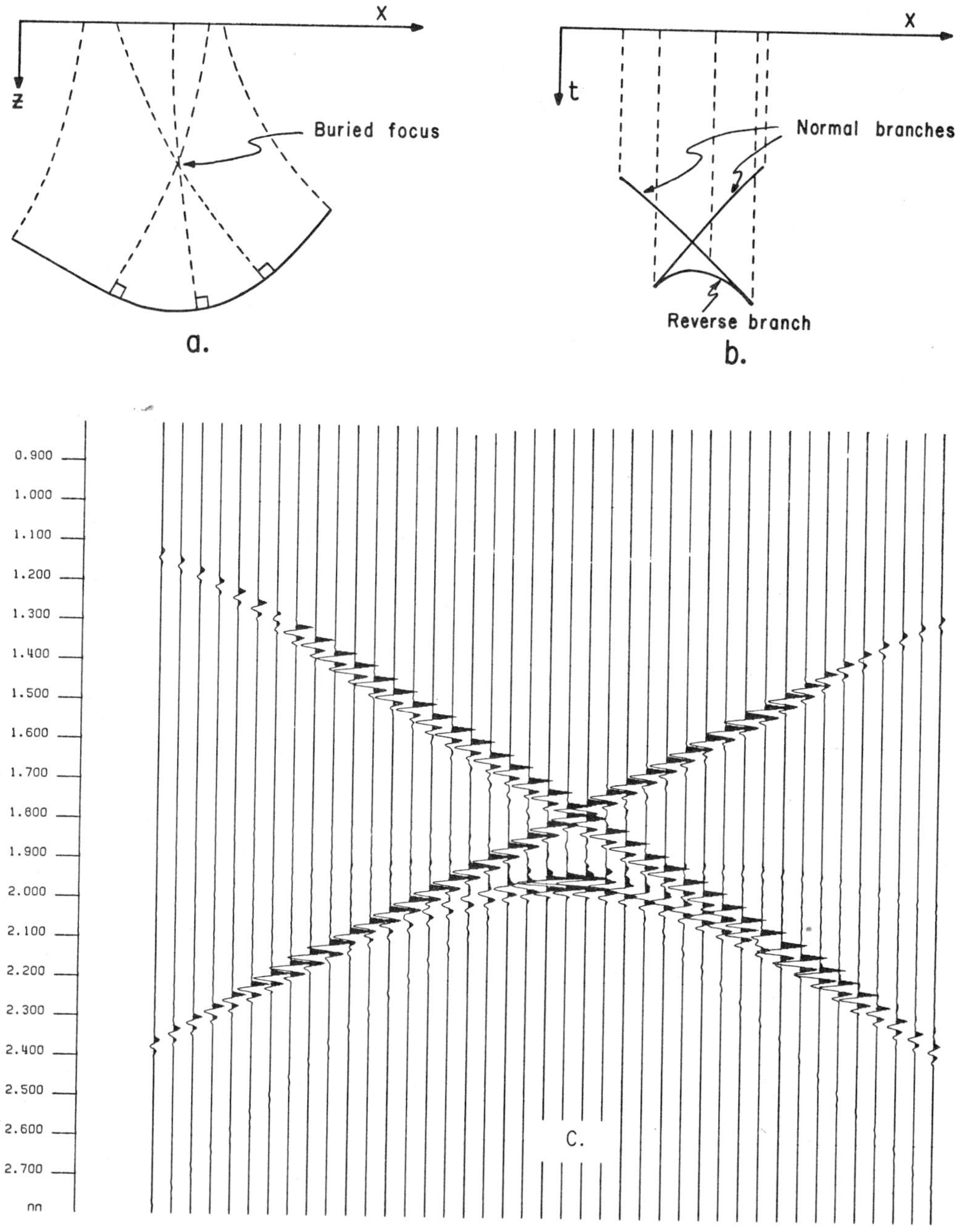

Fig. 28. **Buried focus** reflection. (a) Model of a simple syncline. (b) Schematic time section corresponding to model. (c) Record section showing three branches. Note the 90-degree phase shift in the reverse branch. (Courtesy Chevron Oil Company.)

C

cable: **1.** The assembly of electrical conductors used to connect the geophone or hydrophone groups to the recording truck. See also **streamer. 2.** The assembly of electrical connectors and tensile members used to support a logging sonde or well geophone or bird.

cablebreak: An arrival in a well-velocity survey caused by energy travel in the cable which supports the well geophone. Cablebreak may obscure the desired arrival.

cable drilling: A method of making a hole in hard rock by alternately lifting and dropping the tool.

cable strum: Vibration of a marine streamer produced by sudden tension such as might be caused by pitching of the towing ship or jerks from the tail buoy. A source of noise in marine seismic data.

cable tools: Equipment for drilling a hole by alternately raising and dropping a tool. Occasionally used for seismic shothole drilling in areas of extremely hard surface rock. See **spudder.**

cage: Perforated steel which surrounds a **bubble** energy source (small explosive or air gun), which attenuates the bubble effect by dissipating the energy in turbulent flow of water in and out of the ports in the cage. See Figure 78 **(Flexotir).**

cage shooting: Flexotir.

cake: A deposit of mud which adheres to a borehole wall, especially adjacent to porous formations which absorb much of the mud's moisture.

calibrate (a sonic log): Adjust values so that the integrated traveltime agrees with data from surface shots into a well seis in the borehole.

calibration: **1.** The number of units of the quantity being measured per scale division of the readout device. **2.** A check of equipment readings, made by making a measurement on a known standard.

calibration resistor: A pure resistance of known value used to calibrate the zero-frequency-effect level of a frequency-domain transmitter and receiver.

calibration tails: Calibration records run before and/or after a log run and attached to the logs.

caliper log: A well log which measures hole diameter. Open hole caliper logging tools sometimes have 4 or more arms. Also called *section gauge.* See Figures 117 **(microlog)** and 167 **(sonic log)**. Tools for studing the corrosion of casing or tubing use many "fingers."

camera: A recording oscillograph used to produce a visible pattern representing electrical signals or to make a visible seismic record, usually on photosensitive paper or film. See also **plotter.**

cap: A small explosive designed to be detonated by an electrical current and which in turn detonates another explosive.

capacitance: The ratio of charge *(Q* in coulombs) on a capacitor to the potential across it *(V* in volts) is the capacitance *(C* in farads):

$$C = Q/V.$$

capacitive reactance: Electrical impedance X_C due to capacitance:

$$X_c = 1/(2\pi fC) \text{ ohms},$$

where f is frequency in hertz and C is capacitance in farads.

capacitivity: Permittivity; the property of a material which enables it to store electrical charge. Measured in farads per meter. The ratio of the capacitivity of a material to that of free space is the *dielectric constant"* of the material. Free space has a capacitivity of 8.854×10^{-12} farad/m.

capacitor: A device which adds **capacitance** to an electrical circuit. The standard color code for capacitors is shown in Figure 29.

caprock effect: A sharp positive gravity anomaly superimposed on a broader negative, produced by the dense caprock of a salt dome. Very shallow salt is itself denser than the surrounding sediments, so caprock is not essential in producing this effect.

capture cross-section: **1.** *Atomic capture cross-section* for neutrons is the effective area within which a neutron has to pass in order to be captured by an atomic nucleus. It is a probabilistic value dependent on the nature and energy of the particle as well as the nature of the capturing nucleus. Atomic capture cross-section is often measured in barns (1 barn $= 10^{-24}$cm^2). **2.** *Macroscopic capture cross-section* Σ is the effective cross-sectional area-per-unit volume of material for capture of neutrons; hence it depends on the number of atoms present as well as their atomic capture cross-sections. The unit of measure for Σ is cm^2/cm^3 or *reciprocal cm* (cm^{-1}); Σ is often measured in *capture units* or *sigma units;* 1 cu $= 10^{-3}$ cm^{-1}. **3.** The rate of absorption of thermal neutrons with a velocity v is thus $v\Sigma$.

capture unit: A unit of measure of macroscopic **capture cross-section;** 10^{-3} cm^{-1}. Also call *sigma unit* and abbreviated cu or su.

carbonate-analysis log: A log computed from laterolog, microlaterolog, sonic, density, and sidewall neutron logs which includes a **secondary-porosity index** curve and a **movable-oil plot.**

card: A plug-in printed-circuit module.

cardinal theorem: Sampling theorem.

Carpenter electrode array: Four collinear equispaced electrodes used in electrical surveying. In the Carpenter #1 (or Wenner) array, the two central electrodes are the potential electrodes; in the Carpenter #2 array, one end pair is the potential electrodes; in the Carpenter #3 array, the current and potential electrodes alternate.

cascade: Arrange in series.

casing: Tubes or pipes used in shotholes to keep them from caving in. Usually made in ten-ft lengths which screw together.

casing collar: The coupling between joints of casing.

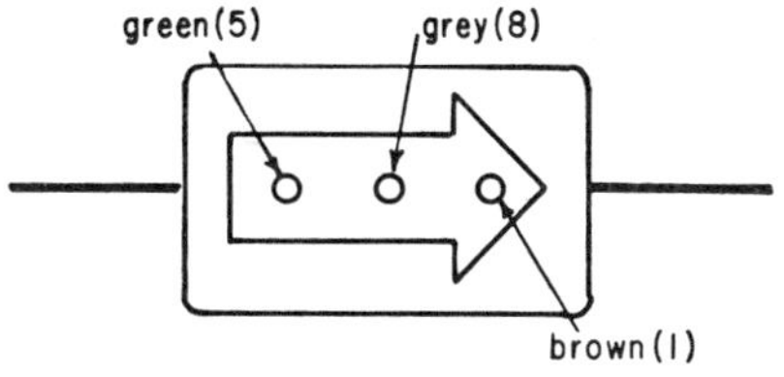

A three-dot 580 µµF mica capacitor

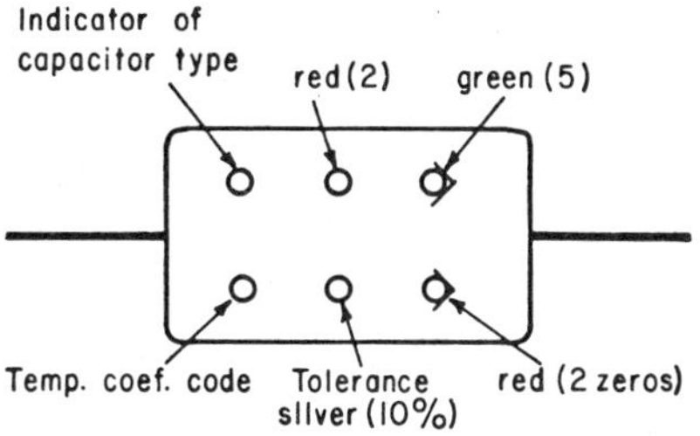

A six-dot 2500 µµF mica capacitor

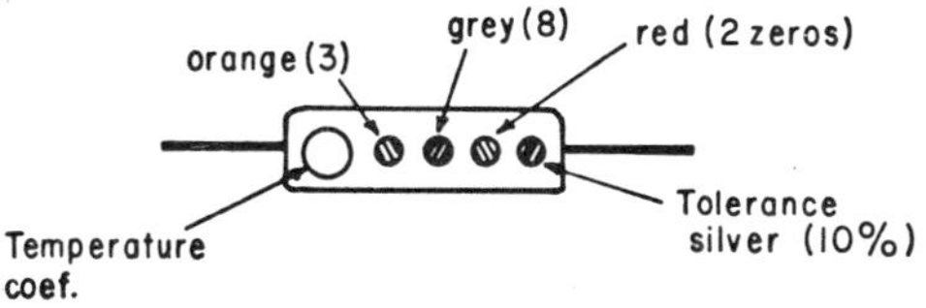

A 3800 µµF ceramic capacitor

Fig. 29. **Capacitor** identification codes. The numerical values for each color are listed in Figure 153 **(resistor)**.

casing-collar locator: A magnetic or scratcher device for locating casing collars, which are used for correlating collars with other logs and which often serve as reference depths in subsequent completion operations.

catalog: Collection of **master curves** or **type curves** for interpreting magnetic, electrical, and electromagnetic survey measurements.

catcher: **1**. A device to prevent cap wires from being blown out of the hole by the shot and endangering personnel by making contact with high-voltage power lines. **2**. A device to catch a Dinoseis gas gun before it falls back to the ground after an explosion, thus preventing a second sharp impact which would complicate the waveform.

cathode: **1**. The electrode at which reduction occurs and electrons are taken up. The negative terminal of an electrolytic cell or the positive terminal of a battery. **2**. The source electrode for electrons inside a vacuum tube or semiconductor diode.

cathode-ray oscilloscope: A device for viewing waveforms (such as voltage as a function of time) and other information. Abbreviated CRO.

cation: A positively charged ion.

cationic membrane: A membrane which permits the passage of cations but not of anions. Shale acts as such a membrane, allowing sodium ions to pass but not chloride ions. Important in generating the **electrochemical SP**.

caved portions: Parts of a borehole where the hole diameter becomes large.

cavitation: The situation where the pressure in a liquid becomes lower than the hydrostatic pressure—it often reaches the vapor pressure of the liquid. The collapse of liquid into the region produces very large pressures and generates a shock wave by "implosion." The outward momentum of water and gases from an underwater explosion (and other marine energy sources) tends to leave behind such a low-pressure region, the collapse of which produces the **bubble** effect.

CBL: **Cement-bong log.**

CDM: **Continuous dipmeter.**

CDP: **Common-depth-point.**

CDPS: **Common-depth-point stack.**

celestial latitude: Angular distance north or south of the ecliptic. Differs from declination.

celestial longitude: Angular distance east of the vernal equinox measured along the ecliptic. Different from right ascension.

celestial pole: The projection of the earth's axis onto the celestial sphere.

celestial sphere: An imaginery sphere of infinite radius concentric with the earth on which all celestial bodies are imagined to be projected.

cementation factor: The exponent *m* in the **Archie Formula.**

cement-bond log: A well log of the amplitude of the acoustic wave which indicates the degree of bonding of the cement to the pipe and formations. See Figure 30. If the casing is poorly cemented, energy which travels through the casing at the fast speed of acoustic waves in steel is strong and little energy travels in the formation; if the casing is well cemented, the casing signal nearly disappears and the formation signal is strong. The log may consist of **1**. an **amplitude log (CBL)** which represents the amplitude of a portion of the longitudinal **acoustic wave** train, or **2**. a display of the **acoustic wave** train such as the **character log,** three-D, microseismogram, VDL, or acoustic signature log.

centipoise: A unit of viscosity, equal to one hundredth of a gram per sec-cm. The viscosity of water at 20°C is 1.005 centipoise.

central induction sounding: An electromagnetic technique in which the vertical magnetic field is measured as a function of frequency at the center of a large horizontal transmitting loop.

central processing unit: The heart of a computer which controls operations and interprets and executes programs. Abbreviated CPU.

CEP: Circular error probability; the radius of a circle such

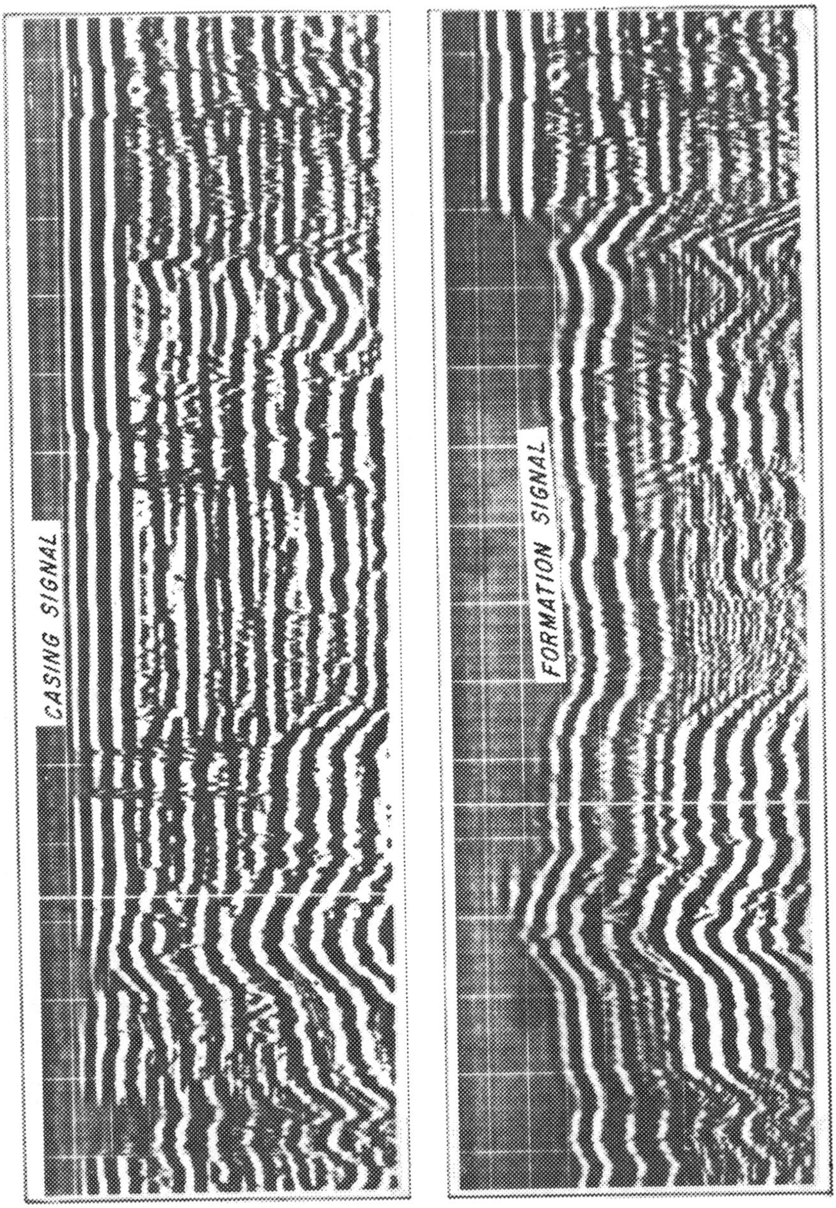

a b

Fig. 30. **Cement-bond** application of microseismogram log. (a) Before squeeze. (b) After squeeze. The disappearance of the casing signal indicates a successful cement squeeze. (Courtesy Welex.)

that half of the measurements fall within the circle.

cesium magnetometer: A type of **optically pumped magnetometer.**

chain: 1. A unit of length equal to 66 ft. **2.** To measure distances directly with a steel tape (chain) as opposed to measuring by stadia or by triangulation. Accuracy of the order of 1:5000 can be achieved with ordinary care. See **chainman.**

chainman: A surveyor's assistant, who helps to measure distances with a chain. Chainmen often work in pairs with a steel tape or wire which is as long as the group length, to measure out seismic spreads. The tail chainman holds one end of the chain on a previously determined group position while the lead chainman walks along the line until the chain becomes taut, backsighting to be sure he is on a straight line. He then "flags" the group and the chainmen move ahead to position the next group.

Chandler wobble: A precession of the earth's rotational axis about its axis of greatest moment of inertia; also called the *free nutation of the earth.* The wobble period is about 435 days and the amplitude is about 0.14 sec of arc.

channel: 1. A single series of interconnected devices through which data can flow from source to recorder. Most seismic systems are 24 channel, allowing the simultaneous recording of energy from 24 groups of geophones. **2.** A localized elongated geological feature resulting from present or past drainage or water action. Such channels often present weathering problems. **3.** An allocated portion of the radio-frequency spectrum. **4.** A time gate during which measurements are made in time-domain electromagnetic surveying. Measurements made during several time gates following source pulses yield several channels of data. **5.** A layer whose velocity is lower than those above and beneath it. See **channel wave** and **SOFAR. 6.** A gate with upper and lower energy thresholds followed by a count rate meter, used in gamma-ray spectrometers.

channel wave: 1. An elastic wave propagated in a layer of lower velocity than those on either side of it. Energy is largely prevented from escaping the channel because of repeated total reflection at the channel boundaries or because rays which tend to escape are bent back toward the channel by the increasing velocity away from it in either direction. The **SOFAR** channel in the oceans is an example of a situation which carries a channel wave. See Figure 31. Energy from a source in the channel, which begins its travel at a small angle with the channel, travels faster than energy along the axis of the channel. Hence an observer in the channel first will hear the energy which begins at the larger angle with the channel, then energy with smaller angles, and finally the energy which travels along the axis. Hence the channel-wave signal from an impulsive source is characterized by a buildup of energy and an abrupt termination. **2.** A channel may consist of the region between two interfaces even without regard to velocity gradients within the region. Energy may be largely or totally reflected at the interfaces so that the energy largely stays within the channel. See also **hidden layer.**

character: 1. The recognizable aspect of a seismic **event;** the waveform which distinguishes it from other events. Usually a frequency or phasing effect, often not defined precisely and hence dependent upon subjective judgment. **2.** A single letter, numeral, or special symbol in a computer system. See also **byte,** with which

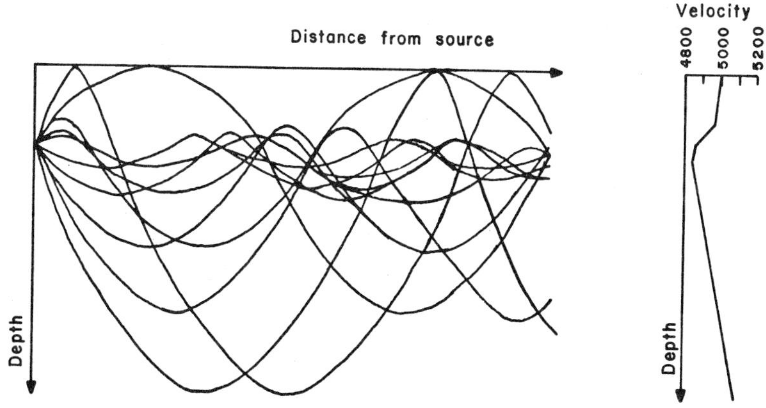

Fig. 31. **Channel waves** in the SOFAR channel. This channel in the ocean is formed by velocity inversion. Because the energy repeatedly gets refracted back toward the velocity minimum, the energy undergoes less divergence than normal. (From Ewing, Worzel, and Pekeris, 1948, G.S.A.Memoir 27, part III.)

it is sometimes used interchangeably. Or it may require several bytes to express a character.

characteristic value: Eigenvalue.

character log: A display of the **acoustic wave** train in wiggle form, as opposed to the similar sort of display in variable density form in the **three-D log,** microseismogram, or VDL log. See also **sonic log, fracture log,** and **cement-bond log.** Similar to *signature log.*

charge: The explosive combination used for a seismic shot, often specified by the weight and type of explosive and sometimes by its length. "Making up a charge" involves assembling the main explosive, primer, and cap so that it is ready to detonate.

chargeability: One of several units of induced polarization in the time domain. Symbol: *M.* **1.** The ratio of initial decay voltage (or *secondary voltage*) to primary voltage. **2.** The dimensionless, induced-polarization parameter of a material in which there is an induced-current dipole moment-per-unit volume P energized by a current density J:

$$\mathbf{P} = -M\mathbf{J}$$

3. The fractional change in resistance measured on a decay curve, as a function of time:

$$M = \Delta\rho(t)/\rho.$$

4. As often measured in the field, the integrated area under an IP decay curve between times t_1 and t_2, normalized by the primary voltage V_p. Units are millivolt-seconds-per-volt.

$$M = \frac{1}{V_p}\int_{t_1}^{t_2} V_t\, dt.$$

For standardization, *on time* and *off time* may be indicated by subscripts; that is $_{33}M_1$ or $M_{(331)}$ means "current on for three sec and decay measured for the first second of a three-sec off time." Field measurements of chargeability are usually calibrated to the $M_{(331)}$ standard, which differs by a factor of about 1000 from the value of *M* given in definitions **1** and **2** above. **5.** The quantity described above multiplied by the conductivity, thus expressed in millifarads/unit length. Also called *specific capacity* when measured in farads/m.

charge anchor: A device fastened to an explosive charge to hold the charge in a fixed position in the shothole and prevent it from floating or moving when another charge in the same hole is detonated.

Chebychev array: A frequency filter or uniformly spaced linear array in which the elements are weighted according to Chebychev polynomials. Such weighting equalizes the height of minor lobes and gives a sharp cutoff. Called *equal ripple filter.* See Figure 32. Also spelled *Tchebyscheff array.*

check: A test of the reliability of data. Check problems with known correct results are run through computers to ascertain correct functioning. **a.** A *summation check* consists of adding a set of figures and using the sum to verify accuracy and completeness. **b.** A *duplication check* requires that an operation be repeated and give identical results. **c.** An *echo check* is used to verify transmitted data; the data are sent back (echoed) to the source station for comparison with the original data. **d.** *Error detecting checks* search data for forbidden combinations. A *parity check* is such a check; the number of bits plus a parity bit must add up to an odd number for odd parity or to an even number for *even parity.* **e.** A *validity check* is a verification that a figure lies within a certain permissible range.

check shots: Shots into a well seismometer (see **well shooting**) to check the results of integrating a continuous velocity or sonic log.

check sum: A *summation check;* see **check.**

chemisorption: Adsorption due to chemical rather than simply electrostatic causes.

chip: A solid-state circuit or circuit element; an integrated circuit.

chirp: A **Vibroseis** signal; a sinusoidal signal of continuously varying frequency. Often implies a linear change of frequency with time.

chi-square: A quantity distributed as $x_1^2 + x_2^2 + \ldots + x_k^2$, where $x_1, x_2, \ldots, x_k$ are independent, Gaussian, with zero mean, and with unity variance. The *chi-square test* is a statistical test of data distribution. if f_i are observed frequencies of occurrence and e_i are the expected number of occurrences, the "goodness of fit" *y* is

$$y = \sum \frac{(f_i - e_i)^2}{e_i}.$$

The goodness of fit can be related (by the use of tables) to the probability that the observed data fit the model with the assumed number of degrees of freedom.

chlorine log: A log based on the counting rate of gamma rays produced by capture of thermal neutrons by chlorine in the formation. By limiting the count to a certain energy range, the tool is made more sensitive

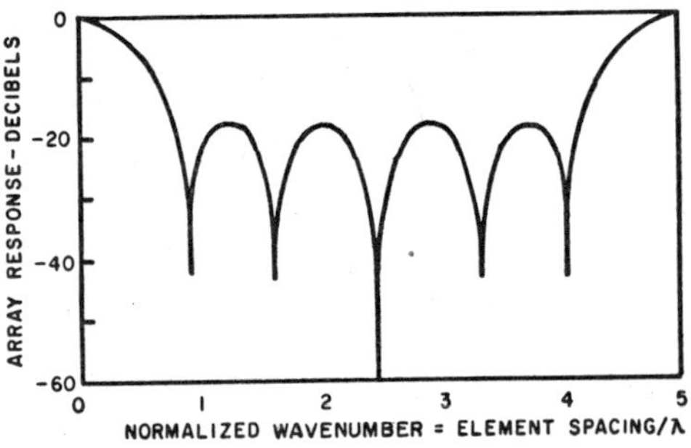

Fig. 32. **Chebychev array** response has all minor lobes of equal height. Response shown is for 6-element array.

to chlorine and relatively insensitive to formation porosity. The chlorine log is now essentially replaced by the **neutron-lifetime log** and thermal-decay-time log.

chopper: An electrical switching device sometimes including an oscillator, such as used to interrupt a dc or low-frequency ac voltage so that it can be measured by an ac voltmeter or amplified by an ac amplifier.

chromatograph: See **partition gas chromatograph.**

circulation: The movement of drilling fluid from a slush pit through pump, drill pipe, annular space in the hole, and back into the slush pit.

circulation loss: Loss of drilling fluid into a porous (or "thief") formation.

cis θ: Euler's identity:

$$\text{cis } \theta = \cos\theta + i\sin\theta = e^{i\theta}.$$

Clarke ellipsoid: The basis for the North American geodetic datum and the reference datum in most of the Western Hemisphere; the Clarke 1866 ellipsoid. (There is also a Clarke 1880 ellipsoid which is used in Africa.) See Table 5 **(Geodetic systems).**

clean: Containing no appreciable amount of clay or shale. Applied to sandstones and carbonates. Compare **dirty.**

clipped: Distorted because amplitude exceeded a maximum permitted amount. Clipping in analog systems usually occurs because of saturation of some element of the system, resulting in distortion of the waveform; see Figure 33. Digital clipping (loss of the first bit) may have variable effects depending on what the first bit represents (i.e., whether the first bit is the sign, the most significant bit, a complement code, etc.); the clipped signal even may bear little resemblance to the unclipped signal.

clipped trace: A display (such as a seismic trace) where amplitudes in excess of a certain amount are not shown.

closure: **1.** The property of a structure whereby it has a closing contour. *Vertical closure* is the vertical distance from the apex to the lowest closing contour; *areal closure* is the area contained within the lowest closing contour. Compare **trap. 2.** The cumulative error around a survey loop; **mis-tie.**

clutter: **1.** Noise on a radar screen resulting from unwanted echoes or scattering such as from a rough sea or from rain. **2.** Coherent interference, in contrast to incoherent nonrepeatable interference.

coastline effect: Distortions in electromagnetic ground waves and natural magnetic fields produced by the contrast in electrical conductivity between land and highly conductive ocean water.

coefficient of anisotropy: See **anisotropy.**

coefficient of variation: Ratio of standard deviation to the mean for a set of quantities. See **statistical measures.**

coercive force: The magnetic field intensity required to reduce the magnetization of a sample to zero. Generally associated with the remanent magnetization acquired from a saturating magnetic field. See **hysteresis** and **coercivity.**

coercivity: **1.** The demagnetizing field intensity required to reduce the induction of a magnetic tape from saturation to zero. Coercivity is used as a figure of merit for magnetic tape. See Figure 99**(hysteresis). 2.** The magnetic hardness of a material, particularly in reference to the distribution of coercive forces of a magnetic system's components. The latter might be different mineral phases, grain sizes, etc.

cofactor: See **matrix.**

coherence: **1.** The property of two wave trains being in-phase. **2.** A measure of the similarity of two functions or portions of functions. If the functions have power spectra P_{ii} and P_{jj} and cross-power spectra P_{ij} (which may be complex), their coherence is

$$P_{ij}/(P_{ii}\,P_{jj})^{1/2}.$$

Also called *coefficient of coherence.* Coherence is the frequency-domain concept which is analogous to **correlation** in the time domain.

coherent: Having a fixed-phase relationship with respect to each other. A set of waveforms is coherent if the phase change from one to the next has a well-defined relationship. For example, seismic reflection events are coherent in a linear way with respect to dip, coherent in a hyperbolic way with respect to normal moveout, coherent with respect to weathering variations in a nonanalytic though systematic way with respect to geophone locations. The principal evidence for a separate seismic event is coherence among the members

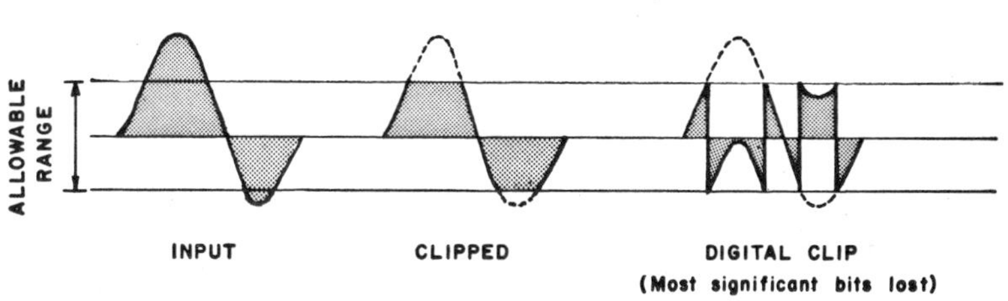

Fig. 33. **Clipping.** Loss of a digital bit can cause various effects depending on how the data are formatted (e.g., the sign bit might be lost).

of a set of seismic traces over a short time interval of the order of 1-1/2 or so cycles of the dominant frequency compared with less coherence elsewhere. Coherence is often measured in a qualitative way in record picking but quantitative measures of coherence are used in automatic picking scheme; see **semblance.** Coherence is used in connection with various types of wave trains, including electromagnetic and seismic waves.

coherent detection: Coherent filtering; a method of noise suppression that filters out the coherent portion of two signals.

colatitude: The complement of the latitude, or 90 degrees minus the latitude.

collar log: Casing-collar locator.

cologarithm: The logarithm of the reciprocal of a number.

colored: Having an arbitrarily shaped amplitude spectrum. Compare **white.**

color mimicry: A technique for correlating the response to different types of measurement for features in two-dimensions. The set of 2 or 3 two-dimensional photos (or maps or other graphical display) are projected through different primary color filters and the images superimposed. See Grossling, 1969, Color Mimicry in Geology and Geophysics: Geophysics, v. 34, p. 249-254.

comb: An infinite sequence of **impulses** $\delta(t - n)$ occurring at unit time intervals. Symbol: $\text{III}(t)$.

$$\text{III}(t) = \sum_{n=-\infty}^{\infty} \delta\,(t - n).$$

Called the *sampling function* because multiplication of a function and a comb gives the sample values at the comb interval. Called the *replicating function* because convolution with a waveform reproduces the waveform at the position of each spike. Also called a *shah.* The Fourier transform of a comb is also a comb:

$$\text{III}(t) \rightarrow \text{III}(f),$$

where the frequency $f = 1/t$, if t is time. See Figure 34. If the impulses are spaced T apart:

$$\text{III}(t) \rightarrow t\,\text{III}(Tf).$$

common-depth-point: 1. The situation where the same portion of the subsurface is involved in producing re-

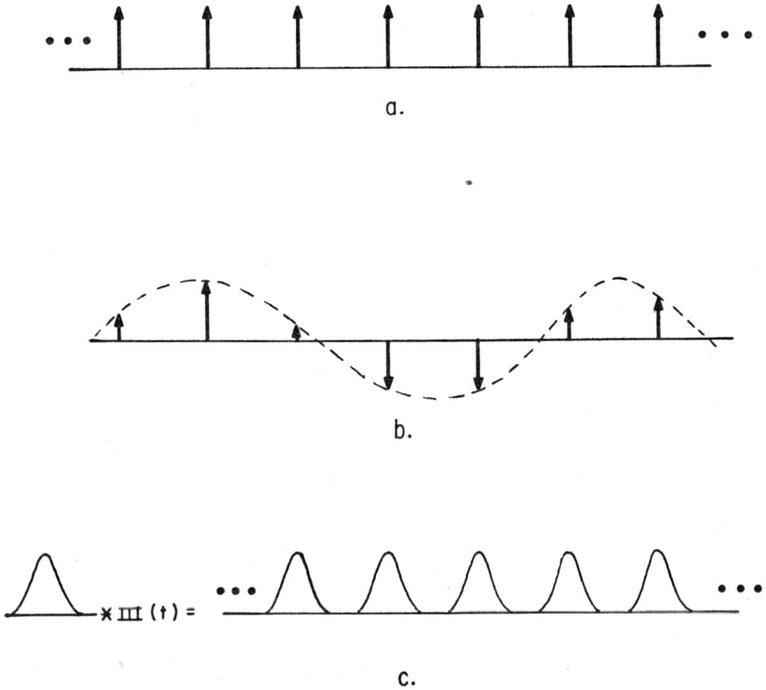

Fig. 34. (a) The **comb.** (b) Multiplication of a function by a comb samples the function. (c) Convolution of a function with a comb replicates the function.

flections at different offset distances on several profiles. **2.** Common-depth-point shooting produces redundant reflection data from which a **common-depth-point stack** can be made. Different shotpoint-geophone combinations are used to record the same reflection from the same subsurface. Also call *roll-along.* See Figure 35. Abbreviated CDP and CRP.

common-depth-point stack: A sum of traces which correspond to the same subsurface reflection point but which are from different profiles and have different offset distances. The traces are corrected for statics and normal moveout before summing (or stacking). The objective is to attenuate random effects and events whose dependence on offset is different from that of normal moveout for primary reflections. Hence multiple reflections which show a different NMO, surface waves, refractions, diffractions, etc., will be attenuated relative to primary reflections. Also called roll-along and horizontal stacking. Petty Geophysical Engineering Company patent.

common-mode rejection: A measure of how well a differential amplifier ignores a signal which appears simultaneously at both input terminals.

common-offset gather: A side-by-side display of traces which have the same shot-to-geophone distance (offset).

common-offset stack: A stack of traces which have the same offset and which are located within a limited range of depth-point locations.

common-range gather: Common-offset gather.

common reflection point: Common-depth-point.

communication: 1. A connecting path bwtween formations so that fluids from one can flow into the other. **2.** The ability to interchange data, as when two computers "have communication with each other."

commutate: To mechanically reverse the direction of a unidirectional electric current by a change of connections, or to reverse every other cycle of an alternating current so as to form a unidirectional current.

compaction correction: An empirical correction applied to porosity values derived from the sonic log in uncompacted formations such as sands at shallow depths or formations under abnormally high pressures. Undercompaction is indicated by abnormally low sonic velocities in adjacent shales.

comparative interpretation: The comparison of survey data with **type curves** which have been calculated for bodies of assumed property contrasts and geometry.

comparator: A circuit which compares two signals and indicates the result of the comparison.

compensated log: A well log made with a sonde designed to correct unwanted effects. The compensated **density log** (FDC) uses the signal from a secondary detector to correct for the effect of mud cake and small irregularities in the borehole wall. The compensated **sonic log** (BHC) uses a special arrangemant of the transducers to correct for irregularities in borehole size and sonde tilt.

competent: A bed which retains its stratigraphic thickness under stress and folds or breaks in comparison with adjacent incompetent beds which tend to flow and so thicken and thin in the **folding** process.

compiler: A program for converting a source program in symbolic language to an object program in machine language independent of the running of the source program. An *interpreter* involves the conversion concurrently with running the program.

complement: The difference between a particular value and full scale. Thus in the decimal system the complement of x is $(10^{\mu} - x)$,where μ is a fixed number; and in binary the complement of x is $(2^{\nu} - x)$, where ν is a fixed number. Adding the complement of x is equivalent to subtracting x; computers often find it easier to generate the complement and add than to subtract. See also **one's complement** and **two's complement.**

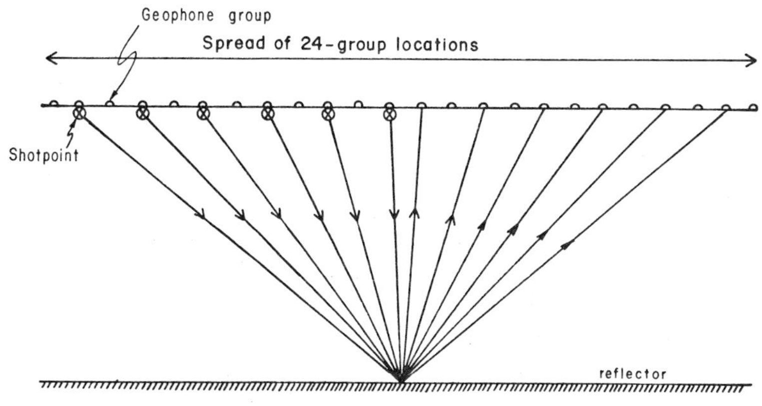

Fig. 35. **Common-depth-point** shooting. In six-fold shooting with 24-geophone groups, the shotpoint is moved two group intervals between successive shots; hence, the same subsurface is sampled six times.

complement of chargeability: An IP time-domain measurement of the area under the decay curve, integrating over the interval between 0.45 and 1.75 sec. Measured on a Newmont-type IP receiver.

complex number: A number with both real and imaginary parts, such as

$$r = x + iy, \text{ where } i = (-1)^{1/2}.$$

[The symbol *j* is also used to indicate $(-1)^{1/2}$.] Complex numbers possess both magnitude (called "modulus") and direction and thus can be used to express two-dimensional vectors. The modulus or magnitude of the above complex number is $(x^2 + y^2)^{1/2}$ and the angle indicating its direction with respect to the real axis is $\theta = \tan^{-1}(y/x)$. θ is in quadrants 1 or 4 if x is positive, in quadrants 1 or 2 if y is positive. A "graph" of a complex function or quantity (such as a frequency spectrum) is shown in Figure 36.

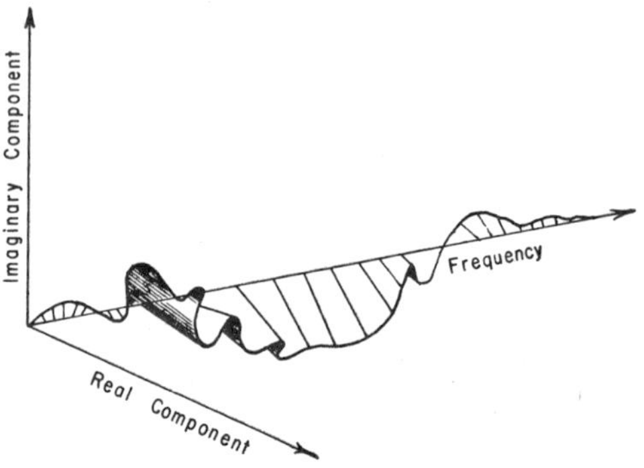

Fig. 36. **A complex function** (such as a frequency spectrum) requires three-dimensional representation. The distance of a point from the frequency axis is the modulus, and the angle with the real plane is the phase.

complex resistance: Impedance.

complex spectrum: See **Fourier transform.**

compliance: The relationship of strain to stress. For an elastic substance, the reciprocal of the elastic modulus; see **elastic constants.** The mechanical or acoustical equivalent of electrical capacitance.

composite: 1. Mix; to combine the energy of different channels without first applying static and normal-moveout corrections. 2. Made up from several different elements, as in *composite decay curve* or *composite time-distance curve.*

composite decay curve: A time-domain decay curve containing more than one component, often with different time constants or even a combination of positive and negative decay curves.

composite time-distance curve: The equivalent of the refraction arrival-time versus offset-distance relationship which would have been obtained from a single shot into an extensive spread of geophones, but actually synthesized from a number of shots and shorter spreads at various locations. See Figure 37.

compositor: A device for **mixing.**

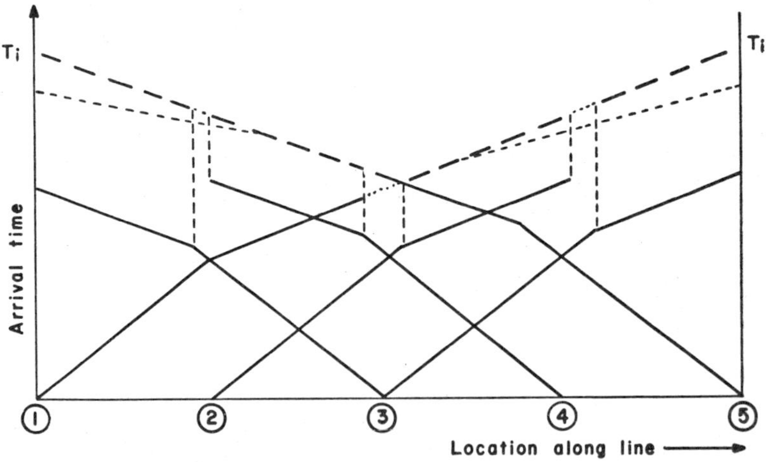

Fig. 37. **Composite time-distance curves** (dashed lines) constructed from first arrival curves from shotpoints 1, 2, 3, 4, and 5 (solid lines). Refraction portions from SP 2, 3, and 4 have been displaced upward to show the "equivalent arrival times" as if the profiles from SP 1 and 5 had been longer. Such displaced curves should overlap for confidence and show the same reciprocal time T_i. Actual time-distance curves from SP 1 and 5 might have been different if another refractor had been reached (e.g., dotted lines).

compressed section: A record section with the horizontal scale compressed so that a large distance is represented, used to present regional aspects of seismic data. Usually associated with considerable vertical exaggeration.

compressional wave: An elastic body wave in which particle motion is in the direction of propagation; the type of seismic wave assumed in conventional seismic exploration. Also call **P-wave,** *dilatational wave,* and *longitudinal wave.*

computer: 1. One who computes corrections for geophysical data. **2.** A *digital computer,* which performs sequences of arithmetic and logical operations as requested by program instructions. (A simple computer program is shown in Figure 79, **flow chart.**) It consists of **(a)** input or "read" portion for receiving program and data to be worked on, **(b)** memory for storing the program and the data, **(c)** a central processing unit (in simplest form, an accumulator or counter) for combining data according to program instructions, and **(d)** an output or "write" portion to deliver the results of the data processing. **3.** An *analog computer,* which operates on quantities in a continuous, smoothly changing, nondiscrete manner. The problem is simulated by analogy. For example, variables in a system of equations may be simulated by currents in an electrical network such that the voltage across a pair of junctions is proportional to the answer to the problem. Instead of electrical quantities, fluid flow, mechanical position, etc., may be used to represent variables.

computer graphics: The visual display of data stored in a computer. Usually refers to a display on a TV tube which is used in **interactive** processing.

computer language: The form in which program instructions can be supplied to a computer for translation into a machine language program which can then be executed. Languages for communicating with computers include *Algol* for scientific applications, *Cobol* for business applications, *Fortran* in several versions for engineering and scientific applications, *PL-1, Basic* for simple problems, and others.

Condep controller: See **depth controller.** Continental Oil Trademark.

condition: To circulate mud in a borehole to make the mud throughout the borehole uniform.

conductance: With direct current, the reciprocal of resistance. With alternating current, the resistance divided by the impedance squared; the real part of the admittance. Measured in mhos.

conduction angle: The number of degrees in a half-cycle ac wave during which a silicon controlled rectifier is turned "on." If ϕ is the *phase control angle,* the conduction angle is (180 degrees $-\ \phi$).

conduction current: Electrical current resulting from the motion of free charges under the influence of an electric field. The density of conduction current **J** at a point in an isotropic medium is

$$\mathbf{J} = \sigma \mathbf{E},$$

where σ is the conductivity and **E** is the electric field. Conduction currents usually are more important than **displacement currents** in electromagnetic prospecting.

conductivity: The ability of a material to conduct electrical current. See **mho per meter.** In isotropic material, conductivity is the reciprocal of resistivity. Sometimes called *specific conductance.*

conductor: A body within which electrical current can flow readily. Often, the "target" of an electromagnetic survey. **(a)** An *electronic conductor* conducts electricity primarily by electron mobility. Most electronic-conducting metallic-luster sulfide minerals are really semiconductor. **(b)** An *ionic conductor* conducts electricity primarily by means of ion mobility. Electrolytes are ionic conductors.

confidence limits: The limits between which a specified percentage of measurements are expected to lie; as in "90 percent confidence limits."

configuration: Arrangement (as of geophones in a group); **array.**

conformable: Two adjacent parallel beds separated by a surface of original deposition, where no disturbance or denudation occurred during their deposition.

conformal mapping: One area can be mapped into another area conformally if there is a continuous one-to-one correspondence of points and if angles are preserved.

conic: A curve defined by a quadratic equation of the form

$$Ax^2 + Bxy + Cy^2 + Dx + Ey + F = 0, \text{ or}$$

$$x^2 + y^2 = e^2 (x + a)^2.$$

For a parabola,

$$e = 1 \text{ or } B^2 = 4AC;$$

for a hyperbola,

$$e > 1 \text{ or } B^2 > 4AC;$$

for an ellipse,

$$e < 1 \text{ or } B^2 < 4AC.$$

conjugate: The conjugate of a complex number is the number with the sign of the imaginary part reversed. Often designated by a superscript asterisk or a superscript bar. Thus if $Z = a + ib$, then $Z^* = a - ib$ or $\bar{Z} = a - ib$.

conjugate points: Pairs of object and image points which correspond to each other.

conjunction: The condition for which an AND gate is used; *intersection.* Each of two (or more) situations must occur. Often written *A* U *B* (read as *"A* and *B").* See Figure 88 **gate).**

connate water: 1. Water trapped in the interstices of the sediments at the time of deposition, as opposed to water which migrated into the formations after deposition. See **interstitial water. 2.** Formation water.

Conrad discontinuity: A sharp increase in the P-wave

velocity in the crust, commonly at a depth of 17-20 km. The velocity below the Conrad discontinuity is of the order of 5.5–6.7 (often 6.–6.7) km/sec (compared to around 8 for the Moho). The Conrad discontinuity is not observed everywhere.

Contact log: A **microlog** in which the sonde is held against the borehole wall. Welex tradename.

contact resistance: The resistance observed: **1.** between a grounded electrode and the ground, **2.** between an electrode and a rock specimen, or, **3.** between electrical contacts.

continuation: Determining from a set of measurements of a potential field made over one surface what values the field will have over another surface (specifically at another elevation). The field at the elevation *z*, *F(x,y,z)* can be found from the field on the surface, *F (x′,y′,0)*, where the surfaces are horizontal and no sources intervene by the *upward continuation* relation:

$$F(x,y,z) = \frac{|z|}{2\pi}\iint \frac{F(x', y', 0)}{R^3}\, dx'\, dy',$$

$$\text{where } R = [(x - x')^2 + (y - y')^2 + z^2]^{1/2}.$$

Interchange of the two fields in this equation gives the *downward continuation* relation. See Peters, 1949, The Direct Approach to Magnetic Interpretation and its Practical Applications: Geophysics, v. 14, p. 290-319. See also **downward continuation.**

continuous profiling: A seismic method in which geophone groups are placed uniformly along the length of the line and shot from holes so spaced that continuous (or 100 percent) subsurface coverage is obtained along the line. Continuous profiling can be accomplished with a variety of spread types. Refraction continuous profiling requires continuous control on the refractor which may require irregular surface layouts.

continuous-velocity log: A **sonic log**; a log of formation velocity against depth. The quantity recorded and graphed is usually the reciprocal of the velocity, the traveltime over a short interval, often expressed in μsec/ft. Abbreviated CVL; CVL is a Birdwell trademark.

contour: A line connecting points of equal value, or representing the locus of a constant value on a map or diagram.

contour interval: The difference in value between two adjacent contour lines. Abbreviated CI.

control: The data on which a map or section are based. Posting of the control allows one to tell which portions result from measurements and which from interpolation and interpretation.

controlled mosaic: A composite aerial photograph fitted to a control plot by rephotographing the component vertical photographs to compensate for scale variations resulting from tilt and variations in flight altitude.

controlled rectifier: An electronic circuit element consisting of a controlled diode or solid-state switch. The controlled diode is usually turned "on" by a small voltage from an external circuit and turned "off" when the voltage is reversed. Used to switch large currents in IP transmitters. When the semiconductor is silicon, these controlled rectifiers are called SCR's.

controller: See **depth controller.**

control station: A point whose position (horizontally and vertically) is used as a base for a dependent survey or as control for adjusting survey errors.

convergence: The effect of computing a survey on the curved surface of the earth as if the surface were plane. The correction from assumed rectangular coordinates to geodetic coordinates allowing for earth curvature is the *convergence correction.*

conversational mode: Interactive mode; a procedure in which each entry from a terminal elicits a response from the computer and vice-versa.

converted wave: Seismic energy which has traveled partly as a P-wave (longitudinal wave) and partly as an S-wave (shear wave), being converted from one form to the other upon reflection or refraction at oblique incidence on an interface. Since mode conversion is small for small incident angles, converted waves become more prominent as offset increases.

converter: A device to perform digital-to-analog (D/A) or analog-to-digital (A/D) conversion.

convolution: Change in wave shape as a result of passing through a linear filter. **1.** A mathematical operation between two functions, *g(t)* and *f(t)*, often symbolized by an asterisk:

$$g(t) * f(t) = \int_{-\infty}^{\infty} g(\tau)\, f(t - \tau)\, d\tau.$$

Convolution is not restricted to one dimension. For example:

$$g(x,y) * f(x,y) = \int_{-\infty}^{\infty}\int_{-\infty}^{\infty} g(\alpha,\beta) \cdot f(x - \alpha, y - \beta)\, d\alpha\, d\beta.$$

2. The act of linear filtering. If a waveform *g(t)* is passed into a linear filter with the **impulse response** *f(t)*, then the output is given by the convolution of *g* with *f*. In discrete form where the input can be thought of as a series of impulses of varying size, each will generate an *f(t)* of proportional amplitude and the output will be the superposition of these. This can be expressed as

$$g_t * f_t = \sum_{k=0}^{L} g_k\, f_{t-k}.$$

This expresses that the output of a linear filter is a weighted linear combination of inputs. *L* is the convolution *operator length* in time divided by the sample interval and (*L*+1) is the number of "points" in the operator. (A simple computer program for this equation is shown in Figure 79, **flow chart.**) The frequency domain operation which is equivalent to convolution in the time domain consists of multiplying frequency-amplitude curves and adding frequency-phase curves. **3.**

The convolution operation is sometimes done (a) by "replacement," where each spike of the input is replaced by a proportionately scaled version of the impulse response and superposition forms the output; (b) by "folding," where the impulse response of the filter is reversed in time and slid past the input, the output for each position of the impulse response being the sum of the products of input and impulse response for corresponding points; (c) by "multiplying the **z-transforms**" of the input and the impulse response to give the z-transform of the output; or (d) by multiplying Fourier or Laplace transforms to give the Fourier or Laplace transform of the output. **4.** Convolution in two dimensions is used with gravity, magnetic, and other data to produce **grid residual, second derivative, continuation,** etc., maps. See Fuller, 1967, Two Dimensional Frequency Analysis and Design of Grid Operators, *in* Mining Geophysics, Vol. 2: Tulsa, SEG, p. 658-708.

convolver: A subcomputer under the control of a main computer which performs large-volume, high-speed multiplications and additions. Used to speed up operations which involve convolution or correlation. Sometimes called a *digital filter box.*

cookbook: A method which is prescribed step-by-step, as opposed to examining data as one goes along and varying methods based on the results of the examinations.

Cooley-Tukey method: A type of algorithm for **Fourier analysis** which considerably reduces computing time. See Cooley and Tukey, 1964, An Algorithm for the Machine Calculations of Complex Fourier Series, *in* Mathematics of Computation: v. 19, p. 297-301.

coordinatograph: X-Y reader.

copy: The degree to which two events or traces look alike.

core: 1. A rock sample cut from a borehole. See **core analysis. 2.** The earth's core is the central portion which will not transmit S-waves. The core is bounded by the Gutenberg discontinuity (2900 km deep) which separates it from the mantle. The core's radius is 3500 km. See Figure 63 **(earth layering)**. Sometimes divided into *outer core* and *inner core,* the radius of the inner core being about 1300 km. **3.** A magnetic core is a device used in rapid-access memory in computers. **4.** Cores which have high-magnetic permeability are often used in the loops which are employed for generating and measuring magnetic fields.

core analysis: Cores from boreholes are analyzed for porosity, permeability, fluid content, fluid identification, lithology, and structure (fractures, cross bedding, etc.). Results are often illustrated on a log or coregraph on which porosity, permeability, water saturation, oil saturation, and lithology are graphed against depth.

coregraph: See **core analysis.**

core slicer: A device using diamond-edged blades which cuts a triangular core about one inch on a side and up to three ft long from the side of a smooth borehole.

Coriband: A computer-produced log of secondary porosity, grain density, water saturation, porosity, fluid analysis, and formation analysis which is calculated from resistivity, density, neutron, and sonic log data. Schlumberger trademark.

Coriolis acceleration: The acceleration (in the rotating coordinate system of the earth) of a body in motion with respect to the earth resulting from the rotation of the earth, as seen by an observer on the earth. A moving gravimeter sees a Coriolis acceleration, which is involved in the **Eötvös effect.**

corner reflector: A radar reflector made of triangles of metal or metal screen at right angles to each other so that one looks into the corner of an open box no matter from which direction it is viewed. Thus it looks like a mirror at normal incidence to radar waves and hence has high reflectivity. Corner reflectors are also used in optics. See Figure 38.

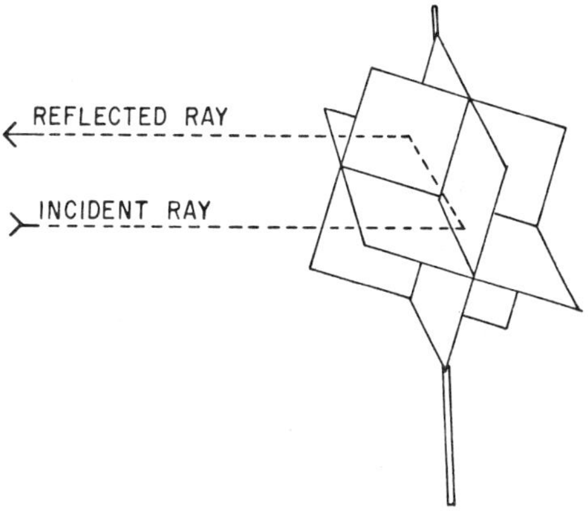

Fig. 38. **Corner reflectors.** After one or more reflections, a ray emerges parallel to the incident ray regardless of the approach direction of the incident ray.

correlation: 1. Identifying a phase of a seismic record as representing the same phase on another record. Indicating that events on two seismic records are reflections from the same stratigraphic sequence, or refractions from the same marker. **2.** The degree of linear relationship between a pair of traces; a measure of how much two traces look alike or the extent to which one can be considered a linear function of the other. The time-domain concept analogous to **coherence** in the frequency-domain. See **autocorrelation** and **crosscorrelation. 3.** The determination of equivalence in stratigraphic position of formations in different wells, based upon similarities in the character of well logs and the occurrence of distinctive features which serve as markers from one well to the next. **4.** The matching of different well logs and other well data either in the same well or in different wells.

correlation filter: A matched filter.

correlation ghost: Nonlinearities in generating a Vibroseis signal introduce second harmonics. The record from the second harmonic will correlate with double the frequency which generated the harmonic to produce a correlation ghost.

correlation method: A seismic method of shooting isolated profiles and correlating events to learn the relative structural positions of reflection horizons. The correlation is often based on similarities in the character of events and in the intervals between events.

correlator: 1. That which one correlates with; either a function or a device. 2. **Matched filter.**

correlogram: Graph of half of the **autocorrelation** function.

COS: **Common offset stack.**

cosine law: In any plane triangle with sides *a, b,* and *c* and the angle *C* between sides *a* and *b*;

$$c^2 = a^2 + b^2 - 2\,ab\cos C,$$

if $C > 90$ degrees, the cosine is negative and the last terms becomes additive. See Figure 162 (**sines law**).

cosine transform: The **Fourier-transform** of the even or symmetrical part of a function. The *sine transform* involves the odd or antisymmetrical part.

cosmetics: The appearance of a record section apart from its information content. *Cosmetic procedures* are designed to improve the appearance of a result rather than its information content.

Coulomb balance: A **torsion balance** used for measuring the remanent magnetism of samples in pulverized form.

Coulomb's law: 1. A force *F* exists between electrical charges Q_1 and Q_2 which are separated by the distance *r*. The force is attractive for charges of unlike sign and repulsive for charges of like sign.

$$F = kQ_1Q_2/r^2.$$

k is a constant which depends on the units used (9×10^9 newton m^2/coulomb^2). 2. A similar relationship between magnetic poles Q_1 and Q_2. *K* is 10^{-7} webers/amp m, where *Q* is in ampere seconds and *r* in meters.

coupled wave: A mode of wave propagation which involves the transfer of energy back and forth between two different wave-propagation modes. Called *C-waves.* The two waves must have the same apparent phase velocity.

coupler: The telephone cradle used in connecting a teletype unit with a telephone line, such as used in time-share computer connections.

coupling: Interaction between systems. 1. Device for fastening together, as the plugs for connecting electrical cables. 2. Aspects which affect energy transfer. Thus the "coupling of a geophone to the ground" involves the quality of the plant (how firmly the two are in contact) and also considerations of the geophone's weight and base area, because the geophone-ground coupling system has natural resonances and introduces a filtering action. 3. The type of mutual electrical relationship between two closely related circuits. *Ac coupling* would exclude dc voltages by employing a series capacitive element. *Dc or direct* coupling may exclude higher frequency signals by using a capacitive element across the inputs or may allow all components to pass. 4. *Capactive coupling* may occur because of mutual capacitive impedance, as between the wires in IP circuits or between a wire and ground. 5. *Inductive coupling* occurs because of mutual inductive impedance, such as between grounded IP transmitter and receiver circuits, especially at higher frequencies, greater distances, or lower earth resistivity. This may give rise to false IP anomalies. Also called *electromagnetic* or *EM coupling.* 6. *Resistive coupling* in IP surveying is due to leakage between wires, between a wire and ground, or through the resistance of the ground itself between two grounded circuits.

covariance: A measure of the common variation between two quantities; a crosscorrelation function which is not normalized. A mean of zero is also implied.

CPU: Central processing unit, the part of a computer which controls operations and interprets and executes programs.

crab: To maintain an angle between a ship's heading and the desired course, such as to compensate for a crosswind or cross sea. Compare **yaw** which is oscillation of the ship's heading.

Cramer's rule: The solution to a set of linear simultaneous equations,

$$\begin{aligned} a_{11}x_1 + a_{12}x_2 + *** + a_{1n}x_n &= b_1 \\ a_{21}x_1 + a_{22}x_2 + *** \quad &= b_2 \\ *** \quad &\quad *** \end{aligned}$$

is the determinant of the coefficients obtained by replacing a row with the b_i's, divided by the determinant of the coefficients (sometimes called "Δ"). Thus,

$$x_1 = \frac{\begin{vmatrix} b_1 & a_{12} & *** & a_{1n} \\ b_2 & a_{22} & & a_{2n} \\ *** & & & *** \end{vmatrix}}{\begin{vmatrix} a_{11} & a_{12} & *** & a_{1n} \\ a_{21} & a_{22} & & a_{2n} \\ *** & & & *** \end{vmatrix}},$$

$$x_2 = \frac{\begin{vmatrix} a_{11} & b_1 & *** & a_{1n} \\ a_{21} & b_2 & & a_{2n} \\ *** & & & *** \end{vmatrix}}{\begin{vmatrix} a_{11} & a_{12} & *** & a_{1n} \\ a_{21} & a_{22} & & a_{2n} \\ *** & & & *** \end{vmatrix}}, \text{ etc.}$$

This is usually not an economical way for computers to solve simultaneous equations.

crater: A large funnel-shaped cavity at the top of a borehole.

crest: **1.** The highest point on a structure. **2.** The **peak** of a seismic (or other) wave.

crew: **Party.**

critical angle: Angle of incidence θ_c for which the refracted ray grazes the surface of contact between two media (of velocities V_1 and V_2):

$$\sin \theta_c = V_1 / V_2.$$

See Figure 38. θ_c has meaning only where $V_2 > V_1$. In general, four critical angles can be defined for the ratios of P- and S-waves in the two media:

$$V_{P1}/V_{P2},\ V_{S1}/V_{S2},\ V_{S1}/V_{P2},\ V_{P1}/V_{S2}.$$

Usually the first is intended unless one of the others is specified.

critical damping: The minimum **damping** which will not allow oscillation.

critical dip: **1.** Dip in the direction opposite to the regional attitude, indicating a closure. **2.** Dip at the most critical point on a closure which is necessary to establish closure.

critical distance: Used in either of two different meanings: see Figure 39. **1.** (The preferred usage) The offset at which the reflection time equals the refraction time; that is, the offset for which the reflection occurs at the **critical angle. 2.** The offset at which a refracted event becomes the first break, the intersection point for the traveltime curves for two refractors. The preferred term is **crossover distance**

critical reflection: A reflection at the critical angle. Amplitude may be exceptionally large in this vicinity. See Figure 39.

CRO: Cathode-ray oscilloscope.

cross: Cross-spread.

crosscorrelation filter: Matched filter.

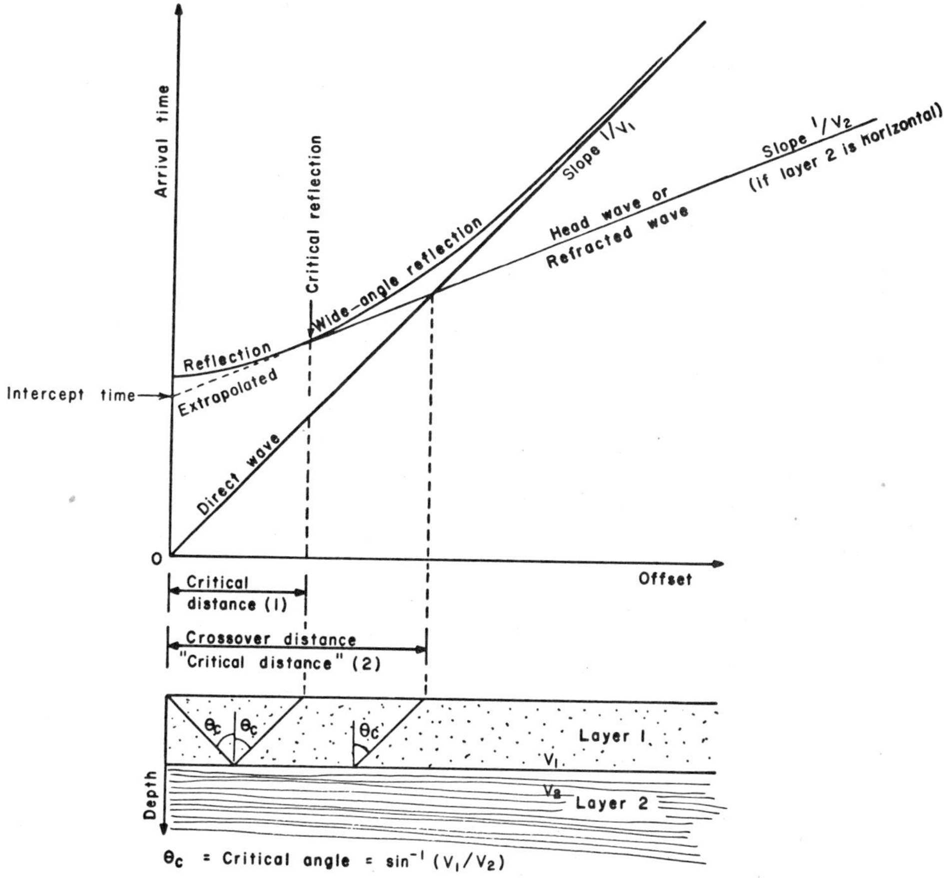

Fig. 39. **Critical distance.** The curvature of the reflection is hyperbolic. The reflection amplitude may be exceptionally large in the vicinity of "critical reflection."

crosscorrelation function: A measure of the similarity of two waveforms, of the degree of linear relationship between them, or of the extent to which one is a linear function of the other. For two waveforms *G(t)* and *H(t)* the normalized crossrelation function $\phi_{GH}(\tau)$ is given by:

$$\phi_{GH}(\tau) = \lim_{T\to\infty} \frac{\int_{-\frac{1}{2}T}^{\frac{1}{2}T} G(t)\,H(t+\tau)\,dt}{\left[\int_{-\frac{1}{2}T}^{\frac{1}{2}T} G^2(t)\,dt \int_{-\frac{1}{2}T}^{\frac{1}{2}T} H^2(t)\,dt\right]^{\frac{1}{2}}}.$$

For digital data this becomes

$$\phi_{GH}(\tau) = \lim_{N\to\infty} \frac{\sum_{k=-N}^{N} G_{(k)}\,H_{(k+\tau)}}{\left[\sum_{k=-N}^{N} G^2_{(k)} \sum_{k=-N}^{N} H^2_{(k)}\right]^{\frac{1}{2}}}.$$

The denominator in the above two expressions is the normalizing factor and is sometimes omitted (as when used in Wiener filtering). When normalized, a crosscorrelation of 1 indicates perfect match, values near zero indicate very little correlation, negative values indicate a degree of match if the phase is inverted. See also **autocorrelation.** Unnormalized crosscorrelation can be accomplished by reversing one function in time and convolving:

$$\phi_{ab} = a(t) * b(-t).$$

The equivalent operation in the frequency domain involves multiplying the amplitudes of common frequencies and subtracting phase-response curves.

cross-coupling effect: The effect in shipboard gravity measurements produced by simultaneous interactive accelerations in two different directions.

cross-equalization: Filtering one channel to match the frequency spectrum of adjacent channels. The matching involves a phase shift as well as adjustment of the amplitude of frequency components. Tends to align coherent events better but may increase short-period reverberations.

crossfeed: Crosstalk. Interference resulting from the unintentional pickup by one channel of information or noise on another channel.

cross information: Information about the direction from which an event approaches the spread, especially the component outside the plane of the section (i.e., outside of a vertical plane which includes the line). The objective is to determine the orientation of the reflector in space; see Figure 40. Cross information is obtained with cross-spreads, from intersecting seismic lines, or in other ways.

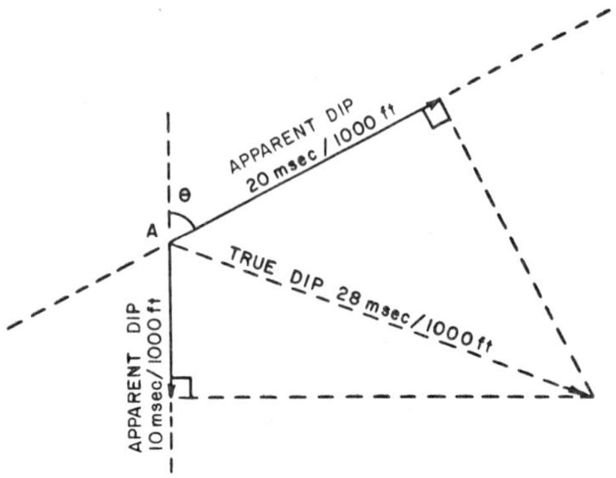

Fig. 40. **Cross information.** Resolved for data from two nonorthogonal lines.

crossover: The reversal of electromagnetic-dip direction over the apex of a conductor. The undisturbed electromagnetic field at stations in the plane of a vertical source loop is horizontal. In the presence of a subsurface conductor, the field will be horizontal over the apex but will have vertical components in opposite directions on either side of the conductor, causing a reversal in dip direction to take place.

crossover distance: That distance at which refracted waves following a deep high-speed marker overtake direct waves or refracted waves which follow a shallower marker. Some refraction-interpretation methods emphasize crossover-distance observation. Sometimes called *critical distance.* See Figure 39 **(critical distance).**

crossplot: A graph to determine the relationship between two different measurements. A crossplot of porosity measured from one type of log against porosity from another type log, for example Figure 41, is used to show secondary porosity which affects the two log types differently.

crosspower spectrum: The Fourier transform of the **crosscorrelation** function.

cross product: **1.** A type of **vector** multiplication. If **i**, **j**, and **k** are mutally orthogonal unit vectors so that two vectors **A** and **B** may be expressed in terms of components in these directions:

$$\mathbf{A} = a_1\mathbf{i} + a_2\mathbf{j} + a_3\mathbf{k} \quad \text{and}$$

$$\mathbf{B} = b_1\mathbf{i} + b_2\mathbf{j} + b_3\mathbf{k}\ ,$$

then the cross product **A** × **B** is orthogonal to both **A** and **B**:

$$\mathbf{A} \times \mathbf{B} = (a_2b_3 - a_3b_2)\mathbf{i} + (a_3b_1 - a_1b_3)\mathbf{j} + (a_1b_2 - a_2b_1)\mathbf{k}.$$

2. The terms in an algebraic multiplication which in-

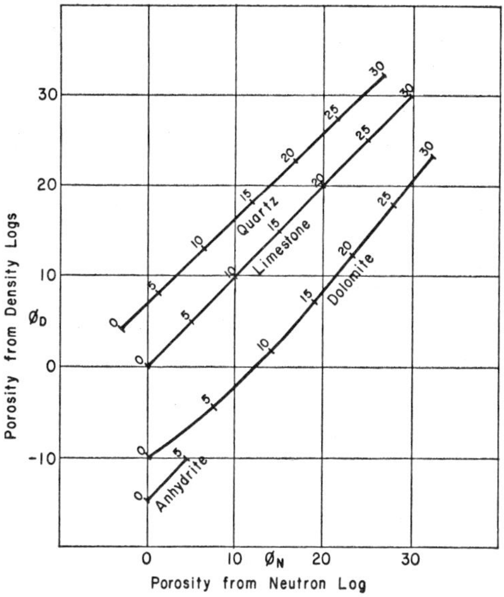

Fig. 41. **Crossplot.** Plot of two different measurements on the same samples. In this example, porosity measured from the neutron log is plotted against porosity measured from the the density log, showing the responses of different rock types. (Courtesy Schlumberger.)

volve elements of different kind; e.g., $2ab$ is the cross-product term in

$$(a + b)^2 = a^2 + 2ab + b^2.$$

cross-section: 1. A plot of seismic reflection events along a seismic line. Events are usually (but not always) migrated and the vertical scale is usually depth (but occasionally time). See **plotted section. 2.** A geological diagram showing the formations and structures cut by a vertical plane. **3.** A concept to represent the probability of collision between particles. A particle has to pass within a certain distance of another particle for the two to interact, the effective distance depending on the type of interaction. Neutron logging is concerned with the capture of neutrons by nucleii, which is expressed by the **capture cross-section.**

cross-spectrum: The expression of the mutual frequency properties of two series. The cross-spectrum is in general a complex-valued function and hence requires a pair of real relationships, such as the amplitude and phase as functions of frequency.

cross-spread: 1. A spread which makes a large angle (often a right angle) with the line of traverse. The objective is to obtain cross information, i.e., information which will allow one to determine the true direction from which the energy reaches the spread so that true position of the reflector in space can be determined. Assume (Figure 40, **cross information**) that a cross-spread (or other cross information) at the angle θ indicates dip rates of 10 and 20 msec/1000 ft; a line to the intersection of perpendiculars from dip components gives the true dip. Note that this construction is different from a vector parallelogram. **2.** A **spread** in the shape of a cross; for example, twelve groups laid out in line and twelve laid out on a line perpendicularly. See Figure 170 **(spread).**

crosstalk: Crossfeed.

cross-track: Perpendicular to a seismic line.

CRP: Common reflection point. Although the inventor (Mayne) used CRP and common reflection point, the more common terms are **CDP** and **common-depth-point.**

crust: The outermost shell of the earth; the portion above the **Moho.** The crust has a P-wave velocity which is usually < 7 km/sec and a mean density of 2.8 to 2.9 gm/cm^3. Crust is commonly used in an imprecise way, sometimes for the lithosphere thus including the upper mantle, sometimes for only continental crust. Continental crust (acidic crust or sial) is granitic to gabbroic; oceanic crust (basic crust or sima) is basaltic. See Figure 63 **(earth layering).**

Curie law: Magnetic susceptibility is inversely proportional to the absolute temperature. This law applies where dipoles are far enough apart that their interaction is small, as in solutions of paramagnetic salts. In paramagnetic solids the susceptibility is inversely proportional to the difference between the temperature and the **Curie point.**

Curie point: The temperature at which a material loses its ability to retain magnetism, that is, changes from ferromagnetic to paramagnetic behavior. Below this temperature atoms interact so that their magnetic moments couple and behave collectively. At the Curie temperature the atom's thermal energy equals the coupling energy, and above this temperature the atomic magnetic moments become decoupled and the substance behaves paramagnetically. The Curie temperature of most rocks is less than 600°C, which is usually reached by about 20 km. The analogous point with antiferromagnetic materials is the "Neel point."

curl: The curl of the vector **V** is given by the **vector** operation:

$$\text{Curl } \mathbf{V} = \nabla \times \mathbf{V},$$

where the operator del is

$$\nabla = \mathbf{i}\frac{\partial}{\partial x} + \mathbf{j}\frac{\partial}{\partial y} + \mathbf{k}\frac{\partial}{\partial z}$$

in rectangular coordinates, where **i**,**j**, and **k** are orthogonal unit vectors. If **V** is expressed as:

$$\mathbf{V} = V_1\mathbf{i} + V_2\mathbf{j} + V_3\mathbf{k},$$

$$\text{Curl } \mathbf{V} = \nabla \times \mathbf{V} = \left(\frac{\partial V_3}{\partial y} - \frac{\partial V_2}{\partial z}\right)\mathbf{i}$$

$$+\left(\frac{\partial V_1}{\partial z} - \frac{\partial V_3}{\partial x}\right)\mathrm{j} + \left(\frac{\partial V_2}{\partial x} - \frac{\partial V_1}{\partial y}\right)\mathrm{k}\,.$$

current density: Current per unit cross-sectional area, determined by the velocity and density of charge carriers. Current density **J** is a vector quantity, measured in amperes per square meter.

current electrode: The *A* or *B* electrode in electrical logging or resistivity or IP surveying. Low electrical resistance at such contacts is desirable to maximize the current into the ground.

curvature: The rate of change of direction of a curve or surface; the inverse of the radius *R*.

$$\text{Curvature} = 1/R = \frac{d^2y}{dx^2}\left[1\left(+\frac{dy}{dx}\right)^3\right]^{-3/2}.$$

Curvature is approximately equal to $2h/S^2$ (Figure 42) for an arc which reaches a height *h* from a chord of length 2*S* (which is small compared to *R*).

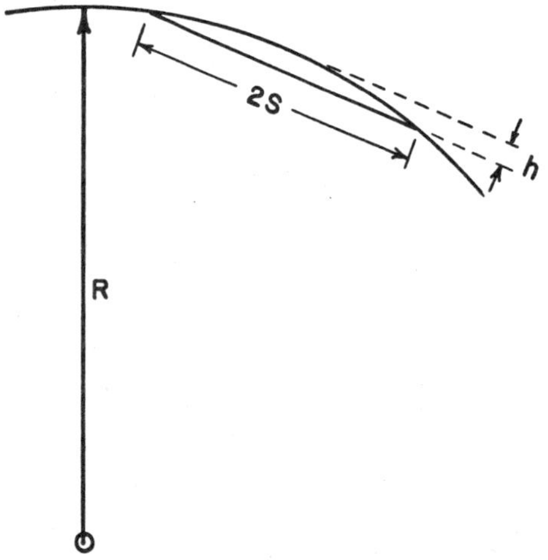

Fig. 42. **Curvature** is the reciprocal of the radius of curvature. For small arcs curvature is approximately $2h/S^2$.

curved path: A seismic raypath which is curved because refraction changes the direction of the ray as the velocity changes with depth. Increase in velocity with depth makes the raypath concave upward.

curve fitting: Finding an analytic equation which approximates a set of data. The most common curve-fitting technique is **least squares** but other methods (such as a **quadratic spline**) are also used.

curve matching: An interpretation method whereby observations are compared with **type curves** (or master curves). Achieving a close fit implies that the actual situation is similar to the model which the type curve represents (which is not necessarily true because of the inherent ambiguity of most geophysical observations).

curve of maximum convexity: A **diffraction curve.** The relation between arrival time and position for energy which has traveled from the observing position to a point and then been diffracted from the point; see Figure 52(**diffraction curve**).The curvature of reflected energy cannot exceed this except for reverse branches and energy which has traveled by a longer path to reach the diffracting point, as with a diffracted reflection. A curve of maximum convexity is specific for a velocity function, and curves of maximum convexity are used in migrating reflections (see Figure 118, **migration**).

cut: To dilute, as may happen to drilling mud if formation water or gas enters the hole.

cutoff: The frequency at which a **filter** response if down by a predetermined amount, such as 3 db. The cutoff points designate the filter; e.g., an 18-57 filter has a low-frequency cutoff at 18 hz and a high-frequency cutoff at 57 hz.

cuttings: Material drilled out of shotholes or drill holes.

CVL: A continuous-velocity log or **sonic log.**

CW: Continuous wave or constant frequency.

cycle skip: **1.** In acoustic or sonic logging, the first arrival is sometimes strong enough to trigger the receiver closest to the transmitter but not the farthest receiver, which may then be triggered by a later cycle resulting in an erroneously high transit time. This situation is called *cycle skipping.* Its onset is characterized by an abrupt deflection corresponding to an added cycle of travel between receivers. *Short cycle skipping* where the near receiver is triggered a cycle too late also can occur, resulting in an abnormally short traveltime. **2.** Jumping a leg in correlating events, as may occur in matching noncorresponding peaks in automatic-statics programs.

cyclographic diagram: A **sterographic projection** showing planes as great-circle intersections of a sphere. Used in three-dimensional structural representation. Also called *beta diagram.* Compare **pole diagram.**

cylindrical: Having symmetry so that shape does not depend on azimuth angle; **two-dimensional.**

cylindrical divergence: Decrease in the amplitude of a wave with distance because of geometrical spreading. As a wavefront expands in a larger circle, the energy becomes spread out and hence the energy density varies inversely as the distance. Surface waves undergo cylindrical divergence whereas body waves undergo **spherical divergence.**

cylindrical hydrophone: A hollow cylinder of piezoelectric material will have a voltage generated between the outside and inside of the cylinder when subjected to radial pressure. Such hydrophones are very stable and durable and have sensitivity independent of operating depth. Most streamer **hydrophones** are of this type.

D

D/A: Digital-to-analog.

damping: A slowing down or opposition to oscillation due to dissipation of the energy of oscillation. **a.** *Critical damping,* μ_c, is the minimum damping which will prevent oscillation from taking place. For a mass, *m* subject to a Hooke's law restoring force with a force constant *k,*

$$\mu_c = 2\,(km)^{1/2}.$$

b. The *damping factor* μ is the ratio of the system friction to that necessary for critical damping, or the quotient of the logarithm of the ratio of two successive oscillations if the system is underdamped. The damping factor is one for critical damping, less than one for an underdamped system (which will oscillate) and greater than one for an overdamped system. See Figure 43. **c.** Increased damping shifts the peak response toward higher frequencies. Electrical damping is increased by decreasing the shunt resistance. Most geophones are slightly underdamped, often having *optimum damping* which is 0.66 μ_c.

darcy: A unit of permeability; the permeability which will allow a flow of one milliliter per second of fluid of one centipoise viscosity through one square centimeter under a pressure gradient of one atmosphere per centimeter. The commonly used unit is the millidarcy or 1/1000 darcy.

Darcy's law: A relationship for the fluid flow rate *q*:

$$q = \frac{kA}{\mu}\,\frac{\Delta p}{\Delta x},$$

where k = permeability, A = cross-sectional area, μ = viscosity, and Δp = pressure differential across the thickness Δx.

dar Zarrouk: The name given by Maillet to resistivity parameters or curves that deal with layered anisotropic materials. The *dar Zarrouk variable* is the transversal unit resistance (ρ_T), the depth integral of the transverse resistivity perpendicular to the strata; the *dar Zarrouk function* is the longitudinal unit conductance ($1/\rho_L$), the depth integral of the conductivity parallel to the strata; and the *dar Zarrouk curve* is a plot of the mean resistivity $(\rho_T\rho_L)^{1/2}$ of the formation down to the depth *Z,* plotted versus the anistropy coefficient times$(Z, (Z\rho_{T}/\rho_L)^{1/2}$. See Maillet, 1947, The Fundamental Equations of Electrical Prospecting: Geophysics, v. 12, p. 529-556.

data set: 1. A set of data. **2.** A device for converting signals from a terminal into a form suitable for transmission and vice-versa.

datum: 1. The arbitrary reference level to which measurements are corrected. **2.** The surface from which seismic reflection times or depths are counted, corrections having been made for local topographic and/or weathering variations. **3.** The reference level for elevation measurements, often sea level.

datum velocity: The velocity assumed beneath the datum surface, often the subweathering velocity.

db: Decibel.

dbm: Decibels less than 1 milliwatt of power. Used, for example, in specifying sensitivity, often in conjunction with input impedance, such as "50 dbm at 5000 ohms."

db/octave: Unit for expressing the slopes of **filter** curves.

dc-pulse method: See **pulse IP method.**

dead: 1. Not electrically connected, as a geophone whose connection to the cable has pulled loose. **2.** Having no signal, as a *dead trace.*

dead reckoning: Determination of position by extrapolation of the track and direction from a previously known point. Inertial navigation and **doppler sonar** are sophisticated versions of dead reckoning. See **positioning.**

debug: To search for and remedy malfunction or errors, as with instruments or computer programs.

Decatrack: A system employing microheads which permits several (usually 10) channels of analog information to be recorded in the space normally occupied by one channel. Mixing or stacking can be accomplished by recording the component channels with microheads and reading them together with a large head which sums them.

decay constant: The time for an exponentially changing voltage to vary by $1/e$, or to change 63 percent from its initial value. Also called *time constant.*

decay curve: A graph of the decay of a quantity as a function of time. An IP voltage decay curve may be characteristic of a particular material. In theory it can be transformed to a **resistivity spectrum.**

Decca: A medium-range **positioning** system involving the comparison of the phase of synchronized CW radio waves transmitted from three (or more) fixed stations. **1.** The *Decca Navigator* is a high-power fixed installation with a range of about 300 miles. Four frequencies in the ratio 5:6:8:9 from 70 to 130 khz are used. *Survey Decca* is a similar system used for temporary installations, with a range of about 100 miles. Both may broadcast different frequencies at one minute intervals to remove lane ambiguity. *Two-range Decca* is a variation of Survey Decca in which the master transmitter is at the mobile station, thus making the system range-range rather than hyperbolic. **2.** Decca *Hi-fix* is a short-range system which involves the broadcasting of a frequency between 1700 and 2000 khz in a time-sharing arrangement between three fixed stations. By locating the master transmitter at the mobile station, the system can be made range-range rather than hyperbolic. Similar to Raydist, Lorac, and Toran. **3.** *Dectra* and *Delrac* are long range systems which operate at 100 khz and at 10 to 14 khz, respectively. They lack the accuracy required for geophysical surveys. See also **lambda.**

decibel: A unit used in expressing power or intensity

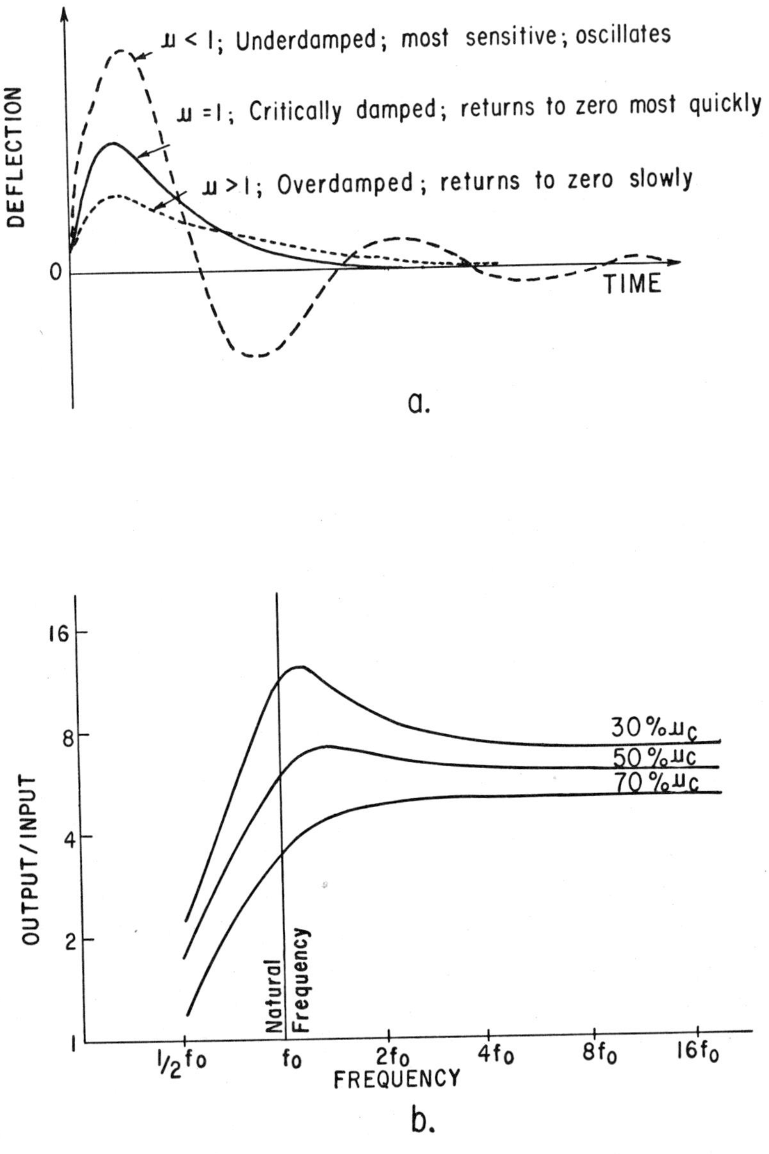

Fig. 43. **Damping.** (a) Effect of damping an impulse. (b) Effect of damping on the amplitude of a periodic input.

ratios: 20 $\log_{10}$ of the amplitude ratio, or 10 $\log_{10}$ of the power ratio. Symbol: db. An amplitude ratio of 2 (which represents a power ratio of 4) is equivalent to 6 db. 1 db = 0.1151 neper.

decimate: To resample digitized data at a longer interval than formerly used. For example, to "decimate from 2 to 4 msec" means to retain only every other sample of data originally recorded every 2 msec, or to "decimate from 2 to 6 msec" means to retain only every third sample. **Alias filtering** must be included in the resampling procedure. Opposite of *reconstitute.*

declination: 1. The angle between geographic north and magnetic north. 2. The angle between the celestial equator and a celestial body. Differs from celestial latitude.

decomposition: 1. Separating effects of different kinds or attributable to different causes. 2. Separating a potential (e.g., gravity) map into regional and residual; **residualizing.** 3. Separating the effects of filtering from a waveform which has been filtered; **deconvolution.** 4. Deriving resistivity stratification from a **kernal function.**

deconvolution: 1. The process of undoing the effect of another filter. Usually an **inverse filter** is designed and convolved with the signal, the objective being to nullify an objectionable effect of an earlier filter action. Recursive filters may also be used to accomplish this result. 2. Deconvolution is sometimes deterministic (i.e., the inverse filter may be known or at least its form may be known) or it may involve assumptions about the statistical nature of the data. 3. Deconvolution is often used to mean any of a number of operations which aim at removing the effects of some previous convolution. Specifically, it may mean (a) *system deconvolution,* in which the objective is to remove the filtering effect of the recording system; (b) *dereverberation* or *deringing,* removing the filtering action of a water layer, see also **Backus filter** and **gapped deconvolution;** (c) removing the filtering action of the near-surface (other than the water layer effects referred to above); (d) *deghosting* and occasionally removing other multiples; (e) **whitening** or equalizing all frequency components within a band-pass in an effort to compensate for filtering actions which have strung out the wavetrain and thereby shorten the reflection pulse length; (f) shaping the amplitude-frequency and/or phase response to match that of adjacent channels; or (g) **predictive deconvolution.** Deconvolution results may vary markedly with different **phase** assumptions, different gate location or width, or different operator length in the filter design. 4. Potential maps and other data sets besides time series may be deconvolved.

Dectra: See **Decca.**

dedicated: Devoted exclusively to; for example, a "dedicated" power generator might supply the power for the seismic recording only, another dedicated generator might power the navigation equipment, both being separate from the ship's normal power supply.

deep seismic sounding: DSS.

deflection of the vertical: The angular difference between a plumb line (the vertical) and a perpendicular to the geodetic ellipsoid. Produced by irregularities in mass distributions (especially the roots of mountain ranges). See Figure 89 **(geodetic latitude).**

degaussing: Demagnetizing; removing the magnetization of a sample by placing the specimen in an oscillating magnetic field, the magnitude of which is gradually reduced to zero. The operation has to be done in a nulled magnetic field (such as may be produced with Helmholz coils) to effect complete demagnetization. This procedure is used to "clean" magnetic tape (remove the data stored on it) so that the tape can be reused; the tape is often rotated during the demagnetization to remove the effect of the earth's magnetic field.

degree: The highest power which a variable assumes. For a differential equation, the power of the highest derivative. *Linear* implies that the degree is one; *quadratic,* that it is two; *cubic* 3; *quartic* 4; etc.

degree of freedom: The minimum number of independent variables which must be specified to define a system.

del: The vector gradient operator; symbol is ∇. In rectangular coordinates:

$$\nabla = \mathbf{i}\frac{\partial}{\partial x} + \mathbf{j}\frac{\partial}{\partial y} + \mathbf{k}\frac{\partial}{\partial z}.$$

∇U is the gradient of the scalar potential field *U.* The operator ∇^2, sometimes called the *Laplacian,* also appears frequently (e.g., in **Laplace's equation):**

$$\nabla^2 = \nabla \cdot \nabla = \frac{\partial^2}{\partial x^2} + \frac{\partial^2}{\partial y^2} + \frac{\partial^2}{\partial z^2}.$$

As an operator on the vector field, **V,** $\nabla \bullet$ **V** is called the **divergence,** and $\nabla \times$ **V** is called the **curl.** Del is also called *nabla.*

delay cap: A cap which detonates a fixed time after an electrical current is applied. Used in achieving a **directional charge.**

delay line: A device capable of retarding a signal by a fixed time interval. (a) A mercury delay line uses transducers to convert electrical signals into ultra-sonic patterns which are then carried across a gap containing liquid mercury to another transducer which reconverts them into electrical signals; (b) electrical delay lines may use lumped capacitative and inductive elements; coaxial cables and transmission lines involve delays by the transit time through the lines; (c) recirculating magnetic tape delays signals by the amount of time needed for the tape to pass from write to read heads.

delay time: 1. In refraction work, the additional time taken for a wave to follow a trajectory to and along a buried marker over that which would have been taken to follow the same marker considered hypothetically to be at the ground surface or at a reference level. See Figure 44. The concept implies that the refractor is nearly horizontal under both shotpoint and detectors. Normally, delay time is assigned separately under the source and under a detector, and is dependent upon the depth of the marker at the incidence and emergence points. *Shot delay time* plus *geophone de-*

lay time equals *intercept time*. **2.** Delay produced by a filter; see **filter correction. 3.** Time lag introduced by a delay cap such as **Elcord. 4.** In induced-polarization work, the time interval between the "off" instant of the charging current and the instant a measuring voltmeter oscillograph is turned "on." Delay times up to 500 or 1000 msec may be necessary to allow dissipation of transient voltages which are not directly related to the polarization decay voltage.

Delrac: See **Decca.**

delta function: An **impulse.**

delta *t* : Moveout, usually written Δt. **1.** The time difference between the arrival times at different geophone groups. **2.** *Dip moveout* is the time difference between groups at maximum offset on opposite sides of a symmetrical spread or at opposite ends of a single-ended spread after correction for normal moveout. The moveout is interpreted in terms of the component of geological dip in the plane of the section barring lateral changes of velocity. See **dip calculation. 3.** Sometimes used for the sum of dip moveout and **normal moveout.** Delta *t* ordinarily does not imply normal moveout unless specifically stated. **4.** Also used in connection with refraction and other events. **5.** Interval transit time, as used with the sonic log.

demodulation: The process of retrieving an original signal from a modulated signal.

deMoivre's theorem: The relationship: $e^{ir\theta} = (\cos\theta + i\sin\theta)^r = \cos r\theta + i\sin r\theta$, where $i = \sqrt{-1}$.

deMorgan's theorems: See **Boolean algebra.**

demultiplex: Separating the individual component channels which have been multiplexed.

densilog: A **density log.** Densilog is a Dresser Atlas tradename.

densimeter: An instrument for measuring density.

density: Mass per unit volume. Commonly measured in $gm/cm^3 = 10^3 kg/m^3$, often without the units being expressed explicitly.

density contrast: The difference in density between two formations.

density log: A well log which records the formation density. The logging tool consists of a gamma-ray source (e.g., Cs^{137}) and a detector so shielded that it records backscattered gamma rays from the formation. This secondary radiation depends on the density of electrons, which is roughly proportional to the bulk density. The source and detector are on a skid which is pressed against the borehole wall. The compensated density-logging tool (FDC) includes a secondary detector which responds more to the mud cake and small borehole irregularities; the response of the second detector is used to correct the readings of the main detector. Density is related to porosity ϕ by the equation

$$\phi = \frac{\rho_{ma} - \rho_{log}}{\rho_{ma} - \rho_f},$$

where ρ_{ma} is the density of the rock matrix, ρ_{log} is the density recorded on the log, and ρ_f is the density of the formation fluid. Sometimes called *gamma-gamma log.* Compare **nuclear cement log** and **photon log.** See Figure 45.

density profile: A line of gravity readings taken over a topographic feature having appreciable relief which is not associated with density variations or structure, the object of which is to determine the best density factor for **elevation corrections.** The most appropriate density is the one that minimizes the correlation of gravity values with elevation. See also **triplets.**

departure curve: Graphs which allow one to correct for measuring conditions which differ from "standard." Such curves, for example, might correct well logs for differences in temperature, hole diameter, mud type, adjacent beds, invasion, etc.

departures: The east or west component of a line expressed in linear units; the difference of the longitudes of the ends of the line measured at a given latitude. For a line directed toward the northeast or southeast quadrant, the departure is positive or *easting;* it is negative or *westing* for a line directed toward the southwest or northwest quadrant. Abbreviated dep

depth controller: A device with moveable vanes which fastens to a marine streamer to maintain it at a predetermined depth; see Figure 46. Air pressure in a tank balances against hydrostatic pressure to move the

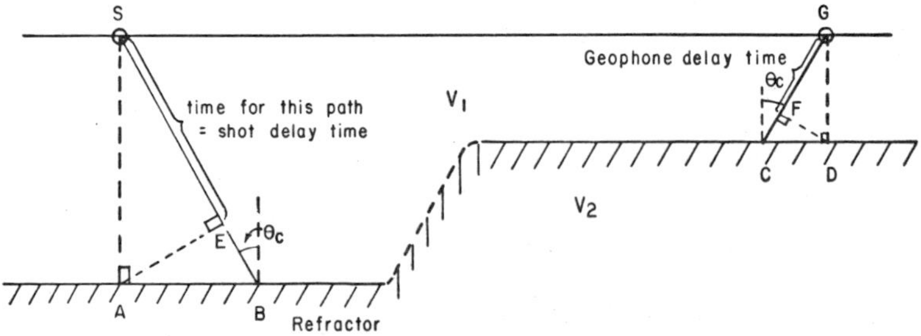

Fig. 44. **Delay time.** Shot delay time $= SB/V_1 - AB/V_2 = SE/V_1$. Geophone delay time $= CG/V_1 - CD/V_2 = FG/V_1$. Shot delay time + geophone delay time = intercept time.

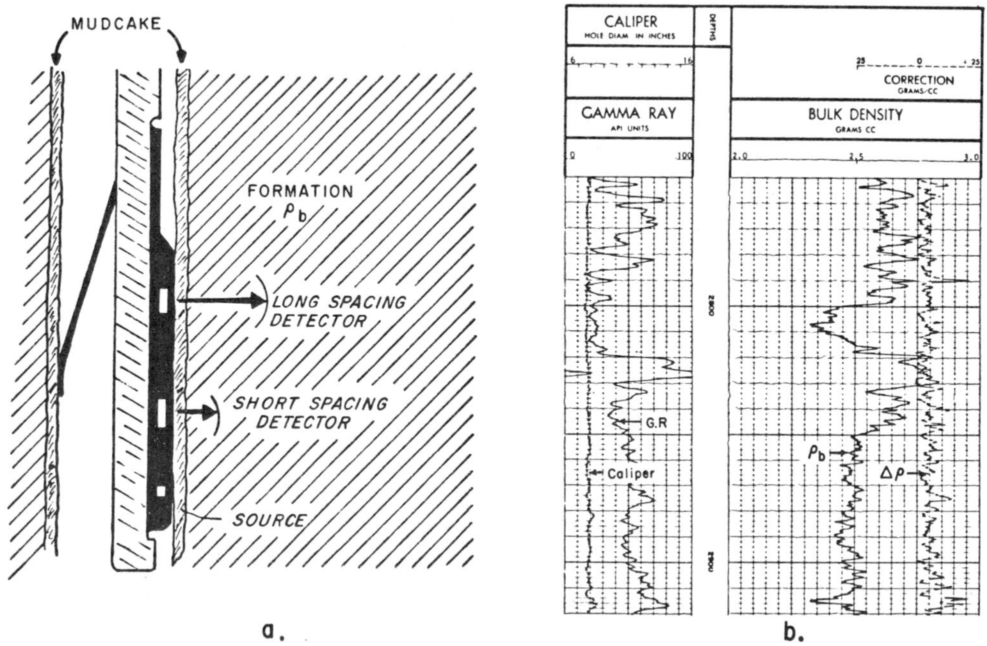

Fig. 45. **Density log.** (a) Compensated density logging sonde schematic. (b) A log showing the density ρ_b and the correction for mudcake, etc., $\Delta\rho$. (Courtesy Schlumberger.)

vanes so that they tend to either depress or elevate the streamer as it is pulled through the water. By varying the air pressure one can vary the depth which the controller is to maintain. Another type balances hydrostatic pressure against the tension of a spring which is remotely controllable.

depth of compensation: The assumed depth at which the pressure due to the overlying crustal elements is constant and below which lateral density variations disappear. Involved in isostatic correction sometimes taken as the top of the asthenosphere. See also **Pratt hypothesis.**

depth of invasion: See **invaded zone.**

depth of investigation: **1.** The depth below the surface to which an exploration system can effectively explore. Depends on array, spacing, property contrast, body geometry, and signal-to-noise ratio. The maximum depth at which interfaces or the sources of anomalies are resolvable considering the signal-to-noise ratio and other measurement considerations. Also see **skin depth. 2.** The radius about a logging sonde within which material contributes significantly to the readings from the sonde.

depth penetration: Depth of investigation.

depth point: **1.** In reflection, work the position midway between shotpoint and geophone, under which the point of reflection is located if the reflector is horizontal and if velocity layering is horizontal. Depth point is usually associated with a spread layout rather than an event (see Figure 47), thus excluding definition 2 below. **2.** Sometimes used for the point of reflection in the subsurface. This usage requires a different depth point for every event, depending on the reflector's dip and

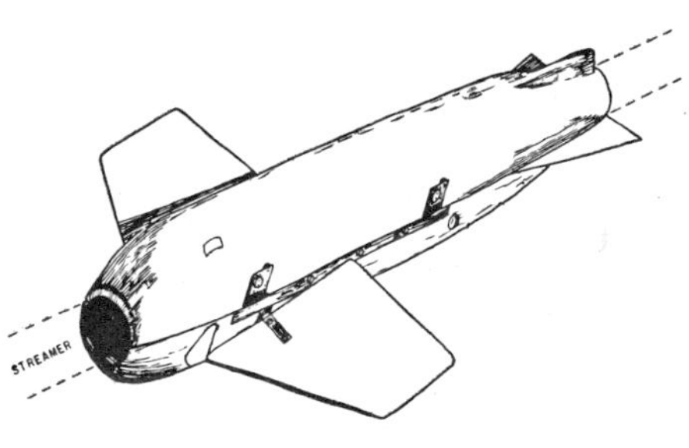

Fig. 46. **Depth controller** is clamped around streamer. When the pressure is less than the setting, the vane tilts causing the controller to sink; when the pressure is greater than the setting, the vane tilts causing the controller to rise. (Courtesy Continental Oil Company.)

depth and the velocity distribution. **3.** In refraction work, the point for which the depth to an horizon has been calculated. Usually the point where the head-wave energy leaves the refractor to travel to the geophone. See Figure 47.

depth probe: 1. A group of refraction profiles for which the offset is successively expanded, designed to obtain information on the velocity-layering pattern in an area. Approximate depths and velocities of refraction markers thus obtained are used to ascertain the nature of the stratigraphic section. Also called *refraction test.* 2. An electrical-surveying technique of exploring vertically down into the earth by employing an orderly horizontal expansion of the interelectrode interval of an array. Data from such a depth probe can be interpreted to give the depth to a resistivity contrast or anomalous IP material if horizontal layering exists. Also called *sounding,* using and *expander, depth profile,* and **vertical electrical sounding** (VES). See **two-dimensional plot.**

depth rule: An algorithm for calculating the depth to an anomalous mass, conductor, or magnetic body. Depth rules are often based on extreme values of an anomaly or its derivatives or on the distances between certain easily determined points on anomaly curves. The simplest depth rule is the half width (the depth to the body is half the width of an anomaly at half the maximum amplitude); see Figure 48. Except for general theorems such as the Bott-Smith relations (Bott and Smith, 1958, The Estimation of the Limiting Depth of Gravitating Bodies: Geophysical Prospecting, v. 6, p. 1-10), most depth rules apply to specific sources such as dikes, point poles (or point masses), etc. Such rules include: **a.** The *slope-distance or Steenland-Vaquier rule* for dike or vertical-prism bodies (depth = the horizontal distance over which the curve approximates a straight line with the maximum slope); **b.** The *Peters rule* for dikes (depth = 5/8 of the horizontal distance between points where the slope is half the maximum slope); **c.** The *Tiburg rule* for poles (depth = 2/3 of the horizontal distance at half the maximum amplitude); **d.** The *Hannel rule for poles (depth = half of* the horizontal distance at a third the maximum amplitude); and other such rules.

depth sounder: Fathometer.

depth sounding: Sounding.

derivative map: A map of one of the derivatives of a potential field, usually the second vertical derivative. The objective of a derivative map is to emphasize anomalies of interest. Second derivative maps are based on Laplace's equation:

$$\frac{\partial^2 \phi}{\partial Z^2} = -\left[\frac{\partial^2 \phi}{\partial X^2} + \frac{\partial^2 \phi}{\partial Y^2}\right].$$

The horizontal derivatives,

$$\frac{\partial^2 \phi}{\partial X^2} \quad \text{and} \quad \frac{\partial^2 \phi}{\partial Y^2}$$

are usually measured from gridded points on the map, often using a residualizing template based on a polar representation of the Laplacian or by two-dimensional convolution with such a template.

det: Detonator; an explosive **cap.**

detail log: A borehole log plotted at a scale larger than 1 inch per 100 ft; specifically, an electric log at a scale of 5 inches per 100 ft.

detection logging: See **exploration logging.**

detector: 1. A device which senses or measures a phe-

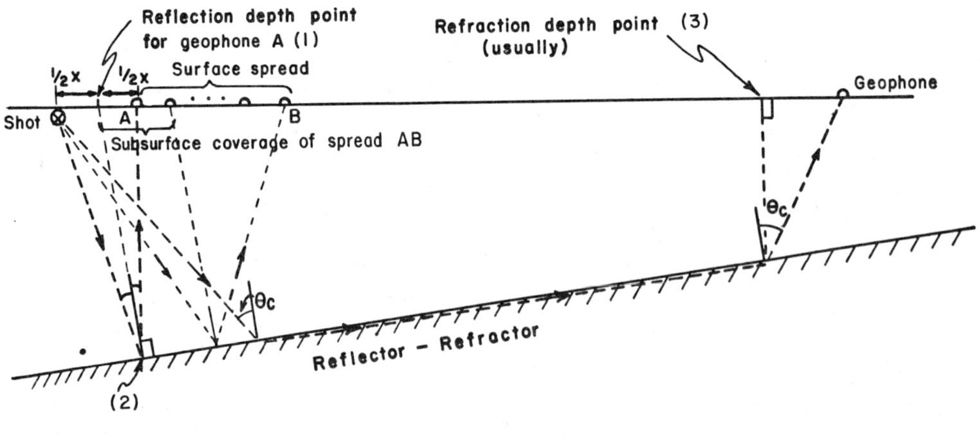

Fig. 47. **Depth point** based on definitions (1), (2), and (3).

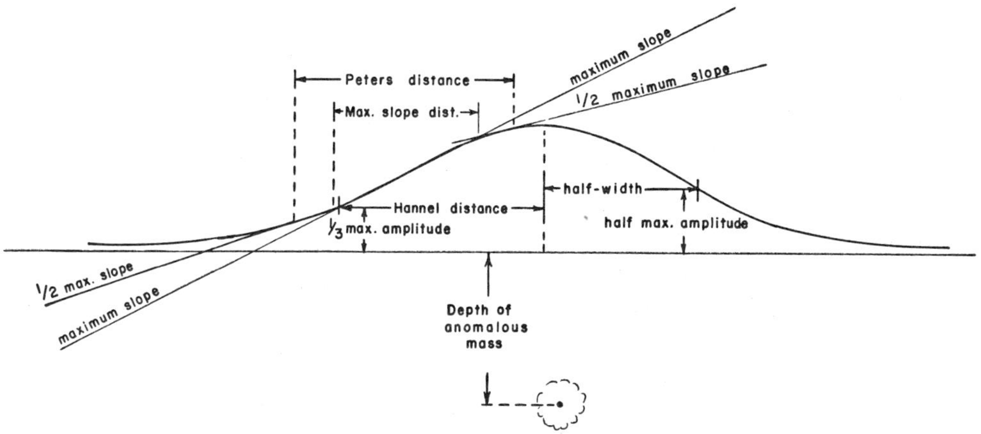

Fig. 48. **Depth rules** showing where various measurements are made on an anomaly curve.

nomenon. 2. A **geophone.**

determinant: A scalar function of a square **matrix** *(A)*:

$$\det A = \sum_{i=1}^{N} a_{ik} A_{ik},$$

where A_{ik} is the cofactor of the element a_{ik}.

deterministic: From a certain set of causes, a unique situation will develop. As opposed to probabilistic, which leads only to the probability that various situations will follow.

detonator: Cap.

deviation: Departure of a borehole from vertical. See **drift, directional survey,** and **rectify.**

dextral: Rotation to the right or clockwise. A dextral strike-slip fault is also called *right lateral.* Opposite is *sinistral.* See Table 3 **(fault).**

diagnostic check: A routine designed to locate malfunctions.

diamagnetic: Having a net negative magnetic susceptibility and a permeability less than that of free space (less than unity in the cgs system). The motion of an electron about a nucleus produces a miniature circular current whose magnetic-moment vector precesses around an applied external field. This additional periodic motion produces a magnetic moment opposite in direction to the applied field. Diamagnetic effects in the earth's field rarely exceed one gamma. The strongest diamagnetic anomalies are probably due to salt domes. Compare **paramagnetic** and **ferromagnetic.**

diamond: A type of geophone or shotpoint array in which the elements are laid out on a grid of lines at about 45 degrees to the seismic line, the pattern having the general shape of a diamond. See Figure 12 **(array).**

diapir: A flow structure whose mobile core has pierced overlying rocks. Salt and shale are the most common sedimentary rocks involved in diapirs. Intrusive rocks can also form diapir-like features but diapiric is usually restricted to plastic flow.

dielectric constant: A measure of the capacity of a material to store charge when an electric field is applied. It is the dimensionless ratio of the capacitivity (or permittivity, the ratio of the electrical displacement to the electric field strength) of the material to that of free space.

dielectric loss: The energy loss per cycle in a dielectric material due to conduction and slow polarization currents or other dissipative effects.

dielectric polarization: The response of a dielectric material to an electric field, producing an induced dipole-moment per unit volume. In an insulating dielectric material, no net electric charge need be transferred by the exciting field. By some definitions, induced polarization is a lossy type of dielectric polarization with a long time constant.

differential: 1. A difference between quantities. Thus, a *differential voltmeter* measures the difference between voltages. 2. A *differential input* on a voltmeter helps reject noise that originates from the ground. See **common-mode rejection.**

differential curvature: For a gravitational equipotential surface, the difference between the curvature of the surface in the direction in which it curves the most to the curvature at right angles to this direction multiplied by the gravitational constant. Measured by the torsion balance.

differential normal moveout: 1. The difference between the normal moveouts of adjacent channels. 2. Sometimes refers to **residual normal moveout,** the normal moveout which remains after an incorrectly assumed amount is removed. 3. Also may refer to the

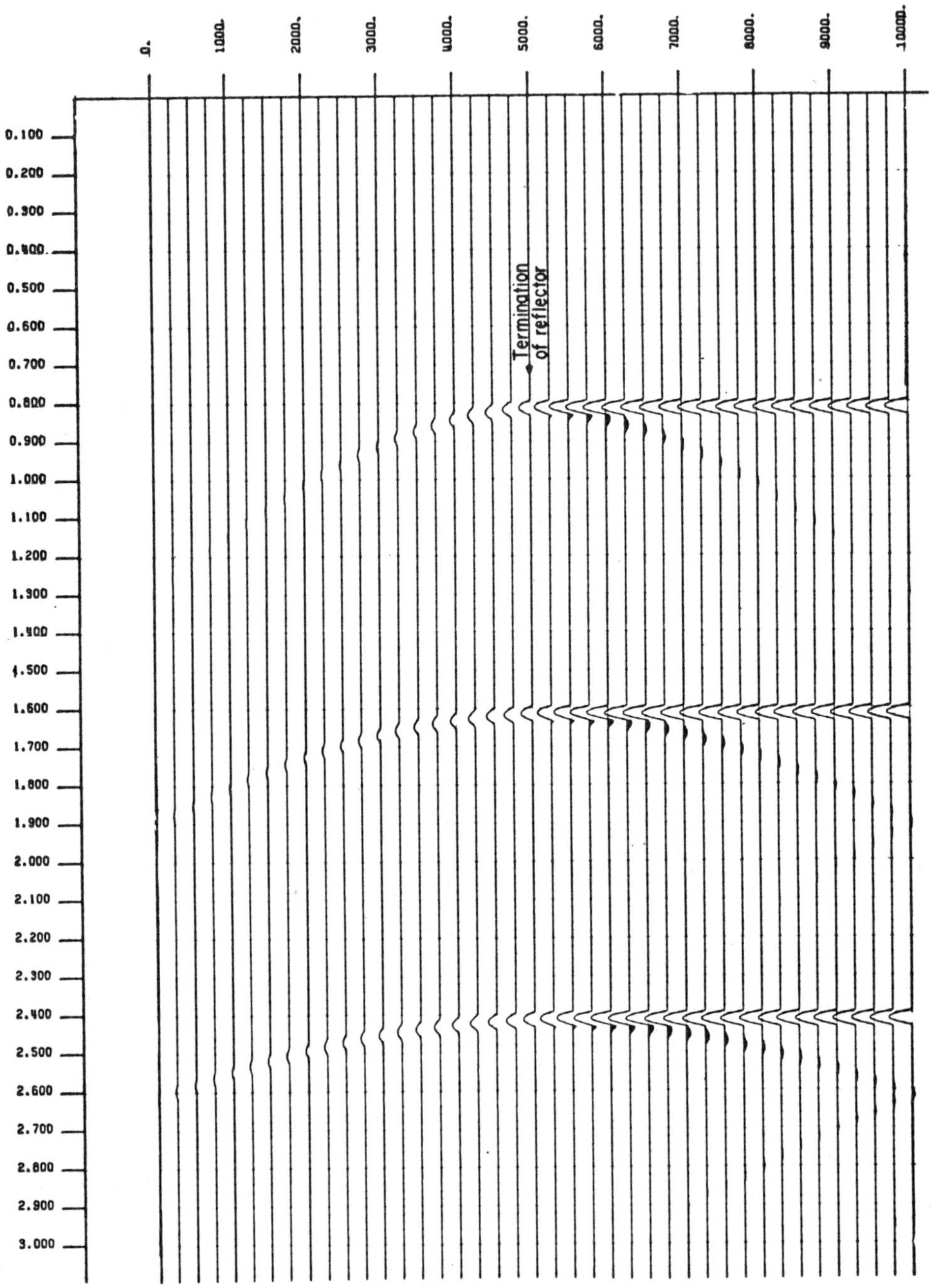

Fig. 49. **Diffractions** generated by termination of three flat reflectors. The curvature of the diffractions becomes smaller with depth. The diffraction branch under the reflection is the inverse of the branch which extends beyond the reflector. The reflection amplitude decreases to one half at the termination point and is continuous with the diffraction. (Courtesy Chevron Oil Company.)

difference between the normal moveout for primary events and that for multiples.

differential Omega: See **Omega.**

diffracted reflection: The diffraction resulting from reflected energy striking a diffracting point; its curvature is appropriate to the diffracting point, not to the arrival time.

diffraction: **1.** A phenomenon common to all waves (light waves, radio waves, seismic waves, surface water waves, etc.). See **Huygens principle. 2.** Scattered seismic energy which emanates from an abrupt discontinuity of rock type, particularly common where faults cut reflecting interfaces. Diffractions from such discontinuities below the seismic line appear along a **diffraction curve** (which depends on the velocity distribution above the diffracting point). Diffractions involving energy which reaches the diffracting point by a longer route than directly (as with a diffracted reflection) have curvature appropriate to the depth of the diffracting point rather than to the arrival time. Diffractions generated by a line source which is not at right angles to the line will appear to have less curvature, becoming flat as the diffracting line becomes parallel to the line of observation. The diffracted energy shows greater curvature than a reflection (except for the reverse branch where there are buried foci). Diffracted energy coming from the termination of a reflection blends with the reflection and obscures (see Figure 49) the location of the termination or becomes confused with dip unless it is correctly recognized. A reflection can be thought of as the interference result of diffractions from the various points lying on the reflector. When correctly migrated, a diffraction should collapse at the location of the diffracting point (Figure 50). **3.** Interference produced by scattering at edges. **4.** The phenomenon by which energy is transmitted laterally along a wave crest. When a portion of a wave train is interrupted by a barrier, diffraction allows waves to propagate into the region of the barrier's geometric shadow. See Figure 51.

diffraction curve: A *curve of masimum convexity;* the relation between arrival time and position for energy which has traveled from the observing position to a diffracting point and then has been diffracted from the point. See Figure 52. One should actually speak of a *diffraction surface* to emphasize the three-dimensional aspect, but "curve" is well entrenched. The curvature of reflected energy cannot exceed this except for reverse branches and certain other situations such as a diffracted reflection. Diffraction curves are specific for a particular velocity function, like a wavefront chart to which they are related and from which they can be constructed. Diffraction curves are used in identifying point diffractions and locating the diffracting points and in migrating reflections (see Figure 118, **migration.** Note that errors in interpreting diffractions can result if the diffracting point lies to the side of the seismic line, if the diffraction event results from a line diffractor which is not normal to the seismic line, or if the diffraction is not the result of the downgoing wavefront (as with reflected energy which is diffracted; a diffracted reflection has curvature appropriate to the vertical time of the diffracting point although it arrives later). See Hagedoorn, 1954, A Process of Seismic Reflection Interpretation: Geophysical Prospecting, v. 2, p. 85-127.

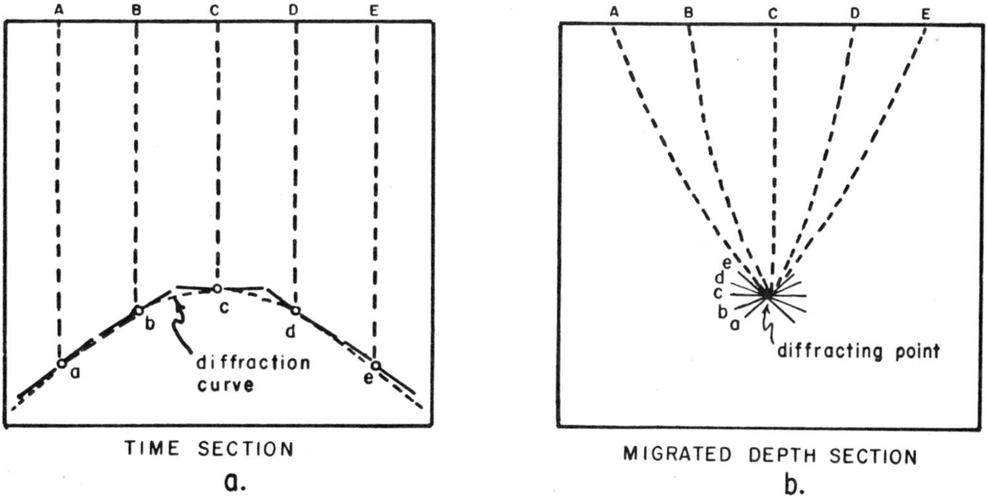

Fig. 50 **Diffraction knot.** (a) A diffraction approximated by line segments. (b) On a depth section the migrated segments form a diffraction "knot."

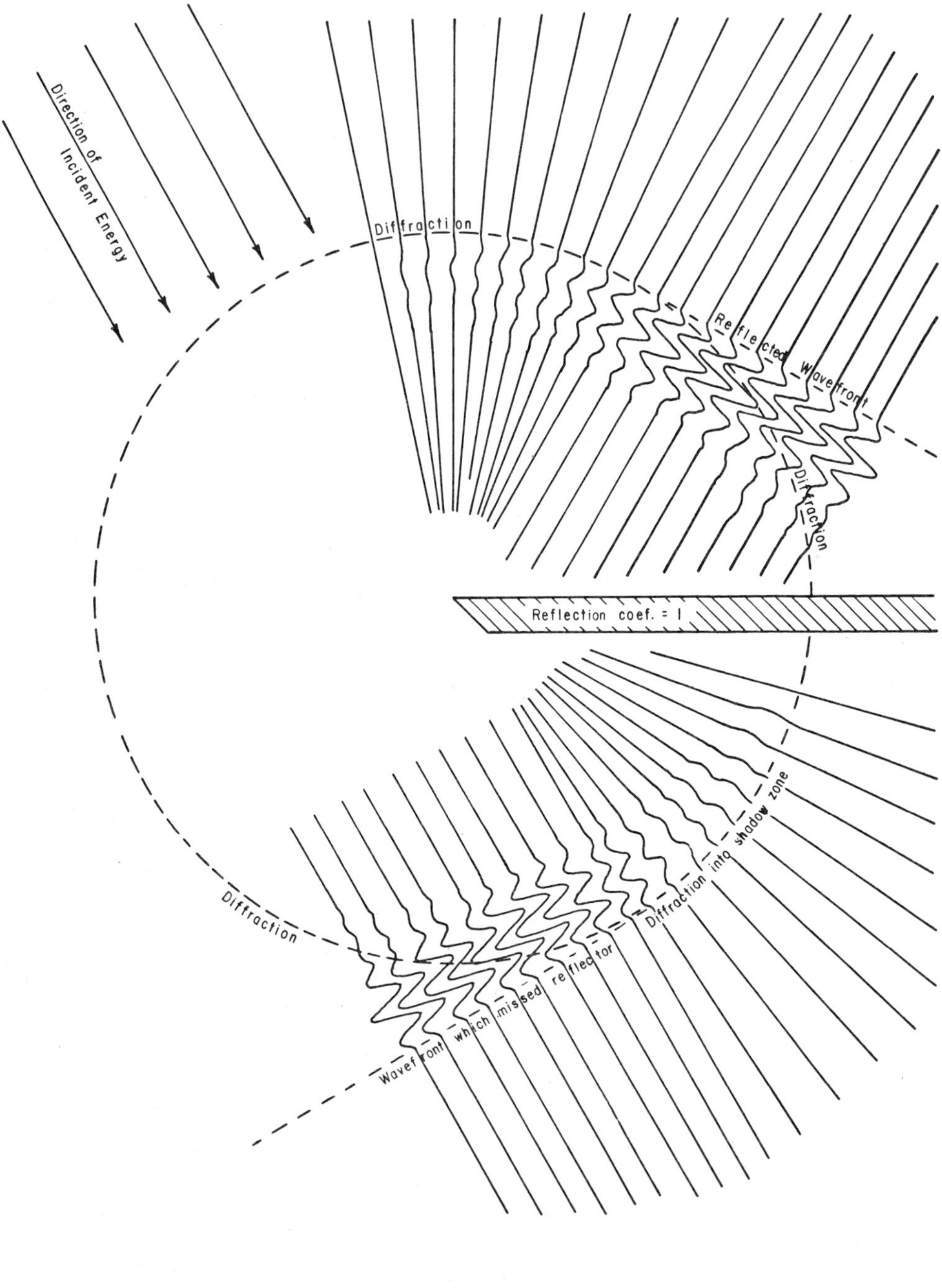

Fig. 51. **Diffraction** of plane wave from a semiinfinite barrier. Schematic diagram of the relative amplitude of the reflected wavefront, of the wavefront which missed the reflector, and of the diffractions from the termination of the reflector. (Courtesy Chevron Oil Co.)

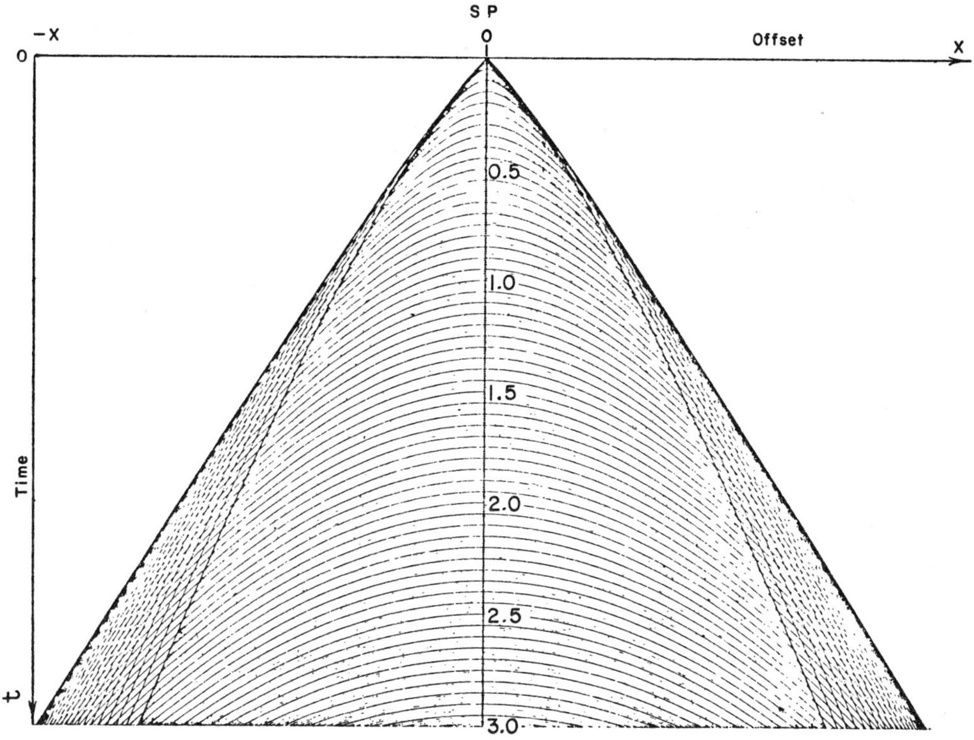

Fig. 52. **Diffraction curves** or curves of maximum convexity. (From Hagedoorn, Geophysical Prospecting: v. 2, no 2, p. 116.)

diffraction function: The function $(\sin \pi x)/\pi x = \text{sinc } x$.

diffraction knot: Where **diffraction** energy is approximated by straight line segments and migrated properly, the migrated segments cluster in an asterisk-like "knot." See Figure 49.

diffraction stack: A weighted stack of all the elements of reflected energy along a **diffraction curve** which yields a migrated section (to the extent that the data were two-dimensional). A method of automatically migrating reflection seismic data.

diffuse layer: The outer, more mobile ions of an electrolyte-solid interface which together with the fixed layer constitutes a **double layer.** Also called *diffuse zone, diffuse double layer,* or *outer Helmholz double layer.*

diffusion: The motion of ions or molecules in a solution resulting from the presence of a concentration gradient.

diffusion impedance: The impedance in a solution produced by a concentration gradient, for instance next to an electrode. See **Warburg impedance.**

digital: Representation of quantities in discrete (quantized) units. A digital system is one in which the information is contained and manipulated as a series of discrete numbers, as opposed to an **analog** system, in which the information is represented as a continuous flow of the quantity constituting the signal.

digital clipping: Loss of the most significant bit of a number, such as produced by overflow. Produces a different result from ordinary clipping and generates spurious high frequencies. See Figure 33 **(clipped).**

digital computer: See **computer.**

digital to ac: Conversion of a digital number to a corresponding ac voltage for navigation gyrocompass compensation.

digital to analog: Conversion of a digital (usually binary) number into a corresponding voltage. Often written D/A.

digitize: To quantize. **1.** To sample a continuous voltage at discrete regular time intervals, quantize the measurements, and record the values as a sequence of numbers in bit combinations on magnetic tape. **2.** The regular sampling of a function. Equivalent to multiplying the function by a **comb. 3.** Conversion of coordinates and other parameters to a form that can be read by a digital computer, as with an **X-Y reader.**

dihedral angle: The angle between two intersecting planes.

dike: A tabular body; igneous rock in the form of a slab of finite thickness which is extensive in other directions, and which cuts across adjacent rock. Vertical and dipping dike models are used in potential field calculations. See Figure 121 **(model)**.

dilatation: Change in volume per unit of volume.

dilatational vave: P-wave.

dimensional analysis: Equating units in a physical relationship so that the dimensions as well as the number values balance.

dimensionless induction number: See **induction number.**

dimensionless units: Ratios which do not depend on the units in which quantities are measured and hence are more general. For example, distance is often measured in terms of wavelengths; frequency is often expressed as a ratio to a natural frequency, etc.

dimple: A shallow velocity anomaly (such as permafrost variation) which depresses or raises all seismic data seen through it and distorts the normal moveout of deeper events by **velocity focusing.**

Dinoseis: A seismic energy source in which a plate is driven against the ground by a confined explosion of gas. An Atlantic Richfield trademark.

dip: 1. The angle which a plane surface makes with the horizontal. **2.** The angle which bedding makes with the horizontal. **3.** The angle which a reflector or refractor makes with the horizontal. **4.** *Apparent dip* is the angle between horizontal and the component of dip in the plane of a section. **5. Pitch** (electromagnetic).

dip calculation: Calculation of the dip or dip component of a reflector from seismic arrival time data; the conversion of dip delta *T* values to dip. May involve resolution of **cross-spread** data. The dip angle for a reflection measured at the surface is usually less than the angle at the reflector because of raypath curvature.

dip log: A **dipmeter** log. Diplog is a Dresser Atlas tradename.

dipmeter: 1. A well log from which the magnitude and azimuth of formation dip can be determined; see Figure 53a. The resistivity dipmeter includes: (a) three or more microresistivity readings made using sensors distributed in azimuth about the logging sonde, (b) a reading of the azimuth of one of these, (c) a reading of the hole deviation or drift angle, (d) its bearing, and (e) one or two caliper measurements. The microresistivity curves are correlated to determine the differences in depth of bedding markers on different sides of the hole and dip calculations are based on such correlations. **2.** Other types of dipmeter use three SP curves, three wall scratchers, etc. **3.** A log showing the formation dips calculated from the above, such as a tadpole plot or stick plot; see Figure 53b.

dipole: 1. A pair of unlike poles, such as a pair of opposite electrical charges or poles of a magnet, that ideally are infinitesimally close together. **2.** In resistivity and IP surveying, a pair of nearby current electrodes which approximates a dipole field from a distance or a voltage-detecting electrode pair. Sometimes called a *bipole*. **3.** In electromagnetic surveying, an electric or

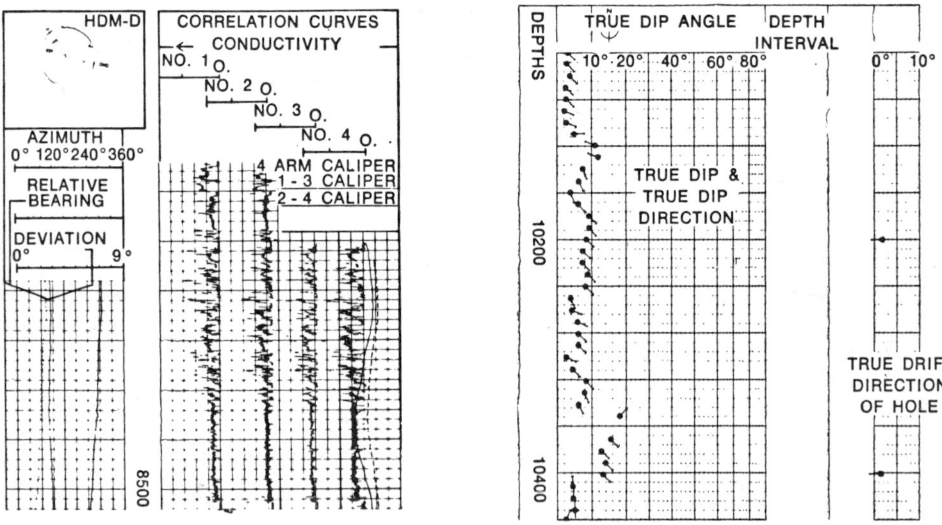

a. b.

Fig. 53. **Dipmeter log.** (a) Log of the field data. (b) Calculated dipmeter log or tadpole plot. (Courtesy Schlumberger.)

magnetic field transmitting or receiving antenna which is small enough to be represented mathematically as a dipole. The near field (electric and magnetic) from a magnetic and electric dipole (respectively) varies as the inverse cube of the distance. **4.** A doublet or two-stick wavelet, that is, *(a, b)*. If $|a| > |b|$, then *(a, b)* is minimum-delay and *(b, a)* is maximum-delay

dipole array: Dipole-dipole **array**.

dipole-dipole array: Electrode **array** used in induced polarization, electrical, and electromagnetic surveying.

dipole moment per unit volume: A measure of the intensity of polarization of a material. Units are ampere-meters per cubic meter.

dip resolution: **1.** Calculation of true dip from **cross-spread** data. **2.** Recognition and separation of two events having nearly the same dip.

dip shooting: A seismic field method whereby structural dips are computed from reflection data (often on **split spreads**) and used to map subsurface horizons. May involve either isolated or continuous profiles or **cross-spreads.**

dip spectrum: **1.** A graph showing the frequency with which different dips occur. **2.** Sometimes used for a **sonograph** which displays coherent energy as a function of apparent velocity (or dip) and arrival time.

dip vector: An arrow on a map pointing in the direction of dip. Sometimes the arrow length indicates (a) the dip magnitude, sometimes (b) the contour spacing, and sometimes (c) the portion of the line over which such a dip direction is evident. Numbers written alongside the arrow may indicate the amount of dip or the rate of dip, and sometimes a grade is also indicated. Used as a method of posting data on a map prior to contouring and also as a grading method to show the distribution and reliability of the data.

Dirac comb: A **comb**; a series of equally spaced delta functions. Also called *shah.*

Dirac function: A **delta function** or impulse.

direct interpretation: Direct mathematical solution of a potential field problem without use of precomputed curves or models. Some used this term for **curve matching.**

directional charge: An explosive charge or charge array in which the explosion front travels at approximately the seismic velocity, so that energy traveling in the desired direction (usually vertically) adds up constructively as opposed to that traveling in other directions. In refraction shooting, a horizontal directional charge detonating at the refractor velocity is sometimes used to concentrate the energy traveling along the refraction path. Sometimes involves the use of broomstick charges, delay caps, impulse blasters, or sausage powder.

directional survey: **1.** Measurements of **drift,** the azimuth and inclination of a borehole with the vertical. A directional survey is often made as part of a dipmeter survey. Sometimes it involves a continuous log with a poteclinometer and sometimes measurements made only at discrete levels with a photoclinometer. **2.** An IP or resistivity-survey method starting from a position such as a drill hole to find the trend direction of an anomalous subsurface body.

directivity graph: **1.** A plot (often in polar coordinates) of the relative intensity versus direction of an outgoing seismic wave such as that resulting from a **directional charge** or from a source pattern; see Figure 54. The directivity results from the interference of the waves from the various components of the pattern. **2.** A plot of the relative response of a geophone pattern or of directivity resulting from mixing; see Figure 54. The response is sometimes plotted in polar form to indicate the relative amplitude resulting from a wave arriving from different directions, but, most commonly, it is plotted as relative response against dip or normalized dip. Sometimes such graphs are given in units of apparent velocity for a certain frequency, apparent wavelength, frequency arriving from a certain direction, apparent velocity, etc. The vertical scale is often logarithmic rather than linear (given in db). Note that the effect on wavelets can be very different from the effect on a sinusoidal wave train so that use of steady-state patterns alone can be misleading. A plot in **f-k space** can be a form of directivity graph showing the response as dependent on frequency also. See also Figure 32 **Chebychev array).**

direct problem: Computation of the effects of a certain model; as opposed to the *inverse problem* of determining the model from observation of effects. See Figure 120 **(model).**

direct recording: Magnetic tape recording in which the magnetization intensity is proportional to the signal strength (or signal strength plus a bias).

dirty: Shaly; containing appreciable amount of shale dispersed in the interstices. Such shale lowers the permeability and effective porosity and affects the readings of many types of logs. A composite electrolyte system of clay and sand can be responsible for membrane polarization effects. Clay particles in the sand act as selective *ion sieves* and surface conduction along the clay minerals causes low resistivity.

disc hydrophone: A piezoelectric hydrophone similar to a **bender.** Two piezoelectric discs are supported around their circumference so that pressure tends to dish them in, causing stresses which generate a voltage across the disc thickness.

disjunction: The logical operation, "either A or B" (written A U B), done in an **OR gate.** Also called *union.* See Figure 88 **(gate).**

dispersion: **1.** Distortion of the shape of a wave train because of the variation of velocity with frequency. The peaks and troughs may advance toward (or recede from) the beginning of the wave as it travels. Leads to the separation of **group velocity** u from **phase velocity** v. Where λ = wavelength and f = frequency,

$$v = f/(1/\lambda)$$
$$u = df/d(1/\lambda).$$

The dispersion of seismic body waves is very small under most circumstances, but surface waves may show appreciable dispersion in the presence of near-surface velocity layering. Figure 55. The disperson of

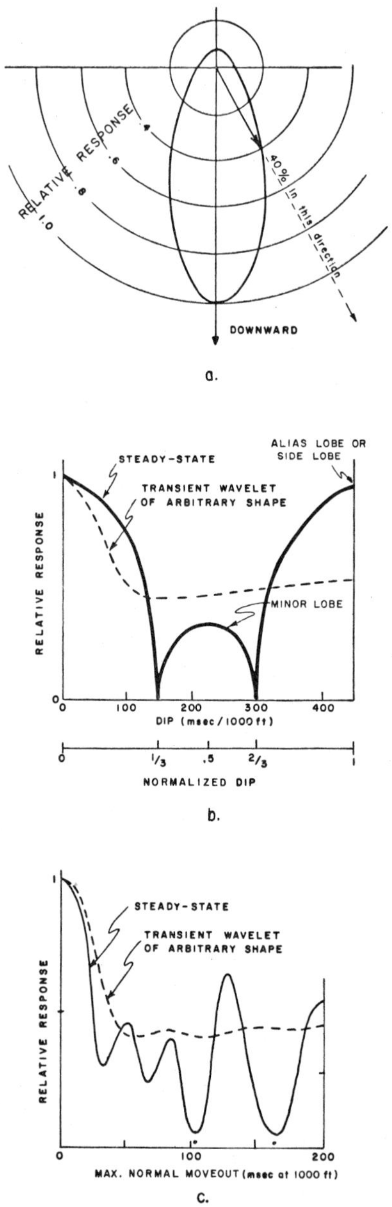

Fig. 54. **Directivity graphs.** (a) Polar plot of the directivity of a directional charge showing the amplitude in different directions. (b) Directivity of three in-line geophones. (c) Directivity of 6-fold CDP stack. (Courtesy Chevron Oil Company.)

electromagnetic body waves is large in most earth materials. **2.** A statistical term for the amount of deviation of a value from the norm.

dispersive filter: A **phase**-shifting filter which does not affect the amplitude spectrum.

displacement: 1. Plotting refraction data with respect to where the refracted energy presumably leaves the refractor rather than under the geophone positions where the arrivals are actually observed. Sometimes called *offset* or *transplacement.* See Figure 130 (**offset**). **2.** Relative movement of the two sides of a **fault.** **3.** The distance a particle is removed from its equilibrium position, as in the ground motion associated with a seismic wave.

displacement current: A current which is proportional to the time rate of change of electric flux density. In most earth materials, displacement currents are negligible compared with conduction currents for the range of frequencies used in electromagnetic methods.

display: A graphic representation of data, especially of seismic data. See Figure 56. The historic display mode for seismic records graphs amplitude as a funtion of time for each geophone group output, giving *wiggle trace* or *squiggle* or *conventional display. Variable area display* in which the area under the wiggle trace is shaded (often involving a bias and trace clipping) has the advantage that coherent events are more evident. *Variable density* has also been used but requires very careful processing to retain a full spectrum of gray shades. *Superimposed modes* or *combined modes* which involve the use of both wiggle trace and variable area or variable density simultaneously retain many of the good features of each type display. Other display modes are also used.

disseminated sulfide mineralization: Sulfide minerals scattered as specks and veinlets through rock and constituting not over 20 percent of the total volume.

distortion: An undesired change in waveform, as opposed to desired changes in waveshape like those from modulation. **a.** *Amplitude distortion* is due to undesired amplitude-versus-frequency characteristics. **b.** *Harmonic distortion* is nonlinear distortion characterized by the generation of harmonics of an input frequency. The percent harmonic distortion is a measure of the fidelity of a system or component. If E_f is the rms voltage of the fundamental and E_n is the rms voltage of the *n*th harmonic, the *percent harmonic distortion* is

$$100\left(\sum E_n^{\,2}\right)^{1/2} / E_f \,.$$

c. *Intermodular distortion* is nonlinear distortion characterized by the appearance, in the output, of frequencies equal to the sum and difference of integral multiples of the component frequencies present in the input. **d.** *Nonlinear distortion* is caused by a deviation from a linear relationship between input and output. **e.** *Phase distortion* results when phase shift is linear with frequency over the bandpass, but where the zero-frequency intercept is not a multiple of π. Where the intercept is a multiple of π, the wave shape is not

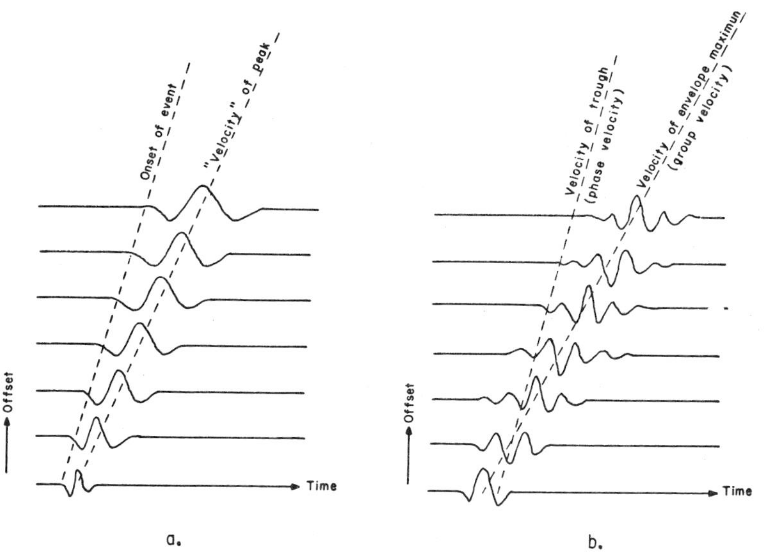

Fig. 55. **Dispersion.** (a) Change in wavelet shape because of dispersion in a Voigt solid. (b) Change in wavelet shape because of energy shifting to later cycles. The axes of time and offset could be interchanged on either graph. (From Balch and Smolka, Geophysics: v. 35, no. 5, p. 745.)

changed, though the entire waveform is delayed. See **linear phase filter.**

distortion point: The input amplitude for which the distortion-generated third harmonic becomes 3 percent or sometimes (in high-fidelity systems) 1 percent. For smaller inputs the system is nearly linear.

distortion tail: A correlation ghost. Harmonic distortion with Vibroseis surface sources produces spurious correlation, especially for the second harmonic (which may be 30 to 100 percent of the fundamental). Such a *distortion tail* follows the correlation for a downsweep and appears as a forerunner for an upsweep.

distributed: Referring to electric circuits, the *smearing out* of resistive, capacitive, or inductive circuit elements such as with a transmission line. Opposite of a **lumped** circuit.

distribution function: A relationship which describes the probability that a quantity will have a value less than a particular value. It is thus the cumulative integral of the probability density (the probability density $P(x)$ of the random variable x is not necessarily a proper function):

$$F(x) = \int_0^x P\{x\}\, dx.$$

diurnals: Phenomena with a periodicity of one day.

diurnal shift: Average daily changes in radio-wave propagation because of diurnal changes in the ionosphere, which affects radio-positioning systems. Published tables allow one to correct for major skywave variations with systems like Omega, but unpredictable local skywave variations and sun-spot effects remain.

diurnal variation: Daily fluctuations of the geomagnetic field, related principally to the tidal motion of the ionosphere. Amplitude and phase vary with season and latitude as much as 40 gammas.

divergence: 1. The decrease in amplitude of a wavefront because of geometrical spreading. With body waves the energy spreads out as the spherical wavefront expands, causing the energy density to vary inversely as the square of the distance *(spherical divergence).* With surface waves the energy density varies inversely as the distance *(cylindrical divergence).* Tube waves do not suffer energy loss because of divergence. Note that energy density decreases for other reasons also. **2.** The divergence of a vector field:

$$\mathrm{div}\ \mathbf{V} = \nabla \cdot \mathbf{V} = \frac{\partial V_x}{\partial x} + \frac{\partial V_y}{\partial y} + \frac{\partial V_z}{\partial z}.$$

divergence theorem: The flux ϕ through a surface (or the integral of the vector flux density g over a closed surface) equals the divergence of the flux density integrated over the volume contained by the surface. Also called *Gauss theorem.*

$$\phi = \int_S \mathbf{g} \cdot d\mathbf{s} = \int_V \nabla \cdot \mathbf{g}\, dV.$$

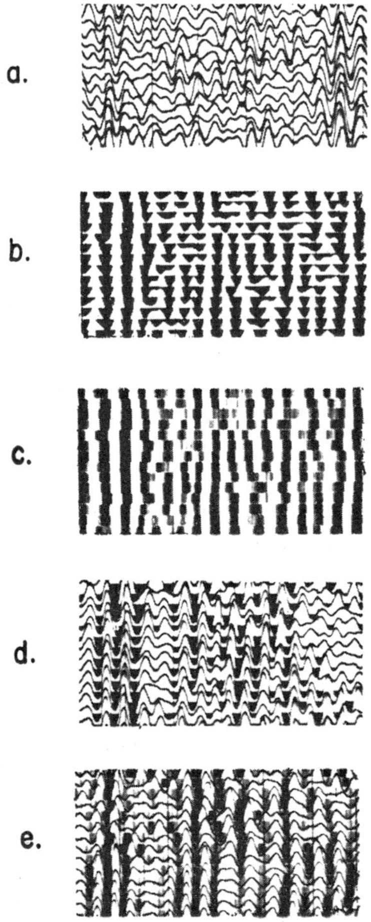

Fig. 56. **Display modes.** (a) Wiggle trace. (b) Variable area. (c) Variable density. (d) Wiggle trace superimposed on variable area. (e) Wiggle trace superimposed on variable density.

diversity stack: 1. A stack in which the components are weighted inversely as their mean power over certain intervals. Used with vertical stacking to prevent occasional large bursts of noise (such as traffic noise) from dominating the stacked record. Also used with common-depth-point stacking to discriminate against ground roll and similar high-amplitude wave trains. **2.** A stack in which amplitudes which exceed some threshold are excluded completely instead of merely given small weight. Amplitudes less than this threshold may not be affected.

diving waves: Refraction in a strong velocity-gradient zone may reverse the component of seismic rays in the downward direction and return them to the surface, where they may be observed (at offset) as refraction arrivals even though they have no appreciable path through a distinctive refractor. See Figure 57. If there are no velocity reversals, diving waves may be used to derive the velocity distribution by means of the Wiechert-Herglotz intergral. See Meissner, 1966, Interpretation of Wide Angle Measurements: Geophysical Prospecting, v. 14, p. 7-16. See also **Blondeau method.**

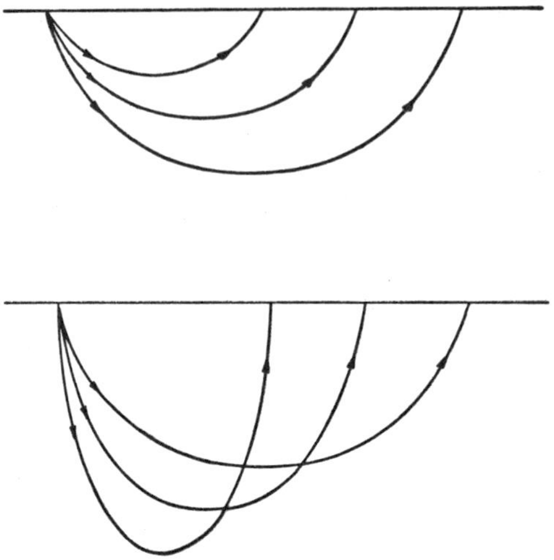

Fig. 57. **Diving waves.** Raypaths are curved because of vertical velocity gradient. **In the lower diagram, the magnitude of the gradient increases rapidly with depth.**

divining: Use of a forked stick which allegedly bends toward water, petroleum, or other sought-for accumulations; used meaning *unscientific.*

Dix formula: For reflections from a sequence of flat, parallel layers, the velocity in the *n*th layer V_n (interval velocity) is given by

$$V_n^2 = (\overline{V}_n^2 t_n \quad \overline{V}_{n-1}^2 t_{n-1})/(t_n \quad t_{n-1}).$$

where $\overline{V}_{n-1}$ and $\overline{V}_n$ are the average velocities from the datum to reflectors above and below the layer and t_{n-1} and t_n are reflection arrival times. This formula is sometimes misused to calculate interval velocities in situations which do not satisfy Dix's assumptions.

D-layer: 1. The innermost layer of the ionosphere. It occurs at heights of 50 to 90 km during daylight hours only, and reflects ELF, VLF, and LF waves, absorbs MF waves, and partially absorbs HF waves. **2.** A layer within the earth; see Figure 63 **(earth layering).**

DNMO: Differential normal moveout.

doghouse: The hut (or cab) which contains **seismic recording instruments** in the field.

dog-leg: An abrupt angular change in direction, as in a survey traverse or in a borehole.

domain: **1.** Associated with the parameter considered to be the independent variable. In the *frequency domain,* frequency is regarded as the variable which can be operated on; in the *time domain* it is record time; in the *f,k domain* or *f,k space* it is frequency and wavenumber; etc. **2.** The set of values which an independent variable may assume. **3.** A region of magnetic polarization in a single direction (magnetic moments parallel) which behaves as a unit during change in magnetization. **4.** The areal extent of a given lithology or environment.

doodlebugger: A geophysicist engaged in field work. Originally used derisively, meaning unscientific.

doppler count: **1.** A count of the number of cycles occurring within a certain time, as in satellite navigation. The satellite emits a constant frequency but the apparent frequency varies because of the velocity of the satellite with respect to the observing station (**doppler effect**). From the counts for several successive time intervals, the instant of closest approach and the distance of closest approach can be calculated. By knowing where the satellite was located at this instant, the position of the observing station then can be determined. **2.** Doppler counting is also used in connection with **doppler radar** and **doppler sonar.**

doppler effect: Apparent change in frequency of a wave caused by motion of the source with respect to the receiver and/or vice versa. Frequencies are increased if the source and receiver are approaching, decreased if they are moving apart. For sound and water waves, the observed frequency f_o is given in terms of the source frequency f_s, the velocity of propagation V, the velocity of the source V_s, and the velocity of the observer V_o (components of velocity toward each other):

$$f_o = \left[\frac{V + V_o}{V - V_s}\right] f_s .$$

The relationship for light and radio waves is slightly different because of relativistic effects and the constancy of the velocity of light irrespective of motion of source and observer:

$$f_o = \left[\frac{V + V_o - V_s}{V - V_o + V_s}\right]^{1/2} f_s .$$

See **doppler navigation.**

doppler navigation: Positioning in which frequency shift because of the **doppler effect** is involved. Usually refers to **doppler radar** (for aircraft) or **doppler sonar** (for ships), occasionally to **satellite navigation** (see **doppler count**).

doppler radar: A radio navigation system used by aircraft. A radio beam transmited from the aircraft and reflected back by various landscape features is received at the aircraft. Because the aircraft transmitter is in motion, the signal undergoes a frequency shift (the **doppler effect**) which is measured and converted to aircraft velocity. A computer on the aircraft determines the plane's position by integrating the velocity over time. Actually, one pair of transmitters beams to the ground ahead of and behind the aircraft to determine the forward speed and another pair of transmitters beams to the right and left of the aircraft to determine the crosstrack speed. The arrangement is similar to that used with **doppler sonar** (Figure 58). The doppler data combined with a compass heading give the direction of the aircraft. Velocity can be measured to about 0.5 percent, somewhat less over water because the surface of the water may itself be in motion.

doppler shift: **1.** Change in observed frequency because of the **doppler effect.** **2.** The shift toward lower frequencies which results from stretching long-offset seismic traces in the removal of normal moveout. Association with the doppler effect only by analogy.

doppler sonar: A sonic location system used by ships, based on the **doppler effect;** similar to doppler radar except utilizing sonic waves. Four transducers send sonar beams fore and aft and to each side, and receive the reflections of these beams from the sea floor; see Figure 58. (The use of four beams allows the effects of the ship's roll, pitch, and yaw to be reduced.) The ship's velocity is computed from the frequency shift of the reflected beams. Velocity measurements together with gyro-compass direction may be integrated to give position. Reflections from the sea bottom (bottom mode) are involved when the water is shallower than 600 to 1000 ft. Returns involve scattering *(volume reverberation* or *water track mode)* when the water is deeper than 600–1000 ft, resulting in considerable loss of accuracy. (A mounting for a doppler-sonar transducer is shown in Figure 158, **sea chest.**)

DOS: Disc **operating system.**

dot chart: A chart used to compute the theoretical gravity (or other potential) effect of a mass distribution; see Figure 59. A dot chart is superimposed on a scale cross-section drawn through the mass; the number of dots which fall within the mass outline multiplied by the anomalous density is proportional to the gravity effect at the origin of the chart. The chart is then moved to a different position on the outline and the number of dots counted again to give the effect at another point, and so on for every point for which the gravity value is to be determined. Most charts assume that the mass distribution extends to infinity perpendicular to the plane of the chart and end corrections have to be applied to remove this restriction. Also called *graticule.*

dot product: The dot product of the vector $X = (x_1, x_2, x_3, \ldots, x_n)$ with the vector $Y = (y_1, y_2, y_3, \ldots, y_n)$ is $X \cdot Y = (x_1 y_1, x_2 y_2, x_3 y_3, \ldots, x_n y_n)$. *The dot product reverse* is $(x_1 y_n, x_2 y_{n-1}, x_3 y_{n-2} \cdots, x_n y_1)$. Compare **cross product.**

double Bouguer correction: The **Bouguer correction** to sea level for measurements made on the ocean floor involves a correction to remove the upward attraction of the sea water above the meter and another correction to replace the sea water with the replacement density. Similar "double" corrections are required for measurements made in mines or in boreholes.

double coverage: **Reversed coverage.**

double layer: The layers of molecular ions and charged dipoles at a solid-to-solution interface. It is electrically

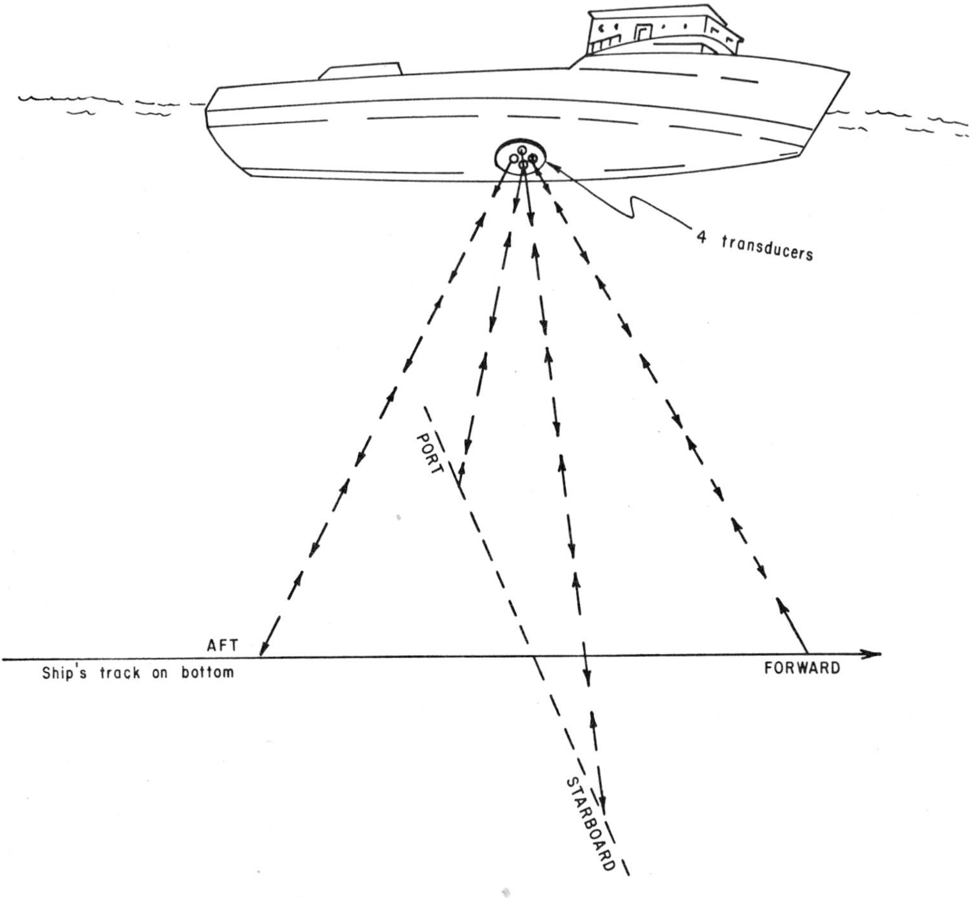

Fig. 58. **Doppler sonar.** (Courtesy Marquardt.)

analogous to a capacitor in that there is charge separation between the solid (electrode) and the charge center of the oriented ions or dipoles. Next to an electrode there may be an adsorbed fixed layer of ions called the *inner Helmholz double layer.* A diffuse layer (*outer Helmholz double layer*) in the electrolyte contains an excess of ions that is usually of the same charge as the electrode but opposite to that of the fixed layer. The thickness of the double layer is less than 100 Angstroms ($10^{-8}m$).

double-layer capacitance: Capacitance due to the presence of the **double layer** of charge at the interface between an electrode and electrolyte. In effect it is in parallel with the **Warburg impedance.**

double-layer weathering: Situation where corrections must be made for two distinctive near-surface low-velocity layers.

double precision: The retention of twice as many digits (bits) to specify a quantity as the computer normally uses.

double-run: A resurvey of a traverse to tie back to the same reference point, to reinforce the certainty that errors have not been made. Used where tie to an independent reference point is not feasible.

doublet: **1.** A two-stick wavelet (a,b). If $|a| > |b|$, (a,b) is minimum phase; if $|a| < |b|$, maximum phase. Some times called *dipole.* **2.** Phase-shift positive followed by phase-shift negative, such as used in **doublet modulation.** **3.** The first derivative of the Dirac delta function.

doublet filter: A digital filter containing only two nonzero values.

doublet modulation: A phase-modulation scheme used to encode the message broadcast in satellite navigation. A phase shift of +60 degrees for 2-1/2 msec

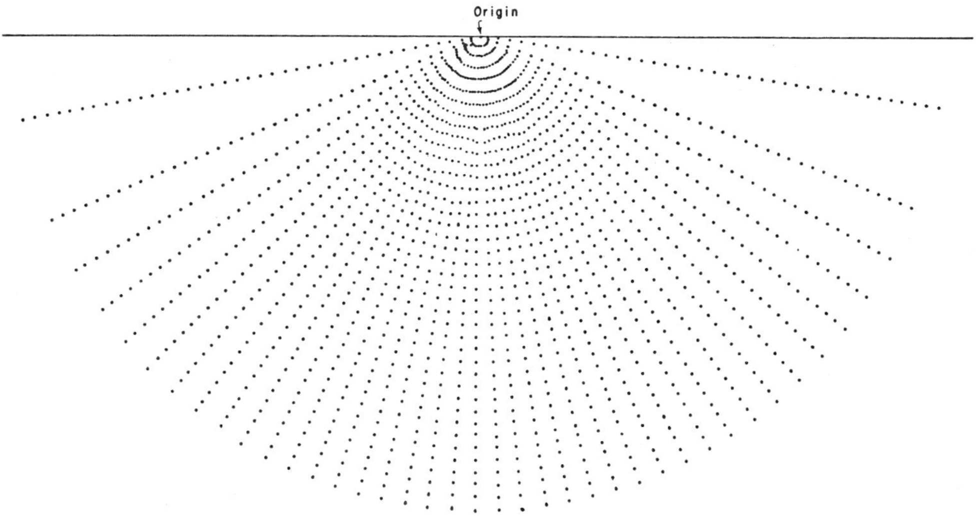

Fig. 59. **Dot chart** (two-dimensional). The gravity effect at the origin of a mass anomaly of given cross-section is $k\rho n$, where ρ is the density contrast, n is the number of dots lying in the anomalous mass, and k is a scale constant.

(Figure 60) is followed by a phase shift of −60 degrees so that there is no net phase shift to introduce error into doppler measurements. A bit consists of two such doublets in opposite sense following each other by 5 msec.

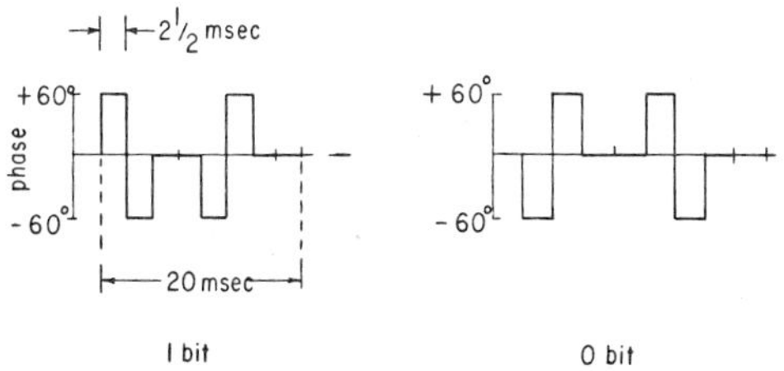

Fig. 60. **Doublet phase-modulation code.** Used for the message from navigation satellites.

Douglas sea state: A scale of sea wave heights. See Table 1 **(Beaufort number)**.

downhole ground: A long electrode often attached to the logging cable some distance (perhaps 100 ft) above an electrical logging sonde or hung just below the casing, used as the reference electrode instead of a reference electrode at the surface. Used in the case of bad SP interference from electrical surface facilities or telluric currents.

downstairs: In the denominator of a fraction.

downsweep: Vibroseis sweep in which the frequency decreases with time.

down-the-hole IP method: Induced-polarization method that explores the region near a drill hole using a single potential or current electrode in the drill hole and other electrodes on the ground surface. Compare **in-hole IP method.**

downward continuation: The process of determining from values measured at one level the value of a potential (e.g., gravitational) field at a lower level. The potential field is not continuous across the boundaries of anomalous masses; as the depth from which anomalies originate is approached, their potential field expression becomes sharper and tends to outline the mass better until the depth of the mass is reached, whereupon the field computed by continuation becomes erratic. Noise in the data often precludes successful application. See **continuation.**

dowser: One who explores for water or petroleum unscientifically.

drag bit: A type of drill bit with fixed blades which drag and cut at the bottom of the shothole as the bit is rotated. See Figure 23 **(bit)**.

drape: Sag in bedding around a feature such as a reef, usually as a consequence of differential compaction, sometimes involving initial dip. Drape also occurs as a result of faulting.

drift: 1. A gradual and unintentional change in the reference value with respect to which measurements are made. If drift is slow and fairly steady in time, the difference produced by drift can be determined by rereading the value of the quantity being measured at the same position where it was previously observed and prorating the difference over other readings made in between. Gravity-meter drift may be caused by gradual heating up of the meter as the day progresses or by "creep" in the spring. Drift is different from "tare," which is a sharp, sudden change in reference value. **2.** The layer of glacial deposits which may vary with position and hence require a variable correction on seismic records. The effect can be the same as that of a weathering layer. This often requires **double-layer weathering** correction, for the entire drift layer and also for the lower-velocity layer of the top part of the drift. **3.** The attitude of a borehole. The drift angle or hole deviation is the angle between the borehole axis and the vertical; the drift azimuth is the angle between a vertical plane through the borehole and north. **4.** A shoran measurement of location with respect to one fixed point. Measurement with respect to a second fixed point is called *rate* or *range.* Both drift and rate values are necessary to establish a fix.

drill: A device for boring holes. Usually means a rotary drill (Figure 61), often truck mounted and mechanically driven, with a means for rotating the drill pipe and with a pump for circulation of a fluid (air or water or **mud)** down through the pipe to wash the cuttings away from the bit and carry them up to the surface in the annular space between the wall of the hole and the drill pipe.

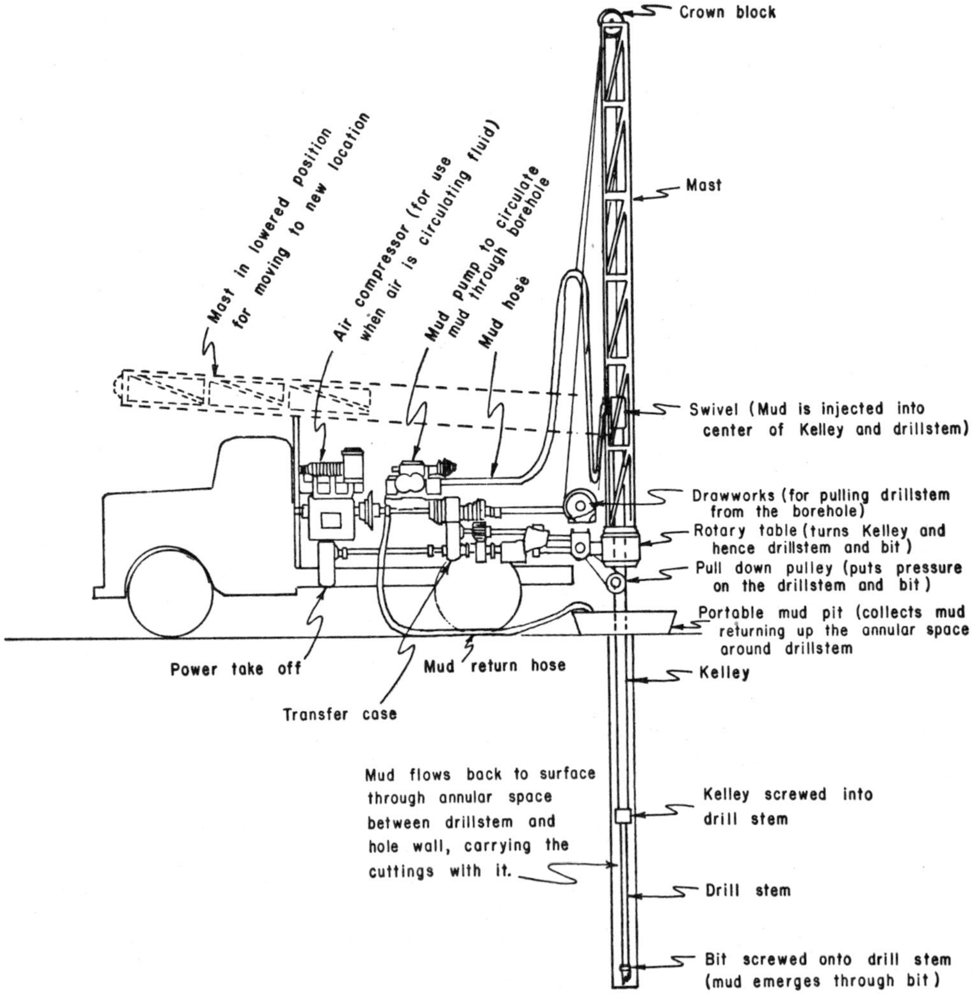

Fig. 61. Rotary **drill.**

Portable drills, water **jets**, augers, spudder or cable tools, and air-blast equipment are also used under certain conditions.

drill collar: A heavy thick-walled length of drill stem which is placed immediately about the bit so as to add weight onto the bit to help it eat into formations faster. See Figure 62.

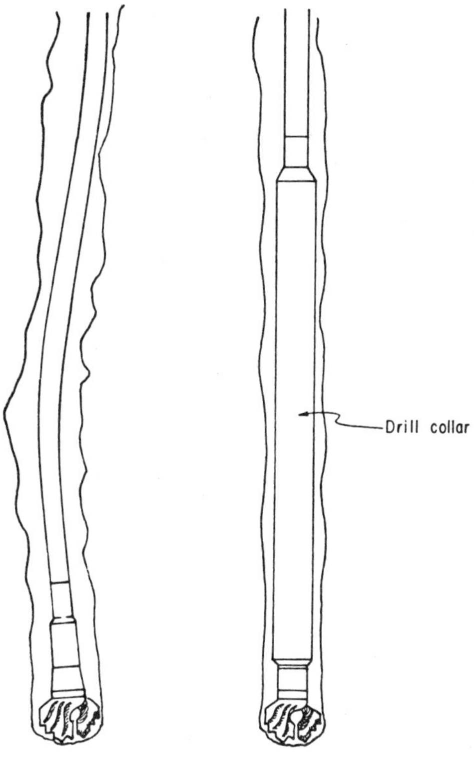

Fig. 62. **Drill collar usage.** Increasing the weight on the drill bit by pushing down on the drill stem may cause the drill stem to bend, resulting in a crooked hole. A heavy drill collar puts weight on the bit without bending the drill pipe.

driller's log: A record of the formations drilled through.

drilling break: An increase in the penetration speed of the drill bit caused by a change in formation, often indicative of penetration into a porous zone.

drilling-time log: A record of the time to drill a unit thickness of formation.

drill pipe: The pipe which is rotated by a drill, to which a **bit** is attached and through which the drilling fluid circulates.

drill-pipe log: A well log which is obtained from a logging instrument which has a self-contained recording mechanism. The log consists of an SP and short and long normals. The tool is lowered through the drill pipe and the flexible electrode assembly is pumped through a port in the bit. The log is recorded by a tape recorder within the tool during the process of coming out of the hole. The tape is played back to obtain the log. Welex tradename.

drop: 1. To drop a weight on the ground in order to generate a seismic wave; see **thumper. 2.** To eliminate.

drop along: Weight drop recording for **common-depth-point** stack.

dropout: A loss of information upon reading or writing on magnetic tape. Usually caused by defects in the magnetic tape or dirt on the tape surface or recording head.

dropped coverage: Portion of a seismic line not shot, perhaps for permit or access reasons or because of danger of doing damage.

drop-point: The surface location where a weight drop occurs, analogous to *shotpoint.* See **thumper.**

DSS: Deep seismic sounding; a long explosion seismic profile (usually refraction) which has the objective of studying the crust, Moho, and upper mantle.

dual: Parallel recording from the same inputs. Parallel or dual recording is often done using different amplifier settings (different gains, filters, etc.). Refraction is often dually recorded with one recording set at high gain to yield sharp first arrivals and the parallel recording at lower gain to allow the picking of secondary arrivals.

dual induction log: An **induction log** consisting of two induction curves with different depths of investigation. Usually run with a focused resistivity device with a shallow depth of investigation, such as a shallow Laterolog.

dummy load: 1. A ground-matching resistance used with pulsed-square-wave transmitters to balance the output load from the transmitter during the "power off" portion of the duty cycle. **2.** Used to load a transmitter during calibration and maintenance when the radiation of energy is unwanted.

dump: 1. To read out for examination the contents of a magnetic tape or other data storage. **2.** Unintentional shutdown of a computer, as by loss of power.

duplex: Simultaneous and independent transfer of data both ways. Compare **half-duplex.**

duty cycle: 1. The proportion of time a switch is "on." **2.** The percent of time in which current is delivered during a complete cycle of a transmitter (such as an IP transmitter).

DWT: A deep well thermometer, a sonde for recording temperature logs.

dynamic corrections: Normal-moveout corrections or corrections which depend on record time.

dynamic correlation: A velocity-analysis operation which involves crosscorrelating traces for different offsets, summing the crosscorrelations for similar pairs of traces over a number of nearby subsurface points, displaying the crosscorrelations for successive differences of offset squared, picking alignments on this display, and computing the residual normal moveout and stacking velocity for such alignments. GSI process.

dynamic range: 1. The ratio of largest recoverable

signal (for a given distortion level) to smallest recoverable signal. The smallest recoverable signal is often taken to be the noise level of the system, and dynamic range is often so defined. However, signals often can be extracted even though they are buried in the noise. The ambiguity can be resolved if the entire signal extraction process is considered in defining dynamic range rather than considering the recording equipment only. **2.** For direct recording magnetic tape, the noise level is for unrecorded tape; bandwidth should be specified because selected narrow bandwidths may give improved dynamic range. The maximum range of standard magnetic tape is about 50 db, of high output tape about 60 db. In digital recording the dynamic range is limited by word length; a 13-bit word represents about 84 db. **3.** The ratio of the maximum reading to the minimum reading (often noise level) which can be recorded by and read from an instrument without change of scale. **4.** Used occasionally for *significance,* the smallest signal which will cause a measurable (significant) change in the presence of a large signal; as shown, for example, by the number of significant bits.

earth holography: Recording of the wave pattern from a constant frequency source along with a reference wave. The resulting hologram can be viewed by light to allow one to "see" the structure which generated the wave pattern. See Hoover, 1972, Acoustical Holography using Digital Processing: Geophysics, v. 37, p. 1–19.

earth layering: Geophysical models of the earth often assume that it consists of horizontal or concentric layers which are homogeneous and isotropic and within which the velocity varies smoothly. The deeper layering is indicated in Figure 63. Raypaths and wavefronts for direct P-waves are shown in Figure 64. Direct S-waves, various reflected body waves, and surface waves will also be observed.

earthquake: Sudden movement of the earth resulting from faulting or volcanism. An earthquake is classified according to the depth of its focus or hypocenter: *shallow* < 60 km, *intermediate* < 300 km, *deep* > 300 km (sometimes > 450 km). The **seismicity** of an area is its likelihood of having earthquakes. Some of the energy released in an earthquake is radiated as seismic energy.

earthquake seismology: Study of earthquakes and their resulting seismic waves as a means of understanding the structure of the earth. A schematic seismic record showing several of the wave arrivals from an earthquake is shown in Figure 65. The scheme for identifying different arrivals is discussed under **wave notation.** Different types of waves may be detected at different distances. The time intervals between different arrivals depends on the distance from the earthquake hypocenter to the observing station. **Intensity** is a measure of the ground motion at a specific locality; **magni-**

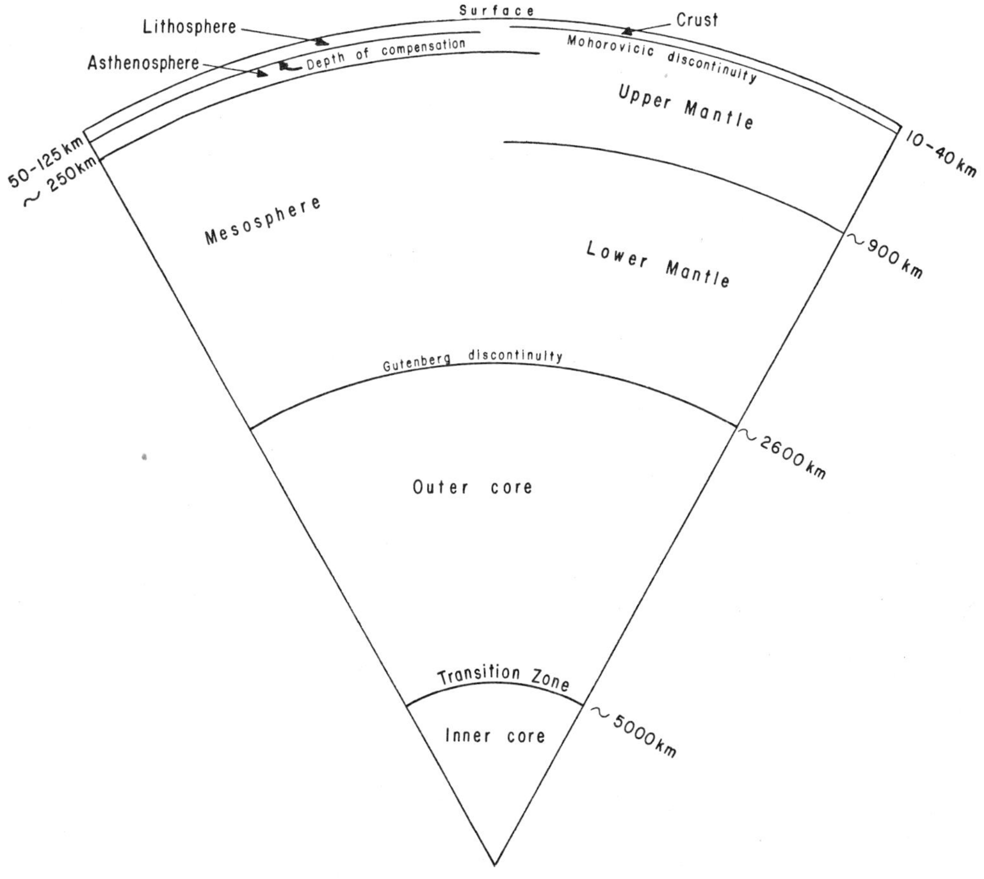

Fig. 63 **Earth-layering** terminology.

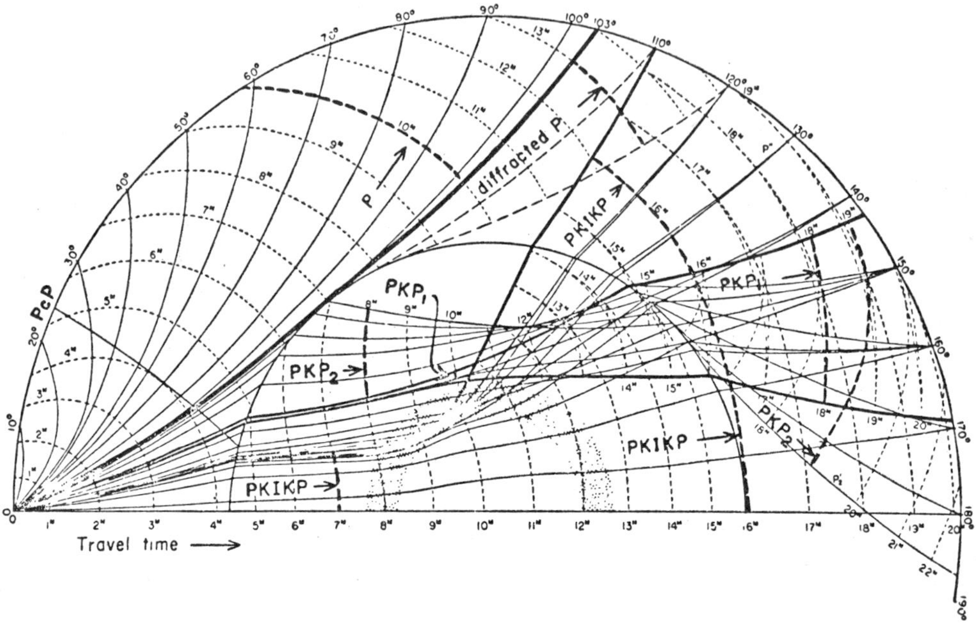

Fig. 64. **Earthquake wavefronts** and raypaths for P-waves. (From Gutenberg, Physics of the Earth's Interior, Academic Press, NY, 1959, p. 104.)

tude is a measure of the energy released in the earthquake.

easting: East component of a survey leg. See **departure.**

eccentering arm: A device which presses a sonde against the borehole wall.

eccentric anomaly: A way of expressing a position on an elliptical orbit (as a parametric equation). See Figure 69 **(ellipticity).**

eccentricity: 1. The condition occurring when the shot is not located at (or perpendicularly offset to) the center of what is intended to be a split spread. This occurs on marine records when a separate shooting boat is employed. **2.** The ratio of the focus-to-center distance to the length of the semimajor axis for an ellipse; see Figure 69 **(ellipticity).**

Eccles-Jordan trigger: A **flip-flop** device.

ecf: Elevation correction factor. See **elevation correction.**

echelon: A staggered arrangement of parallel features; see Figure 66. Sometimes called *en echelon.*

echo sounder: Fathometer.

ecliptic: The apparent annual path of the sun among the stars. *Ecliptic coordinates* refer to celestial latitude and celestial longitude.

eddy current: A circulating electrical current induced in a conductive body by a time-varying magnetic field. Lenz's law states that the direction of eddy current flow

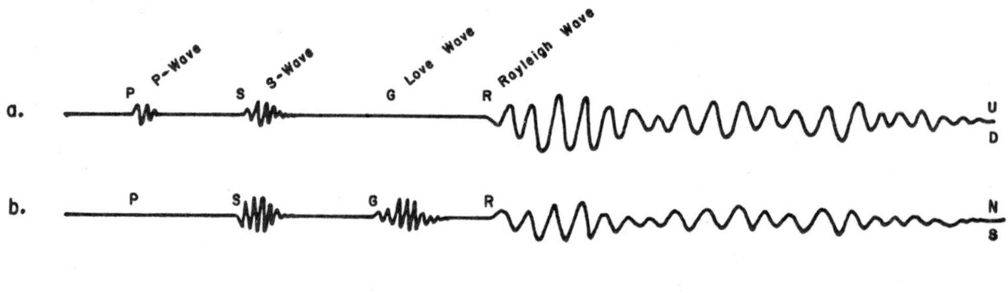

Fig. 65. **Earthquake seismogram** schematic. (a) The vertical component. (b) One horizontal component.

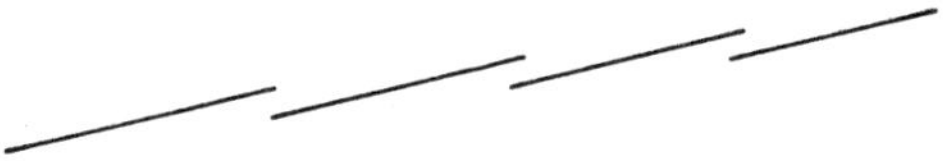

Fig. 66. **Echelon** arrangement.

is such as to produce a secondary magnetic field which opposes the primary field. The secondary field has a quadrature component which depends on the ratio of the resistance to the reactance of the eddy-current path. In electromagnetic prospecting, eddy currents should be distinguished from galvanic currents (which are of electrochemical origin).

edit: The preparation of data for processing. **1.** Specifically, the preparation of a digital tape containing geophysical information. Editing often involves rearranging data ("reformatting"), testing data validity, deleting unwanted data ("killing"), selection of data, insertion of data (such as headers, process parameters, or instructions), breaking the data up into blocks ("gapping"), etc. Editing a digital seismic tape may involve some calculations (such as removing binary-gain effects and correction for spherical divergence). Editing may or may not include static shifts and normal-moveout correction. **2.** With navigation-satellite data, the removal of doppler counts for which the satellite was low in the sky, or because the measurement does not fit with other data.

effort: The number of separate raypaths mixed together to produce the final display: number of energy sources (holes, thumps, pops, etc.) per shotpoint times number of geophones per group times number of records stacked or mixed. Also called *multiplicity.*

eigenvalue: Characteristic root. A square matrix $\mathbf{A} = \|a_{ij}\|$ is characterized by the eigenvalue λ and the eigenvector $\mathbf{q}$, if

$$\mathbf{A}\mathbf{q} = \lambda \mathbf{q}$$

is nontrivial. This eigenvalue equation also can be written as the set of linear equations:

$$a_{11}q_1 + a_{12}q_2 + \cdots a_{1n}q_n = \lambda q_1 \quad \text{and}$$

$$a_{21}q_1 + a_{22}q_2 + \cdots a_{2n}q_n = \lambda q_2, \quad \text{etc.}$$

which also can be written:

$$\begin{Vmatrix} a_{11}-\lambda & a_{12} & \cdots & a_{1n} \\ a_{21} & a_{22}-\lambda & & a_{2n} \\ \cdots & \cdots & & \cdots \\ a_{n1} & a_{n2} & \cdots & a_{nn}-\lambda \end{Vmatrix} = 0,$$

Eigenvalues are used in solving various types of geophysical problems.

eigenvector: See **eigenvalue.** Eigenvectors corresponding to distinct eigenvalues are orthogonal (or independent).

elastic: Returning to its original shape after removal of distorting stress. The return of shape is complete and almost instantaneous rather than gradual.

elastic constants: Elasticity deals with deformations that vanish entirely upon removal of the stresses which cause them. The passage of a seismic wave is an example. The general elasticity tensor relating stress and strain in anisotropic media possesses 21 independent constants. In transversely isotropic media in which properties are the same measured in two orthogonal directions, these reduce to five independent constants. Isotropic media (properties the same measured in any direction) have only two independent elastic constants. For small deformations, Hooke's law holds and strain is proportional to stress. The elastic properties of isotropic materials which obey Hooke's law are specified by *elastic moduli.* These include the following: **a.** *Bulk modulus k*: The stress-strain ratio under simple hydrostatic pressure:

$$k = \frac{\Delta P}{\Delta V/V},$$

where ΔP = pressure change, V = volume, and ΔV = change in volume. $\Delta V/V$ is called the *dilation.* The bulk modulus can be expressed in terms of other moduli as:

$$k = E/3(1 - 2\sigma) = \lambda + 2\mu/3.$$

b. *Shear modulus,* rigidity modulus, or Lame's constant μ; the stress-strain ratio for simple shear:

$$\mu = \frac{\Delta F/A}{\Delta L/L},$$

where ΔF = the tangential force, A = cross-sectional area, L = distance between shear planes, and ΔL = shear displacement. The shear modulus can also be expressed in terms of other moduli as:

$$\mu = \tfrac{1}{2}E/(1 + \sigma).$$

c. *Young's modulus* or stretch modulus E: the stress-strain ratio when a rod is pulled or compressed:

$$E = \frac{\Delta F/A}{\Delta L/L},$$

where $\Delta F/A$ = stress force per unit area, L = original length, and ΔL = change in length. Also,

$$E = \frac{\mu(3\lambda + 2\mu)}{\lambda + \mu} = \frac{9k\mu}{3k + \mu}.$$

$1/E$ is sometimes called *compliance.*

d. *Lame's λ constant:*

$$\lambda = k - 2\mu/3 = \frac{\sigma E}{(1 + \sigma)(1 - 2\sigma)}.$$

For a *Poisson solid,* $\lambda = \mu$.
e. *Poisson's ratio* σ: The ratio of transverse strain to longitudunal strain. When a rod of length L is pulled it is elongated by ΔL and also its width W is contracted by ΔW:

$$\sigma = -\frac{\Delta W/W}{\Delta L/L}.$$

Poisson's ratio varies from 0 to 1/2 (for liquids). Poisson's ratio can also be expressed in terms of elastic moduli:

$$\sigma = \frac{\lambda}{2(\lambda + \mu)} = \frac{3k - 2\mu}{6k + 2\mu} = \frac{E}{2\mu} - 1.$$

Poisson's ratio has the value 1/4 for a *Poisson solid.*
f. *Velocities of P- and S-waves* V_p and V_s: Expressed in terms of the moduli and the density ρ:

$$V_p = [(\lambda + 2\mu)/\rho]^{1/2}$$
$$= [E(1 - \sigma)/\rho(1 - 2\sigma)(1 + \sigma)]^{1/2},$$
$$V_s = (\mu/\rho)^{1/2} = [E/2\rho(1 + \sigma)]^{1/2}, \text{ and}$$
$$V_p/V_s = [(0.5 - \sigma)/(1 - \sigma)]^{1/2}.$$

elastic impedance: Seismic velocity multiplied by density. **Reflection coefficient** depends on changes in elastic impedance. Also called *acoustic impedance.*

elastic wave: An acoustic wave, seismic wave, or sonic wave.

E-layer: 1. The *Heaviside layer,* a layer of air which is ionized by sunlight, about 110 km high in the ionosphere, and reflects HF radio waves back to the earth during the daytime and MF waves at night. See also **F-layer** and **D-layer. 2.** A layer within the earth; see Figure 63 **(earth layering).**

Elcord: A DuPont explosive delay unit, used to make up a **directional charge** which matches explosion velocity to formation velocity. The objective is to send more of the energy vertically downward.

electrical log: 1. A generic term including all electrical borehole logs (SP, normal, lateral, laterologs, induction, microresistivity logs). **2.** Also used for records of surface resistivity surveying; compare **electrical survey. 3.** Electrolog; a borehole log which usually consists of **SP** and two or more resistivity logs, such as short and long **normal** and long **lateral** resistivity logs. Electrolog is a Dresser Atlas tradename.

electrically square: Consisting of the same number of similar elements connected in parallel and in series so that the apparent impedance is the same as that of a single element. See Figure 67.

electrical profiling: An **IP, resistivity method,** or **electromagnetic method** utilizing fixed spacing of electrodes or antennas in which the measuring system is moved progressively along profile lines to detect resistivity changes along the profiles. See **moving source method.**

electrical sounding: An **IP, resistivity method,** or **electromagnetic method** in which electrode or antenna spacing is increased to obtain information from successively greater depths at a given surface location. Electromagnetic sounding can also be done with a fixed spacing by varying the frequency *(time-domain technique).* Electrical sounding is intended to detect changes in resistivity of the earth with depth at this location (assuming horizontal layering).

electrical survey: 1. Measurements at or near the earth's surface of natural or induced electrical fields to map mineral concentrations or for geological or basement mapping. See **electrical profiling, electrical sounding, electromagnetic method, resistivity method, self-potential method, induced-polarization method, telluric method,** and **magnetotelluric method. 2. Electrical logs** run in a borehole.

electric field: A spatial vector quantity measurable as a potential gradient, produced by charged bodies or a time-varying magnetic field. Unit is volts per meter.

electric susceptibility: The ratio of intensity of electric polarization to electric field. A measure of the polarization property of a dielectric. Also called *dielectric susceptibility.* The primary term of the metal factor.

electrochemical SP: The component of the **SP** (self-potential) comprising the sum of the liquid-junction potential and the shale potential, both of which are determined by the ratio of the **activity** of the formation water to that of the mud filtrate. The *liquid-junction* potential is produced at the contact between the invasion filtrate and the formation water as a result of the differences in ion diffusion rates from the more concentrated to the more dilute solution (*concentration cell*). The negatively charged chloride ions have greater mobility than the positive sodium ions and an excess negative charge tends to cross the boundary, resulting in an emf. The *shale potential* results because the shale bed acts as a cationic membrane permitting the sodium cations to flow through it but not the chloride anions. The liquid-junction potential and shale potential are additive. See also **SSP** and **electrokinetic potential.**

electrode: A piece of metallic material that is used as an electrical contact with a nonmetal. Can refer to a grounding contact used for field surveying, to the metallic minerals in a rock, or to electrical contacts in

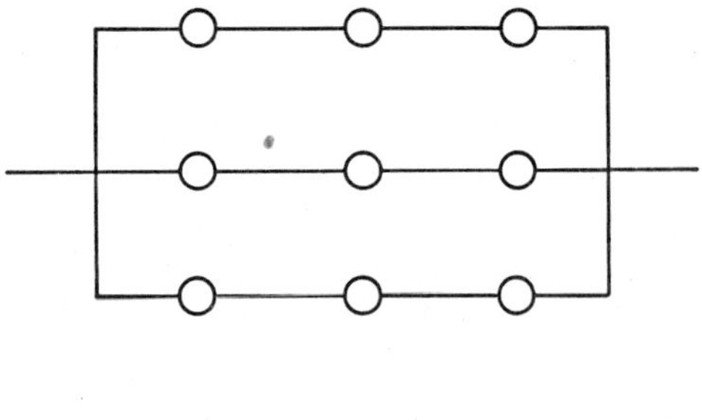

Fig. 67. **Electrically square** geophone arrangement. The diagram refers to electrical connection, not surface disposition. The array of 9 geophones has the same impedance as a single geophone.

laboratory equipment.

electrode array: See **array.**

electrode equilibrium potential: The reversible equilibrium potential (no energy loss) across the interface between an electrode and an electrolyte, measurable when no current passes through the interface. Measured as the voltage difference between a reference electrode and the electrode in question. It is primarily due to the free energy of the electron-transfer process.

electrode impedance: **1.** In electrochemistry, the total impedance across the interface between an electrode and an electrolyte. The equivalent-circuit model includes solution resistance, capacitances in the fixed and diffuse layers, and Warburg impedance. **2.** In electrical-circuit theory, either the self impedance of a single electrode or the mutual impedance between electrodes.

electrode polarization: **1.** In electrochemistry an electrode is polarized if its potential deviates from the reversible or equilibrium value. **2.** Polarization also can be induced due to passage of current through an interface or a change in ion concentration at an electrode surface. The amount of extra polarization is the overvoltage or **induced polarization** of the electrode.

electrode potential: See **electrode equilibrium potential.**

electrode resistance: The electrical resistance between an electrode and the immediate surroundings; sometimes called *contact resistance, self resistance,* or *mutual resistance,* depending on the situation.

electrodialysis: Migration of charge through a membrane in an electric field.

electrodynamic geophone: Moving-coil geophone.

electrokinetic potential: A component of the SP (self-potential) produced as a result of movement of the invasion fluid through the mud cake. Variations in ion concentration produced by the processes which generate the **electrochemical SP** tend to be swept along by the movement of the invading fluid, and this flow of charge results in the electrokinetic potential. The electrokinetic potential is usually small compared to the electrochemical SP and is often neglected when estimating formation water resistivity from the SP. Also called *streaming potential* and *electrofiltration potential.*

electrolog: Electrical log.

electrolyte: **1.** A material in which the flow of electric current is accompanied by the movement of matter in the form of ions. **2.** Any substance that dissociates into ions.

electrolytic tank: A container holding a conductive solution in which electrical model experiments can be carried out. See **analog modeling.**

electromagnetic method: A method in which the magnetic or electrical fields associated with artificially generated subsurface currents are measured. In general, electromagnetic methods are those in which the electric and magnetic fields in the earth satisfy the diffusion equation (which ignores displacement currents) but not **Laplace's equation** (which ignores induction effects) nor the wave equation (which includes displacement currents). One normally excludes methods which use microwave or higher frequencies (and which consequently have little effective penetration) and methods which use dc or very low frequencies in which induction effects are not important **(resistivity** and **IP** methods). Some methods such as Afmag which employ natural energy as the source are usually classified as electromagnetic methods whereas other methods using natural energy such as the **magnetotelluric method** are not.

electromagnetic spectrum: The range of frequencies or wavelengths of electromagnetic radiation. Such radiation is associated with periodically varying electric and magnetic fields and travels at the speed of light. See Figure 68.

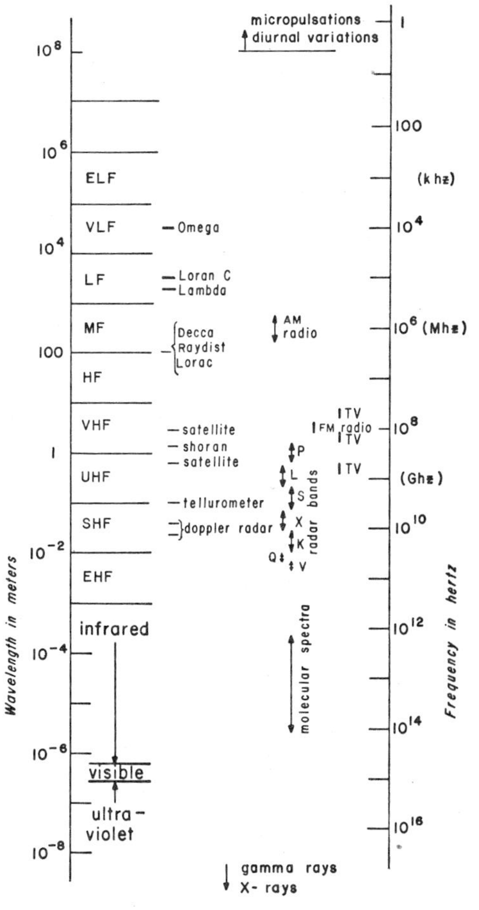

Fig. 68. **Electromagnetic spectrum.**

electron density: See **density log.**

electronic: Relating to devices, circuits, or systems in which conduction is primarily by electrons moving through a vacuum, gas, or semiconductor.

electronic conductor: A material such as a metal that conducts electricity by virtue of electron mobility.

electron-transfer reaction: An electrode surface phenomenon involving an oxidation-reduction reaction, generating a faradaic current.

electro-osmosis: The phenomenon whereby an electric field moves a fluid through a membrane. Also, charge separation in an electrolyte by osmotic action.

electrotape: An electronic surveying instrument similar to the **tellurometer.**

elevation angle: In satellite navigation, the angle measured from the horizontal at the receiver location to the satellite. Satellite passes are often designated by the elevation angle at closest approach.

elevation correction: **1.** The correction applied to reflection or refraction time values to reduce observations to a common reference datum. **2.** In gravity, the sum of the free air and Bouguer corrections. The elevation correction is obtained by multiplying the difference between station and reference elevation by the elevation correction factor (ecf).

$$\text{ecf} = (0.0941 - 0.01276\rho)\ \text{mgal/ft}$$
$$= (0.03086 - 0.004185\rho)\ \text{mgal/m},$$

where ρ = density.

ELF: Extremely low frequency; electromagnetic energy between 30 and 300 hz. Natural energy in this band originates mostly from lightning strikes. See **sferics** and Figure 68.

ellipse area: A quantity measured in **telluric surveys** which can be interpreted in terms of variations in earth resistivity.

ellipse of polarization: See **polarization ellipse.**

ellipsoid: A solid figure for which every plane cross-section is an ellipse. An ellipsoid of revolution can be generated by rotating an ellipse about one of its axes. An ellipsoid of revolution is usually used to approximate the **geoid.** Some of the standard ellipsoids are listed in Table 5 **(geodetic systems).**

ellipsoid of anisotropy: The equipotential surfaces about a point current source in an anisotropic, homogeneous medium. Such surfaces are ellipsoids of revolution, flattened by the ratio which is the coefficient of **anisotropy.**

ellipticity: **1.** The ratio of the major to minor axes of an ellipse (Figure 69). **2.** Specifically in electromagnetic surveying, the ratios for the **polarization ellipse** which can be determined by measuring the amplitudes and phases of two orthogonal components of the magnetic field.

Eltran: An early electrical transient exploration method similar to time-domain EM methods, which used a large-spaced dipole-dipole electrode array for petroleum exploration.

EM: Electromagnetic; see **electromagnetic methods.**

empty hole: A borehole which is filled with air or gas.

emu: The cgs-electromagnetic system of units, which is similar to the cgs-esu system except that "practical" electrical units (except for the abampere) are used.

end-of-file mark: Machine-readable mark on digital magnetic tape indicating the end of a complete group of data on the tape such as a seismic record.

end-on: Seismic field arrangement wherein the shot is at or near the end of the geophone spread. See Figure 176 **(spread).**

en echelon: **Echelon.**

energy: **1.** The sum of the square of the amplitude elements in a wavelet (times a proportionality constant which is often omitted). Thus if

$$b_t = (b_0, b_1, b_2, \cdots b_n),$$

the energy of b_t is

$$k(b_0^2 + b_1^2 + b_2^2 + \cdots + b_n^2).$$

2. The capacity to do work.

engineering geophysics: Use of geophysical methods to get information for civil engineering. Bathymetry, seismic reflection and refraction, side-scan sonar, gravity, magnetic, electrical, and sampling methods are employed. Shallow refraction is commonly used to give the depth to bedrock. In water-covered areas high-power fathometers, sparkers, gas guns, and other seismic reflection methods employing high frequencies (up to 5 khz) are used to obtain reflections from shallow interfaces so that bedrock and the nature of fill material can be diagnosed. Such methods are also used to

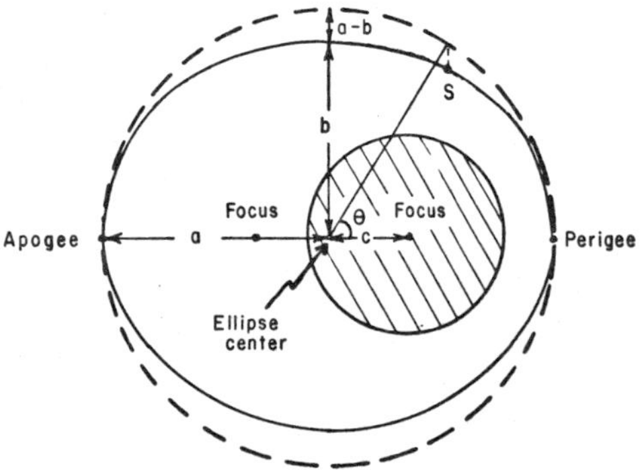

Fig. 69. **Ellipse** terminology. An ellipse is the locus of points for which the sum of the distances from the two foci is constant. A satellite follows an elliptical path about a body at one focus. a = semimajor axis. b = semiminor axis. Eccentricity = $e = c/a = (2f - f^2)^{1/2} = [1 - (1/E)^2]^{1/2}$. Ellipticity = $E = a/b = 1/(1 - f) = (1 - e^2)^{-1/2}$. Flattening = $f = (a - b)/a = 1 - 1/E$. θ = the eccentric anomaly when the satellite is at S.

locate large pipe lines on, or buried in, the sea bottom by the prominent diffractions which they generate.

enhancement: Improvement of data by a filtering or noise-rejection process.

entropy: **1.** A thermodynamic quantity which measures the energy which is not available. Higher entropy represents increased disorder and the entropy always increases in any reaction, according to the second law of thermodynamics. **2.** A measure of the uncertainty in a message. If $P(m_i)$ is the probability that the message m_i has been transmitted, then the entropy H where there are i possible messages, is given by

$$H = -\sum P(m_i) \log_2 P(m_i).$$

The entropy of a situation with no uncertainty is zero. Entropy is related to the information content of a message.

envelope: The low-frequency curves encompassing or bounding deflections of higher frequency. The curves are usually drawn by smoothly connecting adjacent peak amplitudes. Figure 70.

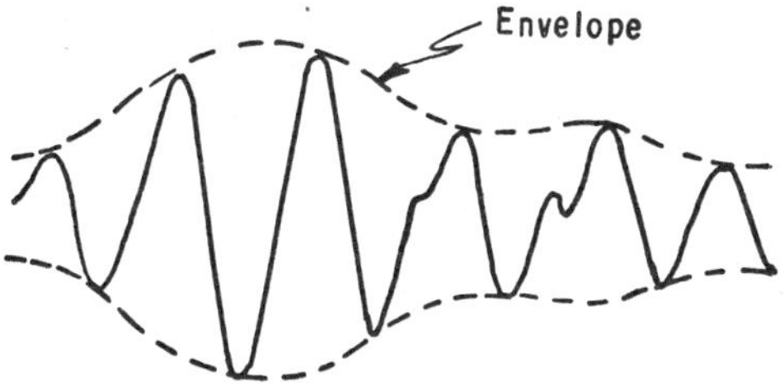

Fig. 70. **Envelope** of a wavetrain.

Eötvös effect: The vertical component of a Coriolis acceleration observed when measuring gravity while in motion. The meter's velocity over the surface adds vectorially to the velocity due to the earth's rotation, varying the centrifugal acceleration and hence the apparent gravitational attraction. The Eötvös correction in mgal for a meter whose velocity is V knots at an azimuth angle α and latitude ϕ is

$$E = 7.503\ V \cos\phi \sin\alpha + 0.004154\ V^2.$$

The Eötvös uncertainty dE in terms of direction uncertainty $d\alpha$ and velocity uncertainty dV is

$$dE = (7.503\ V \cos\phi\ \cos\alpha)\ d\alpha + (7.503\ \cos\phi\ \sin\alpha + 0.008308\ V)dV$$

See Glicken, 1962, Eötvös Corrections for a Moving Gravity Meter: Geophysics, v. 27, p. 531-533.

Eötvös unit: A unit of gravitational gradient or curvature; 10^{-6} mgal/cm.

epeirogenic: See **tectonic types.**

ephemeral data: Part of the data broadcast from a navigation satellite which varies with each broadcast. The ephemeral data includes the time since the hour or half hour and corrections to the satellite location; see Figure 107 **(Kepler coordinates).**

ephemeris: A table showing the position of a body (such as a navigation satellite) at various times.

epicenter: The location on the earth's surface below which the first motion in an earthquake occurs. Compare **hypocenter.**

epilog: A **synergetic** log.

epithermal neutron: See **neutron log.**

equalization: **1.** *Trace equalization* involves adjusting the gain of different channels so that their amplitudes are comparable. **2.Cross-equalization** involves matching the frequency spectrum of different channels to each other or to a predetermined curve, including adjustments because of phase difference. **3.** Filtering to correct for the linear-with-frequency response inherent in magnetic-tape pickup from direct analog recordings, for the geophone response, or for other frequency discrimination in recording or playback.

equatorial array: A configuration of electrodes used in resistivity surveying. See Figure 13 **(array).**

equilibrium conditions: **1.** A condition of balance at a state of minimum energy where energy is neither produced nor consumed. **2.** A condition predicted by the law of mass action where the velocity of forward and reverse reactions of a reversible process are equal.

equinoctial: The celestial equator; the intersection of the plane of the earth's equator and the celestial sphere.

equipotential-line method: A resisitivity-surveying method in which a current is caused to flow through the earth between two electrodes (one of which may be remote). The line of points on the surface of the ground which constitutes the locus of a given voltage difference from one of the electrodes is mapped with a third "probe" electrode. Similar to pole-pole array. Also called *equipotential survey* and *applied-potential method.*

equipotential surface: The continuous surface which is everywhere perpendicular to lines of force. No work is done against the field when moving along such a surface. Mean sea level is an equipotential surface with respect to gravity.

equivalent circuit: An electrical circuit which has the same input-output relationship.

equivalent width: The width of a boxcar with the same peak amplitude which contains the same energy.

ergodic: Having the same statistical properties throughout the ensemble. An ergodic system will return sometime arbitrarily close to any prior state. The statistical properties measured over sufficiently long intervals anywhere throughout the ensemble will be the same.

error: A deviation from the correct value. See round-off error and **truncation error; random** and **systematic; accuracy** and **probable error.**

error control: A system for detecting errors and sometimes for correcting them also. See **check.**

error function: The error function erf(t) is

$$\mathrm{erf}(t) = \frac{2}{\sqrt{\pi}}\int_0^t e^{-y^2}\, dy.$$

The probability that an error lies between $\pm a$ is erf(*ha*), where $h =$ precision index, for data which have a Gaussian distribution.

error message: A message from a computer about an incompatibility in program instructions.

ES: Electrical survey.

Euler's identity:

$$e^{i\theta} = \cos\theta + i\sin\theta = \text{cis}\,\theta,$$

$$\cos y = (e^{iy} + e^{-iy})/2 = \cosh(iy), \text{ and}$$

$$\sin y = (e^{iy} - e^{-iy})/2i = -i\sinh(iy).$$

even function: A function which has the same value when the sign of the variable is changed; i.e., $f(x) = f(-x)$. The Fourier transform of an even function is the cosine transform; its frequency-domain representation is zero phase. Any function can be separated into an even part and an odd part. Aeromagnetic anomalies are sometimes separated into even and odd parts for interpretation.

event: A lineup on a number of traces which indicates the arrival of new seismic energy, denoted by a systematic phase or amplitude change on a seismic record; an *arrival.* May be a reflection, refraction, diffraction, or any other type of wavefront.

exaggeration: Use of a different vertical than horizontal scale. See Figure 190 (**vertical exaggeration**).

EXCEPT gate: A circuit with multiple inputs which functions when signal is present on one input and absent on other inputs. Differs from OR gate, which does not have the restriction of absent signal on the other inputs. Also called *exclusive OR.* The EXCEPT truth table for inputs *A* and *B* is:

A	B	EXCEPT
0	0	0
1	0	1
0	1	1
1	1	0

See Figure 88 (**gate**).

excess three: See **binary-coded decimal.**

exchange current: A term in electrochemistry for the reversible electric current at an electrode that is in equilibrium with an electrolyte.

exclusive filter: A filter with a very sharp, narrow passband.

exclusive OR: See **EXCEPT gate.**

executive: An **operating system** or supervisor.

expand: 1. To break down into elements, as to expand in terms of frequency components. **2.** To increase the gain, as with a **gain control. 3.** To extend a spread to longer offsets.

expander: 1. Expanding spread; specifically, a depth probe. **2.** An IP and resistivity-surveying technique in which the electrode-separation interval is successively expanded so as to achieve greater depth of exploration. Also called *sounding* or *probing.* Data from an expander can be interpreted to give the depth to horizontal layers with contrasting physical properties, if such exist.

expanding spread: 1. A **spread** moved to greater offsets for successive shots from the same location so that the equivalent of many geophone groups is obtained from a single shot. Used in refraction work, in noise analysis, and in velocity analysis. **2. Depth probe. 3.** A *Dix-type expanding spread* is used in the field with the $X^2 - T^2$ velocity determination. See Dix, 1955, Seismic Velocities from Surface Measurements: Geophysics, v. 20, p. 68-86.

expanding-spread vertical-loop technique: Electromagnetic-survey method using a fixed transmitter and movable receiver, often along lines at about 45 degrees to the anticipated strike of the conductor sought.

expectation: 1. An operator which denotes the mean, or weighted mean where values are not equally probable. Usually denoted by brackets, $E[\ldots]$. **2.** In seismic processing usage it indicates the sum if for finite wavelets (b_t) or the mean if for an infinite time series (u_t):

$$E\{b_t\} = b_0 + b_1 + b_2 + \cdots + b_n \text{ and}$$

$$E\{u_t\} = \lim_{n\to\infty} \cdot (u_n + \cdots + u_0 + u_1 + \cdots + u_n)/(2n+1).$$

expected value: Expectation.

exploding wire: A marine seismic energy source in which an electrical arc discharge is initiated over a wire which vaporizes during the discharge. Called *WASSP.*

exploration logging: The use of widely spaced in-hole electrodes, perhaps combined with surface electrodes, to expand the exploration effectiveness of a drill hole. Also called *detection logging.*

exponentiate: 1. To introduce an empirically derived exponential time-dependent gain in seismic playback. Often refers to additional gain adjustment after removal of gain effects during recording and correction for spherical divergence. **2.** To magnify variations in input data so that the output is proportional to a constant raised to an exponent of the input value.

extended-range shoran: Extremely sensitive **shoran** system which depends on energy refracted around the earth's curvature and hence is not line-of-sight limited, as ordinary shoran is. Called *XR shoran* and has a range of the order of 150 miles.

extended spread: In-line-offset spread.

extrinsic conduction: The low-temperature ionic or electronic conduction in solid electrically conducting materials, due to weakly bonded impurities or defects. Also called *structure-sensitive* conductivity. See **intrinsic conduction.**

eye ball: To make an educated guess after casual visual inspection without actually measuring, or calculating.

Table 2. Fault types

FAULT TYPE	STRESS DIRECTION MINIMUM	 MAXIMUM	CHARACTERISTICS	RELATED TERMS
NORMAL 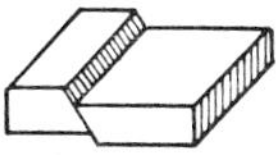	HORIZONTAL (Tension)	VERTICAL (Gravity)	Dip usually 75° to 40°	TENSION FAULT GRAVITY FAULT SLIP FAULT LISTRIC FAULT (CURVED FAULT PLANE)
REVERSE 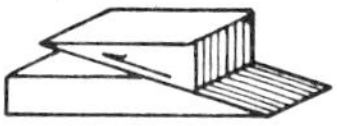	VERTICAL	HORIZONTAL (Compression)	Fault plane may disappear along bedding	THRUST FAULT LOW ANGLE (dip < 45°) HIGH ANGLE (dip > 45°)
STRIKE - SLIP 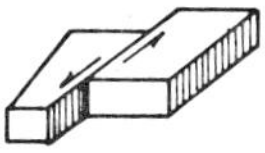	HORIZONTAL	HORIZONTAL	Fault trace often 30° to maximum stress	TRANSCURRENT FAULT TEAR FAULT WRENCH FAULT RIGHT LATERAL  (Dextral) LEFT LATERAL  (Sinistral)
ROTATIONAL 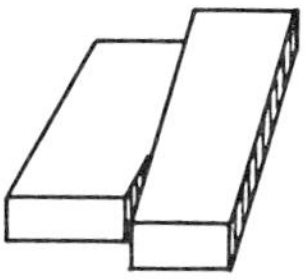			Throw varies along fault strike; may vary from normal throw to reverse.	SCISSORS FAULT HINGE FAULT
TRANSFORM 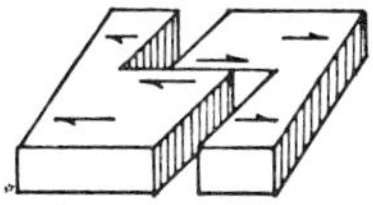	HORIZONTAL		Associated with separation or collision of plates New material fills rifts between separating plates or one plate rides up on another if plates collide.	DEXTRAL 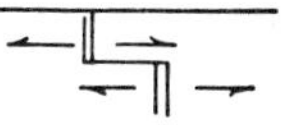 SINISTRAL 

factor analysis: A method of finding a mathematical relationship among values. Assume that a set of observations $X_1, \ldots, X_k$, can be accounted for by a set of linear relationships involving the factors $Z_1 \ldots, Z_p$, where $p < k$:

$$X_i = a_{i1}Z_1 + a_{i2}Z_2 + \cdot \cdot \cdot + a_{ip}Z_p + E_i,$$

where E_i is an error term (also called the "specific factor"). Factor analysis is a method of finding the parameters a_{ij}. Also called *multivariant analysis.*

fade: To attenuate the early part of long offset traces in stacking; to *mute.*

fairing: **1.** Adjusting to produce a regular surface of correct form. Specifically, smoothing to remove errors of measurement or noise in the data, so as to reveal the significant portion. **2.** Irregularities which induce turbulent flow. The drag on a cable pulled transversely through the water is less for turbulent flow than for nonturbulent flow. Irregularities such as short strings tend to create small vortices which reduce fluid friction. If the cable were pulled without fairing it would tend to shed large vortices alternately and then have spurts of motion, causing vibration and inducing noise.

fairway: **1.** The region within which effort is to be concentrated, such as a band of possible velocities within which one searches for velocity picks from normal-moveout measurements.**2.** A trend of hydrocarbon accumulations.

FAL: **Formation-analysis log.**

false color: The use of colors to represent different frequency bands (or other measurable characteristics) where the colors are not those naturally characteristic of the frequencies. Used to make infrared differences evident and in other applications.

faltung: **Convolution.**

fan-filter: **Velocity filter.**

fan shooting: An early use of the refraction seismograph to indicate the presence of salt domes within a thick low-velocity section. Detectors are located in (widely spaced) fan-like arrays radiating from shot locations. A **normal traveltime curve** (versus distance) is established from a *calibration* or *normal profile* shot where no dome is present. An early arrival **(lead)** with respect to the normal indicates part of the travel path is at an abnormally high velocity, signifying the presence of a dome of high-velocity salt. Figure 71.

faradaic: Pertaining to an electrochemical electron-transfer reaction as in the oxidation-reduction process in an electrolytic cell.

faradaic path: Passage of current at an electrode by the conversion of atom to ion or vice-versa; i.e., as the result of an electrochemical reaction. See **Warburg impedance.** A nonfaradaic path involves the ionic layers which are adjacent to the electrode acting as a condenser.

far-field: Field remote from the source. In electromagnetic phenomenon, far-field represents radiated energy. If the distance from the source is R and the wavelength is λ, then far-field implies that $R >> \lambda$. Compare **near-field.**

fast Fourier transform: One of several algorithms (such as the Cooley-Tukey method) which accomplishes the **Fourier transform** in a high-speed digital computer more rapidly than direct evaluation of the Fourier integral. Most involve iterative methods.

fathom: 6 ft or 1.8288 m.

fathometer: A device for measuring water depth by timing sonic reflections from the water bottom; an echo sounder.

fatigue: See **hole fatique.**

fault: A displacement of rocks along a shear surface. See Figure 72. The surface along which the displacement occurs is called the *fault plane* (often a curved surface and not "plane" in the geometric sense). The *dip* of the fault plane is the angle which it makes with the horizontal; the angle with the vertical is called the *hade.* The *trace* of a fault is the line which the fault plane makes with a surface (often the surface of the ground or a bed). Faults are classified (see Table 2) as *normal, reverse,* or *strike-slip,* depending on the relative motion along the fault plane. A fourth type of fault associated with plate tectonics is a *transform fault.* A *hinge* or *scissors fault* is produced by rotation of the blocks across the fault about an axis perpendicular to the fault plane so that throw varies systematically along the fault trace. Primary faults may have faults of a different type associated with them. Thus thrusting may produce tensions which cause secondary normal faults. Faulting and folding commonly are associated, both being responses to the same stresses; see Figure 82 **(fold).** Faulting during sediment deposition may cause stratigraphic associations with faulting. Thus beds may abruptly thicken and become more sandy across a normal "growth" fault. Faults evidence themselves in seismic data principally by: **a.** abrupt termination of events, **b.** diffractions, **c.** changes in dip, flattening or steepening, **d.** distortions of dips seen through the fault, **e.** cut-out of data beneath the fault or in the shadow-zone of the fault, **f.** changes in the pattern of events across the fault, and **g.** occasionally a reflection from the fault plane.

Faust's equation: **1.** Empirical relationships between seismic velocity V in feet/sec, geologic age T, true formation resistivity ρ in ohm feet, and depth of burial Z in feet:

$$V = K(ZT)^{1/6}$$

$$V = 2 \times 10^3 (\rho Z)^{1/6} .$$

Z is sometimes taken to be the maximum depth to which the formation has ever been buried, not necessarily its present depth. See Faust, 1953, A Velocity Function Including Lithologic Variation: Geophysics, v. 18, p. 271-297. **2.** The reciprocal of the second equation above is often written in more general terms with three constants (k,a,b) to be evaluated empirically from sonic-log transit time ΔT:

$$\Delta T = k \rho^{-a} Z^{-b} .$$

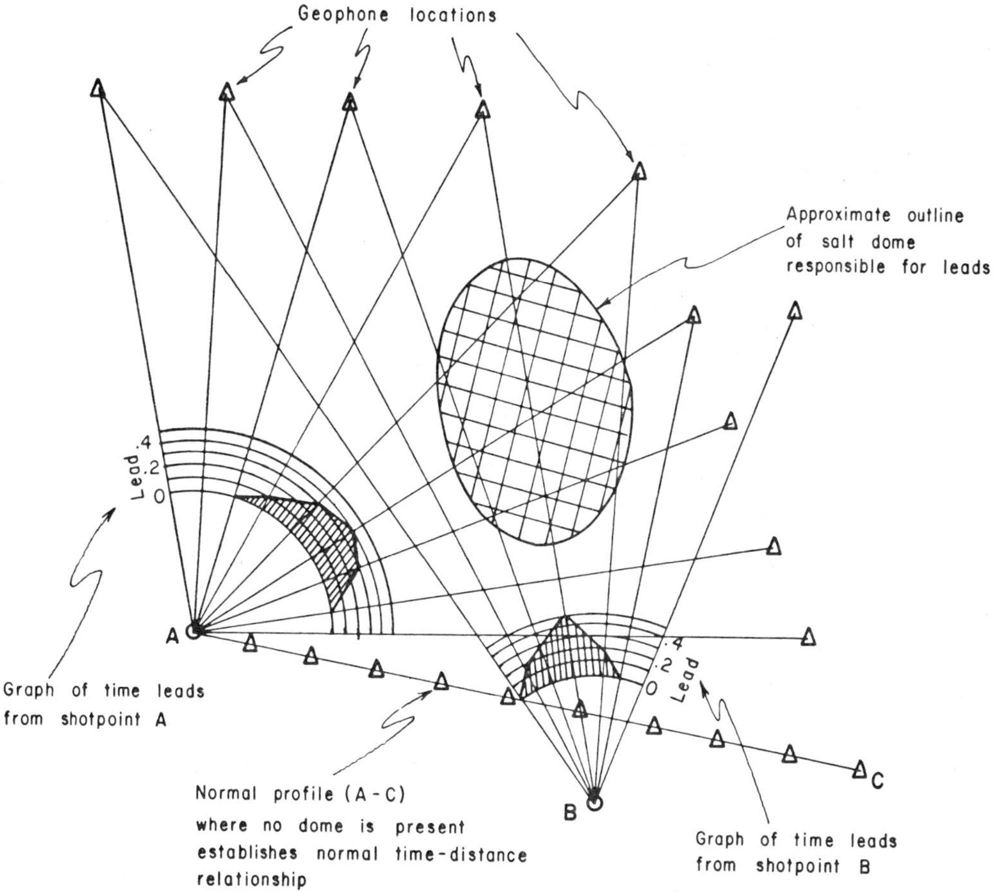

Fig. 71. **Fan shooting.** (From Nettleton, Geophysical Prospecting for Oil, 1940, New York, McGraw-Hill, p. 277.)

FDC: Compensated formation-density log; see **density log.**

FDL: Formation-density log; see **density log.**

feathering: En-echelon arrangement of successive spreads, as produced in marine shooting when a cross-current causes the cable to drift at an angle to the seismic line. Figure 73.

feather pattern: A weighted or tapered pattern of geophones within a group (or of shotholes or weight-drop points) such that the contribution of the elements of the pattern decreases with distance from its center.

feedback: The use of part of the output of a system as a partial input. *Negative feedback,* where part of the output is fed out-of-phase into the input, is used to attenuate variations such as for self-correcting or control purposes. *Positive feedback,* where a portion of the output is fed in-phase into the input, is used to produce oscillations.

feedback filter: Recursive filter.

fence diagram: Network of simplified sections, usually displayed on a map in vertical isometric projection, to illustrate variations in the third dimension. A photograph of a fence-diagram model is shown in Figure 74.

fence effect: An IP, resistivity, or electromagnetic anomaly produced by the presence of a nearby grounded conductor, such as a fence. Fences are indicated on IP maps as wooden post fence, steel post fence, etc.

Fermat's principle: The raypath which seismic energy will follow between two points is that for which the first order variation with all neighboring paths is zero. It is sometimes phrased as that path for which the traveltime is a minimum (or, in certain cases, a stationary

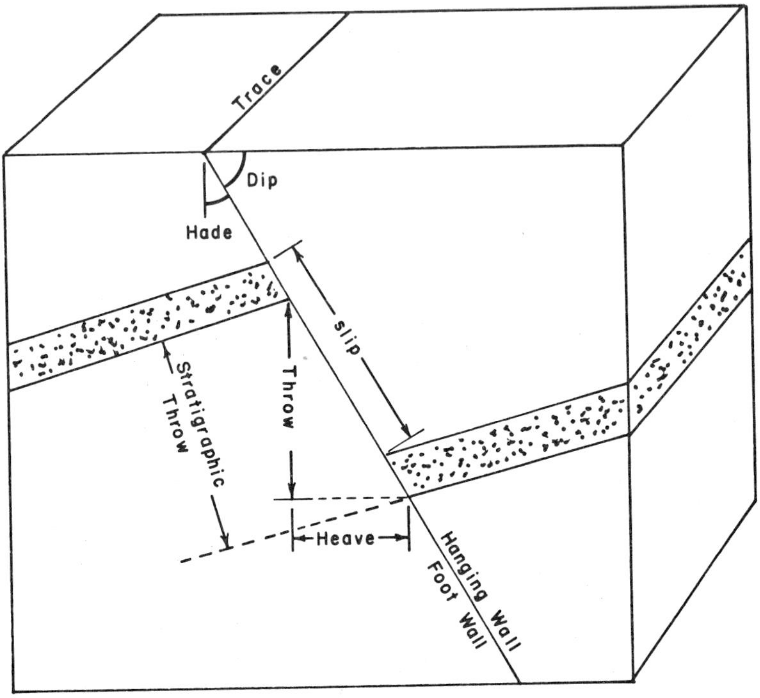

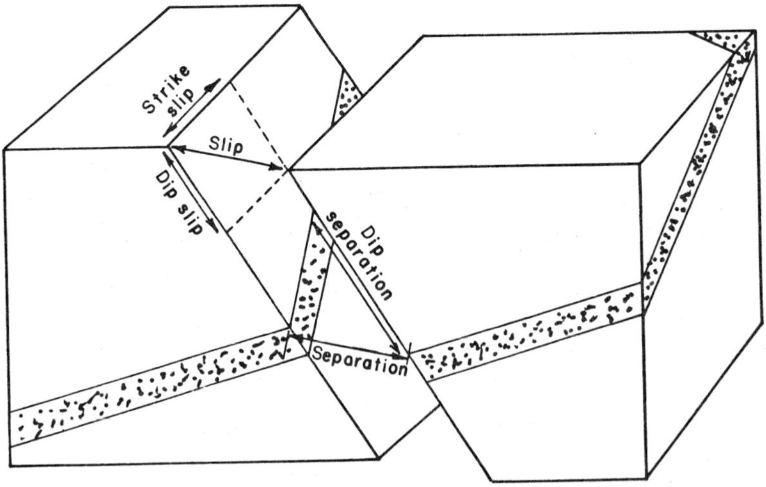

Fig. 72. **Fault** nomenclature.

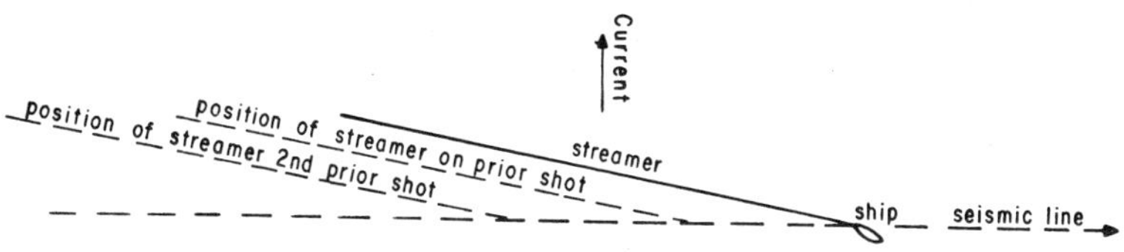

Fig. 73. **Feathering** of a marine streamer.

value or a maximum) compared with all neighboring paths. If the intervening medium contains parts having different speeds, the path will not be straight, but will be such that the over-all traveltime is minimized (usually). The resulting raypath will be the **least-time path** or *brachistochrone.* **Snell's law** follows from Fermat's principle.

ferrimagnetism: Property of some spinel-structured ferrites which show both ferromagnetic and antiferromagnetic properties because the ionic interaction favors both parallel and antiparallel alignment of group magnetic moments. Ferrimagnetic substances include distinct ferromagnetic sublattices which couple antiferromagnetically so that the observed magnetism is the difference.

ferromagnetic: Having positive and relatively large susceptibility and generally large hysteresis and remanence. In ferromagnetic materials the atoms interact and atomic magnetic moments couple so that groups of atoms behave collectively and orient in a parallel configuration. As the temperature of such materials rises to the Curie point, the thermal energy of the atoms becomes sufficient to overcome the coupling energy and the material behaves paramagnetically. See also **diamagnetic** and **paramagnetic.**

Fig. 74. **Fence diagram** of seismic record sections. (Courtesy Chevron Oil Company.)

FET: Field effect transistor.

FFI: Free fluid index; see also **nuclear-magnetism log.**

FFT: Fast Fourier transform; an algorithm for rapidly approximating the Fourier transform, such as the Cooley-Tukey method.

Fick's laws: The diffusion rate laws of electrochemistry: The time rate of change of diffusion flux is proportional to the concentration gradient and the activity of the electrode.

fiducials: 1. Points accepted as fixed bases of reference. **2.** Marks which indicate points of simultaneity; e.g., a mark on a magnetic-intensity record showing which point corresponds to a point on an altimeter record and to a point on the map or navigation records which were made at the same time. **3.** Time marks on a seismic record. **4.** In satellite navigation, the number of two-minute intervals in the Julian day.

field: 1. A region of space influenced by some force-producing agent. See **potential field** and **solenoidal field. 2.** The outdoors where geophysical surveys are made.

field balance: See **Schmidt field balance.**

field continuation: See **continuation.**

field-effect transistor: A semiconducting device which uses as input a transverse electric field to vary its conductance and thus control its output current. Ideally it is a voltage-controlled current source. Abbreviated FET.

field intensity: Force per unit. For a magnetic field, force per unit magnetic pole or current per unit length; for a gravitational field, force per unit mass (or acceleration); for an electrical field, force per unit charge or potential gradient.

field tape: A magnetic tape containing geophysical data recorded in the field. As opposed to a processed tape on which the data have been modified by processing in a computer.

field timing: The timing signals recorded in the field, usually displayed on the field monitor record as timing lines, on the field tape as 100 hz timing signal, and on digital tape as timing words or words counted from the time break. Commonly originates from a tuning fork or from a crystal-controlled oscillator. See Figure 149 **(record).**

figure of the earth: Geoid.

file: A collection of related records or program instructions treated as a unit.

file protect: An interlock which prevents writing on a

Table 3. Filtering

Band-pass filtering	Attenuates high and/or low frequencies.
Notch filtering	Attenuates a very narrow band of frequencies (e.g., 60 hz).
Deconvolving Whitening Optimum filtering	Shape the frequency spectrum according to some knowledge of prior filtering actions or according to some assumptions about signal characteristics or desired output.
Hi-line balancing	Adjusts resistive/reactive impedance to balance out high line.
Velocity filtering	(Multichannel filter) attenuates certain apparent velocities.
Coherence filtering	(Multichannel filter) attenuates where certain coherence tests are not satisfied.
Picking	(Multichannel filter) eliminates data which fail certain coherence and amplitude tests.

magnetic tape and thus destroying information already stored there.

filter: 1. That part of a system which discriminates against some of the information entering it. The discrimination is usually on the basis of frequency, although other bases such as wavelength or moveout (see **velocity filter**) or coherence may be used; see Table 3. **2.** Linear filtering is called **convolution.** Linear filters may be characterized by their **impulse response** or more usually by their frequency-domain transfer characteristics (amplitude and phase response). Filter characteristics are often designated by specifying the frequencies at which their amplitude is down by a given amount, often 3 db (70 percent or half power), and by the slope of their cutoff. Typical seismic filters are shown in Figure 75. **3.** Some passive electrical filters are shown in Figure 76. Simple RC or RL filters have slopes of 6 db/octave, but they may be cascaded to

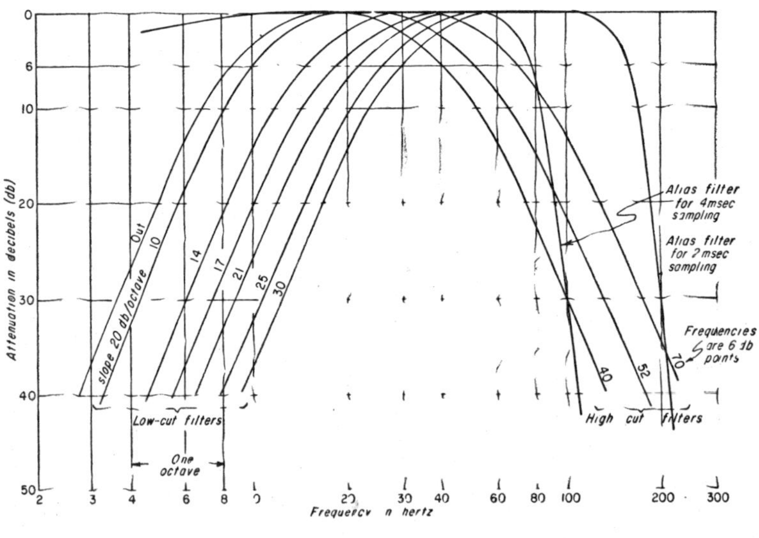

Fig. 75. Filter curves for typical seismic amplifier.

achieve higher slopes. Thus a double-section simple filter may have a slope of 12 db/octave. A *K*-type filter may have a slope of 18 db/octave. These may also be cascaded; hence a *KK* filter has a 36 db/octave slope. See also **high-cut filter** (= low-pass filter) and **low-cut filter** (= high-pass filter). **4. Alias filters** are very sharp high-cut filters designed to prevent aliasing. **5.** *Band-pass* filters are often specified by listing successively their low-cut and high-cut component filters. Hence, 2–21, 1–55 might mean a double-section simple low-cut filter down to 70 percent at 21 hz and a single-section simple high-cut filter down to 70 percent at 55 hz; or a 21 *KK*–55*K* might mean two cascaded low-cut *K* filters and a single high-cut *K* filter. The order of the specification is sometimes reversed. **6.** *Notch filters* reject sharply at a particular frequency. The *M*-derived filter (Figure 76) is such a type; it is usually used in conjunction with another filter, e.g., *MK;* otherwise it would partially pass frequencies above the notch as well as those below the notch. **7.** *Digital filters* provide a means of filtering data numerically in the time domain by summing weighted samples at successive time increments. Digital filtering can be the exact equivalent of electrical filtering but digital filtering is more versatile and permits one to filter easily in accordance with arbitrarily chosen characteristics which might prove difficult or impossible to achieve with electrical components. **8.** Specific types of filters used in electric circuits include the *Butterworth filter,* a low-pass filter design with flat response, and the *Tchebyscheff filter* (= *Chebychev filter),* a band-pass filter with a steep rolloff, characterized by a uniform ripple in the passband. (Figure 32, **Chebychev array,** shows such filtering action in the wavenumber domain.) **9.** Filtering can be accomplished by optical methods as well as by electrical and digital methods; see **laserscan.** See also **inverse filter** and Robinson and Treitel, 1964, Principles of Digital Filtering: Geophysics, v. 29, p. 395-404.

filter analysis: Use of a series of narrow band-pass filters to see the effects of different passbands on various events. Also called *frequency slicing.*

filter cake: Mud cake.

filter correction: Correction of record times made necessary when filters with different time delays are used in recording different portions of data that must be correlated.

filtrate: Fluid which has passed through a filter. Specifically, the drilling fluid which has passed through the **mud cake.**

first arrival: First break.

first break: The first recorded signal attributable to seismic-wave travel from a known source. First breaks on reflection records are used for information about the weathering. Much refraction work is based principally on first breaks, although secondary (later) refraction arrivals are also used. Also called *first arrival.*

fish: A sensor which is towed in the water, such as used with side-scan sonar.

fishtail bit: A drilling bit used to penetrate soft formations. Also called **drag bit.**

fish trap: Conservation agent attached to a seismic crew (usually a marsh or water crew).

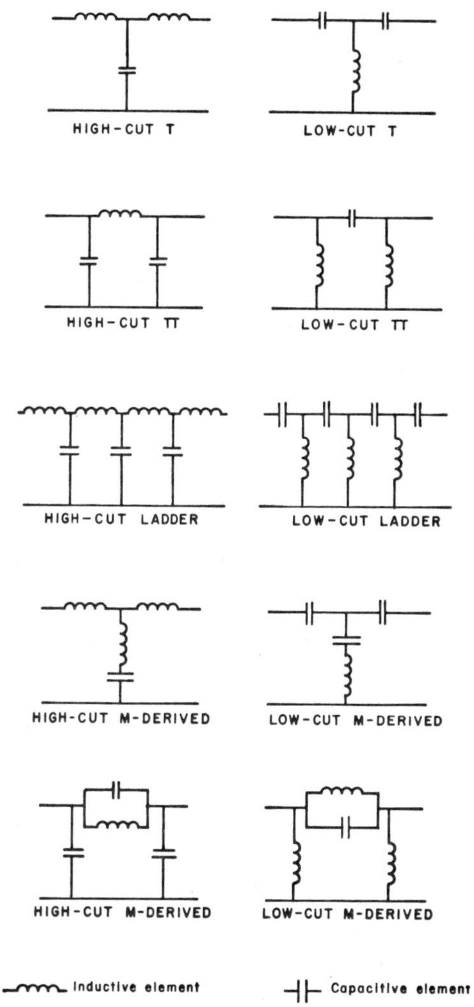

Fig. 76. **Filters** made of passive electrical elements.

fix: A determination of location, as by the intersection of two lines-of-position. A fix is made without reference to a former position. See Figure 113 **(line of position)** and **positioning.**

fixed data: Data which are constant for a record or similar collection of data and thus need not be input to a computer more than once.

fixed layer: A compact layer of ions and molecules held in place on an electrode or solid by chemical or electrostatic adsorption forces. Also called the *inner Helmholz double-layer.*

fixed-layer capacitance: Capacitance due to the presence of **fixed-layer** ions.

fixed point: **1.** A method of calculating in which the computer does not consider the location of the decimal point. Data must have values within a certain range which is the responsibility of the programmer. Compare **floating point. 2.** Sometimes implies integer operation.

fixed-source method: A geophysical profiling method in which the source of energy is stationary and the receiver is moved about to explore the area. Most electromagnetic sounding techniques are fixed-source methods. See **moving-source method.**

f,k space: A domain in which the independent variables are *frequency (f)* and *wavenumber (k)*. Seismic data analysis sometimes involves *f-k* plots in which energy density within a given time interval is plotted and contoured on a frequency-versus-wavenumber basis. Sometimes used in velocity-filter design. Figure 77.

flag: **1.** A bit attached to a computer word to indicate the boundary of a field. **2.** An indicator to tell a later part of a computer program about a condition which occurred earlier in the program.

flagging: Strips of plastic, cloth, or paper used to mark instrument or shot locations.

flattened section: **1.** A seismic record section in which a particular record event has been made flat (or sometimes merely smooth) by introducing arbitrary static shifts. Sometimes used in determining static corrections. If flattened on an event which represents a bed which was deposited horizontally, the flattened section resembles a paleosection which shows the attitude of deeper events at the time of deposition of the flattened bed (assuming that data migration and nonprimary reflection events do not cause problems). Useful in studies of variations in the time interval between events, such as used in locating reefs. **2.** A **paleosection.**

flattening: The ratio of the difference between the major and minor axes of an ellipse to the major axis; see Figure 69 **(ellipticity)**. The flattening of the earth is about 1/298.25.

flat-topping: Reducing values which exceed some chosen value to the chosen value; clipping. May be caused by loss of sensitivity due to saturation of some part of the measuring system. Compare **digital clipping** and see Figure 33 **(clipped).**

F-layer: One of the layers of ionized air in the ionosphere which reflects back radio waves up to about 50 Mhz. In the daytime the F-layer subdivides into two layers, the lower of which (F_1) is usually 175 to 250 km

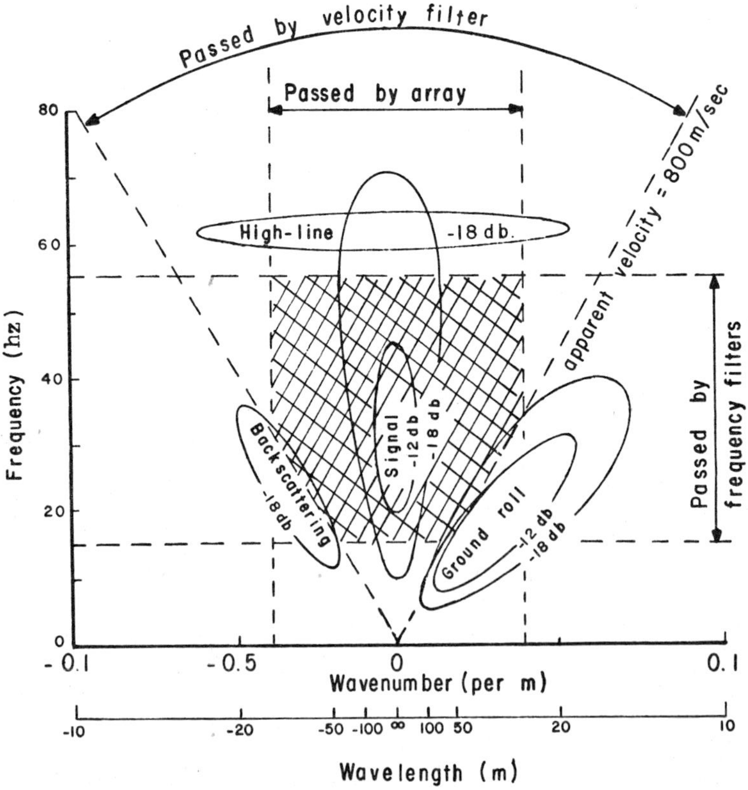

Fig. 77. ***F,k* plot.** The region passed by all three filters (array, frequency filter, and velocity filter) is cross-hatched. Radial lines through the origin represent constant apparent velocity *(V = f/k)*.

high and is found only during daylight. The higher F_2 layer at 250 to 400 km is present both night and day and is the principal reflector of HF radio waves at night.

Flexichoc: An implosive energy source for marine shooting. Two plates are separated by compressed air until they lock into position. The air between them is then pumped out, creating a vacuum. On signal the plates are unlocked and the water pressure forces them together, creating a shock wave in the water. See Figure 78a. Institut Francais du Petrole trademark.

Flexotir: A seismic method for marine shooting whereby small charges (about 2 ounces) are propelled through a rubber hose by water under pressure into a steel cage, where they are detonated. Also called *cage shooting.* See Figure 78b. Holes in the cage allow the water repelled by the explosion to flow out and in, thus dissipating some of the energy in the bubble of waste gases and thus attenuating the bubble effect. This permits charges to be fired much deeper than otherwise (at about 40 ft) which greatly increases the shot efficiency. Institut Francais du Petrole trademark.

flip-flop: A bistable oscillator; a device with two and only two stable states. Used to store one bit of information.

floating: Not electrically connected to **ground** or the system reference voltage.

floating charge: A seismic charge which is not as deep as intended and thought to be. Characterized by early uphole time and delayed reflection times.

floating datum: A reference surface used in areas of rough topography. The elevation of the floating datum is varied to minimize the illusion of subsurface structure which would be caused by surface relief if a constant-elevation datum were used. The topography may produce velocity variations because of the increased loading or the topography may merely reflect structure which also affects velocity distribution. A *one-third floating datum* is sometimes used in which the reference reflects one-third of the surface elevation above a flat plane. Tilted datums are also used sometimes. The floating datum is merely one way of introducing static corrections so that the velocity distribution beneath the datum is simplified, usually by assuming no horizontal variation. More elegant methods of dealing with lateral velocity variations are now used, thus lessening reliance on floating datums.

floating point: A number expressed by a certain number of significant figures times a base raised to a power. Thus 139000 might be written as 1.390×10^5 to indicate four significant figures. Writing numbers in floating point format prevents the loss of significant figures in case the number becomes too small or too large for a fixed register. Computers usually use the base 2 rather than the base 10.

flow chart: A diagram showing the operations involved in a process. Often used to show a computer program or computer processing steps and their sequence. See Figure 79 and also Figure 141 **(processing).** The emphasis is often on where and who does what.

flowmeter: A device which measures the flow of fluid. Sometimes the flowmeter is lowered through the flow stream in a borehole; sometimes it is set in one spot with a packer. See **spinner survey.**

fluid sampler: See **formation tester.**

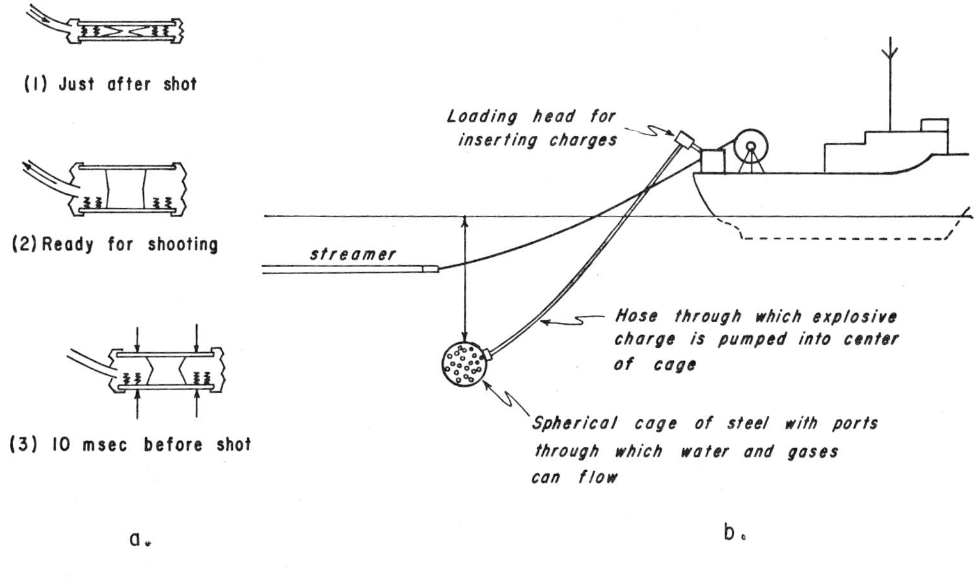

Fig. 78. **Flexichoc** (a) and **Flexotir** (b). (Courtesy Compagnie Générale de Géophysique.)

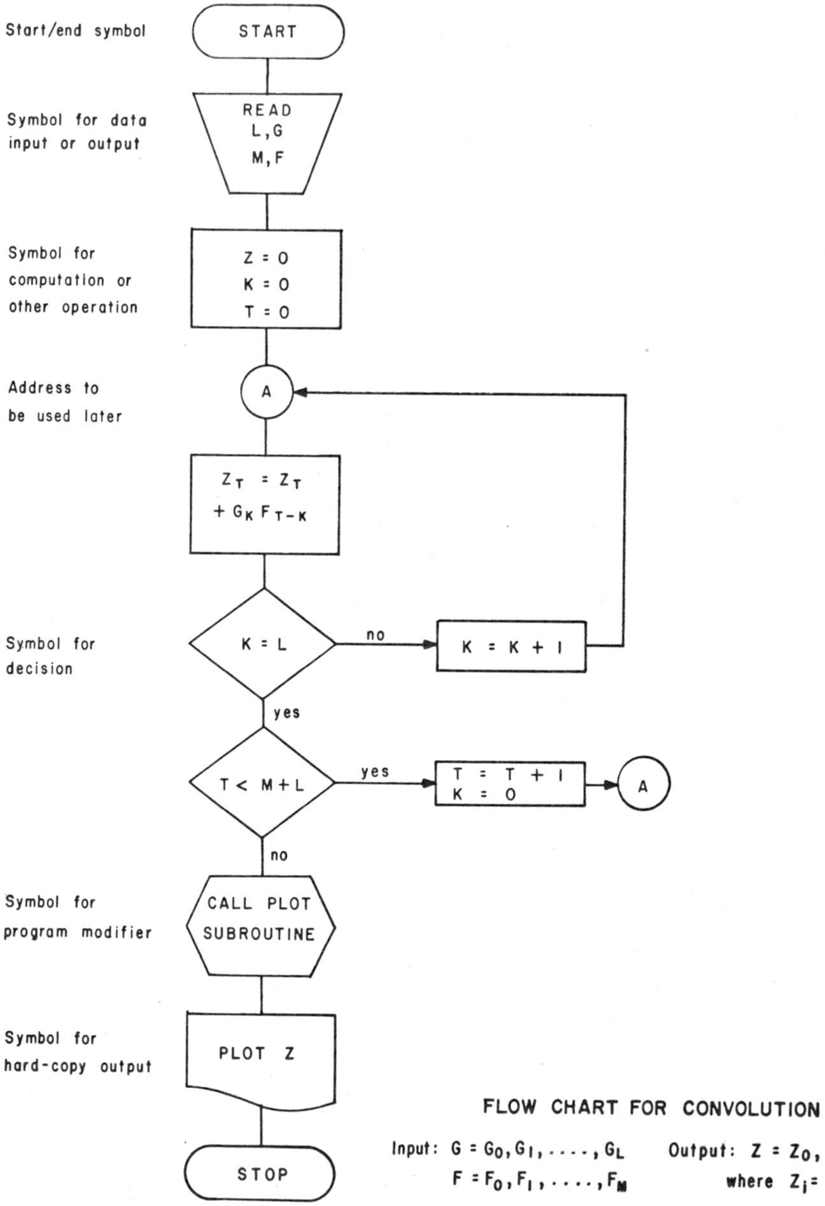

Fig. 79. **Flow chart** for the convolution of G with F. Input: $G = [G_O, G_1, \ldots, G_L]$; $F = [F_O, F_1, \ldots, F_M]$. *Output:* $Z = [Z_O, Z_1, \ldots, Z_{L+M}]$, where $Z_i = \sum_{k=0}^{L} G_k F_{i-k}$.

fluid-travel log: A recording of the flow of fluid past a logging tool which is held stationary. A radioactive slug is injected into the fluid stream through ports in the logging tool and the radioactive counting rate is measured as a function of time at two detector positions on the tool. Used to locate leaks.

flushed zone: See **invaded zone.**

flute: **Mute.**

flutter: Noise caused by variations in speed of magnetic tape; **wow.**

fluxgate magnetometer: An instrument capable of detecting changes in the geomagnetic field of the order of one gamma. Contains two cores of highly permeable material which approaches saturation in the weak magnetic field of the earth (Figure 80). A 1000-hz cyclic field is superimposed by a coil around the core. This field completes saturation of the core. The place in the energizing cycle at which saturation is reached is a measure of the strength of the earth's field. A secondary coil detects the changes in flux. Two parallel cores are used with windings in opposite directions so that the difference is measured. A current through an additional winding nullifies most of the background magnetic field so that the magnetometer is sensitive to small changes in the field. The magnetometer measures the magnetic field component along the axes of the cores and must be oriented with the field if the total intensity is to be measured. With the Gulf magnetometer this is accomplished by three mutually perpendicular fluxgate instruments and servomechanisms which minimize the magnetic field in two of these, thus maximizing the field in the third. Compare **proton magnetometer** and **optically pumped magnetometer.**

flyer: A number of geophones permanently connected at intervals along a short cable. A typical flyer might contain, for example, six geophones connected in series at 20 ft intervals, with clips at one end for connecting it to the main cable which carries the signal to the recording equipment.

FM: **Frequency modulation.**

focus: The location of the first rupture in an **earthquake;** the *hypocenter.*

focused log: A **laterolog** or guard log; sometimes refers to an **induction log,** FoRxo, microlaterolog, or proximity log.

fold: Common-depth-point multiplicity. Where the same CDP point is sampled at 12 offset distances, e.g., it is referred to as "12-fold." See also **folding.**

fold-back: A cable which is doubled back on itself such that two or more geophone stations are located at the same position on the ground. Used for making comparisons (for example, different types of geophone arrays) on the same shot.

folding: **1.** Frequency folding as a consequence of sampling produces **alias** problems. **2. Convolution. 3.** Bending of geological strata. *Primary folding* is response to deep seated forces with a strong horizontal component. *Gravitational folding* results from downward sliding and flow, which are secondary results of uplift and tilting. Local folding can be caused by compaction or by the upwelling of salt or igneous rocks. Folding can also result from superficial sliding or flow, as with slumps and slides. The upraised part of a fold forms an anticline, the downwarped portion a syncline. The long axis of an anticline or syncline is the **b-axis.** Other fold terminology is shown in Figure 81. Different materials respond differently to the same stresses (and stress duration) and hence cleavage, flow, and faulting

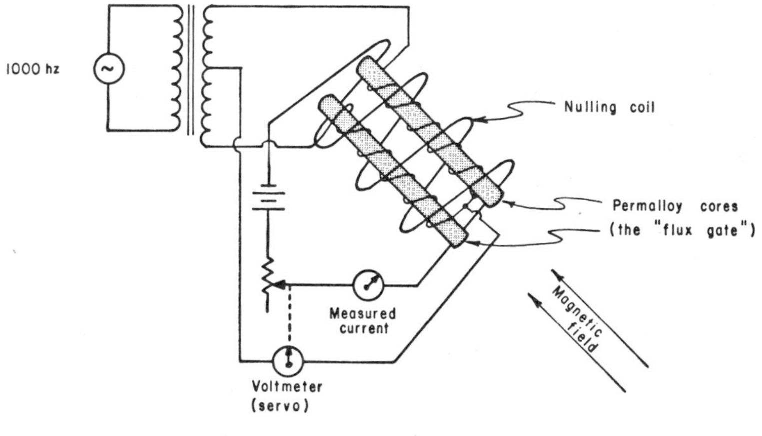

Fig. 80. **Fluxgate magnetometer.** The two permalloy cores are magnetically saturated by the 1000-hz current. Because their solenoids are oppositely wound, the earth's magnetic field unbalances them and causes a voltage across the servo (shown as a voltmeter) which changes the current in the nulling coil to produce a magnetic field which balances out the earth's field. The current through the nulling coil is a measure of the magnetic field strength.

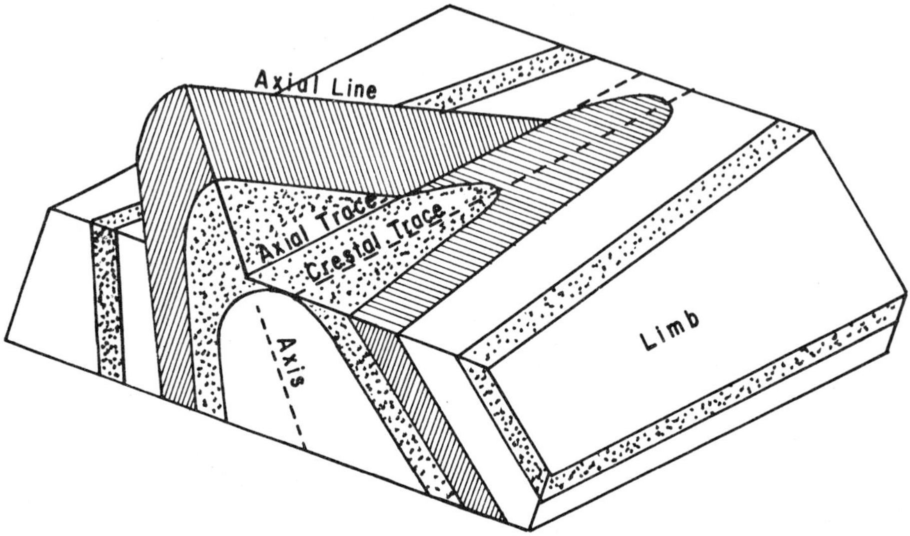

Fig. 81. **Fold** terminology. The plunge of the anticline is the angle which the axial line makes with the surface. The b-axis is the direction of the axial line.

usually are associated with folding; see Figure 82. *Competent beds* tend to retain their thickness in folding and govern the folding wavelength (fold wavelength is of the order of 25 times the thickness of the most competent member). *Incompetent beds* flow in response to folding stresses, but the distinction with competent beds is gradational. In *concentric* or flexural-slip folding (Figure 83), competent beds slip along the bed-

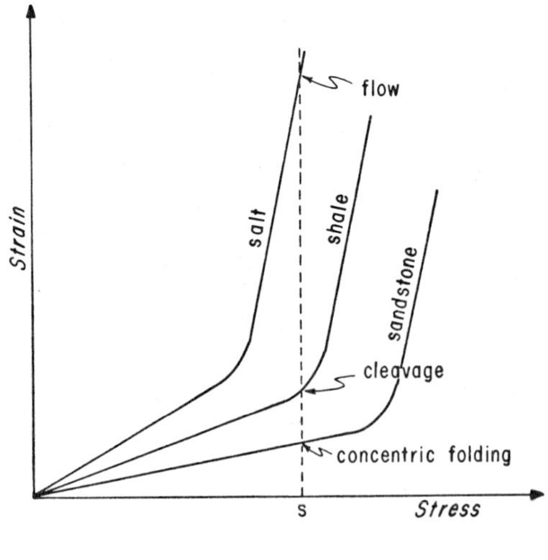

Fig. 82. **Folding** or cleavage or flow may result from the same stress on different rocks. The stress duration is involved as well as the stress magnitude. (From deSitter, 1964, Structural Geology: New York, McGraw Hill, p. 68.)

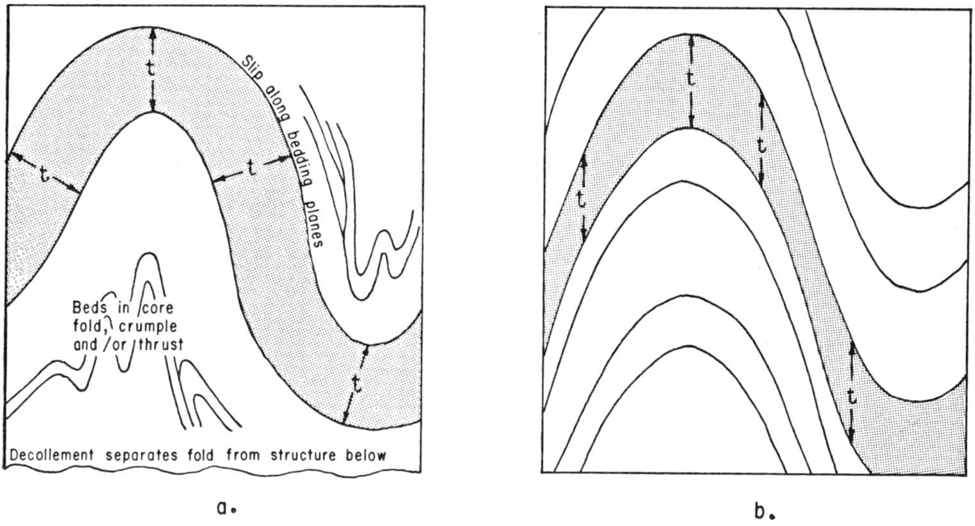

Fig. 83. **Folding** types. (a) In concentric folding, competent beds tend to retain constant thickness perpendicular to the bedding, whereas incompetent beds fold and fault and separate the fold from deeper structure. (b) In similar folding, beds tend to retain their vertical thickness, thus thinning on the sides of the fold.

ding planes with respect to each other and the curvature becomes progressively smaller for higher beds in an anticline as the intensity of the folding decreases. The core of such a fold is an incompetent bed which has been squeezed out. In *similar folding,* successive beds show essentially the same curvature. Vertical thickness is preserved but beds are thinner on the flanks as measured perpendicular to the bedding. Folding is *disharmonic* where the folding of one bed is not related to the folding of nearby beds, the beds being separated by incompetent beds.

folding frequency: Nyquist frequency; see **alias.**

foreign: 1. Not part of a system. **2.** Not part of one's own company.

foresight: A surveying measurement with the objective of determining the position and elevation of the stadia rod with respect to the survey instrument.

format: The arrangement of data (as on a magnetic tape) involving the placement of bits of different significance, number of bits per byte and bytes per word, parity, multiplexing arrangement, timing and gain information, record identification and other auxiliary data, gaps, start-stop codes, etc. Standard formats have been specified by the SEG (Northwood et al, 1967, Recommended Standards for Digital Tape Formats: Geophysics, v. 32, p. 1074–1084; and Meiners et al, 1972, Recommended Standards for Digital Tape Formats: Geophysics, v. 37, p. 36–44). Specific computer systems may use their own format for processing so that the first step with new data may be to reformat it.

formation: A lithological unit which has some distinction which permits identification. It is not necessarily a time unit. May be composed of several "members" and be part of a larger "group."

formation-analysis log: A computed log of apparent fluid resistivity and apparent porosity based on induction log and either sonic- or density-log data. Abbreviated FAL; also called *R_{wa}-analysis log.*

formation-density log: See **density log.** Abbreviated FDL.

formation evaluation: The analysis and interpretation of well-log data, drill-stem tests, etc., in terms of the nature of the formations and their fluid content. The objectives of formation evaluation are to ascertain if commercially productive hydrocarbons are present and, if so, the best means for their recovery and to derive lithology and other information on formation characteristics for use in further exploration. See Table 11 **(well log).**

formation-resistivity factor: The ratio of the resistivity of a formation to the resistivity of the water with which it is saturated. While formation factor is, strictly speaking, defined only for clean brine-saturated sands (for which it is nearly independent of the fluid resistivity), it is often used without this restriction. Apparent formation factor is a function of porosity, pore structure, pore-size distribution, and clay content. Usually abbreviated F or F_R. See **Archie's formula.**

formation tester: A tool run on a wireline to obtain samples of formation fluid. Hydrostatic, flow, and shut-in pressures are recorded.

Table 4. Fourier transform pairs

The time functions on the left are Fourier transforms of the freauency functions on the right, and vice-versa. Many more transform pairs could be shown.

	Time Function		Frequency Function
Boxcar	$G(t) = \begin{cases} 1, & \lvert t \rvert < \tau/2 \\ 0, & \lvert t \rvert > \tau/2 \end{cases}$ 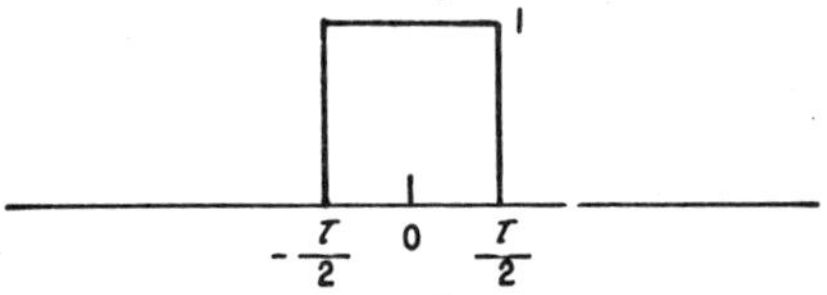	Sinc	$S(f) = \tau \, sinc(f\tau) = (1/\pi f) \sin(\pi f t)$ 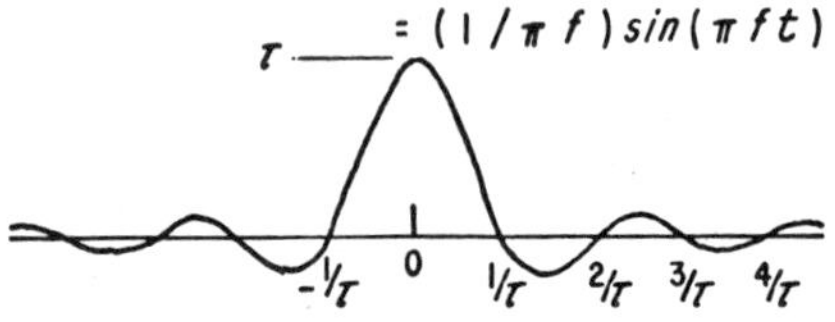
Triangle	$G(t) = \begin{cases} 1-\lvert t \rvert/\tau, & \lvert t \rvert < \tau \\ 0, & \lvert t \rvert > \tau \end{cases}$ 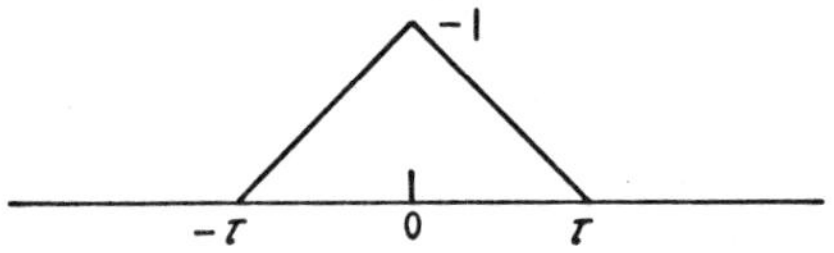	Sinc ^{2}	$S(f) = \tau \, sinc^2(ft) = (1/\pi^2 f^2 \tau) \sin^2(\pi f t)$ 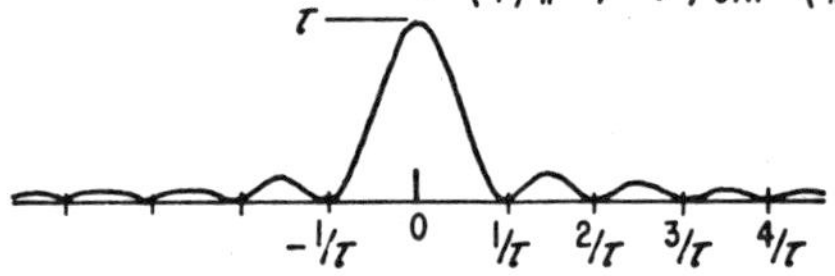
Gaussian	$G(t) = e^{-1/2 t^2}$ 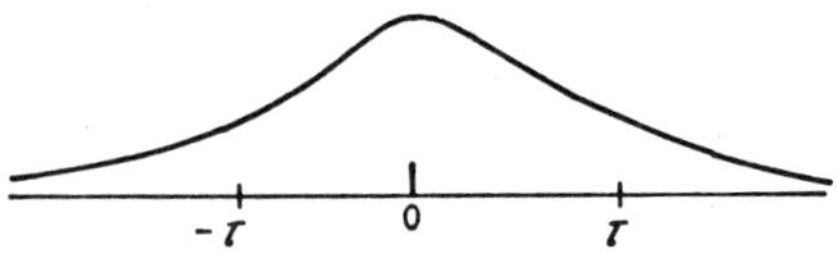	Gaussian	$S(f) = \tau (2\pi)^{1/2} e^{-(\pi f \tau)^2}$ 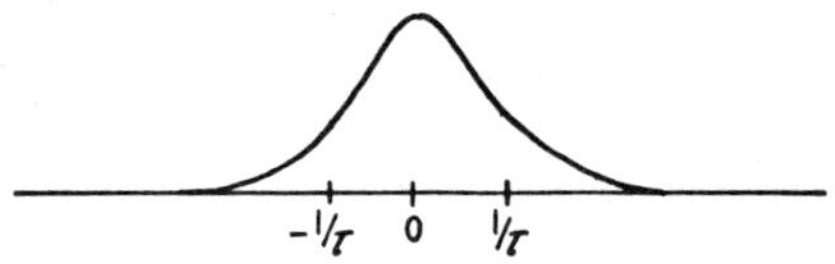
Impulse	$G(t) = \delta(t) = 0, \quad t \neq 0$ 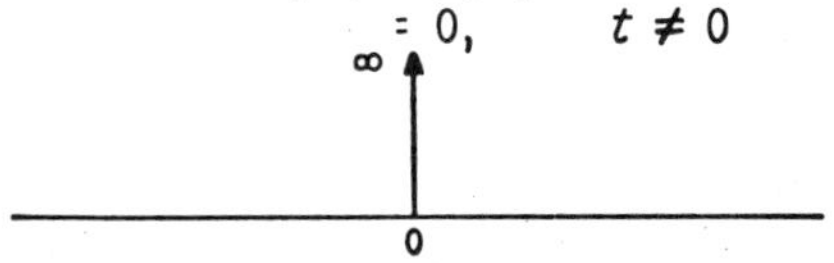	DC Shift	$S(f) = 1$ 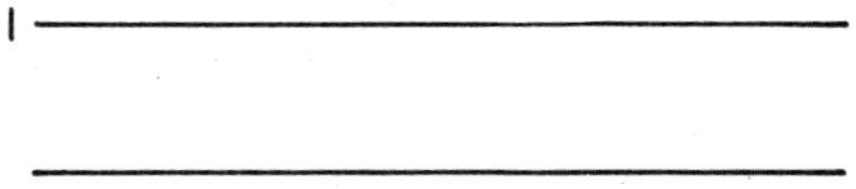
Sinusoid	$G(t) = \cos \omega_0 t$ 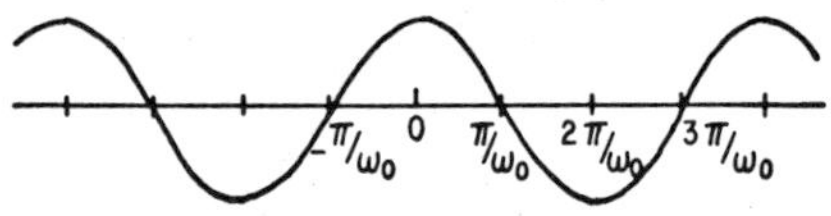	Single Freq.	$S(f) = \tfrac{1}{2}(\delta(f+f_0) + \delta(f-f_0))$ 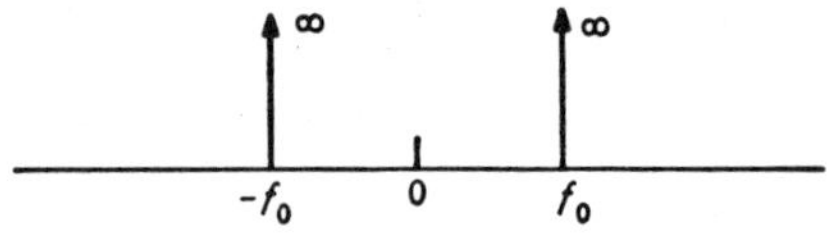
Comb.	$G(t) = Ш(t) = \sum_{-\infty}^{\infty} \delta(t-n\tau)$ 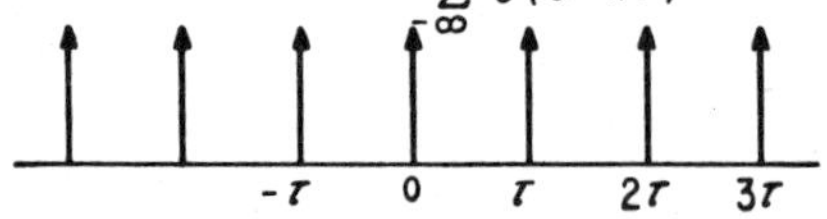	Comb.	$S(f) = \sum_{-\infty}^{\infty} \delta(f-n/\tau)$ 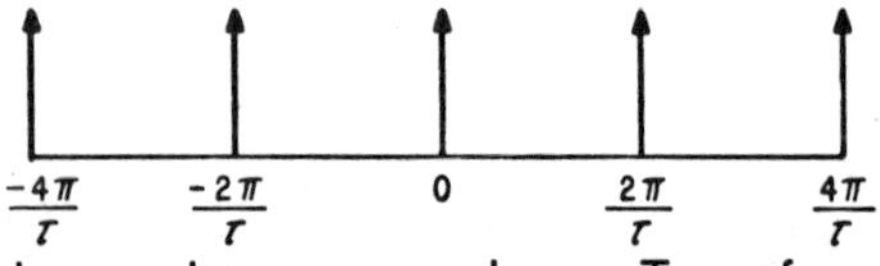

The above are all even functions and hence have zero phase. Transforms for real odd functions are imaginary, i.e., have a phase shift of $\pm \pi/2$. Transforms of functions which are neither odd nor even involve variation of phase with frequency.

Table 4. (Continued)

Doing the "time operations" is equivalent to doing the "frequency operations" on the transform of the data.

Time operation	Freq. operation	Significance
Linear addition $ag(t) - bh(t)$	Linear addition $aG(f) - bH(f)$	Linearity and superposition apply in both domains. The spectrum of a linear sum of functions is the linear sum of their spectra (if spectra are complex, rules of addition of complex quantities apply). Any function may be regarded as a sum of component parts and the spectrum is the sum of the component spectra.
Scale change $g(kt)$	Inverse scale change $\frac{1}{k} G\left(\frac{f}{k}\right)$	Time-Bandwidth invariance. Compressing a time function expands its spectrum in frequency and reduces it in amplitude by the same factor. The amplitude reduces because less energy is spread over a greater bandwidth. The case where $k=-1$ reverses the function in time, interchanging positive and negative frequencies; so for real time functions, reversing the phase.
Convolution $g(t) * h(t)$	Multiplication $G(f)H(f)$	The spectrum of the convolution of two time functions is the product of their spectra. If the spectra are complex, multiplication is equivalent to multiplying amplitude soectra and adding phase spectra.
Multiplication $g(t)h(t)$	Convolution $G(f) * H(f)$	The spectrum of the product of two time functions is the convolution of their spectra.
Delay $g(t - t_o)$	Linear added phase $e^{-j2\pi f_o t} G(f)$	Delaying a function by a time t_o multiplies its spectrum by $e^{-j2\pi f_o t}$, thus adding a linear phase $\theta = -2\pi f t_o$ to the original phase. Conversely, a linear phase filter produces a delay of $-d\theta/df = 2\pi t_o$.
Complex modulation $e^{j2\pi f_o t} g(t)$	Shift of spectrum $G(f-f_o)$	Multiplying a time function by $e^{j2\pi f_o t}$ shifts its spectrum to center about f_o rather than zero frequency.

form factor: 1. Geometric factor, the geometric multiplying factor which depends on the type of electrode array and interval being used. **2.** The type curve for a profile across an idealized body, e.g., a sphere.

Fortran: FORmula TRANslation; a language designed to simplify programming for digital computers.

forward solution: Determining measurements which would result from a given model, such as predicting the electric potential for a given model of resistivity distribution and a given distribution of current sources, for use in comparing with observed data. Also called the *direct problem;* see Figure 120 **(model)**.

FoRxo: See **microlaterolog**. Welex tradename.

Foucault current: See **induction log**.

Fourier analysis: The analytical representation of a waveform as a weighted series of sinusoidal functions. Determining the amplitude and phase of sine or cosine waves of different frequencies into which a waveform can be decomposed. The Fourier analysis relations are part of the **Fourier transform** relations. Opposite of Fourier synthesis. See Figure 86. **(Fourier transform)**.

Fourier pairs: Operations and functions which **Fourier-transform** into each other. See Table 4. Fourier pairs can be generalized into more dimensions, such as illustrated in Figure 84 (where the alternate domains might be offset-time versus frequency-wavenumber or *f-k* space).

Fourier plane: **Frequency domain.**

Fourier synthesis: Superimposing the sine and/or cosine waves represented by a spectrum so as to construct the waveform (or time-domain representation). See Figure 85. The Fourier synthesis relations are part of the **Fourier-transform** relations.

Fourier transform: The formulas which convert a time function (waveform-seismic record trace, etc.) *G(t)* into the frequency-domain representation *S(f)* and vice versa. An example is

$$S(f) = \int_{-\infty}^{\infty} G(t)\, e^{-j2\pi ft}\, dt$$

$$= \int_{-\infty}^{\infty} G(t) \cos(2\pi ft)\, dt - j\int_{-\infty}^{\infty} G(t) \sin(2\pi ft)\, dt$$

and the inverse.

$$G(t) = \int_{-\infty}^{\infty} S(f)\, e^{j2\pi ft}\, df$$

$$= \int_{-\infty}^{\infty} S(f) \cos(2\pi ft)\, df + j\int_{-\infty}^{\infty} S(f) \sin(2\pi ft)\, df.$$

Finding *S(f)* from *G(t)* is called *Fourier analysis* and finding *G(t)* from *S(f)* is called *Fourier synthesis;* see Figure 86. *S(f)* is the *complex spectrum,* the real part being the *cosine transform* and the imaginery part, the *sine transform* whenever *G(t)* is real. Another expression for *S(f)* is

$$S(f) = |A(f)|e^{j\phi(f)},$$

where the functions *A(f)* and *φ(f)* are real and are, respectively, the *amplitude spectrum* and the *phase spectrum* of *G(t)*.

$$A(f) = [(\text{Real part of } S)^2 + (\text{Imaginary part of } S)^2]^{1/2}.$$

$$\phi(f) = \tan^{-1}\left[\frac{\text{Imaginary part of } S}{\text{Real part of } S}\right].$$

φ is in the 1st or 2nd quadrant if the imaginary part is positive, in the 1st or 4th quadrant if the real part is positive. A record trace *H(t)* which extends only from 0 to t_1 may be assumed to be repeated indefinitely and so expanded in a Fourier series of period t_1:

$$H(t) = \sum_{n=0}^{\infty} a_n \cos(2\pi nt/t_1) + \sum_{n=1}^{\infty} b_n \sin(2\pi nt/t_1)$$

$$= \sum_{n=0}^{\infty} S_n\, e^{j2\pi nt/t_1},$$

where $a_n = (2/t_1)\int_0^{t_1} H(t) \cos(2\pi nt/t_1)\, dt,$

$b_n = (2/t_1)\int_0^{t_1} H(t) \sin(2\pi nt/t_1)\, dt,$

and $S_n = |A_n|e^{i\phi_n}.$

$A_n = (a_n^2 + b_n^2)^{1/2}.$

$\phi_n = \tan^{-1}(b_n/a_n).$

The same rules for quadrants apply to ϕ_n as expressed for *φ(f)*. a_o is the zero frequency component (or *dc shift*). The frequency spectrum is discrete because the function is periodic. If H_t is a sampled time series sampled at intervals of time t_2, then we can stop summing when $n > 2t_1/t_2$ (see **sampling theorem**). In this case a_n and b_n can be expressed as sums:

$$a_n = (2/t_1) \sum_{t=0}^{t_1} H_t \cos(2\pi nt/t_1) \text{ and}$$

$$b_n = (2/t_1) \sum_{t=1}^{t_1} H_t \sin(2\pi nt/t_1).$$

See also **phase response** and **fast Fourier transform** The Fourier-transform relations can be generalized for more than one dimension.

TIME DOMAIN FREQUENCY DOMAIN

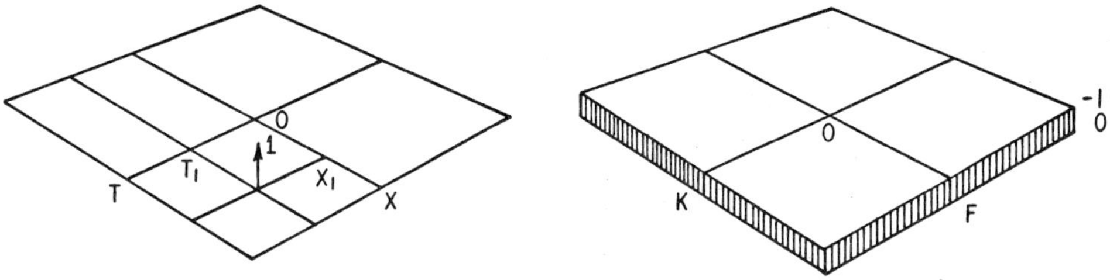

An impulse, $\delta(X-X_1, T-T_1)$, transforms into a dc shift plus a phase shift.

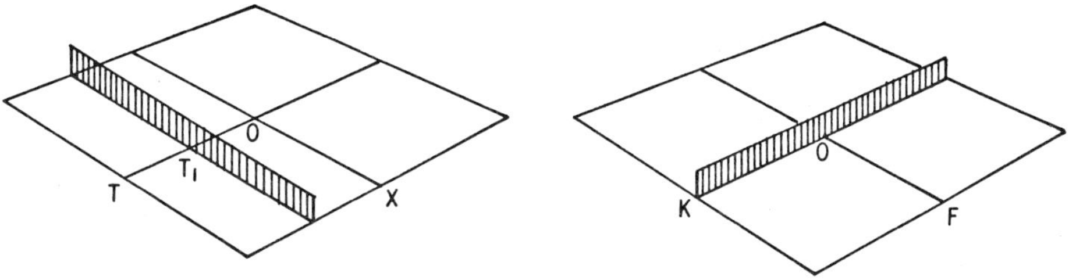

A one-dimensional impulse, $\delta(T-T_1)$, transforms into a one-dimensional impulse, $\delta(F)$, plus a phase shift.

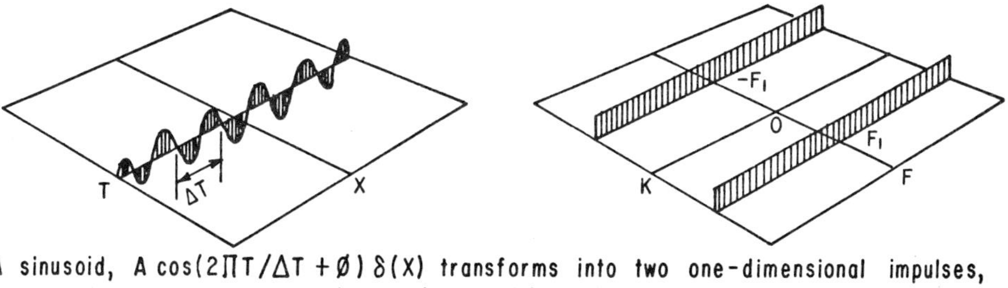

A sinusoid, $A\cos(2\pi T/\Delta T + \emptyset)\,\delta(X)$ transforms into two one-dimensional impulses, $\delta(F-F_1)$ and $\delta(F+F_1)$, where $F_1 = 1/\Delta T$.

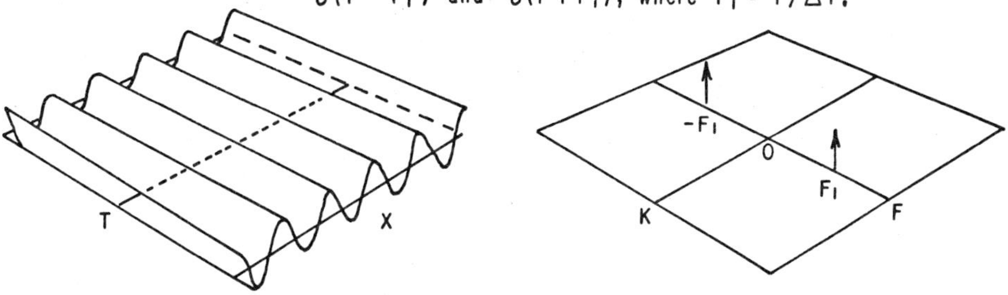

A sinusoidal surface transforms into two impulses, $\delta(F-F_1,K)$ and $\delta(F+F_1,K)$

Fig. 84. **Fourier-transform pairs** in two-dimensions. (From Lindseth, 1970, Recent Advances in Digital Processing of Geophysical Data: Calgary, Computer Data Processors, p. 10.18 and 10.19.)

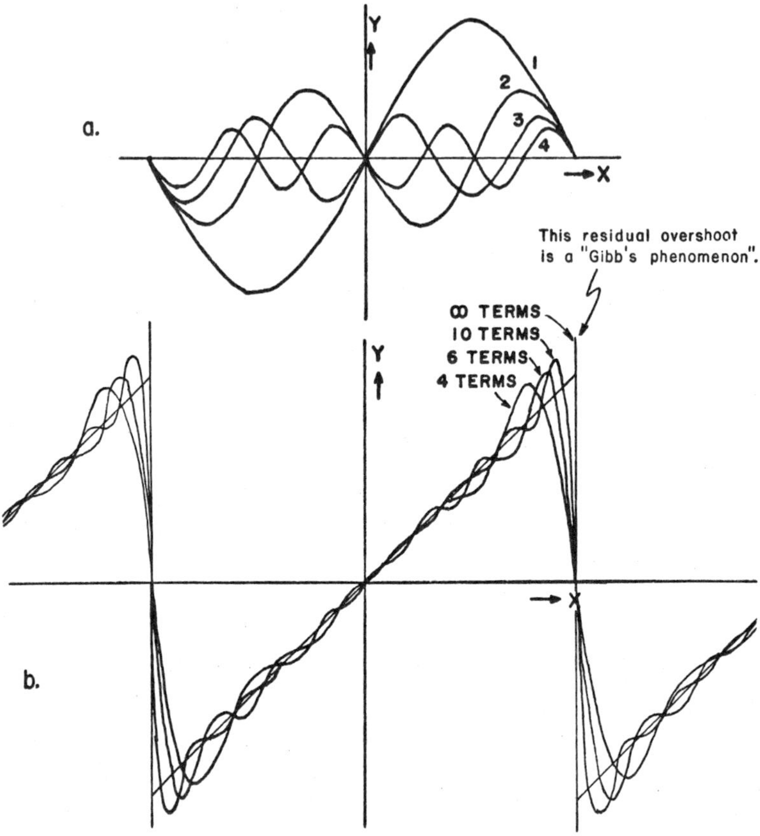

Fig. 85. **Fourier synthesis** of a sawtooth. (a) The first four Fourier components of a sawtooth are: sin x, $-(1/2)$ sin $2x$, $(1/3)$ sin $3x$, and $-(1/4)$ sin $4x$. (b) Superposition of the Fourier components reconstitutes the sawtooth.

four-way dip: Dip calculated from a cross-spread (especially where both in-line and cross-spreads are splits). See Figure 40 **(cross-spread)**.

fracture log: A well log of the cumulative amplitude of the wave arrivals from a sonic logging tool during a certain gate time. A fracture zone attenuates the acoustic energy.

free-air anomaly: Gravity data which have been corrected for latitude and elevation **(free-air correction)** but not for the density of the rock between datum and

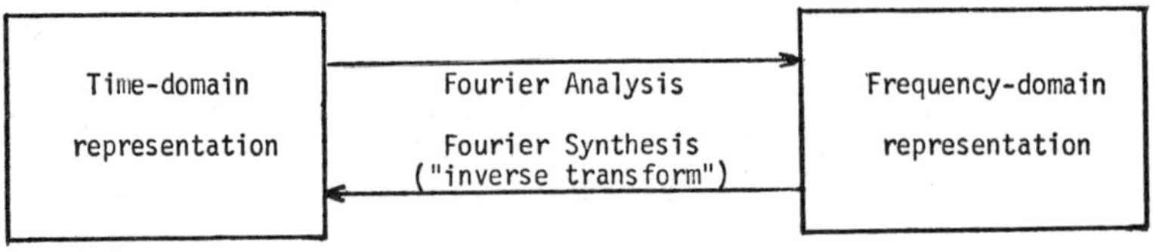

Fig. 86. The **Fourier transform.**

the plane of measurement (**Bouguer correction**).

free-air correction: 1. A correction for the elevation of a gravity measurement required because the measurement was made at a different distance from the center of the earth than the datum. The first term of the free-air correction is 0.09406 mgal/ft or 0.3086 mgal/m. **2.** In Turam, normalizing a ratio of successive measurements by dividing by the ratio of the calculated free-space vertical magnetic field. Compare **normal correction.**

free-fluid index: The percent of the bulk volume occupied by fluids which are free to flow as measured by the **nuclear-magnetism log.** Abbreviated FFI; gas gives a low FFI.

free oscillation of the earth: A simple change-of-shape oscillation of the whole earth. The period is 53 minutes in the lowest mode. See Figure 87.

free-space field: The field about an antenna in the absence of nearby conductors. See **primary field.**

frequency: 1. Symbol, *f.* The repetition rate of a periodic waveform, measured in "per second" or hertz. The reciprocal of **period.** Compare **spatial frequency. 2.** *Angular frequency* ω, measured in radians/sec, is to frequency, *f*, as 2π:

$$\omega = 2\pi f.$$

3. The *dominant frequency* of wavelets refers to an approximate repetition (the reciprocal of the peak-to-peak time interval) even though the entire wavelet does not repeat. See Figure 192 (**wave**). **4.** The *frequency content* of a waveform refers to the amplitudes of the sinusoidal components into which the waveform can be decomposed by **Fouier analysis,** even where there is nothing repetitive about the waveform itself. **5.** For mathematical symmetry, the Fourier integral is usually written for frequencies from $-\infty$ to $+\infty$. Negative frequencies sometimes disturb those who like to connect physical meaning with concepts; negative frequency can be thought of as the repetition rate where one counts backward in time.

frequency domain: 1. Measurements as a function of frequency or operations which involve frequency. The Fourier transform variable when transforming from time. **2.** The frequency method of making induced-polarization measurements.

frequency effect: The difference between resistivity (or voltage) measurements (ρ_1 and ρ_2) made at two frequencies f_1 and f_2, divided by the resistivity (or voltage) at one of the frequencies. Abbreviated FE.

$$FE = (\rho_2 - \rho_1)/\rho_1$$

Percent frequency effect (PFE) is often used.

frequency modulation: Modulation in which the instantaneous frequency of the modulated wave differs from the carrier frequency by an amount proportional to the instantaneous value of the modulating wave. Abbreviated FM. The amplitude of the modulated wave is usually constant. See Figure 122 (**modulation**).

frequency response: The characteristics of a system as a function of frequency. See **Fourier transform.**

frequency spectrum: The characteristics of a wavelet as a function of frequency. See **Fourier transform.**

fresh: Very low in dissolved salts. Sometimes used comparatively with respect to normal sea water (which is 35,000 parts of dissolved salts per million), sometimes used comparing formation water with mud filtrate. "Fresh water" has less than 2000 ppm dissolved salts.

FT: Field tape; a magnetic-tape record of geophysical data made directly "in the field" as opposed to a processed tape which is made from information which originates on another tape.

full-wave rectifier: An electronic device which uses both positive and negative polarities of alternating current to produce direct current.

fundamental: The lowest frequency of a periodic function.

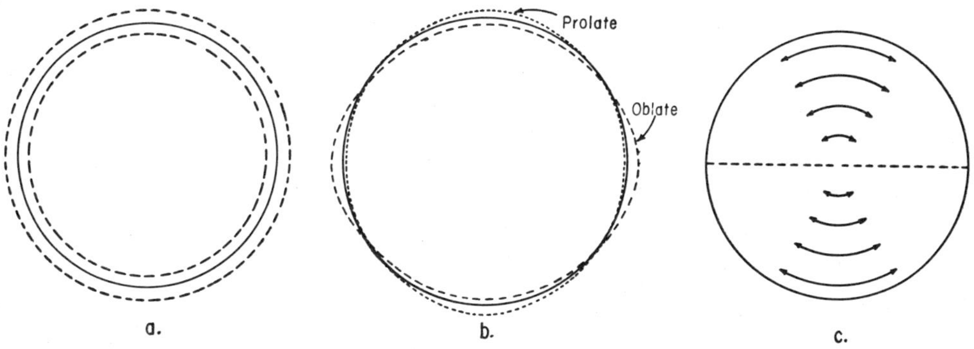

Fig. 87. **Free oscillation** of the earth. The simplest modes are (a) radial oscillation, (b) spheroidal oscillation, and (c) toroidal oscillation. (From Stacey, Physics of the Earth, 1969, John Wiley, New York, p. 110.)

G: Designation for **Love waves.** Also designated L_Q. See also **wave notations** and Figure 65 **(earthquake seismology).**

GAG: Geophysical Analysis Group; a research project at Massachusetts Institute of Technology during 1952-57, which applied communications-theory ideas to seismic analysis. Many digital data-processing techniques developed from these studies. See Flinn, Robinson, and Treitel (editors), 1967, Special Issue on MIT Geophysical Analysis Group Reports: Geophysics, v. 32, no. 3.

GAGC: Ganged automatic **gain control.**

gain: An increase (or change) in signal power (or amplitude) from one point in a circuit or system to another.

gain control: 1. Variation of the amplification or attenuation of an amplifier to compensate for variations in input signal strength. Gain controls often use a feedback loop whereby the output level is used to control the gain so as to keep the output level within certain limits. Gain control is used in many types of amplifiers. At maximum gain the amplifier has maximum sensitivity. At lower gains the amplifier will accept larger inputs without distortion. **2.** Types of gain control used in seismic amplifiers include: **a.** *Automatic* gain control in which the gain of each channel is controlled automatically and independently of other channels. **b.** *Ganged automatic* gain control, in which the gain of all channels is the same although automatically determined, being based either on an average of the energy levels of a number of the channels or on a single channel. **c.** *Preset* or "programmed" gain control, in which the gain as a function of record time is determined arbitrarily prior to the shot. **d.** *Binary* gain control, in which gain is allowed to vary only by factors of two but the time at which the gain changes is determined automatically. May be ganged or individual for each channel. The times at which the changes occur are recorded so that the gain effects can be removed in processing. See Figure 22 **(binary gain).** **e.** *Quaternary* gain control, in which gain is allowed to vary only by factors of four.

gain trace: A trace on a seismic record which indicates the gain (amplification) used on one or more channels. See Figure 149 **(record).**

gal: A unit of acceleration, used in gravity measurements. One gal = 1 cm/sec^2 = 10^{-2} m/sec^2. The earth's nominal gravity is 980 gal.

galvanic contact: An actual electrical contact with the ground, as opposed to causing electric current flow by induction. Called a *conduction contact* and an *ohmic contact* if linear and rectification is not involved.

galvanometer: A device to measure small currents. A coil suspended in a constant magnetic field rotates through an angle proportional to the electrical current flowing through the coil. A part of many seismic cameras. Often abbreviated "galvo."

gamma: A unit of magnetic field. A gamma = 10^{-5} gauss = 10^{-9} tesla = 1 nanotesla.

gamma-gamma log: See **density log.**

gamma-ray log: A well log which records the natural radioactivity. **1.** In sediments the log mainly reflects shale content because minerals containing radioactive isotopes (the most common of which is potassium) tend to concentrate in clays and shales. Volcanic ash, granite wash, and some salt deposits also give significant gamma-ray readings. The log often functions as a substitute for the SP for correlation purposes in cased holes, in conductive muds in open holes, and for thick carbonate intervals. See Figures 45 **(density log),** 112 **(laterolog),** and 126 **(neutron log).** **2.** Used in exploration for radioactive minerals such as the uranium minerals.

gamma-ray surveying: Measurements of naturally occuring gamma rays in the search for radioactive materials. Portable gamma-ray spectrometers incorporate energy-discrimination ability which permits distinguishing between radiomaterials. Thus one might measure all gamma rays, only those above 1.3 Mev (for potassium + uranium + thorium), only those above 1.6 Mev (for uranium + thorium), or only those above 2.5 Mev (thorium).

ganged gain control: **Gain control** in which the gain of several channels varies in the same way.

gap: 1. *Shotpoint gap,* a larger interval between geophone groups on either side of the shotpoint than between other groups. **2.** *Interrecord gap,* an interval of space on a digital magnetic tape during which no information is recorded which serves as a sentinel to indicate the beginning of a new record or a new block of data. Gaps facilitate the input of data into a computer in quantized blocks of information.

gapped deconvolution: Deconvolution with a **gapped operator.**

gapped operator: A method of deconvolving water bottom reverberations in deep water, as with a **Backus filter** where groups of nonzero filter elements are appreciably separated from each other. See Kunetz and Fourmann, 1968, Efficient Deconvolution of Marine Seismic Records; Geophysics, v. 33, p. 412-423.

Gardner method: A refraction interpretation method which involves separating intercept time into constituent **delay times** associated with the shot end and the geophone end of the trajectory. The separation is not unique unless some assumptions are made. Gardner's concept of delay time is central to other refraction methods. See Gardner, 1939, Areal Plan of Mapping Subsurface Structure by Refraction Shooting: Geophysics, v. 4, p. 247-259.

gas exploder: A seismic energy source in which a mixture of propane or butane with oxygen or air is exploded under the water. Also called *gas gun.*

gas gun: **Gas exploder.** (Different from *air gun).*

Gassmann equation: Seismic P-wave velocity through a minimum-volume packing of uniform spheres is proportional to

$$[E^2 z/(1-\sigma^2)^2]^{1/6},$$

where E = Young's modulus, z = pressure (or depth where proportional to pressure), σ = Poisson's ratio. See Gassmann, 1951, Elastic Waves Through a

Packing of Spheres: Geophysics, v. 16, p. 673–685.

GASSP: Gas source seismic profiler; a **gas exploder** system in which gas is exploded in a long rubber tube which expands because of the explosion. Shell Development trademark.

gate: **1.** The interval of record time over which a function (such as an **autocorrelation** or **crosscorrelation**) is evaluated. Also called **window**. **2.** A circuit with several inputs and one output which is used in digital logic (computer design). Truth tables for inputs A and B to these gates (assuming 0 indicates "no signal" or **false** and 1 indicates "signal" or true) is:

A	B	$\bar{A}$	AND	OR	EXCEPT	NOR	NAND
0	0	1	0	0	0	1	1
1	0	0	0	1	1	0	1
0	1	1	0	1	1	0	1
1	1	0	1	1	0	0	0

(A superscript line ($\bar{A}$) indicates INVERSION, which is also indicated in circuit diagrams by a small circle.) See Figure 88 and **Boolean algebra**.

gather: A display of the input data to a stacking process rearranged so that all the seismic traces corresponding to some criterion are displayed side by side. Used for checking corrections and evaluating the components of the stack. A *CDP gather* displays data for the same reflecting point after correction for normal moveout and statics. A *common-range gather* displays data for the same offset but a number of nearby reflection points.

gauss: The cgs-emu unit of magnetic induction (or flux density) **B**. It is a measure of the number of magnetic lines of force per unit area. 1 gauss = 1 maxwell/cm^2 = 10^5 gammas = 10^{-4} tesla = 10^{-4} weber/m^2. "Gauss" is also used in the cgs system as a unit for magnetization, or dipole moment per unit volume. In cgs units, a magnetizing force (in oersted) gives rise to a flux density or field (in gauss), with the units being equal in magnitude if in free space.

Gauss error function: See **error function.**

Gaussian distribution: A "normal" distribution. A set of values so distributed about a mean value *m* that the probability $\epsilon(\Delta a)$ of a value lying within a small interval Δa centered at the point *a* is:

$$\epsilon\,(\Delta a) = (e^{-\frac{1}{2}(a-m)^2/\sigma^2})\,\Delta a/(2\pi)^{\frac{1}{2}}\,\sigma,$$

where σ is the standard error. $\epsilon(\Delta a)$ is called the *error function.*

Gauss law: The total flux ϕ through any closed surface is equal to $4\pi k$ times the source strength *m* enclosed by the surface:

$$\phi = 4\pi km = \oint_S \mathbf{g}\cdot d\mathbf{s} = 4\pi k\int_V \rho\, dV = \oint_S \nabla U \cdot d\mathbf{s}.$$

(The 4π is often deleted in the mks system.) *k* is a constant which depends on the units of measure. This can also be expressed in terms of the flux density or field strength **g**, the source density ρ, and potential *U*. ϕ may be electrical flux if *m* is electrical charge; in the mks system, ϕ is in webers if *m* is in coulombs and k $\approx 9 \times 10^9$. Or ϕ may be gravitational flux if *m* is mass, in which case $k = -\gamma$, where γ is the gravitational constant. Or ϕ may be magnetic flux if *m* is magnetic poles.

GCT: Greenwich civil (mean) time. Usually written **GMT**.

gelatin: An explosive, usually of the dynamite type.

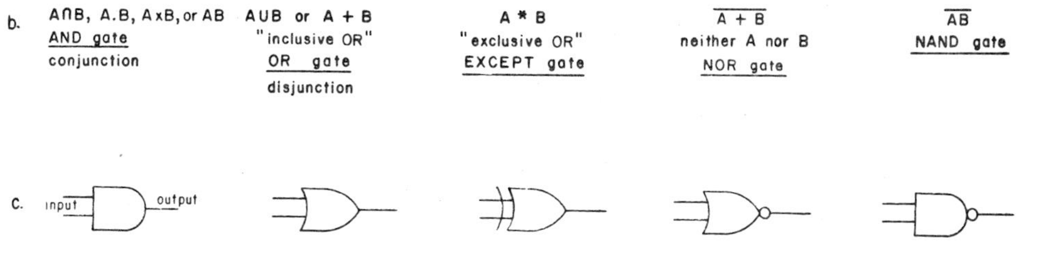

Fig. 88. **Gate types.** (a) Venn diagram representations. (b) Symbol and word representations. (c) Circuit diagram representations. Gates may have more than two inputs.

geocentric latitude: The angle between a line through the center of the earth and a plane through the equator. Compare **geodetic latitude.**

geodesic: The shortest distance between two points subject to some constraint, such as being on a given surface, such as the surface of the earth.

geodesy: The study of the earth's form and gravitational field. Involves the location of points on the earth with respect to reference systems.

geodetic latitude: Ordinary latitude, the angle between a tangent to the ellipsoid which approximates the earth's shape and the earth's axis; see Figure 89. Geodetic latitude differs from "geocentric latitude" because of the ellipticity of the earth. If the geocentric latitude is ψ and the geodetic latitude is ϕ, then $\phi - \psi = 11.7 \sin 2\phi$ (approximately) in minutes of arc. The maximum difference (at 45 degrees) is about 21.5 km. Geodetic Latitude also differs from "astronomic latitude" in places where mass distribution causes a vertical line to point other than perpendicularly to the ellipsoid. *Reduced latitude* is sometimes used; the location is projected parallel to the earth's axis onto a sphere whose radius is the ellipsoid's major axis and the angle between a radius to this point and the axis is the *reduced latitude.*

geodetic system: Reference for latitude and longitude. Location systems in different parts of the world are based on different reference positions and ellipsoid assumptions. The major systems are listed in Table 5.

Geoflex: A seismic land source in which an explosive cord (such as Primacord) is buried by a plowlike device. Imperial Chemical Industries trademark. Compare **aquaseis.**

Geograph: Thumper or the weight-drop method. Trademark Mandrel Industries.

geographic latitude: geodetic latitude

geoid: The sea-level equipotential surface to which the direction of gravity is everywhere perpendicular. An oblate ellipsoid of revolution (the "spheroid") which approximates the geoid is the reference for **geodetic latitude** determinations.

geoidal height: The distance between the reference ellipsoid and the geoid; height of geoid above ellipsoid.

geomagnetic pole: See **magnetic pole.**

geomagnetic reversal: Change in the polarity of the earth's magnetic field from its present polarity, which occurred during several periods in the earth's history.

geomagnetic-variation method: Studies of deep conductivity variations in the earth (particularly lateral variations) by simultaneously observing variations in the three components of the magnetic field at different stations, as the field varies naturally.

geometric factor: 1. The geometry-dependent weighting factor for determining how the conductivities of each medium in the vicinity of a sonde affect the apparent-conductivity measurement, especially used with the induction log. **2.** A numerical factor used to multiply the voltage-to-current ratio from measurements between electrodes to give apparent resistivity. Geometric factor is dependent on the type of electrode array and spacing used. Also called *geometric constant, array factor,* or *form factor.*

geometric sounding: An electromagnetic depth sounding in which the geometry is varied with the frequency held constant, as opposed to frequency sounding (sometimes called *parametric sounding)* in which the geometry is held constant and the frequency is varied.

geophone: The instrument used to transform seismic energy into an electrical voltage; a *seismometer,* a *jug,* or *pickup.* Most land geophones are of the moving-coil type (Figure 90). A coil is suspended by springs in a magnetic field (the magnet often is integral with the case of the instrument). A seismic wave moves the case and the magnet but the coil remains relatively stationary because of its inertia. The movement of magnetic field with respect to the coil generates a voltage across the coil which is proportional to the difference between

Table 5. Major geodetic systems

Datum	Ellipsoid	Equatorial radius	1/Flattening	Comments
North American "Meades Ranch"	Clarke 1886	6,378,206m	295	Used throughout Western Hemisphere
European "Pelmert Tower, Potsdam"	International 1910	6,378,388m	297	Used in most of Eurasia and Africa
Russian "Pulkovo"	Bessel 1841 Krassouski 1938	6,377,397m 6,378,245m	299.2 298.3	
Tokyo	Bessel 1841	6,377,397m	299.2	
Indian "Kolimpur"	Everest 1830 International 1910	6,377,276m 6,378,388m	300.8 297 293.5	
	Clarke 1880 Heiskanen 1926 Hough (US Army) 1956 APL (Applied Physics Lab)	6,378,249m 6,378,397m 6,378,260m 6,378,144m	193.5 297.0 297.0 298.23	Used in Africa Navigation satellite

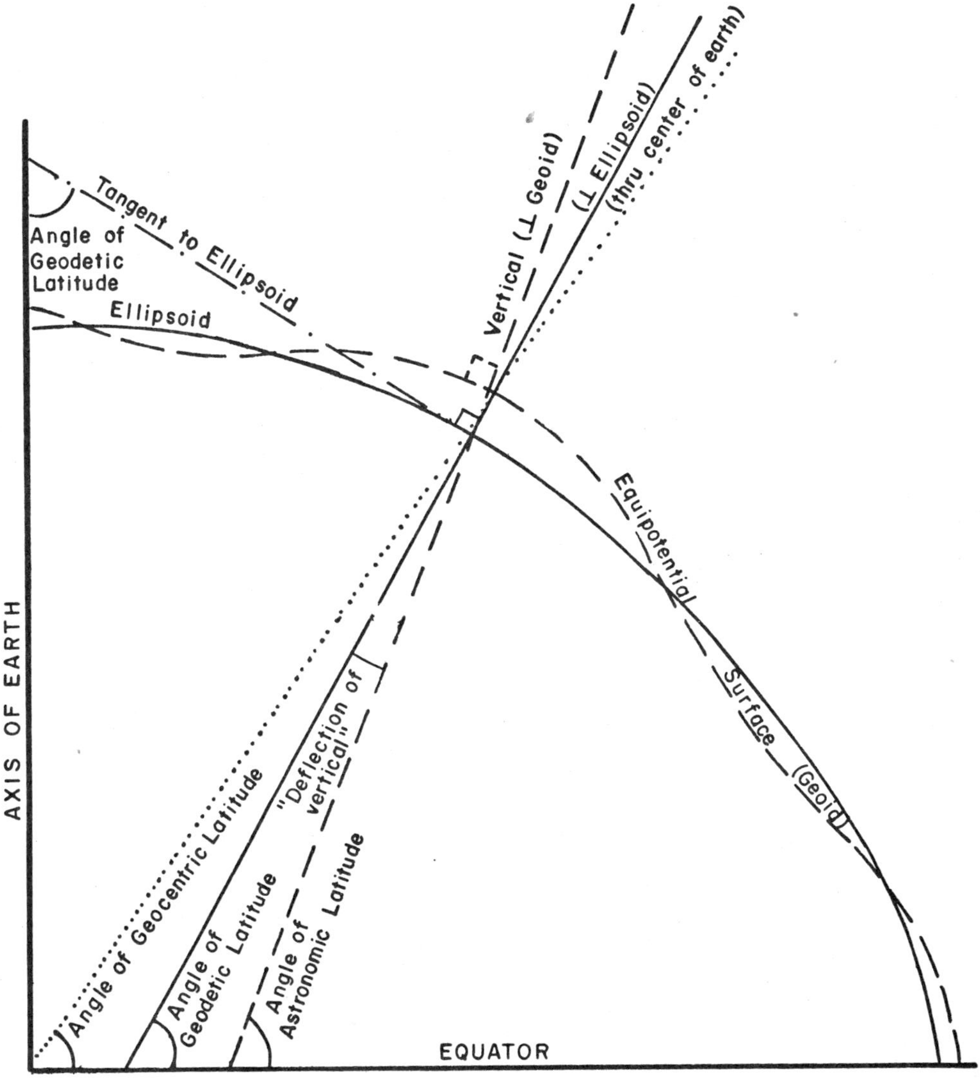

Fig. 89. **Geocentric**, geodetic, and astronomic latitude.

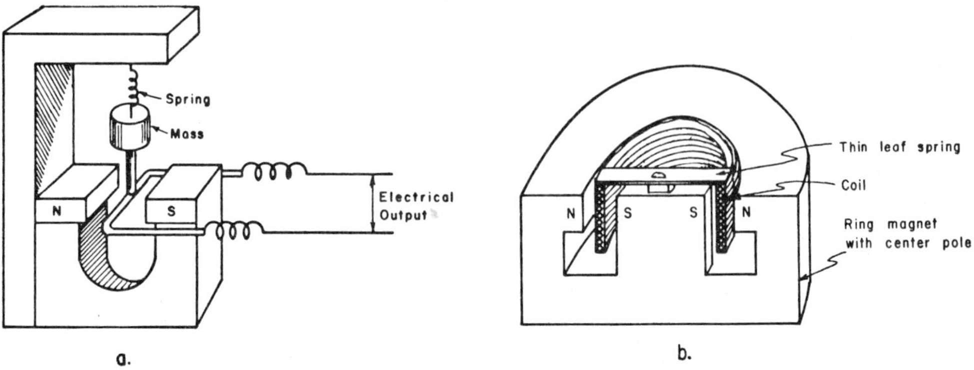

Fig. 90 **Geophone.** (a) Schematic of essential elements. (b) Half of a moving-coil geophone.

the velocity of the coil and the magnet. Thus over a region which is above the natural frequency, the voltage output is proportional to the velocity of the seismic wave motion and relatively insensitive to frequency. At frequencies below the natural frequency of the geophone, the output (for input of constant velocity of magnet motion) is proportional to frequency and hence to the acceleration involved in the seismic wave. See Evenden, Stone, and Anstey, 1971, Seismic Prospecting Instruments: Berlin, Gebruder Borntraeger, v. 2, p. 1–30. See also **dampling** and compare **hydrophone** and **streamer.**

geophone interval: **1**. The distance between adjacent geophones within a group. **2**. Sometimes used for **group interval,** the separation between the centers of adjacent geophone groups.

Geophysical Analysis Group: See **GAG.**

geophysical survey: A program to make geophysical measurements over an area with the objective of learning about the geology. The objective of many surveys is to locate accumulations of hydrocarbons, minerals, water, etc. See **electrical survey, geothermal prospecting, gravity survey, magnetic survey, reflection survey, refraction survey, remote sensing,** and **well log.**

geophysics: **1**. The study of the earth by quantitative physical methods, especially by seismic reflection and refraction, gravity, magnetic, electrical, and radiation methods. **2**. The application of physical principles to studies of the earth. Includes the branches of (a) seismology (earthquakes and elastic waves), (b) geothermometry (heating of the earth, heat flow, and volcanology and hot springs), (c) hydrology (ground and surface water and sometimes including glaciology), (d) oceanography, (e) meteorology, (f) gravity and geodesy (the earth's gravitational field and the size and form of the earth), (g) atmospheric electricity and terrestrial magnetism (including ionosphere, Van Allen belts, telluric currents, etc.), (h) tectonophysics (geological processes in the earth), and (i) exploration and engineering geophysics. Geochronology (the dating of earth history) and geocosmogony (the origin of the earth) are sometimes added to the foregoing list. Enthusiasts in particular branches are inclined to appropriate the word "geophysics" to their own branch exclusively, whether that branch be ionospheric studies or exploration for oil. **3. Exploration geophysics is the use of seismic, gravity, magnetic, electrical, electromagnetic, etc., methods in the search for oil, gas, minerals, water, etc., for economic exploitation.**

geostatic pressure: Pressure due to the weight of the overlying rock, which will generally be different from the pressure of fluids in the rock.

geosyncline: A deeply subsiding area of extensive sediment and/or volcanic accumulation; a more-or-less continually sinking area. Usually more-or-less linear.

geothermal gradient: The rate of change of temperature as one goes deeper into the earth. In the earth's crust, it averages about 25° C/km.

geothermal heat flow: See **HFU.**

geothermal prospecting: Measurements of variations in earth temperature which are not attributable to the daily variations in solar heating. The objectives may be to locate sources of geothermal energy, geological features which affect heat flow (salt domes, dikes, faults, etc.), or ground water variations. Diurnal temperature variations penetrate to about one meter and annual temperature variations to up to 20 m, so measurements made at 1-1/2 to 2 meters over a relatively short time period reflect variations of heat flow. Corrections have to be applied for the type of vegetation cover which affects readings to the extent of 2 or 3 degrees. Measurements may also be made below the bottom of lakes and marsh. See Poley and van Steveninck, 1970, Geothermal Prospecting; Delineation of Shallow Salt Domes and Surface Faults: Geophysical Prospecting, v. 18, p. 666-700.

geothermic surveying: See **geothermal prospecting.**

g factor: An expression used in refraction calculations. If v_1 and v_2 are velocities above and below the interface involved in refraction, and i_c is the critical angle, $g = V_1/2 \cos i_c = V_1 V_2/2(V_2^2 - V_1^2)^{1/2}$. See Stulken, 1967, Constructions, Graphs, and Nomographs for Refraction Computation, *in* Seismic Refraction Prospecting: A. W. Musgrave, editor, Tulsa, SEG, p. 304—329.

ghost: 1. Energy which travels upward from the shot and then is reflected downward, such as occurs at the base of the weathering or at the surface. Ghost energy usually joins with the down-traveling wave train, changing its wave shape and adding a tail. Ghost energy sometimes is separated sufficiently from the main wave train to form a separate wave and separate spurious reflection events on a record. Correcting for the change in waveshape which ghosting produces is the objective of **uphole stacking.** Sometimes called *secondary reflection* (which is also applied to multiples). 2. A **correlation ghost** which results from harmonic distortion when using the Vibroseis. 3. A reflection of light from the front side of a mirror as well as from the silvered back side.

Gibb phenomenon: When a waveform which includes a discontinuity (or whose derivatives are discontinuous) is Fourier synthesized, the fit is poor near the discontinuity. As the number of frequency components included in the synthesis approaches infinity, the region of poor fit becomes vanishingly narrow. This poor fit sometimes is called *Gibb's ears.* See Figure 85 **(Fourier synthesis).**

giga-: A prefix meaning 10^9.

GIGO: Garbage in, garbage out. A "principle" in data processing which emphasizes that meaningful input signal must be present if a meaningful result is to be obtained.

gimbals: A device for supporting an instrument so that it will remain essentially horizontal even when the support tips.

Gish-Rooney method: 1. An electrical-surveying method in which the polarity of both current flow and potential-measuring electrodes is reversed frequently to cancel out the effects of polarization at the electrodes. 2. A resistivity-interpretation method.

GMT: Greenwich mean (civil) time. The international reference time, the time at the observatory at Greenwich, England.

gnomonic projection: See **map projection.**

goodness of fit: See **chi square.**

GOR: Gas-to-oil ratio. The ratio of the gas to the oil which a well produces. A well is usually classed as an oil well if its GOR $< 15{,}000$ ft^3/bbl, as a gas well if GOR $> 150{,}000$ ft^3/bbl, or as a condensate well if intermediate.

Goupillaud medium: A medium of parallel layers of such thickness that the traveltime perpendicularly through each is equal. Used in computing synthetic seismograms. Goupillaud, 1961, Approach to Inverse Filtering of Near-Surface Layer Effects from Seismic Records: Geophysics, v. 26, p. 754-760.

graben: A down-dropped block bounded by normal faults, often relatively long and narrow.

gradient: 1. The rate of change of a spatially dependent function in the steepest direction. See **gradiometer.** 2. The operation which finds the gradient from a potential function:

$$\text{Gradient of } U = \text{grad } U = \nabla U = \left(\mathbf{k}\frac{\partial}{\partial x} + \mathbf{j}\frac{\partial}{\partial y} + \mathbf{k}\frac{\partial}{\partial z}\right)U.$$

3. The component of the gradient in an arbitrary direction, as the *vertical gradient* of gravity.

gradient array: See **array.**

grading: Indicating the relative reliability of data or their interpretation. An important part of interpretation; usually a subjective process. Coherency-measuring criteria are sometimes used to make grading quantitative. 1. Seismic event picks are usually graded by letters: *vg* = very good, *g* = good, *f* = fair, *p* = poor, *vp* = very poor, and *?* = questionable. Sometimes a double letter grade is used, the first letter being a measure of the certainty that the pick is a reflection event and the second letter being a measure of the precision with which it is measured. 2. Correlations are often graded by letters similar to those for seismic picks. 3. Contours are often graded by varying the type of line: solid contour = reliable; dashed contour = less reliable or interpolated between data; and dotted contour = speculative.

gradiomanometer: A device for determining the density of the wellbore fluid by measuring the pressure on two bellows and therefore the vertical pressure gradient.

gradiometer: A device for measuring the gradient of a potential field. 1. An arrangement of two magnetometers, one above the other, so that the difference in their readings is proportional to the vertical gradient of the magnetic field. 2. A three-arm torsion balance which is sensitive to gravity gradients but not to curvature.

grand slam: A combination of logs or a computation procedure for calculating the depth of invasion and the resistivity of both invaded and uncontaminated zones, based on a dual-induction laterolog and a proximity log or microlaterolog.

graphic log: Strip log.

graticule: 1. A template for graphically calculating gravity or similar data. See also **dot chart** and **zone chart.** 2. A grid network such as lines representing parallels and meridians on a plotting sheet.

gravimeter: An instrument for measuring variations in gravitational attraction; a gravity meter. Most present gravimeters are of the unstable or astatic type. The gravitational force on a mass in the meter is balanced by a spring arrangement, and a third force is provided which acts when the system is not in equilibrium. This third force intensifies the effect of changes of gravity and increases the sensitivity of the system. In the LaCoste-Romberg gravimeter (Figure 91), the main spring which balances out the gravitational pull on the weight is a "zero-length spring" inclined at an angle. A zero-length spring has a stress-strain curve which passes

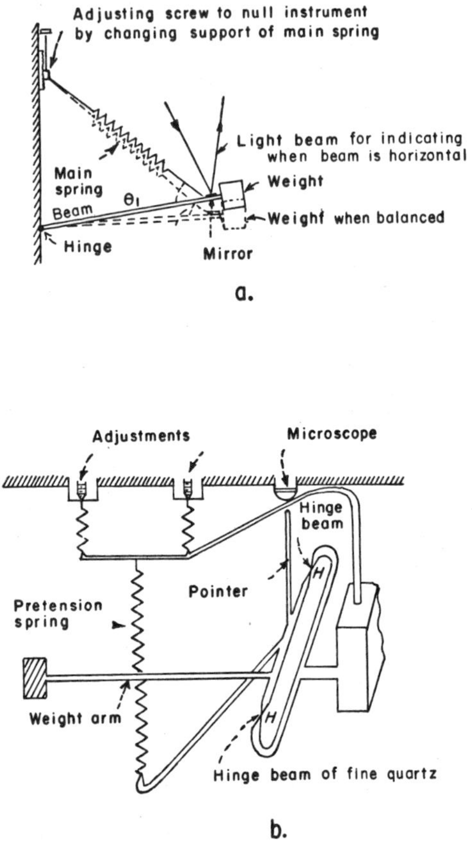

Fig. 91. **Gravimeters.** (a) La Coste-Romberg schematic. (b) Worden schematic.

through zero length when projected back to zero strain. (For example, a spring which requires an initial stress before the coils begin to separate.) Zero-length springs have very long periods and high sensitivity. The Worden gravimeter uses a horizontal torsion fiber or "hinge" as the main spring. This and the accessory measuring springs and supports are made of quartz and are very small and hence very light in weight. The Worden meter incorporates temperature-compensating features and is usually encased in a thermos bottle to further reduce its temperature sensitivity. See also **shipboard gravimeter.**

gravitational constant: The proportionality constant γ in Newton's Law of Universal Gravitation. The gravitational force F between two point masses m_1 and m_2 can be related to the distance r between them:

$$F = \gamma m_1 m_2 / r^2.$$

γ has the value 6.670×10^{-11} newton m^2/kg^2. The gravitational acceleration g is the force per unit (face on the mass which is in the gravimeter):

$$g = F/m_1 = \gamma m_2 / r_2.$$

gravity: The force of attraction between bodies because of their mass. Usually measured as the acceleration of gravity. See **gravitational constant.**

gravity anomaly: 1. Bouguer anomaly. 2. The difference between the gravity which is observed and that expected from a model. **3. Residual** gravity.

gravity meter: Gravimeter.

gravity reduction: Applying **Bouger elevation, free-air, isostatic, latitude,** or **terrain corrections** to gravity measurements.

gravity standard: The value of gravity at the Pendelsaal of the Geodetic Institute in Potsdam, East Germany, is defined as 981.27400 gal. (This standard value may be about 13 mgal too large.) See also **international gravity formula.**

gravity survey: Measurements of the gravitational field at a series of different locations over an area of interest. The objective in exploration work is to associate variations with differences in the distribution of densities and hence of rock types. Occasionally the whole gravitational field is measured (as with a pendulum) or derivatives of the gravitational field (as with a torsion balance), but usually the difference between the gravity field at two points is measured (as with a gravimeter). Gravity data usually are displayed as Bouguer or free-air anomaly maps.

gravity unit: A unit of gravitational acceleration, equal to 0.1 mgal or 10^{-6} m/sec^2. Sometimes called *G unit.*

gray code: A binary number system code in which successive numbers differ by only one bit; See Table 8 **(number systems).** The gray code is used in error minimization because the number of bit changes are the same for a step change regardless of the magnitude of the quantity.

Green theorem: A form of the divergence theorem relating volume integrals to surface integrals. If F and G are two scalar functions, then

$$\oint_V (F\nabla^2 G - G\nabla^2 F)\, dV = \oint_S (F\nabla G - G\nabla F) \cdot \mathbf{dS}.$$

greenwich: 1. Longitude measured with respect to the *prime meridian* which passes through the Royal Astronomic Observatory at Greenwich, England. **2.** The time at the Greenwich observatory; **GMT,** sometimes GCT.

grey level: In variable-area and variable-density record sections, the overall "greyness" of the section is an important psychological factor affecting the section's interpretability.

grid effect: 1. A result of systematic errors in interpolating values at grid points where no data observations exist. Such errors tend to create false anomalies along certain orientations. **2.** Where grid residual maps are made with templates which are not azimuthally symmetrical, false anomaly trends may be created.

grid residual: A method of emphasizing anomalies of a certain size in a potential-field map. A grid (usually square or trigonal) is drawn on a contour map and values are determined at the grid intersections by interpolation. The residual at one of the grid intersections is the value at that point less the average at other intersections a fixed distance away. Averages at several distances may be used and weighted to approximate **second-derivative** values or other operators. The process of making grid residuals is also called **map convolution** because it represents map data convolved with a residualizing operator (or template). See also **residualize.**

grid smoothing: A method of smoothing sharp irregularities in potential-field measurements which arise from very shallow disturbances. A grid is drawn on a contour map and the smoothed value at a grid intersection is the average of values a fixed small distance away.

ground: A point in a circuit used as a common reference point, often the conducting chassis on which the electrical circuit is physically mounted. Frequently, but not necessarily, connected to the earth by a low-resistance conductor.

ground mix: The use of a pattern of shots or a pattern of geophones distributed over a sizeable surface area. The objective is to have vertically traveling reflection energy add up in-phase while horizontally traveling energy partially cancels. The term is sometimes reserved for situations where adjacent geophone or source patterns actually overlap.

ground roll: Surface-wave energy which travels along or near the surface of the ground. Usually characterized by relatively low velocity and low frequency but high amplitude. Ground roll tends to mask desired signals; shot and geophone patterns, filtering, and stacking are used to discriminate against it. Rayleigh waves are usually the main source and ground roll is sometimes called *pseudo-Rayleigh wave.* See Figure 147 **(Rayleigh wave).**

ground truth: 1. Data obtained on the ground concerning the significance of anomalies observed in **remote sensing** to help interpretation. **2.** Data from a ground monitor to show that extraneous events did not occur during the acquisition of airborne data, as use of a ground magnetometer to ascertain the absence of disturbing magnetic storms.

ground unrest: Background or ambient noise, such as produced by wind, microseisms, etc.

group: The various geophones which collectively feed a single channel. The number of geophones may vary from one to several hundred. A large group is sometimes called a *patch.* See **array.**

group interval: The horizontal distance between the centers of adjacent geophone groups.

group velocity: The velocity with which the energy in a wave train travels. In dispersive media where velocity varies with frequency, the wave train changes shape as it progresses so that individual wave crests appear to travel at a different velocity (the phase velocity) than does the envelope of the wavetrain. The velocity of the envelope is the group velocity. See also Figure 55 **(dispersion)** and compare **phase velocity.** See Brilloin, 1960, Wave Propagation and Group Velocity: New York, Academic Press, Inc.

gu: Gravity unit; 0.1 **milligal.**

guard electrodes: Extra electrodes whose function is to focus the current flow from a measuring electrode into the formation. Also called *bucking electrodes.* See **laterolog.**

guard log: A log made with **guard electrodes. A laterolog** or *focused log.*

guest: A mineral found in association with another mineral or mineral assemblage.

guided wave: 1. An interface wave or **surface wave. 2.** A **channel wave.**

Gulf magnetometer: A type of **fluxgate magnetometer.**

gun: 1. A **gas gun,** a seismic energy source in which an explosive gas mixture is detonated. **2.** An **air gun,** a seismic energy source from which a bubble of highly compressed air is released. **3.** A device for obtaining **sidewall cores. 4.** A device used to **perforate** or open holes in casing so that fluid can flow into the borehole.

G unit: Gravity unit or 0.1 mgal.

Gutenberg discontinuity: The boundary between the earth's mantle and core. See Figure 63 **(earth layering).**

G-wave: A **Love wave** usually restricted to an oceanic path.

gyrocompass: A gimbal-mounted gyroscope incorporating unbalanced masses which make the axis of rotation precess about true north. A gyroscope will maintain its orientation in space in the absence of disturbing forces (Figure 92). If a torque tries to change the plane of rotation of the gyroscope, the gyroscope axis will rotate about an axis which is perpendicular to both the gyroscope's axis of spin and the torque; this is called *precession.* If a weight is applied to the vertical circle of a horizontally spinning gyroscope, tilt of the gyroscope axis because of rotation of the earth will tend to raise the weight, thus exerting a torque which will make the gyroscope precess unless the axis is aligned north-south. This "meridian seeking" ability is the essence of a gyrocompass.

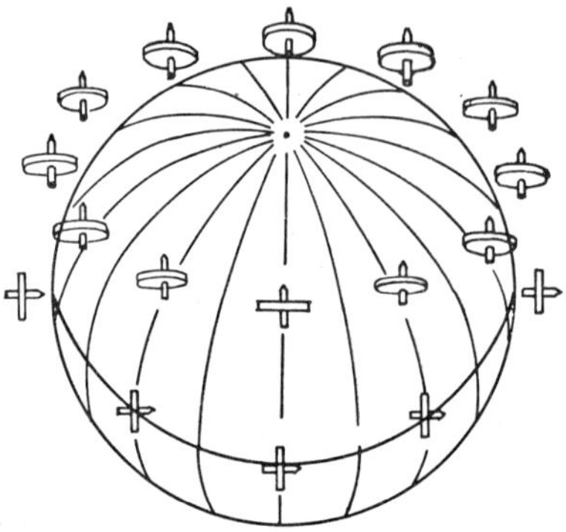

a.

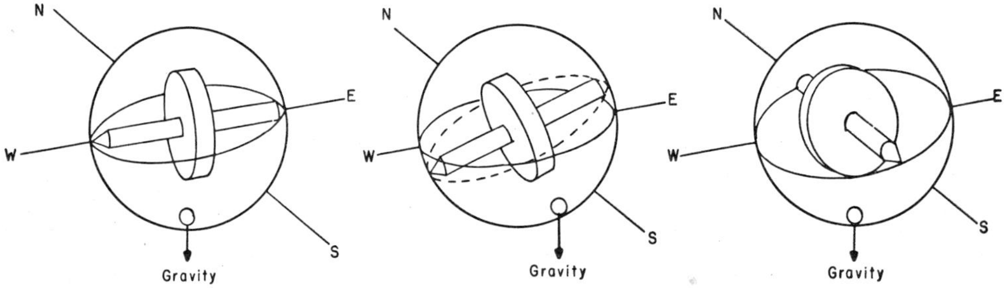

b.

Fig. 92. **Gyrocompass.** (a) Gyroscope tends to maintain its orientation in space, making it appear to change direction as seen from the rotating earth. (b) A weight on the vertical circle makes a gyroscope into a gyrocompass. Gravitational pull on this weight as the earth rotates produces a torque which tends to make the rotor axis precess about the north-south direction.

hachure: A short line or mark along a contour or fault trace which points in the down-dip direction or toward smaller values.

hade: The complement of dip. See Figure 72 **(fault)**.

Hagedoorn method: Plus-minus method of refraction interpretation.

Hales method: A graphical refraction interpretation method, particularly useful where the refractor changes depth markedly, such as where there is considerable relief or over large faults. Hales's method assumes constant velocity above the refractor. Reversed time-distance curves are plotted with the data from one direction plotted with respect to the reciprocal time (see Figure 93). Lines such as CD are drawn with the slope $V_o^2/V_R \cos \phi$, where ϕ is the dip of the refractor ($\cos \phi$ is usually so close to unity that it is ignored). From C', the projection of C onto the horizontal, the line $C'E$ is drawn at the critical angle i with the horizontal ($i = \sin^{-1} V_o/V_R$). The vertical projection of the midpoint of CD on this line locates the "center of the profiling circle" F. The radius of the circle is given by $R = V_o \Delta T/2 \cos(i)$, where ΔT is the time distance between C and D (actually, $t_C + t_D - t_{AB}$ because of the inverted time scale from the one shotpoint). The arc of radius R centered at F is tangent to the refractor. The refractor configuration is found by plotting many such arcs and finding their common tangent. See Hales, 1958, An Accurate Graphical Method for Interpreting Seismic Refraction Lines: Geophysical Prospecting, v. 6, p. 385-294.

half adder: A circuit with two inputs (A and B) and two outputs, sum and carry (S and C). Its truth table is:

A	B	C	S
0	0	0	0
1	0	0	1
0	1	0	1
1	1	1	0

C is an AND gate and S is an EXCEPT **gate.**

half adjust: A kind of rounding in which the value of a particular digit determines whether a one shall be added to the next-higher significant digit.

half-duplex: A system in which transmission can occur in only one direction at any time. Transmissions in opposite directions have to alternate. Compare **duplex**

half-maximum distance: half-width.

half-space: A mathematical model bounded only by one plane surface; i.e., the model is so large in other dimensions that only the one boundary affects the results. Properties within the model are usually assumed to be homogeneous and isotropic, though other models are also used.

half-width: A distance measurement on magnetic and gravity profiles made in the process of estimating the depth to the center of mass (for gravity data). On anomalies attributed to point or line masses (such as spherical or cylindrical masses, Figure 94), the half-width is half the width at half amplitude. If the anomaly results from a horizontal cylinder perpendicular to the line of profile, the depth to the center of the cylinder is the half-width; if the anomaly results from a sphere

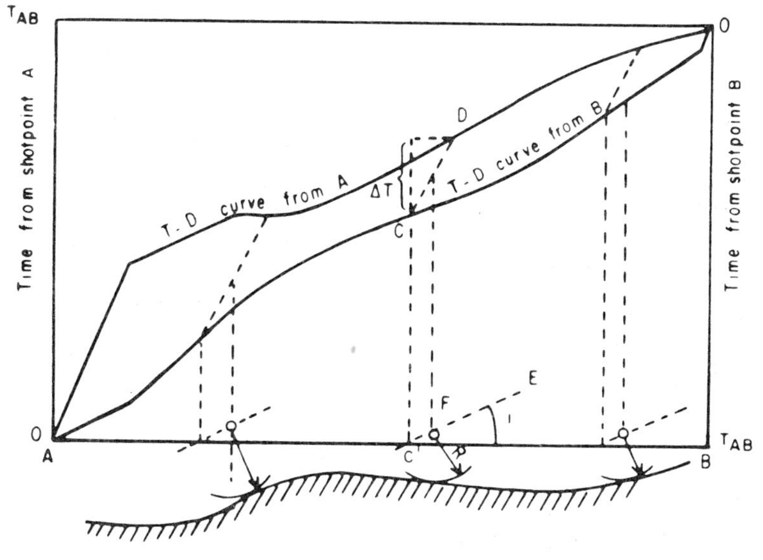

Fig. 93. **Hales refraction method.** (From Hales, Geophysical Prospecting: v. 6, no. 3, p. 292.)

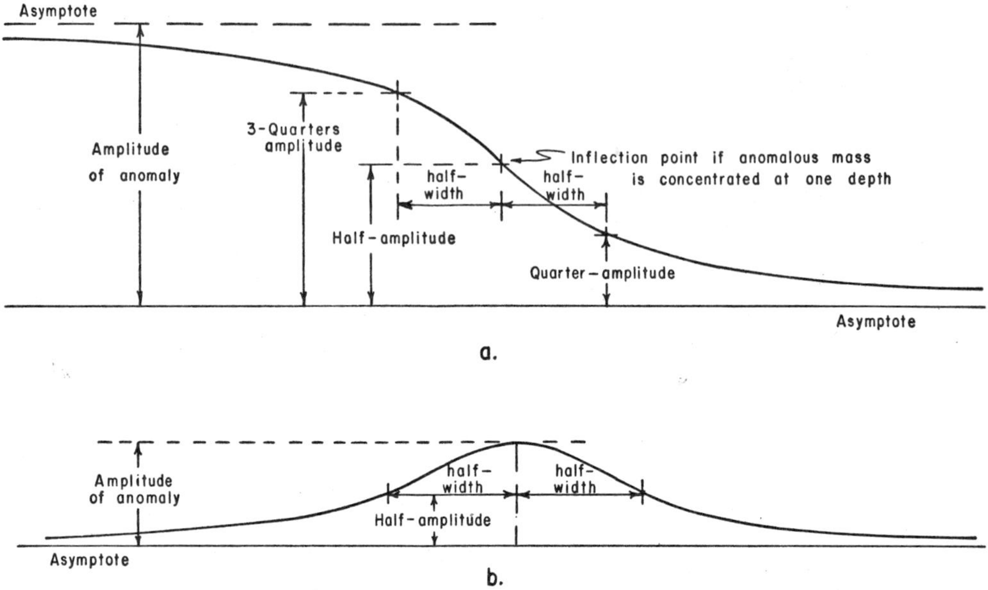

Fig. 94. **Half-width.** (a) Gravity anomaly due to a simple fault (semiinfinite slab). Depth to center of anomalous mass = half-width. (b) Gravity anomaly due to simple sphere or horizontal cylinder. Depth to center of sphere = 1.305 × half-width. Depth to center of cylinder = half-width.

beneath the line of profile, the depth to the center is 1.305 times the half-width. On fault-type anomalies, half the width between points where the anomaly is one-quarter and three-quarters amplitude. If the anomaly results from a simple step perpendicular to the line of profile, the depth to the middle of the step is the half-width. See also **depth rule.**

Hall effect: A transverse potential develops across a semiconductor or strip of metal which carries a current when located in a strong magnetic field. Used in analog correlators for Vibroseis.

halo effect: 1. Many residual and second-derivative methods produce a ring or *halo* of opposite sign around an anomaly, reflecting the opposite curvature of the field which exists around the periphery of the anomaly. Halos do not represent anomalous masses. Halos can be reduced or eliminated by biasing. **2.** A ring anomaly, claimed to be characteristic of certain electromagnetic or geochemical effects of structures or hydrocarbon accumulations.

hammer: A hammer striking a steel plate is used as seismic source for shallow refraction measurements. The hammer incorporates a switch which starts a timer when the hammer strikes; the timer is stopped when energy received by a geophone reaches some threshold value.

Hamming: Smoothing with weights 0.23, 0.54, 0.23.

Hamming function: A function used to shape a **window** or **gate** in data processing in order to avoid undesirable truncation effects such as produced by the discontinuities in a boxcar-shaped window. The Hamming function (different from the **Hanning function** which is also used as window taper) is

$$\begin{cases} 0.54 + 0.46 \cos \phi & -\pi < \phi < \pi \\ 0 & \phi < -\pi \text{ or } \phi > \pi, \end{cases}$$

where ϕ increases linearly from $-\pi$ to π across the window. For the *k*th element out of *n* elements within the window, $\phi = 2\pi k/(n+1) - \pi$. See also Figure 198**(window).** (Named after R. W. Hamming.)

Hanning: Smoothing with weights 0.25, 0.50, 0.25.

Hanning function: A function used to shape a **window** in data processing to avoid undesirable effects of sharp truncation. The Hanning function is

$$\begin{cases} 1/2 + 1/2 \cos \phi & -\pi < \phi < \pi \\ 0 & \phi < -\pi \text{ or } \phi > \pi. \end{cases}$$

See also Figure 198**(window).** (Named after J. von Hann.) Compare **Hamming function.**

hard copy: A printed (or otherwise displayed) copy of data in human-readable form.

hardware: Equipment, especially computing-machine equipment.

harmonic: A frequency which is a simple multiple of a fundamental frequency. The third harmonic, for example, has a frequency three times that of the fundamental. Two frequencies are *harmonically related* if they are each harmonics of a common fundamental.

harmonic distortion: **Distortion** characterized by the generation of harmonics of input frequencies.

harmonic function: A solution of **Laplace's equation.**

Haskell matrix: See **Thomson-Haskell method.**

haversine: Haversine $\alpha = (1 - \cos \phi)/2$.

HDT: High-resolution dipmeter. A **dipmeter** which records four high-resolution microresistivity curves and has an additional electrode on one pad which yields another curve at displaced depth. The displaced-depth curve is used to correct for variations in sonde velocity.

head: A device which reads from or writes on a storage medium. A **magnetic head** is used with magnetic tape, perforating or reading fingers to punch or read data on cards, etc.

head-check pulse: An impulse applied to all channels simultaneously so that the alignment of the magnetic heads can be checked. Usually appears on a record before the time break. See Figure 149 **(record).**

header: The identification information and tabulation of parameters which precedes data, as on magnetic tape.

head wave: A refraction wave or Mintrop wave; a wave characterized by entering and leaving the high-velocity medium at the critical angle. See Figure 39 **(critical distance)** and Figure 95. Particularly the wave which gives a refraction first break.

heat flow unit: HFU.

Heaviside function: A **step function.**

Heaviside layer: E-layer.

height of instrument: The elevation of a survey-instrument eyepiece with respect to the datum. Abbreviated **HI.** Used with theodolites, levels, alidades, etc.

Helmholtz coil: A pair of coaxial coils with their distance apart equal to their radius, which permits an accurate calculation of the magnetic field between the coils. Used in calibration of magnetometers and nulling an ambient magnetic field for magnetic measurements.

Helmholtz double layer: See **double layer**

HEM: **1.** Helicopter electromagnetics. **2. Horizontal-loop** electromagnetic method. See **slingram method.**

Hermitian matrix: A matrix which equals the transpose of its conjugate:

$$(A^*)^T = A \ .$$

It has real eigenvalues and can be diagonalized by similarity transformations.

herringbone: A pattern of systematic deviations of contours on a contour map produced when one line of data is systematically off, e.g., if it contains a systematic bias or is mislocated. Figure 96.

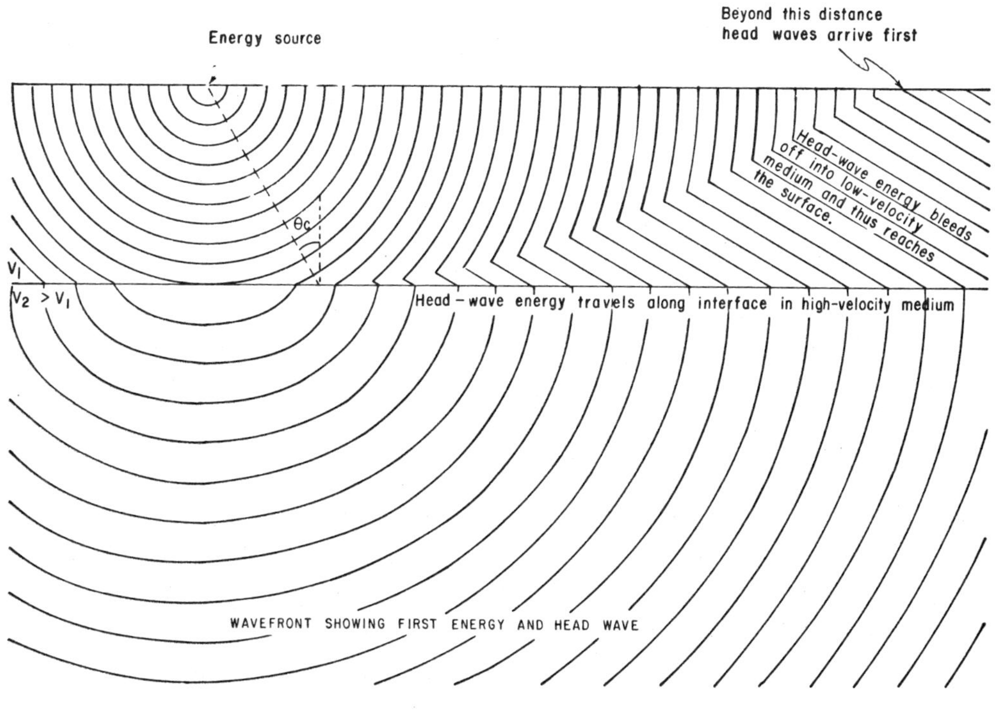

Fig. 95. **Head waves** and wavefronts showing first energy.

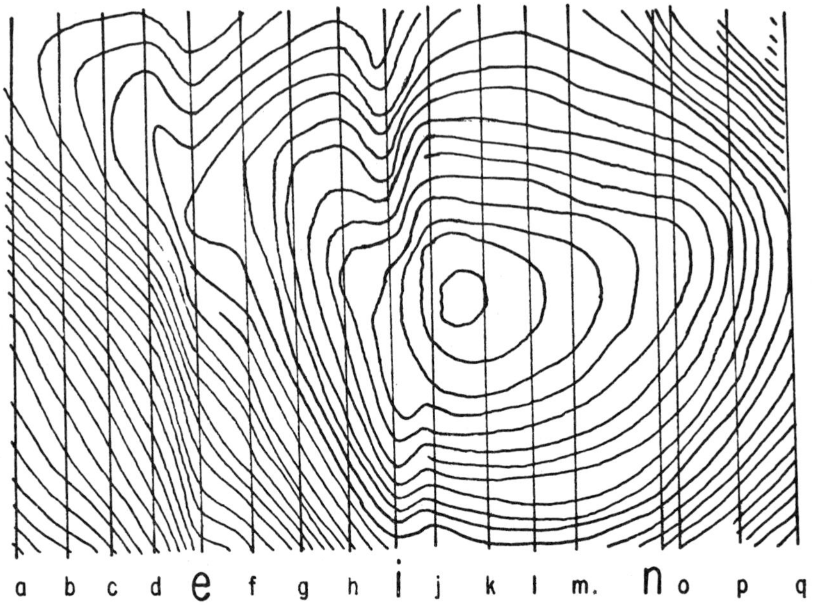

Fig. 96. **Herringbone effect.** Flight lines *e, i,* and *n* are not located in their actual positions (*e* and *i* are displaced southward, *n* is displaced eastward), thus producing fictitious anomalies. The effect on the contours is called "herringbone."

hertz: A unit of frequency. Hertz = hz = cycles per second = cps.

heuristic: Pertaining to a "trial and error" method.

HFU: Heat flow unit; 10^{-2} cal/m^2 sec. The mean heat flow of the earth is 1.2–1.5 HFU. Heat flow ranges from about 0.9 in shield areas to over 2 HFU in Cenozoic volcanic areas. Mid-ocean ridge values reach 8 HFU.

HI: Height of instrument.

hidden layer: A layer which cannot be detected by refraction methods. See Figure 97. **1.** A low-velocity layer lying beneath a high-velocity layer. **2.** A layer which is too thin or has insufficient velocity contrast to give a distinct arrival (sometimes insufficient to give a first arrival). See **blind zone.**

hi-fix: See **Decca.**

high-cut filter: A **filter** that transmits frequencies below a given **cutoff** frequency and substantially attenuates all others. The same as low-pass filter.

high-line: Small voltages induced in seismic cables or instruments by high voltage electric transmission lines. Characterized by the frequency of standard power transmission (60 or 50 hz) or its harmonics. Coupling may be by capacitive linkage, by electromagnetic induction (especially if the transmission lines are unbalanced), or by leakage currents from ground-return systems. The problem is most severe when the seismic cables and geophones are not well insulated or are close to and/or parallel to power lines. High-line eliminators commonly are a part of seismic recording equipment. These may consist of bridges which balance the voltage across the input transformers with respect to a center tap at ground potential, the assumption being that voltages with respect to ground are noise. Such bridges usually have two adjustments, one for the resistive component and another for the reactive component of the high-line-induced voltages. High-line interference is also sometimes reduced with a 60 hz (or 50 hz) notch filter which removes a narrow band of frequencies around the high-line frequency. See also **humbucking.**

high-pass filter: A filter that passes without significant attenuation frequencies above some cutoff frequency while attenuating lower frequencies. The same as *low-cut filter.*

high-resolution seismic work: HR.

high-resolution thermometer: A small-diameter fast-response thermometer for logging open or cased boreholes with a temperature resolution of 0.5°F.

high-speed layer: A layer in which the speed of wave propagation is greater than in the layer above it and which therefore carries head-wave (refraction) energy.

Hilbert transform: Given an $h(t)$ which is nonsingular at $t = 0$ and which is a causal response so that $h(t) = 0$ for $t \leq 0$, then its Fourier transform $H(f) = R(f) + j\,X(f)$ has the corresponding special property

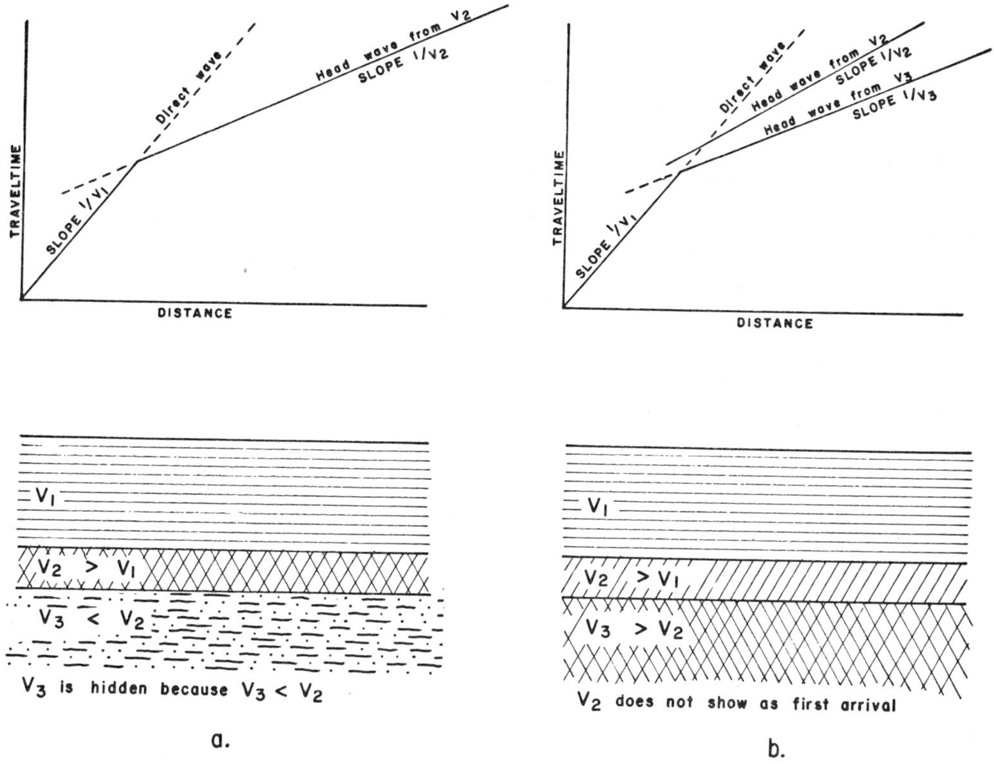

Fig. 97 **Hidden layers.** (a) A layer involving a velocity inversion does not produce a head wave. (b) The head wave from a layer without sufficient velocity contrast may not produce a first-arrival.

known as the Hilbert transform, expressed by

$$X(f) = -R(f) * \frac{1}{\pi f} = -P\frac{1}{\pi}\int_{-\infty}^{\infty}\frac{R(f')df'}{f-f'}$$

$$\text{and } R(f) = X(f) * \frac{1}{\pi f} = P\frac{1}{\pi}\int_{-\infty}^{\infty}\frac{X(f')df'}{f-f'},$$

where P denotes the Cauchy principal value at $f = f'$. This follows from $h(t) = h(t) \cdot u(t)$ which implies

$$H(f) = H(f) * \left(\frac{1}{2\pi jf} + \frac{\delta(f)}{2}\right).$$

This relation between R and X is also known as the *dispersion relation.* Similarly, given an $H(f)$ which vanishes for $f < 0$, its Fourier transform $h(t) + jx(t)$ has $h(t)$ and $x(t)$ forming a pair of Hilbert transforms. h and x have the same amplitude spectrum but differ in phase by 90 degrees. $h + jx$ is called the *analytic signal* belonging to h, and x is the *quadrature signal* corresponding to h.

Hilbert-transform technique: A technique for determining the phase of a minimum-phase function from its power spectrum. The principal use is in computing a deconvolution operator. Given the power spectrum $P(f)$ and that the wavelet is minimum phase, the wavelet's frequency-domain representation $W(f)$ is

$$W(f) = A(f)\, e^{j\phi(f)} = |P(f)|^{1/2}\, e^{j\phi(f)}.$$

The amplitude $A(f)$ is the square root of the power spectrum. The phase $\phi(f)$ must be computed. Taking the logarithm of both sides,

$$\ln W(f) = \tfrac{1}{2}\ln P(f) + j\phi(f)$$

splits the function into real and imaginary parts. To be minimum phase, the function must be analytic in the lower half plane. Then the Hilbert transform can be used to find the phase $\phi(f)$ from $\ln P(f)/2$:

$$\phi(f) = (\tfrac{1}{2})\ln P(f) * 1/(\pi f).$$

Since the amplitude and phase are known, the Fourier transform can be computed and the time-domain expression for *W(t)* is determined.

hi-line: See **high-line.**

hinge: 1. The part of a fold where the curvature is greatest. 2. In a *hinge fault* the blocks across the fault have rotated about an axis perpendicular to the fault plane. Thus the throw varies along the fault strike. Also called *scissors fault.* See Table 2 **(fault).**

hiran: High-precision shoran. Essentially ordinary shoran with minor improvements.

hodograph: The figure described by the terminus of a moving vector. See Figure 147 **(Rayleigh wave).**

hole: A mobile vacancy in the electronic valence structure of a semiconductor (an atom with less than its normal number of electrons). *P*-type semiconductors have an excess of holes. An electron from a neighboring atom can fill the hole so that in effect the hole moves to the neighboring atom. The apparent movement of holes in an electric field is equivalent to a current.

hole blow: 1. Ejection of water, mud, and sometimes rocks from the shothole as a result of the shot explosion. 2. Noise on a seismic record caused by such ejection.

hole fatigue: The effect involved where there is a delay between the detonation of a shot and the initiation of the seismic impulse from it because of changes in the shot environment (usually formation of a cavity) produced by an earlier shot in the same hole.

hole logging: Drill-hole IP or resistivity surveying. See **hole probe.**

hole noise: Noise from the shot, caused by hole blow or discharge of the gases resulting from the explosion. Hole noise may last for several seconds. It often is excessively strong on geophone groups near the shothole and attenuates rapidly with distance.

hole plug: A device used to plug a shothole after shooting. Usually the plug is pushed far enough into the hole to prevent its being dislodged and earth is shoveled over it level with the surrounding ground. Also called *bridge plug.*

hole probe: A drill-hole IP or resistivity survey in which closely spaced in-hole electrodes are used to determine the electrical properties of rock near the drill hole. Also called *electric logging, IP logging, resistivity logging,* and *hole logging.*

holography: Recording of the intensity of the interference pattern which results from the scattering of coherent radiation and a reference beam. Optical holography usually involves photographing the interference from laser light (in order to achieve a coherent source); subsequent illumination of the photographic plate allows one to "see" the photographed object in three dimensions. See **earth holography.**

homocline: A region of broadly uniform dip.

homogeneous: The same throughout. Uniformity of a physical property throughout a material.

Hooke's law: Stress is proportional to strain. See **elastic constants.**

hop: Travel of a radiowave to the ionosphere and back to earth.

horizon: 1. The surface separating two different rock layers. Where such a surface (even though not itself identified) is associated with a reflection which can be carried over a large area; a map based on the reflection event may be called a *horizon map,* sometimes contrasted to a **phantom** map. 2. A line which indicates the horizontal direction.

horizontal cylinder: A model used in calculating potential field effects. A horizontal cylinder infinitely long so that the ends of the cylinder do not produce any effects. The model is equivalent to a line whose mass is ρA per unit of length when ρ is the density contrast and A is the cross-sectional area of the cylinder.

horizontal-dipole sounding: Electromagnetic sounding configuration using a transmitting coil with its axis horizontal.

horizontal-loop method: An electromagnetic survey method wherein the planes of the transmitter and receiver coils are horizontal and the coils are carried a fixed distance apart. In-phase and quadrature components are measured by the receiver. Also called *slingram.*

horizontal mixing: 1. Horizontal stacking or **common-depth-point stacking.** 2. Ocassionally, **ground mixing.**

horizontal profiling: See **profiling.**

horizontal stacking: Common-depth-point stacking.

horst: A block raised up by normal faulting; usually long compared to its width. Compare **graben.**

hot shot: To carry out a short rush program, often without moving the field camp or crew headquarters.

hot-wire analyzer: A device used to detect hydrocarbon gases returned to the surface by the drilling mud. Basically a Wheatstone bridge, two arms of which are kept at a high temperature. Hydrocarbon gases become oxidized as they pass over one arm, which increases its temperature, changes its resistance, and unbalances the bridge. The hot wire analyzer response is sometimes plotted as a well log.

hour angle: Angular distance of a body west of the projection of a meridian onto the celestial sphere. *Local hour angle* is the angle between a body and the projection of the observer's meridian onto the celestial sphere. *Greenwich hour angle* is with respect to the projection of the Greenwich meridian. *Sidereal hour angle* is angular distance west (unlike the others which are measured east) of the vernal equinox; it is thus the supplement of right ascension.

housekeeping: Administrative, accounting, or overhead operations for a computer.

HR: High-resolution; recording of seismic frequencies (up to 500 hz, but especially 75 to 150 hz) above the normal exploration range with the objective of improving resolution, especially of shallow events.

H-type section: A **three-layer resistivity model** in which the middle layer is more conductive than the layers above or below it.

hum: Electrical interference or noise occurring at the power-line frequency or its harmonics. See **high-line.**

Humble formula: A special form of Archie's formula, applicable in granular (sucrosic) rocks:

$$F = 0.62\ \phi^{-2.15},$$

where F = formation factor and ϕ = porosity.

humbucking: An arrangement to reduce electromagnetic pickup, especially from power lines. Reverse-wound geophones have their coils wound in opposite direction from normal-wound ones so that electromagnetic-pickup voltages have opposite polarity while seismically induced voltages are in-phase (because the magnets also are reversed in polarity). Humbucking geophones incorporate both normal and reverse elements in the same case.

hundred-percent section: A seismic record section which does not utilize redundant data (CDP or similar duplication).

hunting: Successively overshooting in opposite directions; following a desired course in an oscillatory manner. In "hunting a course" one veers slightly to the right of the desired course, then slightly to the left, etc., and so is never very far off position. A characteristic of an underdamped servo-system.

Huygens principle: The concept that every point on an advancing wavefront can be regarded as the source of a secondary wave and that a later wavefront is the envelope tangent to all the secondary waves. Figure 98.

H-wave: **Hydrodynamic wave.**

hybrid scale: A scale used with laterologs which is linear with resistivity for low resistivities and linear with conductivity (the reciprocal of resistivity) for high resistivites. See Figure 112 (**laterolog**).

hydrocarbon saturation: Fraction of the pore volume filled with hydrocarbons.

hydrodynamic wave: **1.** H-wave; a seismic surface wave similar to a Rayleigh wave except that it moves in the opposite sense (that is, forward at its "up" position). **2.** Fluid waves.

hydrogen index: Hydrogen per unit volume compared with that in fresh water. **Neutron log** response depends mainly on the hydrogen index.

hydrophone: A pressure detector, a detector which is

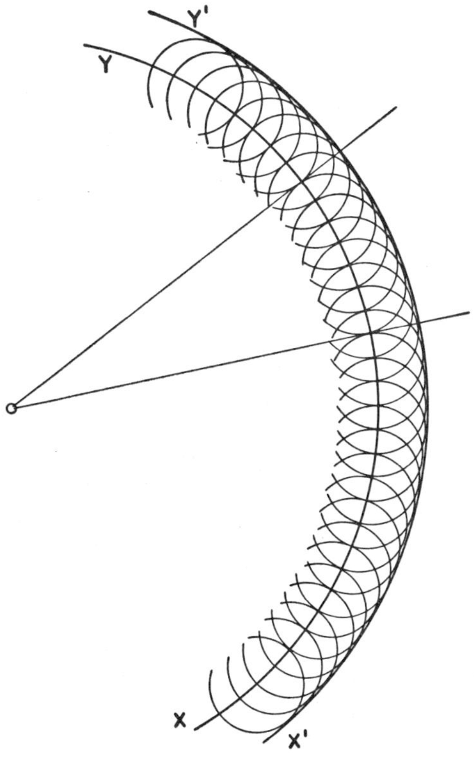

Fig. 98. **Huygens principle.** *XY* = wavefront at *T*. *X'Y'* = wavefront at $T + \Delta T$. Radii of small circles = $V\Delta T$.

sensitive to variations in pressure, as opposed to a geophone which is sensitive to motion. Used when the detector can be placed below a few feet of water as in marine or marsh work or as a well seismometer. Some hydrophones operate because of **magnetostriction.** Pairs of motion-sensitive geophones in an opposition arrangement (so that translational motion affects each equally but in opposite sense whereas pressure effects add) also have been used; but most hydrophones are **piezoelectric.** Piezoelectric hydrophones include **benders,** disc hydrophones, and **cylindrical hydrophones.** The sensing element is usually a piezoelectric ceramic material such as barium titanate, lead zirconate, or lead metaniobate. Piezoelectric hydrophones are high-impedance devices and signals from hydrophones or hydrophone arrays may be passed through preamplifiers or impedance-matching transformers before transmission through the streamer to the recording instruments.

Hydrosein: A marine seismic source utilizing the implosion which results as two plates are driven suddenly apart, creating a void between them into which water rushes. The implosion of the water generates a seismic shock wave. Western Geophysical Company trademark.

hyperbolic line of position: A **line of position** determined by measuring the difference in distance to two fixed points.

hyperbolic search: A search for coherency among traces in a comon-depth-point gather along a hyperbolic trajectory such as normal moveout should produce. The objective is a measure of the best value of normal moveout. Used in many **velocity-analysis** methods.

hypocenter: An earthquake *focus;* the point at which the first motion in an earthquake originates. The projection on the surface of the earth is the *epicenter.* Hypocenters of earthquakes often lie along the interface of plates of the earth's crust where one plate is overriding another; see **plate tectonics.**

hysteresis: **1.** A phenomen exhibited by a system or material in which the response depends nonlinearly on past responses; the failure of a property which has been changed to return to the original state after the cause of the change has been removed. **2.** Especially the effect where the magnetization produced by an applied field lags behind the field, which involves energy loss. See Figure 99. When the applied field returns to zero, the residual magnetism which is retained is called *remanent magnetization.* The magnetic field intensity required to reduce the remanent magnetization to zero is the *coercive force.*

hz: Hertz, a unit of frequency; cycles/sec.

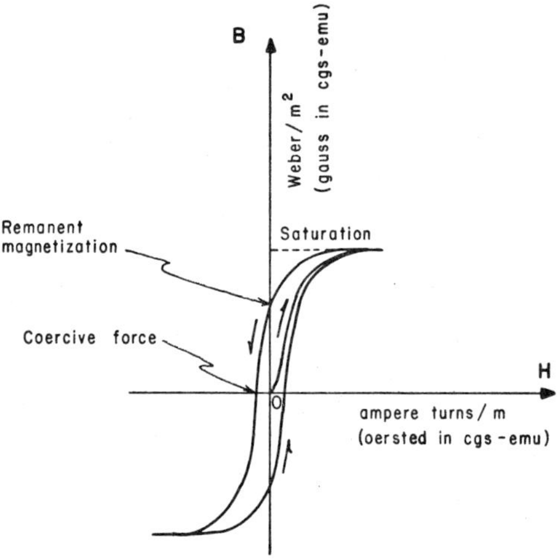

Fig. 99. **Hysteresis** loop.

i: **1.** Symbol to indicate $(-1)^{1/2}$. *j* is also used. **2.** A unit vector in the *x*-direction; as in the operator **del.**

ic: Integrated circuit; a solid-state circuit or circuit element.

ice-bridge effect: In cold areas an ice plug may form in the top of a shothole, thus confining the gases from the explosion and producing secondary shocks similar to **bubble pulses.**

ice-noise: Seismic noise resulting from expansion and contraction caused by solar heating. Also noise generated by differential movement of ice floes.

ideal body: A simple **model** (see Figure 121) such as a point mass, line mass, cylinder, sphere, vertical step, etc.

ideal polarized electrode: A metal-to-electrolyte contact at which no charge crosses the interface. Charge accumulates and the electrode interface behaves like a capacitor without leakage. No chemical reaction takes place and there is no exchange current or faradaic process. This condition is approximated when high-overvoltage, nonreactive metals are at equilibrium with an electrolyte.

IES: Induction electrical survey: A borehole log which usually includes **SP,** 16-inch normal, and deep-investigation **induction logs.** Also abbreviated IEL.

iff: If and only if.

IGY: International Geophysical Year; the period from July 1, 1957 to December 31, 1958.

II(t): Unit **boxcar.**

III(t): **Comb** or shah.

ILd: Deep investigation **induction log.**

ILm: Medium investigation **induction log.**

image: **1.** A method used in tracing raypaths through plane interfaces. An image of the source is located beneath the reflector so that straight lines from the image appear the same at the geophones as would reflected energy from the source (Figure 100). **2.** A method used in electrical modeling whereby boundaries are replaced by source images so that the potential and current distribution in the zone of interest is unchanged. The objective is to facilitate computation of the potential distribution.

imaginary: The part of a complex number that involves the factor $(-1)^{1/2}$. The **out-of-phase** component. In impedance, the reactive component.

imaging: See **remote sensing.**

impedance: **1.** The apparent resistance to the flow of alternating current, analagous to actual resistance in a dc circuit. The impedance is (in general) complex, of magnitude $|Z|$ with a phase angle θ. These can be expressed in terms of the resistance *R* (in ohms), the inductive reactance $X_i = 2\pi fL$ and the capacitive reactance $X_c = 1/2\pi fC$:

$$Z = [R^2 + (X_i - X_c)^2]^{1/2} \text{ ohms}$$

$$\theta = \tan^{-1}\left(\frac{X_i - X_c}{R}\right).$$

f is frequency in hertz, *L* is inductance in henrys, and *C* is capacitance in farads. **2.** See **acoustic impedance. 3.** In magnetotelluric exploration, if E_x is the electric-field component in an arbitrary horizontal direction and H_y is the magnetic-field associated with it (in the orthogonal horizontal direction), the *wave impedance* (or *field impedance* or *surface impedance*) Z_s is:

$$Z_s = E_x/H_y\,.$$

The surface impedance is a complex number because of phase differences between E_x and H_y. Measurements of surface impedance versus frequency can be interpreted in terms of the electrical properties of the subsurface.

impedance matching: Making the impedances of two connecting circuits be complex conjugates of each other. Impedance matching gives maximum transfer of power and other benefits.

implosion: Collapse into a region of very low pressure. The creation of such a region under water (as with the Hydrosein or Flexichoc) causes the water to rush in with great force and the collision of the inrushing water on itself generates a seismic shock wave. The outrush of water propelled by a bubble of high-pressure gas (as from an underwater explosion or from an air gun) leaves behind a region at the vapor pressure of the water, into which the water subsequently collapses, causing the **bubble effect.**

impulse: **1.** The limit of a pulse of unit area as its width approches zero and its height approaches infinity. Also called *Dirac function* and *delta function* and symbolized by $\delta(t)$. It has a value at only one instant and unit energy content:

$$\delta(t-a) = 0, \text{ if } t \neq a \text{ and}$$

$$\int_{-\infty}^{\infty} \delta(t)\, dt = 1.$$

(The impulse is sometimes called the *unit impulse* because of the unity in the above equation.) The essential

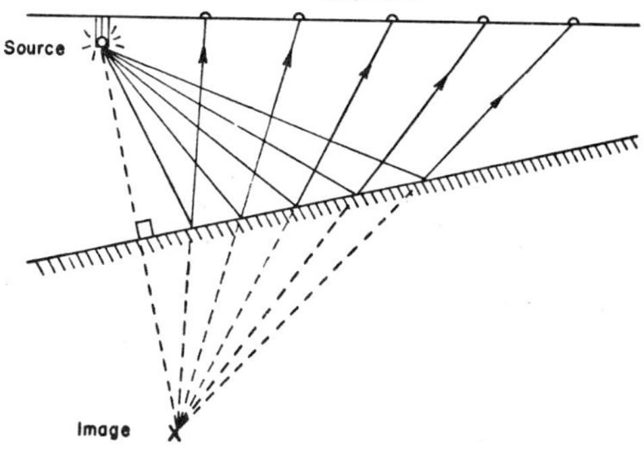

Fig.100. **Image.** Reflected rays reaching the geophones appear the same as direct rays from the image.

characteristic of an impulse (which is sometimes used as a definition) is expressed by

$$\int_{-\infty}^{\infty} f(t)\ \delta(t-a)\ dt = f(a).$$

An impulse contains all frequencies in equal proportions at zero phase. In digital form an impulse δ_t is:

$$\delta_{t-a} = \begin{cases} 1, & \text{if } t = a \\ 0, & \text{if } t \neq a. \end{cases}$$

See **impulse response. 2.** A very sharp wave of very short duration, such as might result from an explosion.

impulse blaster: A device which fires an electrical blasting cap when it senses a shock wave. Used to achieve a **directional charge** by firing a second charge when the shock from the initial explosion reaches it so that the charge effects add up in-phase.

impulse response: The response of a system to an **impulse** input. Also called *memory function.* The impulse response characterizes a linear system and contains the same information as the frequency-domain **transfer function,** which is its Fourier transform. The output of a linear system is given by the input to the system convolved with the system's impulse response.

incident angle: The angle which a raypath makes with a perpendicular to an interface, which is the same as that which a wavefront makes with the interface in isotropic media.

inclination: 1. The angle between a line's direction and the horizontal; e.g., magnetic inclination, the angle at which magnetic lines of force dip. **2.** The dip of a plane (bed, fault, or other tabular body) measured from the horizontal. The attitude of the plane may be characterized by the direction of a line normal to it and the inclination of the plane by the angle between its normal line and the vertical.

inclinometer: 1. A device for measuring hole inclination and azimuth. See **directional survey. 2.** A device for measuring the pitch and roll of a ship. Usually either pendulous or gyroscopic. **3.** A surveying instrument which measures the angle between the horizontal and the line of sight.

incompetent: See **folding.**

independent: Not expressable in terms of each other. Two quantities are statistically independent if they possess a joint distribution such that knowledge of one does not give information about the distribution of the other.

index: A symbol or number to identify one element out of a set, such as an element in a matrix.

index factor: A constant which when multiplied by certain measurements made on potential-field anomalies gives an estimate as to the depth to the anomalous mass (sometimes the maximum depth at which the anomalous mass could be located). Used in magnetic and gravity interpretation. See **depth rule.**

induced-current dipole moment per unit volume: A vector parameter describing induced-polarization properties **P** as a function of chargeability *(M)* and current density **(J)**:

$$\mathbf{P} = -M\mathbf{J}.$$

induced polarization: 1. Usually abbreviated IP. An exploration method involving measurement of the slow decay of voltage in the ground following the cessation of an excitation current pulse (time-domain method) or low-frequency (below 100 hz) variations of earth impedance (frequency-domain method). Also known as the *overvoltage* method. Refers particularly to electrode polarization (overvoltage) and **membrane polarization** of the earth. Also called induced potential, overvoltage, or interfacial polarization. Various electrode configurations are used; see Figures 13 and 14 **(array). 2.** The production of a double layer of charge at mineral interfaces or of changes in such double layers as a result of applied electric or magnetic fluids.

inductance: The capability of an electric circuit or of an adjacent circuit to induce an electromotive force (emf) within the circuit. Proportional to the ratio of total flux linking the circuit to current flow. Measured in henrys.

induction: 1. Causing a voltage by varying the magnetic flux. According to Faraday's Law of Induction,

$$e = -\frac{d\phi}{dt} = -L\frac{di}{dt} \quad ,$$

where e = voltage, $d\phi/dt$ = time rate of change of magnetic flux in webers/sec, L = inductance in henrys, and di/dt = time rate of change of current in ampere/sec. The vector directions are shown in Figure 101. **2.** See **magnetic induction** and **induced polarization.**

induction log: An electrical conductivity/resistivity well log based on electromagnetic-induction principles. See Figure 102. A high-frequency alternating current of constant intensity induces current flow in the formation. This *Foucault current* flowing in the formation *ground loop* causes an alternating magnetic field which produces a current in a receiving coil. The receiving-coil current is nearly proportional to the conductivity of the formation. Induction sondes may have several transmitting and receiving coils to produce a highly focused log. An induction log can be recorded where the borehole fluid is conductive or nonconductive, as in oil-base muds or gas. A dual induction log measures different depths of penetration (**ILd** and **ILm,** deep and medium).

induction method: An electromagnetic method in which eddy currents are induced in the earth by a time-varying magnetic field. Usually the term is applied only to those electromagnetic methods for which the receiver is in the **near-field** or induction zone of the transmitter.

induction number: A dimensionless parameter which determines the response of an electromagnetic system. It is equal to the product of the wavenumber and a characteristic length of the system. As examples, the response of a two-loop system on a homogeneous earth is given by $(\sigma\mu\omega)^{1/2} R$ and on a thin layer by $(\sigma\mu\omega t R)^{1/2}$, where σ is the conductivity, μ the magnetic permeability, ω the angular frequency, t the thickness

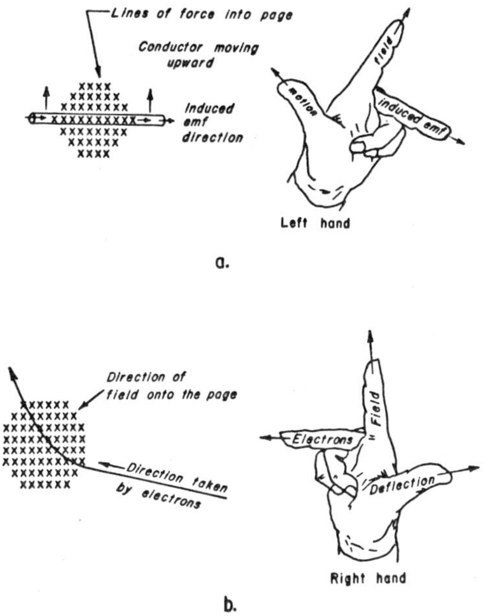

Fig.101. **Induction** (a) and motor rules (b).

of the sheet, and *R* the source-detector loop separation. A full-scale electromagnetic system and a scale-model analog have the same response if their induction numbers are equal.

inductivity: Magnetic permeability; the three-dimensional inductance of a material. Free space (and nonmagnetic rocks) has an inductivity of 1.257×10^{-6} henrys per meter.

inertial navigation: A dead-reckoning method of determining position in which accelerations are measured with very sensitive accelerometers and doubly integrated to give position. Inertial systems tend to accumulate error and hence need to be updated frequently. Used in conjunction with other positioning systems rather than on a stand-alone basis.

infinite electrode: A (usually) fixed, remote electrode which is a great distance from the successively moved field-surveying electrodes. Ideally it is located far enough from the measurement electrodes to have negligible effect on the measurements.

inflection-tangent-intersection method: A magnetic interpretation method which involves making a number of measurements of shape features on a profile across a magnetic anomaly and making the best fit of these measurements to theoretical values for vertical dike. Both distance measurements and amplitude measurements may be made as shown in Figure 103. These values are plotted on a logarithmic scale which is slid along a graph of the theoretical values until the best match is achieved. The fit permits one to determine depth, width, magnetic inclination, and magnetization in a plane perpendicular to a vertical dike. Also called *ITI method* and *Naudi method.* See Naudi, 1970, Une methode d'analyse fine des Profils Aeromagnetiques: Geophysical Prospecting, v. 18, p. 56-63.

information retrieval: A system for storing and searching large quantities of data for wanted information.

infrared: The part of the electromagnetic spectrum with wavelengths between those of visible light and 10^{-3}m. See Figure 68 (**electromagnetic spectrum**). *Thermal infrared* implies wavelengths longer than those of normal heat radiation.

in-hole IP method: Technique for measuring near-hole IP and resistivity properties, using at least one potential and one current electrode in the hole. Compare **down-the-hole IP method.**

inhomogeneity: Lack of spatial uniformity of a physical property. Also called *heterogeneity.*

initial blanking: **Mute.**

initialize: To set the initial value. Quantities being altered in iterative processing (as might happen in a Fortran "DO loop") must have a starting value assigned them by an "initializing" statement before the loop (as in Figure 79, **flow chart**). Some systems assume initial values of zero and so do not require initializing statements.

initial suppression: Attenuation at the beginning of a seismic recording to prevent high apparent noise on traces prior to the first breaks, or to prevent overload from the high-amplitude first-break energy. Also called *presuppression.*

injection: Insertion of orbit information into a navigation satellite. Satellites are observed by fixed tracking stations and the orbit data are updated every 12-16 hours.

in-line: Along a survey line. An in-line geophone array has all the geophones located on the seismic line. An in-line electromagnetic configuration has the source and receiver on the traverse line.

in-line offset: A **spread** which is shot from a shotpoint which is separated (**offset**) from the nearest active point on the spread by an appreciable distance (more than a few hundred feet) along the line of the spread. See Figure 171 (**spread**).

in-phase: Electrical signal with the same phase angle as that of the exciting signal.

input: **1.** Data to be processed. **2.** To transfer data into a computer. **3.** The current, voltage, or driving force applied to a circuit, system, or device. **4.** The terminals where the input to a system is applied.

input impedance: The impedance across the input terminals of an electrical circuit. Where the input impedance of a measuring device is much higher (say, by 100 times) than the impedance being measured, the measurement will not be altered greatly by the presence of the device.

Input system: A time-domain electromagnetic-survey system in which measurements are made during the off-periods between source pulses. Barringer trademark.

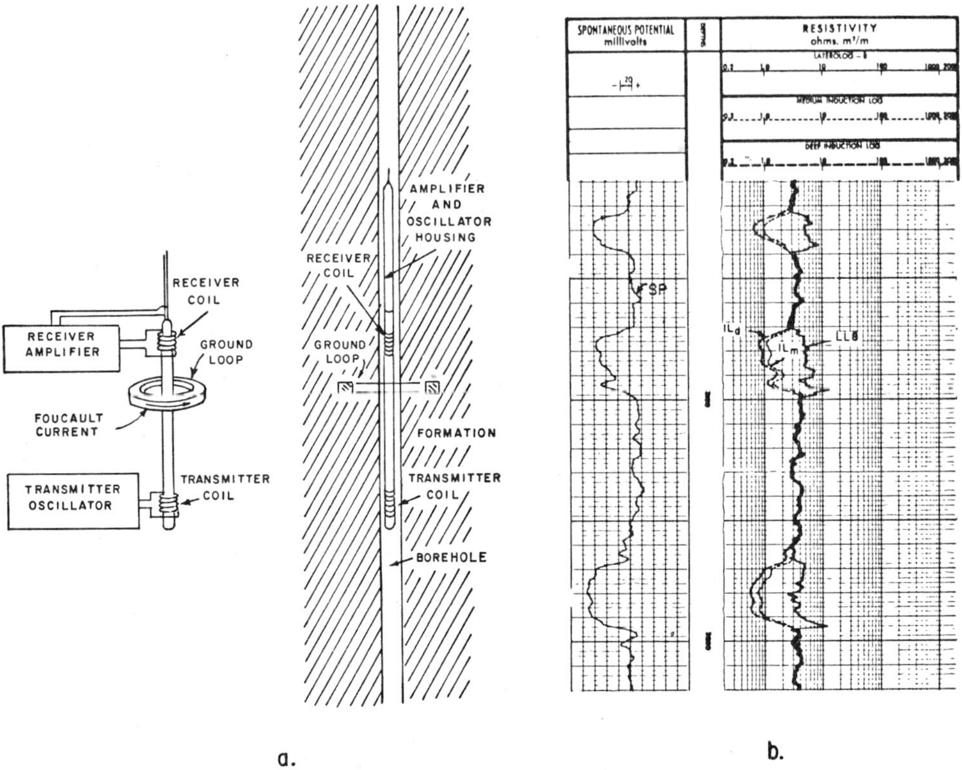

Fig.102. **Induction log.** (a) Schematic two-coil induction logging sonde. (b) Dual-induction log. IL_d is the deep-induction log, IL_m the medium-induction log. *LL8* is a Laterolog-8 and SP is an SP (spontaneous potential) log run at the same time. (Courtesy Schlumberger.)

in situ: Material in its original location. Used in connection with measurements of properties of the material which do not involve altering or moving the material. For example, in-situ velocity measurements would be made on rock in place in contrast to measurements on a sample of the rock in a laboratory.

insulator: A nonconductor of electricity.

integrated circuit: A single solid-state electrical-circuit element replacing sometimes a number of individual components. Complete amplifiers, gates, oscillators, and other modules may be made as integrated circuits. Abbreviated IC.

integrated navigation system: A combination of **positioning** systems in a synergetic manner. Specifically, the combination of **satellite navigation** with **doppler sonar** and **gyrocompass** (and other subsystems) into a system. The doppler sonar and gyrocompass provide location information between satellite fixes and also provide the velocity information required for an accurate satellite fix. The satellite fixes are used to give the reference locations and remove cumulative-error effects in the doppler sonar.

integration of chargeability: Measurement of the area under an IP decay curve by integrating the decay voltage with time. When normalized by dividing by the primary voltage, this measurement is one definition of *chargeability.* The areas under several successive decay curves can be averaged to improve the precision of the measurement.

intensity: **1.** The rate of flow of seismic energy through a unit area perpendicular to the direction of the seismic wave. **2.** The degree of shaking at a specified place because of an earthquake. The rating uses the descriptive Mercalli scale with grades from I to XII. See Table 7 **(Mercalli scale)**. Compare **magnitude,** which is independent of location. **3.** Often refers to **electric field** or magnetic **field intensity.**

intensity of magnetization: Magnetic moment per unit volume (occasionally, per unit mass).

interactive: Permitting decisions to be made in dialog

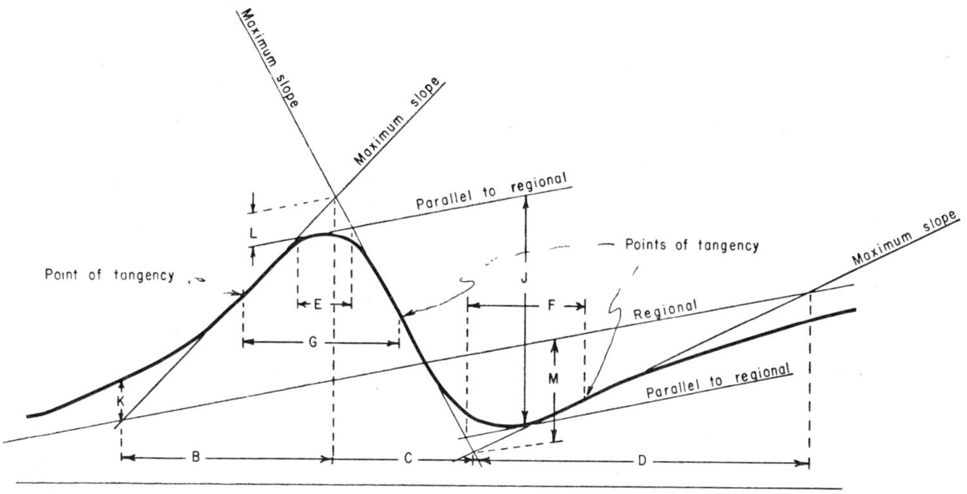

Fig.103. **Inflection-tangent-intersection method.** The letters indicate the various parameters which are measured. (From Naudy, Geophysical Prospecting: v.18, no. 1, p. 59.)

back and forth during processing. The selection of parameter values and other decisions may be based on inspecting intermediate results. Often involves display on a TV tube of data in the computer *(computer graphics).* Also permits rapid evaluation of trial choices of parameters. As opposed to processing where decisions are made before the processing begins.

intercept time: The time obtained by extrapolating the refraction alignment on a refraction time-distance *(T-X)* plot back to zero offset. See Figure 181 **(time-distance curve).**

interface: 1. The common surface separating two different media in contact. **2.** The contact or connecting element between two computing machines or components by means of which information is passed between the two. Interface devices are used to reformat data (such as convert serial data bits to parallel bits or convert from analog to digital form or vice-versa), to hold data until it is needed (buffer function), and sometimes to do simple operations like summing (counting). **3.** To make the output of one device acceptable as input of another device.

interfacial polarization: See **induced polarization.**

interference: 1. The superposition of two or more waveforms. Interference is *constructive* where peaks add to peaks, *destructive* where peaks cancel troughs. **2.** Signals from another source (e.g., atmospheric static) which tend to obscure a desired signal.

interlocking: 1. Containing energy which has traversed the same raypath, usually in opposite directions. Two seismic records are interlocking if the end (or some) geophone group of the one record occupies the shotpoint location of the interlocking record, and vice versa. Reflection events on interlocking traces should have the same arrival time. **2.** Interdependent controlling of several units whose functions has to be coordinated for proper (or safe) operation.

intermediate storage: Use of magnetic tape, discs, or drums for temporary storage of seismic data during digital or analog processing.

International Gravity Formula: The *theoretical gravity* at the latitude ϕ is 978,049 $(1 + 0.0052884 \sin^2\phi - 0.0000059 \sin^2 2\phi)$ mgal. [A better value may be 978,031.8 $(1 + 0.0053024 \sin^2\phi - 0.0000058 \sin^2 2\phi)$ mgal.] See **latitude correction.**

interpreter: 1. One who determines the geological significance of geophysical data. **2.** A machine which reads coded information (such as punched cards) and prints out the translation. **3.** A computing-machine routine which translates from one computer language into machine language at execution time, as with programs written in Fortran. Compare **compiler.**

interrupt: A signal which advises that some operation needs to be done. Interrupts allow peripheral on-line equipment to advise a computer's control that they have data to be read or require instructions or data.

intersection: Conjunction.

interstitial water: Water in the interstices or pore spaces in a formation.

interval time: The time difference between two reflection events. Changes in interval time are used as indications of reefing and other stratigraphic and structural variations. Interval time may vary because of changes in velocity from one material to another or changes in thickness such as produced by differential compaction.

interval transit-time: The traveltime of a compressional sonic (seismic) wave over a unit distance, hence proportional to the reciprocal of P-wave velocity. Measured in the **sonic log**, usually in microseconds per foot.

interval velocity: Seismic-wave **velocity** measured over a depth interval. **1.** In sonic log determinations the interval may be 1 to 3 ft, in well shooting it may be 1000 ft or more. Usually refers to compressional velocity and usually implies measurements perpendicular to the bedding. **2.** Often used for velocity calculated by the **Dix Formula** from velocities measured from normal moveout.

intrabasement anomaly: A local anomaly caused by a unit wholly within the basement complex.

intrinsic conduction: Conduction due to major components of the material, as opposed to conduction due to impurities or imperfections. At high temperatures intrinsic conduction dominates other conduction modes.

intrinsic IP: The true induced polarization of a material, in contrast with its apparent value.

intrinsic thermal-neutron decay-time: The true decay-time of the formation as opposed to the measured decay-time which is subject to hole and diffusion effects. See **neutron-lifetime log.**

invaded zone: The portion about a wellbore into which drilling fluid has penetrated, displacing some of the formation fluids. Invasion takes place in porous, permeable zones because the pressure of the mud is kept greater than that of the formation fluids. See Figure 104. A *mud cake* builds on the formation wall, limiting further flow of mud fluid (*filtrate*) into the formation. Directly behind the mud cake is a *flushed zone* from which almost all of the formation water and most of the hydrocarbons have been displaced by filtrate. The invasion process alters the distribution of resistivities and other properties and consequently the value which logs read. The *depth of invasion* is the equivalent depth in an idealized model rather than the maximum depth reached by filtrate. In oil-bearing intervals, the filtrate may push a bank of formation water ahead of it to produce a relatively low-resistivity **annulus** which is especially important with deep-investigation induction logs.

inverse filter: A filter with characteristics complementary to another filter so that when used in series with the other filter no frequency-selective filtering occurs (except for overall time delay). See also **deconvolution.**

inverse Fourier transform: See **Fourier transform.**

inverse matrix: The inverse of a square **matrix** A is the matrix which yields the identity matrix when multiplied by A: $A^{-1}A = I$. The inverse matrix can be found by dividing the adjoint (or adjunct) of A by the determinant of A. A nonsquare matrix B can be squared up by multiplying by its transpose and the inverse can then be found:

$$[B^{T} B]^{-1} B^{T} B = I.$$

inverse problem: The problem of gaining knowledge of the physical features of a disturbing body by making observations of its effects (e.g., fields and potentials); finding the model from observed data. In contrast to the *direct, forward,* or *normal* problem, calculating what would have been observed from a given model. See Figure 120 **(model).**

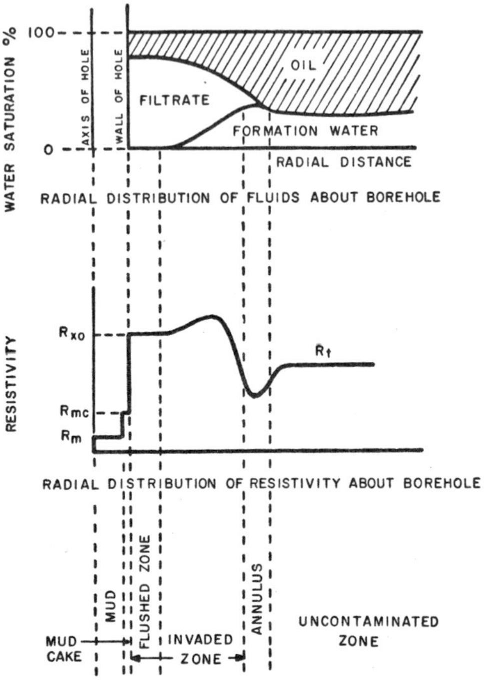

Fig. 104. **Invaded zone.** (Courtesy Schlumberger.)

inverse square law. 1. A potential field surrounding a unit element has a magnitude inversely as the square of the distance from the element.

$$\text{Field} = km/r^2.$$

Applies to a gravitational field where the element is mass and k is the gravitational constant, to a magnetic field and where the element m is a magnetic pole (k is 10^{-7} webers/ampere meter, where pole strength m is in ampere meters) and applies to an electrostatic field, where the element m is an electrical charge (k is 9×10^9 newtons/coulomb, where m is in coulombs). Also called *Coulomb's law.* **2.** The energy density of a seismic body wave varies inversely as the square of the distance from a point source. Also called *spherical divergence.* **3.** The intensity of electromagnetic energy (light or radio) varies inversely as the square of the distance from a source (neglecting reflection effects if a reflector is present).

I/O: Input/output; refers to the input or output functions of a computing machine.

ion exchange: The property of some minerals (particularly clays) which enable them to absorb certain anions and cations and retain them in a state whereby they can be exchanged for other anions and cations in solution. Ion exchange is a diffusion process and its rate depends on ion mobility.

ion mobility: Ease of movement of ions in an electric field, measured by the ratio of ion velocity to electric-field strength.

ionosphere: That part of the earth's atmosphere which includes several layers of ionized gas at a height of 50 to 250 miles, which bend certain radiowaves back toward the earth. See **D-layer, E-layer, F-layer,** and **refraction correction.**

IP: **Induced-polarization method.**

IP susceptibility: A term used as a measure of induced polarization, implying an analogy with other types of polarization such as induced magnetic effects.

irreducible water saturation: The fraction of the pore volume occupied by water in a reservoir at maximum hydrocarbon saturation. It represents water which has not been displaced by hydrocarbons because it has been trapped by adhering to rock surfaces, trapped in small pore spaces and narrow interstices, etc. Irreducible water saturation is an equilibrium situation. It differs from *residual water saturation,* the value measured by core analysis, because of filtrate invasion and the gas expansion that occurs when the core is removed from the bottom of the hole to the surface.

IR survey: A **remote sensing** method using either infrared radiation reflected from the surface after being beamed downward from the aircraft **(thermal infrared)** or occasionally natural infrared radiation. Infrared surveys are usually flown at night to reduce the natural radiation and solar effects.

ISD: **Intermediate storage** device.

iso-: A prefix meaning "equal," used in conjunction with other words to denote contour lines through points on maps at which the measured value is the same. Examples include the following: **a.** *isoanomaly:* A line on a map connecting points of equal anomaly, used especially for maps showing magnetic or gravity anomalies. Also spelled "isanomaly." **b.** *isobar:* Line joining places of equal barometric pressure. **c.** *isobath:* Line of equal water depth. **d.** *isochore:* Line of equal interval between two beds or two seismic events. Differs from *isopach* in that it may express variations in many units and the effects of unconformities, whereas isopach implies variations within a single unit. Term now falling into disuse. **e.** *isochron:* Line of equal reflection time, or equal time difference, or equal delay-time; isotime. **f.** *isocline:* Line of equal magnetic inclination. **g.** *isogal:* Line of equal gravity anomaly. **h.** *isogam:* Line of equal magnetic intensity. **i.** *isogon:* Line of equal magnetic declination. **j.** *isohyet:* Line of equal amounts of rainfall. **k.** *iso-J:* Line of equal values of the Jacobian or relative ellipse area used in telluric surveying. **l.** isopach: See seperate entry. **m.** *isopor:* Line of equal secular change, such as equal annual change of isogonic or isoclinic lines. **n.** *isorad:* Line of equal radioactivity or lines of equal gamma radiation. **o.** *isoseismal:* Line between regions of different intensity for a given earthquake. Such a line might indicate the boundary between zones IV and V on the Mercalli scale, for example. **p.** *isotherm:* Line of equal temperature. **q.** isotime: See separate entry.

isopach: **1.** A contour which denotes points of equal thickness of a rock type, a formation, a group of formations, etc. Usually used for *isochore.* **2.** A contour denoting equal vertical distance, thus not necessarily corrected for the dip of the bedding. **3.** Sometimes incorrectly used for a contour denoting equal time difference between reflections. Compare **isotime.**

isostasy: The gravitational balance of large portions of the earth's crust as though they were floating on a denser underlying layer (the asthenosphere); it thus accounts for major topography. The Pratt hypothesis assumes density variations so that areas of less-dense crust rise topographically above areas of more-dense crust (see Figure 105). The Hayford modification of the Pratt hypothesis requires that the pressure be balanced at the "depth of compensation." The Airy hypothesis

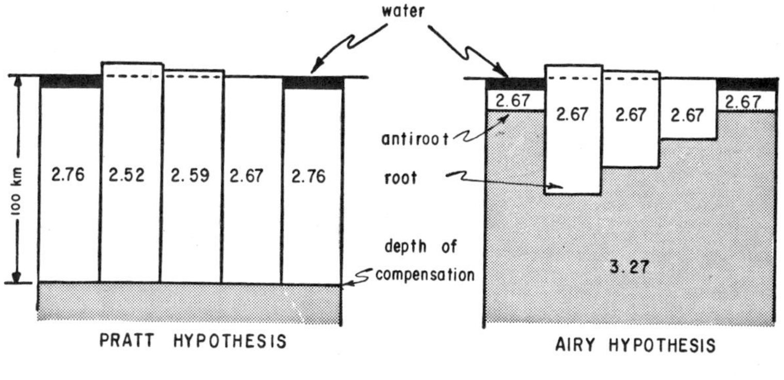

Fig. 105. **Isostasy** concepts.

varies the thickness of crustal blocks of constant density so that the thicker parts ride higher; thus mountainous areas are compensated by deep "roots" and deep ocean basins by "antiroots." Ocean basins thus may have antiroots at 6-8 km with the roots of mountains extending to 50-60 km. The Heiskanen modification permits density to vary but compensates 2/3 of the topography with roots. The Vening Meinesz hypothesis allows some of the balance to be accommodated laterally by the surrounding region rather than achieve balance in the vertical direction only; the "radius of regionality" is of the order of 200 km.

isostatic correction: A correction to gravity data to compensate for lateral density or thickness variations between large blocks of the earth's crust. The correction must assume a model and is made from elevation data and waterdepth data using zone charts. See Heiskanen and Vening Meinesz, 1958, The Earth and Its Gravity Field: New York, McGraw-Hill Book Co., Inc., p. 159-170.

isotime: **1.** The time interval between two reflections, mapped in order to locate stratigraphic changes in the interval, reef buildup, variation in salt thickness, etc. The isotime may vary because the velocity or the thickness varies or both. **2.** Contours of reflection time, time intervals, etc.; *isochron.*

isotropic: Having the same physical properties regardless of the direction in which they are measured. Strictly, applies only to an arbitrarily small neighborhood surrounding a point and to single properties, such as "isotropic electrically." Compare **anisotropic.**

isovelocity surfaces: Surfaces at which the seismic velocity is constant. In the absence of structural uplift and lateral variations of lithology, isovelocity surfaces are apt to be horizontal planes. But where structural uplift occurs, isovelocity surfaces tend to follow structure but with less relief than the structure. Changes in velocity bend seismic rays and hence alter apparent structure. Curvature of velocity surfaces can act toward seismic rays like lenses toward light rays, distorting both structure and measurements such as velocity calculated from **normal moveout**. Knowledge of how the velocity varies is essential for correct interpretation.

iterative: **1.** A procedure which repeats until some condition is satisfied. See **loop.** **2.** Processing by successive approximations, each based on the preceding, in such a way as to converge onto the desired solution, such as used in computing a satellite-navigation fix.

ITI method: **Inflection-tangent-intersection method.**

J: **Jacobian.**

j: **1.** Symbol to indicate $(-1)^{1/2}$. *i* is also used. **2.** Multiplication by *j* rotates a complex vector by 90 degrees. **3.** A unit vector in the y-direction, as in the operator del.

jack: The receptacle into which an electrical plug fits.

Jacobian: For a set of transform equations $y_i = y_i(x_1, x_2, \ldots, x_n)$, the Jacobian matrix **J** is:

$$J = \left\| J_{ik} \right\| = \left\| \frac{\partial y_i}{\partial x_k} \right\|.$$

Vanishing of the determinant of the Jacobian shows that relations are not independent. For the 2 × 2 matrix,

$$\begin{Vmatrix} a & b \\ c & d \end{Vmatrix},$$

independence is shown by $ad - bc \neq 0$.

Jacob's staff: A pointed stick about 5 ft long used to support a surveyor's compass.

JCL: **Job control language.**

Jeffreys-Bullen curves: Relationships between the arrival time of various modes of seismic waves and the distance (angular distance) from the epicenter of the earthquake which generated them. See Figure 106 and **wave notation** for explanation of the labeling of the curves.

jet: To drill a shothole by pumping water (or mud) down the drillstem or drill casing, where the water flow rather than the grinding action of a bit is the main force for removal of the material.

job-control language: A method of telling a computer what jobs to do, in what order to do them, what tapes to use, and what to do with the results.

Johnson noise: Noise resulting from the random thermal energy of conduction electrons. Also called *thermal noise.* For an electrical element, the mean square voltage due to Johnson noise is proportional to the absolute temperature *K,* the bandwidth *B* (in hertz), and the resistance *R* (in ohms):

$$\text{rms noise voltage} = 7.4 \times 10^{-12}\,(KBR)^{1/2}$$

JOIDES: Joint Oceanographic Institutions for Deep Earth Sampling. A program to obtain cores of the sediments in the deep oceans. Holes drilled from the ship *Glomar Challenger* did much to prove **plate tectonics** and hence has had tremendous impact on geology and geophysics.

joint probability: The probability of simultaneous occurrence of values of two or more quantities.

jug: **Geophone.**

jug hustler: One who lays out and picks up the seismic spread and geophones.

jug line: **1.** Cable connecting geophones to instruments. **2.** The **spread.**

jug planter: **Geophone planter.**

Julian day: The number of a day within a calendar year. Referred to Greenwich.

jump a leg: To miscorrelate one or more cycles. See also **leg.**

jump correlation: See **correlation.**

jumper: **1.** A relatively short electrical connection. **2.** A connection different from the ordinary; for example, an extra cable to connect with the main spread cable when it is otherwise difficult to reach because of access problems. **3.** A temporary electrical connection used to by-pass part of a circuit. **4.** A seismic cable without takeouts for geophone connections.

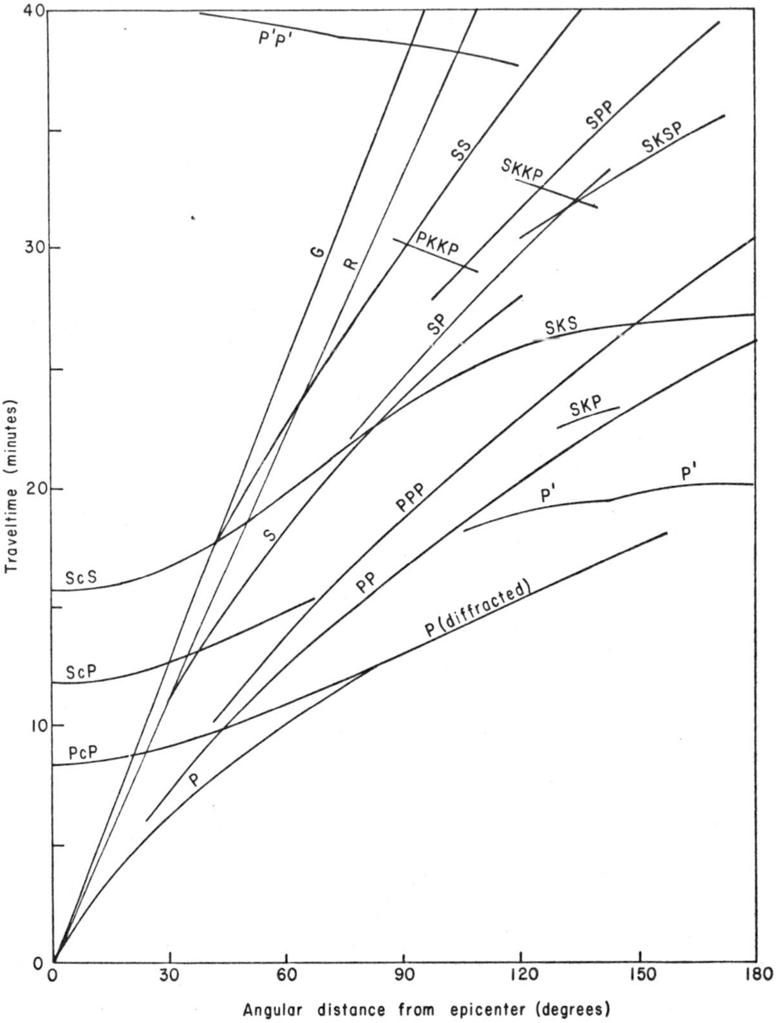

Fig. 106. **Jeffreys-Bullen traveltime curves** for some waves from a shallow earthquake. (From Seismological tables, British Assn. for Advancement of Science, London, 1940.)

k: **1.** Wavenumber; the number of waves per unit distance; the reciprocal of wavelength.

$$k = 1/\lambda = f/v,$$

where λ = wavelength, f = frequency, and v = velocity. (Some authors use $k = 2\pi/\lambda$ instead of the above.) **2.** Seismic usage often implies "apparent" wavenumber, λ being the apparent wavelength and v the apparent velocity. Because apparent velocity is used, k for a particular energy train varies with the angle between the raypath and the line of measure (the line of the spread, usually). **3. Resistivity-contrast factor. 4.** A unit vector in the z-direction, as in the operator del.

Kalman filter: A recursive filtering scheme applicable to nonsteady-state linear systems. The system is described by a model of differential equations involving orthogonal state variables. The errors in each measurement and in the model description are assumed to be independent and Gaussian. The filter estimates the error based on prior measurements and incorporates each new measurement in future error estimates. Kalman filtering is used in real-time reduction of integrated satellite-navigation data and in some seismic-filtering schemes. See Bayless and Brigham, 1970, Application of the Kalman Filter: Geophysics, v. 35, p. 2-23.

K-band: Radar band (10.0 - 36.0 Ghz) which is sensitive to vegetation, used in remote sensing and doppler radar.

kelly: The unit at the top of the **drill** pipe which transmits the rotary motion of the rotary table to the drill pipe. The kelly is usually a square, grooved, or hexagonal member supported at the upper end by the swivel and secured into and supporting the drill pipe at the lower end.

Kepler coordinates: The quantities with respect to the celestial sphere which describe the elliptical orbit of a satellite. Navigation satellites broadcast their Kepler coordinates and ephemeral information which allows the calculation of their location at any given instant (see Figure 107).

Kepler's laws: *First law:* Every planet follows an elliptical path with the sun at one focus. *Second law:* A line from the center of the sun to the center of a planet sweeps the same area in the same time; hence the planet travels fastest when closest to the sun. *Third law:* The square of the period of revolution of a planet is proportional to the cube of its mean distance from the sun.

kernel function: 1. A mathematical function of resistivity and depth which can be calculated from apparent-resistivity data, from which one tries to derive the resistivity stratification. Koefoed (1965, Direct Methods of Interpreting Resistivity Observations: Geophysical Prospecting, v. 13, p. 568-591) derives the kernel function from Schlumberger-configuration data and Paul (1968, Direct Interpretation of Resistivity Profiles for Wenner Electrode configuration: Geophysical Prospecting, v. 16, p. 159-162) from Wenner-configuration data. The derivation of resistivity stratification from the kernal is shown by Pekeris (1940, Direct Method of Interpretation in Resistivity Prospecting: Geophysics, v. 5, p. 31-46) and Vozoff (1956, Numerical Resistivity Analysis–Horizontal Layers: Geophysics, v. 23, p. 536-556). The electric potential V at the surface of a horizontally layered earth due to a dc point-source which is also located at the surface was expressed by Stefanesco in 1930 as a Hankel integral,

$$V = \frac{C}{r} + 2C\int_{-\infty}^{\infty} K(\lambda)\, J_o(\lambda r)\, d\lambda,$$

where r is the distance from the point-source to the observation point, $J_o(\lambda r)$ is a Bessel function, λ is a phantom variable of integration, C is a constant, and $K(\lambda)$ is the kernel function. Also called *Stefanesco function.* **2.** For any linear transform

$$F(\xi) = \int K(X,\xi)\, f(x)\, dx, \quad K(X,\xi) \text{ is the "kernel."}$$

key bed: A reflection with sufficient distinguishing characteristics to make it easily identifiable in correlations.

keyseat: Where a borehole changes direction so that the drilling pipe "feels" the side of the borehole. The effect is to wear a groove in the borehole wall. The drill pipe may fit in this groove but the groove may be smaller than the couplings between units of drill pipe, making it difficult to pull the drill pipe.

kick: 1. Break or *onset.* **2.** Loss of drilling fluid because fluid is penetrating a porous formation. **3.** Gas entering the drilling fluid from a porous formation.

kill: To make a trace equal to zero.

K-index: A measure of the average intensity of magnetic disturbances in time, such as magnetic storms, but excluding diurnal and lunar time variations.

Kirchoff diffraction equation: An expression of the wave amplitude at a point P which is several wavelengths λ from a diffracting aperture in the plane B. Referring to Figure 108, if α and β are the angles between the normal to B and the rays from Q and P to B, the lengths of the rays being r and s, then the wave amplitude u at P can be expressed as an integral over the aperture area, $d\sigma$:

$$u = -\frac{i}{2\lambda}\iint \frac{1}{|rs|} e^{i(r+s)/\lambda} (\cos\alpha - \cos\beta)\, d\sigma.$$

Kirchoff's laws: *First law:* The vector sum of all currents into any junction point is zero. *Second law:* The vector sum of all voltages around a closed loop is zero. Kirchoff's laws apply to three-dimensional materials such as the earth as well as to conventional electrical circuits.

Klauder wavelet: The autocorrelation of a Vibroseis sweep. The Vibroseis process of injecting a sweep of frequencies into the ground and then correlating with the sweep pattern to yield a seismic record is equivalent to convolving with the autocorrelation so that the Klauder wavelet is in effect the seismic waveform.

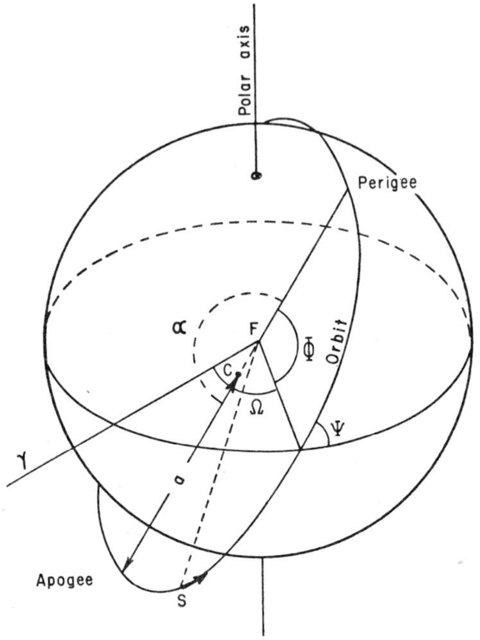

Fig. 107 **Kepler coordinates** and satellite ephemeral information. ♈ is the first line of Aries or the vernal equinox.

C = Center of the elliptical orbit.
F = Focus = center of earth for navigation satellites.
S = Satellite.

To define plane of orbit:
- Ω = Right ascension of the ascending node of the orbit.
- Ψ = Inclination of the orbital plane; approximately 90 degrees for polar orbit. (Cos Ψ and sin Ψ are broadcast.)

To define shape of orbit:
- Φ = Argument of perigee.
- a = Semimajor axis of orbit.
- e = Eccentricity = $\overline{CF}/a$.

To locate satellite in orbit:
- α = Angular position of satellite in orbit; calculated from (1) Greenwich time of perigee and (2) period of satellite.

To locate rotating earth:
- Right ascension of Greenwich at time of perigee.

To allow for precession:
- $d\Omega/dt$ = Rate of change of right ascension.
- $d\Phi/dt$ = Precessional rate of perigee.

Also broadcast are:
- Time of last injection.
- Satellite frequency.

Ephemeral information:
- Greenwich time.
- Correction to the eccentricity anomaly.
- Correction to the semimajor axis
- Distance out of the orbital plane.

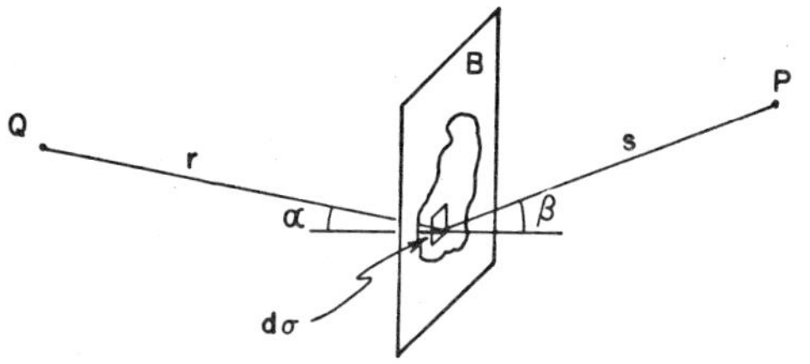

Fig. 108. **Kirchoff diffraction equation.**

knot: A nautical mile per hour = 1.1508 statute mile per hour = 1.852 km/hour = 101.27 ft/min = 0.5148 m/sec.

Knott equations: Equations governing the partition of energy between reflected and refracted plane P- and S-waves at a plane interface, as a function of the velocities, elastic constants, and densities in the two media and of the incident angle. See Muskat and Meres, 1940, Reflection and Transmission Coefficients for Plane Waves in Elastic Media: Geophysics, v. 5, p. 115-155. See also **Zoeppritz's equations.**

Koenigsberger ratio: The ratio of the remanent magnetization to the product of susceptibility and the earth's field strength; symbol: Q.

Kronecker delta: A unit tensor, whose diagonal elements are unity and whose other elements are zero. Usually written as δ_{ij}. Compare **impulse.**

K-type section: A **three-layer resistivity model** in which the middle layer is more resistive than the layers above or below it.

kurtosis: A measure of the sharpness of a population distribution. If x_{10}, x_{25}, x_{75}, and x_{90} are the values for which 10, 25, 75, and 90 percent of the population is smaller, respectively, then

$$\text{kurtosis} = \frac{x_{75} - x_{25}}{2\,(x_{90} - x_{10})}.$$

k-wave: An earthquake wave which has traveled in the earth's core. See **wave notation.**

L: 1. A surface wave. *(L* stands for *long waves.)* L_Q denotes a Love wave, L_R a Rayleigh wave. See **wave notation.** 2. The area under a decay curve from 0.45 to 1.75 sec (induced-polarization usage).

ladder network: A step-type electrical network composed of H, L, T, or pi sections connected one after the other. Often used so that one can pick off different levels, such as different voltages at the same impedance. Ladder filters are shown in Figure 76 **(filter).**

lag: 1. A difference in the time of two events. 2. A delay in the arrival time of seismic events. Lagging refraction or reflection arrivals may indicate subsurface structure or delay caused by weathering variations, phase shifts in filtering, shothole fatigue, etc. Negative of **lead.**

lagged product: The product of two values corresponding to different times.

Lagrange interpolation formula: A method of calculating a polynomial for interpolating between a set of values which are not necessarily equally spaced. Given $y_1 = y(x_1), y_2 = y(x_2), \ldots y_n = y(x_n)$:

$$y(x) = \frac{(x-x_2)***(x-x_n)}{(x_1-x_2)***(x_1-x_n)}y_1$$

$$+\frac{(x-x_1)(x-x_3)***(x-x_n)}{(x_2-x_1)(x_2-x_3)***(x_2-x_n)}y_2$$

$$+***+\frac{(x-x_1)(x-x_2)---(x-x_{n-1})}{(x_n-x_1)(x_n-x_2)---(x_n-x_{n-1})}y_n.$$

lambda: A medium-range positioning system in which the mobile station is the master and which involves the transmission of two frequencies (100 to 200 Khz) to remove lane ambiguity. Two small fixed stations use reception from the mobile master to control phase-locked transmitters. The use of phase-locked oscillators permits continuous transmission despite short losses of signal such as might result from sky-wave interference.

Lambert conic projection: A conformal secant conic map in which the earth's features are projected radially from the earth's center onto a cone which intersects the earth along two standard parallels; see Figure 116 **(map projections).** Parallels are thus the arcs of circles and meridians are straight lines. Lambert projections are used as the standard map references in some states. *Lambert coordinates* on such a map are the rectangular grid coordinates with respect to an arbitrary reference point. The Lambert coordinates thus are not oriented precisely north-south (except along the reference meridian) nor east-west. (The *Lambert equal-area* map is an azimuthal projection.)

Lamé constants: See **elastic constants.**

lane: The unit of measuring position with radio-positioning systems. In phase-comparison (CW) systems, the lane is the distance represented by one cycle of the standing-wave interference pattern resulting from two radiated waves. Its linear distance is not constant but depends on the position within the network. Phase-comparison systems permit location within a lane but do not necessarily determine in which lane. The lane ambiguity has to be resolved either by counting the number of lanes which have been crossed since a known location or by some additional measurement, such as measurements at a different frequency. Lane shape depends on the configuration of the system, but circular, hyperbolic, and occasionally elliptical systems are used, as are combinations of these. See Figure 113 **(line of position).**

language: A method of communicating instructions to a computer. Computer languages are Algol (algorithmic language), Cobol (common business oriented language), Fortran (formula translation) in several versions, Basic, PL1, etc.

Laplace's equation: The relationship showing that the Laplacian $\nabla^2 U$ of a potential function U vanishes in space which contains neither sources nor sinks. (∇ is the operator del.) In rectangular coordinates

$$\nabla^2 U = \frac{\partial^2 U}{\partial x^2} + \frac{\partial^2 U}{\partial y^2} + \frac{\partial^2 U}{\partial z^2} = 0.$$

Compare **Poisson's equation.**

Laplace transform: The linear transforms

$$F(s) = \int f(t)e^{-st}\,dt \quad \text{and}$$

$$f(t) = \frac{1}{2\pi i}\int F(s)e^{st}\,ds.$$

s is a complex number and t is a real one. When the limits of integration are $\pm\infty$, the transform is "two-sided." The two-sided Laplace transform becomes identical with the Fourier transform when s is purely imaginary. More often the "one-sided" transform is used, especially in the study of transient waveforms. In this case the integral is

$$F(s) = \lim_{h\to 0}\int_h^\infty f(t)e^{-st}\,dt \quad \text{and}$$

$$f(t) = \frac{1}{2\pi i}\int_{c-i}^{c+i} F(s)e^{st}\,ds.$$

The one-sided transform is often written with the limits 0 to ∞, the limit being implied. Laplace transforms may not exist for all values of s, and, hence, many Laplace transforms are limited to *strips of convergence,* the range of values for the real part of s for which the above integrals are finite. The Laplace transform domain is often called the *s-plane.*

Laplacian: See **Laplace's equation.**

Larmor frequency: The frequency with which gyromagnetic moments precess in a magnetic field. Atoms and nuclei possess magnetic moments because of their spin and precess like a small gyroscope about the direction of an externally applied steady magnetic field (such as the earth's field). Radio-frequency energy at right angles to the steady field will be absorbed because of resonance when the RF frequency equals the precession frequency. This principle is involved in **proton-resonance** and **optically pumped magnetometers** and in the **nuclear-magnetism log.**

LASA: Large-aperture seismic array; a geophone array in Montana set up to detect nuclear explosions and distinguish them from earthquakes. LASA consists of 21 subarrays of 25 detectors each, the subarrays being of the order of 7 km in diameter and the LASA itself 200 km in diameter. Figure 109.

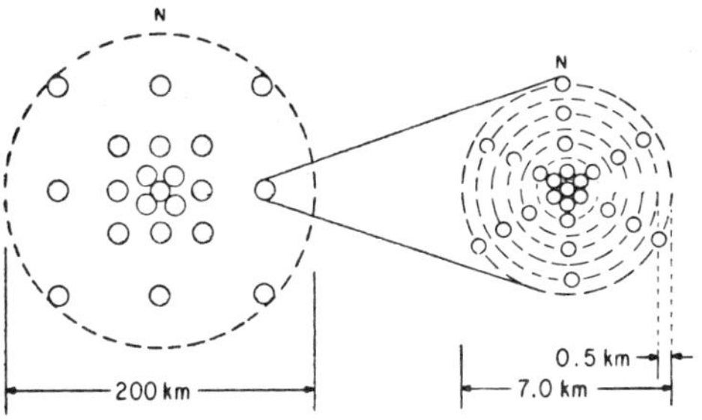

Fig. 109. **Large-aperture seismic array** (LASA). Each subarray is composed of sensors as shown to the right. (From White, Geophysics, v. 34, no 1., p. 1.)

LaserScan: An optical process whereby a seismic record section undergoes two Fourier transformations to make another record section out of it. A lens accomplishes the Fourier transform; see Figure 110. If a grating is present in the object plane at the focal point of the lens, parallel rays will emerge and constructive interference will occur only at I_0, I_1, I_2, etc., separated by a distance Z which depends on the grating spacing. In optics, one usually thinks of white light passing through a uniform grating and separating out into its frequency components; in the Laserscan monochromatic light passed through a nonuniform "grating" (e.g., a seismic record section) separates out into the spacing components of the record section (like f,k space). The lens thus accomplishes a Fourier transform of the record section. A second lens synthesizes the components back into a record section. By blocking certain components in the transform plane their effect can be eliminated in the image plane. This can accomplish frequency or velocity filtering. Trademark of Conductron Corporation. See Dobrin, Ingalls, and Long, 1965, Velocity and Frequency Filtering of Seismic Data Using Laser Light: Geophysics, v. 30, p. 1144-1178.

lateral: A type of resistivity log. A constant current is passed between an electrode A in the borehole and a remote electrode while the potential difference is measured across two electrodes, M and N, located on the sonde; see Figure 111 and compare with the **normal** arrangement (Figure 129). The MN distance is small compared to the *spacing,* which is the distance be-

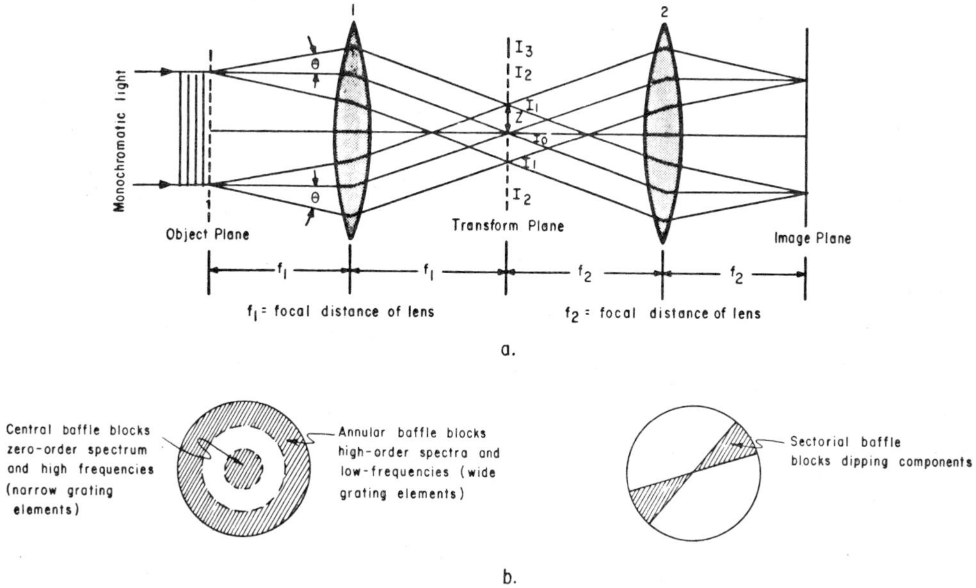

Fig. 110. **Laserscan** principle. (a) Each lens in effect Fourier transforms the data. (b) Baffles to accomplish filtering when inserted in transform plane.

tween the current electrode and the midpoint between the potential electrodes, usually 16 to 22 ft, often 18 ft 8 inches (5.69m). A *short lateral* sometimes uses a spacing of 6 to 9 ft (1.8 to 2.7m). The potential electrodes are usually below the current electrode, but on the *reciprocal sonde* the functions are interchanged so that the potential electrodes are above the current electrode.

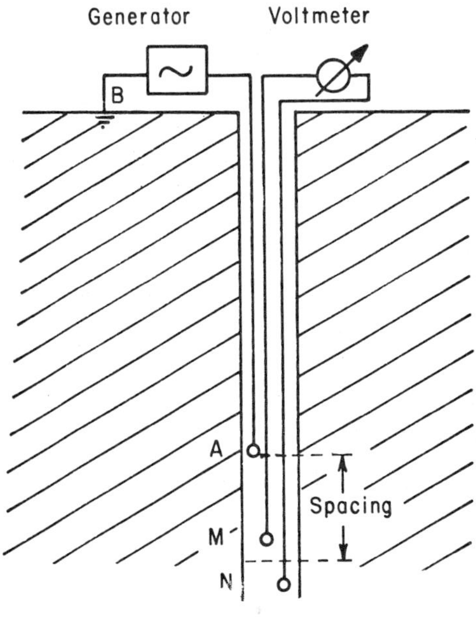

Fig. 111 **Lateral** electric-log configuration.

lateral variations: Changes in a horizontal direction.

laterolog: A resistivity log made with a sonde that is focused by use of guard or bucking electrodes which force the "surveying current" to flow nearly at right angles to the logging sonde. Also called *guard log* or *focused log.* One type uses three electrodes plus guard electrodes. In the laterolog-7 (Figure 112), sufficient current is fed into the bucking electrodes A_1 and A_2 so that the current from electrode A flows out nearly at right angles to the sonde (cross-hatched area in Figure 112). Sensing electrodes M_1, M_1', M_2, and M_2' are used to adjust the bucking-electrode currents until this flow condition exists. The laterolog-3 and guard log use long bucking electrodes above and below the current electrode. See also Figure 102 **(induction log)**. Laterolog is a Schlumberger trademark.

latitude: 1. Distance north (positive) or south (negative) of a reference point or of an east-west reference line. Used in the latitude-**departure** survey method, where the distances are called *northing* or *southing.* 2. The angle between a tangent to the ellipsoid and the earth's axis; **geodetic latitude.** 3. A smoothed magnetic inclination; see **magnetic latitude.**

latitude correction: 1. A correction of gravity data because of variation in centrifugal force resulting from the earth's rotation as the distance to the earth's axis varies with latitude θ and for variation of the earth's radius because of polar flattening. The International Gravity Formula for such variation is

$$g = 978{,}049\,(1 + 0.0052884 \sin^2\theta - 0.0000059 \sin^2 2\theta) \text{ mgal.}$$

The latitude correction amounts to 1.308 sin 2θ mgal/mile = 0.813 sin 2θ mgal/km. 2. A gyrocompass correction for the rotation of the horizontal north vector as a function of latitude. (The horizontal north vector is tangent to the earth and hence the rotation is the result of earth curvature.)

law of reflection: The angle of reflection = angle of incidence, for the same wave mode in an isotropic medium. This is a special case of Snell's law. The more general form of Snell's law must be used for the wave generated by mode conversion upon reflection. If i_p is the incident angle for an incident P-wave (or i_s for an incident S-wave), then the angles of the reflected P-wave will be r_p and the angle of the reflected S-wave will be r_s:

$$\frac{\sin i_p}{V_P} = \frac{\sin i_s}{V_S} = \frac{\sin r_p}{V_P} = \frac{\sin r_s}{V_S},$$

where V_P is the P-wave velocity and V_S is the S-wave velocity.

law of refraction: The sine of the angle of incidence is to the sine of the angle of refraction as the velocity of the incident wave is to the velocity of the refracted wave. Usually called *Snell's law.* See Figure 165 **(Snell's law).**

layer: See **earth layering.**

layout chart: 1. A diagram showing subsurface reflecting positions assuming flat reflectors (or the points midway between shotpoint and geophone group centers) for a number of profiles, such as the component profiles for a common-depth-point stack. Also called *stacking chart.* 2. A diagram showing the shotpoint and geophone positions on the surface of the ground. See Figure 172 **(stacking chart).**

L-band: Radar frequencies between 390 and 1550 μhz; see Figure 68 **(electromagnetic spectrum).**

lead: 1. An indication of interesting structural or other geological conditions. 2. Time arrival by which one event is ahead of another, or by which an arrival is ahead of its "normal" arrival time. Negative of **lag.** 3. The phase angle by which current is ahead of emf in a capacitive electrical circuit. 4. An electrical conductor for connecting to electrical devices.

leakage: Low electrical resistance to ground where there should be high resistance, as in a wet seismic cable.

leaking modes: 1. Head-wave energy which travels through a refractor. The amplitude of head waves within a refractor (for energy incident at the critical angle or greater) decreases exponentially away from the interface. If the refractor is thin some of the energy will "leak" through the refractor and appear as seismic

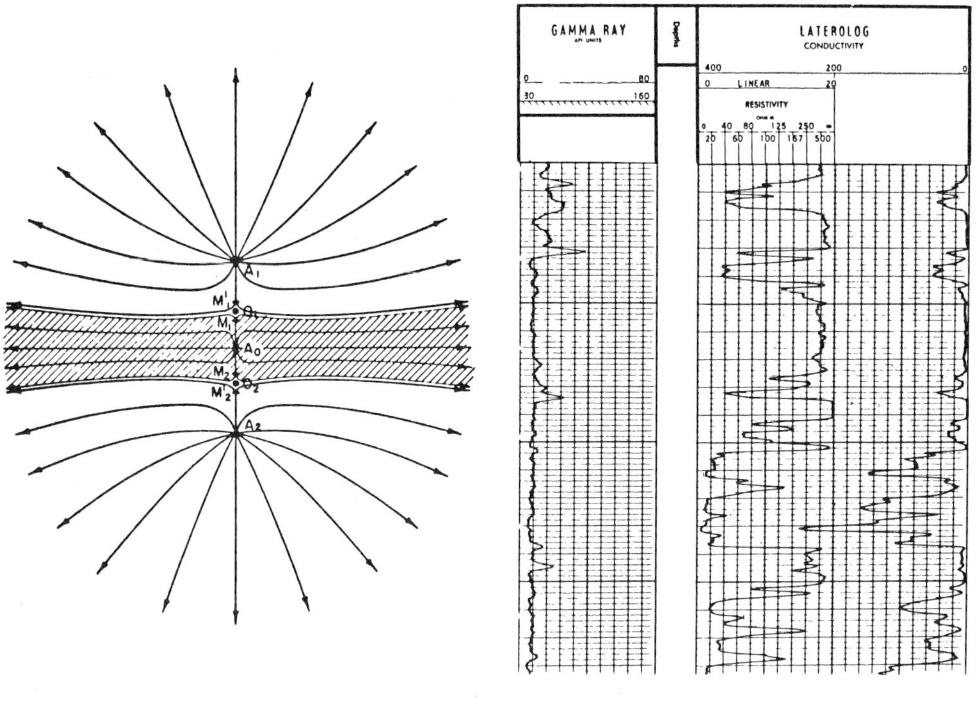

a. b.

Fig. 112. **Laterolog.** (a) Schematic of laterolog-7 showing current flow lines. (b) Laterlog plotted on hybrid scale. (Courtesy Schlumberger.)

waves below the refractor. **2.** Propagation in imperfect wave guides.

least-squares fit: An analytic function which approximates a set of data such that the sum of the squares of the "distances" from the observed points to the curve is a minimum. (Distance usually implies vertical distance or along paths x = constant; other criteria such as shortest distance are sometimes used.) One must determine the functional form of the fit (whether linear, quadratic, etc.) and what is to be minimized in order to define the problem. For example, different velocity functions will result depending on whether seismic time-depth data or velocity-depth data are fitted or if the data are weighted or differently distributed in depth. Least-squares fitting is used in computing satellite navigation fixes.

least-time path: The path between two points which takes the least time to traverse (subject to certain constraints). The path which a seismic ray takes according to Fermat's principle (though Fermat's principle may dictate a path other than the least-time). Also called *minimum-time path* or *brachistochrone.* The raypath will generally be curved or bent because of velocity variations. Also applicable to light rays, etc.

ledge: The first consolidated bedrock.

Lee partitioning method: A variation of the Wenner electrode array in which an additional electrode is placed between the potential electrodes and the potential between it and each of the other potential electrodes is measured.

left-hand rule: A rule which gives the direction of an induced emf in a conductor which cuts a magnetic field. See Figure 101 **(induction)**.

leg: A cycle of more-or-less periodic motion. When following a seismic event from trace to trace or from record to record, one usually concentrates on a particular trough or peak of the energy. If an erroneous correlation is made into an adjacent trough or peak, one has "jumped a leg."

Legendre theorem: For a sherical triangle which is small compared with the spherical radius, the sides bear the same length relationship as a plane triangle whose corresponding angles are smaller by a third of the **spherical excess.**This theorem is used in correcting plane surveying for earth curvature.

leg function: A segmented curve.

leggy: Tailing; the character of a wave train which includes a number of cycles as opposed to a few. Legginess is produced by too narrow a filter bandpass.

level: **1.** A survey device used to ascertain which point on a survey rod is at the same elevation as the instrument. **2.** Amplitude, as in "a potentiometer controls the voltage level." **3.** A track on digital magnetic tape. Thus in "7-level recording," seven magnetic heads spaced across the tape width record seven bits of information at one time (i.e., in one byte).

Levinson algorithm: An algorithm used to solve the normal equations for a Wiener optimum filter (the discrete Wiener-Hopf equation). The equations are of the form $RF = G$ where $R = r_{ij}$ is an autocorrelation matrix, usually the autocorrelation of the input to the filter; $F = ||f_i||$ is the unknown filter; and $G = ||g_i||$ is usually the crosscorrelation of the input and desired output of the filter. The time required by the algorithm goes as the square of the length of the filter as opposed to usual matrix inversions which go as the cube. This is possible because R is a symmetric Toeplitz matrix where the entire matrix is determined by knowledge of the first row or column. See Levinson, 1960, Appendix B in Wiener, Extrapolation, Interpolation, and Smoothing of Stationary Time Series: New York, MIT Technology Press and John Wiley and Sons, Inc.

Lg-wave: "Long" surface waves including both **Rayleigh waves** and **Love waves.**

library: Programs and data which a computer system has available to use.

line: **1.** A linear array of observation points, such as a seismic line. **2.** Equipment which is directly connected to and controlled by a central controller. See also **off-line.** **3.** A channel. **4.** In a power spectrum, the contribution of a single frequency. Physically, the contribution of a very narrow frequency band.

lineagenic: See **tectonic types.**

linear: Having a straight line relationship: $x = a + by$. See also **linear system.**

linear filter: See **convolution.**

linearity of IP: **1.** The proportional relationship between polarization and current density, in which chargeability is a constant. **2.** The symmetrical identity between voltage versus time of IP charge and decay curves. **3.** The proportionality between decay-curve amplitude and polarizing voltage.

linearly independent: Not being expressable as a linear combination of other elements. Thus a set of equations (or solutions) is linearly independent if none of them can be expressed as a linear combination of the others. A set of linear functions, $f_i(x_1, \ldots, x_n) = \sum a_{ij} x_j$, is linearly independent if the determinant does not vanish (i.e., if $|a_{ij}| \neq 0$). See **Jacobian.**

linear-phase filter: A filter which shifts phase proportional to frequency, thus introducing a constant delay without change of waveshape. A linear-phase filter is a mixed-phase filter which shifts component frequencies proportional to the frequency. The phase shift versus frequency graph is thus linear over the band-pass and the intercept must be a multiple of 2π so that each component may be delayed equally. Also called *delay filter.* See Figure 119 **(minimum phase).** Such a filter produces no phase distortion. If the intercept is an odd multiple of π, it will invert a wavelet. Linear-phase filters are used to achieve zero-phase filtering by shifting the time scale after filtering. Sometimes used as part of quadrature or polarization filtering, for filtering the horizontal or vertical components of motion before combining them so as to emphasize one mode of wave propagation with respect to another, as to enhance P-waves with respect to Rayleigh waves.

linear sweep: Vibroseis sweep where the vibrator frequency varies linearly with time and hence the argument of the cosine is quadratic. Thus the sweep amplitude is $[A \cos (w_o t + bt^2)]$ for upsweep or $[A \cos (w_o t - bt^2)]$ for downsweep. See Figure 190 **(Vibroseis).**

linear system: **1.** A system whose output is linearly related to its input. If a linear system is excited by an input sine wave of frequency f_1, the output will contain only the frequency f_1; the amplitude and phase may be changed, however. The rules of scaling (if input of A results in an output of B, then input of kA results in output of kB) and superposition (if A results in B and C results in D, then $A + B$ results in $C + D$) apply. **2.** An electrical circuit whose impedance is independent of applied voltage (or current).

line-mile: A unit of measure for geophysical work such as indicating continuous coverage over one mile of seismic line or a mile of aeromagnetic data.

line of force: A curved line in a potential field such that a tangent anywhere along the line is in the direction of the force on a charge (pole, mass, etc.) at that location. The density of lines of force is called the *flux density* or *field strength.* Used in connection with electric fields, magnetic fields, gravitational fields, etc. See **Gauss law.**

line of position: The locus of equal values measured with a positioning system, such as a line of equal phase difference with Raydist or Omega. The distance between two adjacent lines of position is called a lane. The intersection of two lines of position determines a location *fix.* Lines which represent constant range from a reference location are circles (Figure 113); lines which represent constant difference in ranges are hyperbolas; lines which represent constant sum of ranges are ellipses. Positions may be determined by the intersection of two families of circles, by the intersections of two families of hyperbolas, by combinations of circles and hyperbolas, etc. Abbreviated LOP.

line of sight: The straight-line distance from an object to an observer. High-frequency (short-wavelength) radio waves travel such raypaths so that the curvature of the earth limits the range which can be achieved. (The "effective" earth curvature differs somewhat from the actual earth curvature; see **radio earth.**) Line-of-sight range R is given in terms of the height of transmitting and receiving antennas h_t and h_r:

$$R = k(h_t^{1/2} + h_r^{1/2}).$$

The constant k is 1.22 nautical miles when h is given in feet, 4.08 km when h is given in meters. Refraction will increase the effective range of radio waves depending on their wavelength and the sensitivity of the detecting system will also affect the range.

line printer: A computer output device which prints a line of characters simultaneously.

line source: A source of energy which can be treated mathematically as though it were condensed into an infinitely long line in three dimensions or a point in two dimensions. In the seismic method, for example, anomalous pulses which appear to emanate from a line, such as a diffraction from a fault. In the gravity and magnetic methods, line sources represent a concentration of mass or magnetized matter into a point of two dimensions (such as a horizontal cylinder treated as having the mass concentrated along the axis of the cylinder). In electrical methods, a long current-carrying wire can often be regarded as a line source. Usually the current is taken to be equiphase along the wire. In the near-field zone the magnetic field varies inversely with distance for a line source, inversely with distance squared for an electric dipole, and inversely with distance cubed for a magnetic dipole.

lineup: In-phase alignment across the traces of a seismogram; **coherent** energy.

liquid-junction potential: Liquid-boundary potential; see **electrochemical SP.**

Lissajous figure: The steady pattern on an oscilloscope when periodic waves which are harmonics of a common frequency are applied to the horizontal and vertical plates. Such patterns are used to ascertain that the frequencies are exactly the ratios of small integers and to determine their relative phase relationship.

list: Overall average tilt of ship to starboard or port (about an axis in the principal direction of motion). As opposed to roll, which is ship rotational motion.

listening period: The time between periodic inputs. In time-domain IP surveying, responses are measured during the "listening periods" between periods of current application. In doppler-sonar navigation, the frequency is measured during the listening periods between transmissions.

listric surface: A curved fracture (fault) surface which becomes less steep as one goes deeper, becoming nearly horizontal at some depth. Because of the curvature of the fault plane, rotation accompanies slippage along the fault.

lithologic log: A log showing lithology as a function of depth in a borehole. Sometimes a strip log based on samples, sometimes interpreted from other borehole logs.

lithosphere: The upper 100 km (approximately) of the earth which is relatively rigid, thus including the earth's crust and upper mantle. Characterized by relatively

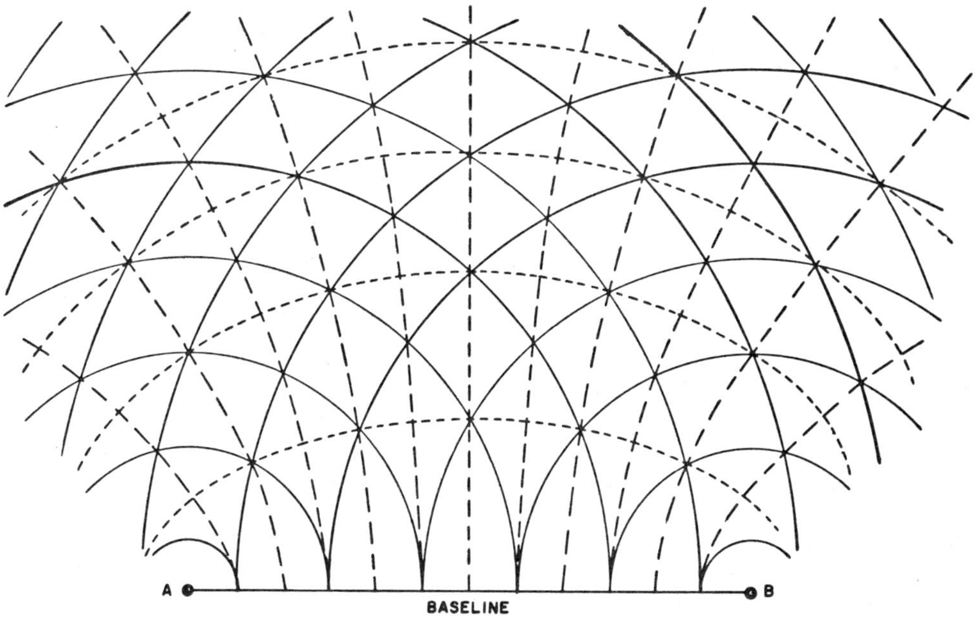

Fig. 113. **Lines of position** (LOP) for positioning systems. The solid lines indicate the range from A (or B) and are circles The dashed lines indicate constant difference in range from A and B and are hyperbolas. The short dashed lines indicate that the sum of the ranges from A and B is constant and are ellipses.

low attenuation of seismic waves (high Q). Underlain by the asthenosphere and the mesosphere. See Figure 63 **(earth layering)**.

little slam: Small slam.

live: Responsive. A geophone is live if it is responsive and properly connected. A channel is live if it is responsive to input energy. As opposed to dead.

LLI: Log-level indicator.

ln: Natural logarithm; logarithm to the base *e*. Ln x = 2.3026 $\log_{10}x$.

load: 1. The power which a device consumes or delivers. **2.** An impedance connected across an output.

loading pole: A pole (usually in 10-ft sections) for placing explosive charge in a shothole.

load-point marker: A marker (such as a band of metal foil affixed on a digital magnetic tape) which indicates to the computer the beginning or end of data storage on the tape.

lobes: Passbands in a directivity graph. See Figure 54 **(directivity graph)**. A bulge in a spectral window. The main window is called the *main lobe* and smaller bulges to the side are called *side lobes.* Used in connection with seismic directivity, radio antenna, patterns, etc.

local gravity: Bouger value from which the regional has been subtracted; **residual.**

local magnetic anomalies: Anomalies of restricted areal distribution caused by the magnetization of units in the uppermost parts of the earth's crust.

lock on: Establishment of phase agreement between an oscillator in a receiver and a received radio signal. Phase-lock loops are used in satellite navigation receivers.

locus: The set of all points which satisfies a given requirement. Thus a circle or sphere is the locus of points equidistant from a particular point. See Figure 113 **(line of position).**

log: 1. A record of measurements or observations, especially those made in a borehole. See **well log. 2.** An instrument for measuring a vessel's speed or distance traveled or both.

logarithmic contour interval: Plotting of data on a logarithmic scale is sometimes used where properties of materials vary by many orders of magnitude. Resistivity and IP data often are contoured in intervals which are approximately logarithmic (or geometric), such as 1, 2, 5, 10, 20, 50, 100; or 1, 3, 10, 30, 100, 300.

logging: Measuring the physical properties of the material around a borehole. See **well log.**

log-level indicator: A trace which indicates the log of the gain of one amplifier channel which is under AGC control, thus roughly indicating energy level. Abbreviated LLI. See Figure 149 **(record).**

log-normal: A statistical distribution which, when plotted logarithmically, has the appearance of a normal Gaussian-distribution curve.

longitudinal: Along the major axis of a feature; as opposed to transverse. The longitudinal axis of a structural feature (such as an anticline) is its *b axis.*

longitudinal wave: P-wave or compressional wave.

long normal: A **normal** resistivity log made with the A and M electrodes in the sonde (usually) 64 inches apart (1.63m).

long-path multiple: A seismic reflection whose travel path is much longer than required for a primary reflection from the deepest interface reached. A long-path multiple tends to appear as a separate event rather than blending into the tail of the primary. For example, the energy might be reflected by a deep reflecting interface, then at or near the surface, and again by the same or another deep interface. See **multiple.**

long wave: Surface wave.

loop: 1. Field observations which begin and end at the same point with a number of intervening observations. Obtaining data in loops is useful in correcting for drift in gravity-meter observations or diurnal variation in magnetometer surveys, and in detecting faults or other cause of misclosure in seismic work. **2.** An electrical circuit which provides feedback, as an *AGC loop.* **3.** A part of a computer program in which the last instruction can repeat the preceding series of operations (with or without modification) until some particular condition is reached. **4.** Transmitting or receiving coil used in electromagnetic surveying.

LOP: Line of position. The line (which may be hyperbolic, circular, straight, etc.) indicated by a positioning measurement along which one is located. The intersection of two lines of position determines a location *fix.* See Figure 113 **(line of position).**

Lorac: A medium-range surveying system involving the phase comparison of beat frequencies of radio waves transmitted over different paths, similar to Raydist, Decca, and Toran. Lorac uses a fixed *reference station* which observes the transmissions of the center station and "red" and "green" end stations and broadcasts a different frequency which is amplitude modulated by information about the phase differences of the CW stations. See Figure 114. Seismograph Service Corporation trademark.

loran: Long-range navigation. One of several government-maintained long-range pulse-type electronic positioning systems. Hyperbolic lines of position are determined by measuring the difference in the time of reception of synchronized pulse signals from fixed transmitters at known geographic positions. **1.** *Loran A* ("standard" loran) involves the transmission of pulses of 1850, 1900, and 1950 khz energy from fixed transmitters. A mobile receiver times the differences between the arrival of these pulses to give a hyperbolic line of position. **2.** *Loran B* is an experimental loran which has been superseded. **3.** *Loran C* operates at lower frequencies and hence has longer wavelength and longer range; it also allows radiofrequency cycle-matching to provide increased accuracy; see Figure 115. The loran C master station transmits groups of pulses at a group-repetition period which characterizes a network. The repetition period and the 100 khz transmission frequency are controlled by ultrastable cesium frequency standards. Two slave stations transmit independently generated pulses which are also ultra-stable and synchronized with the master station. At a user's mobile station the time differences between the receipt

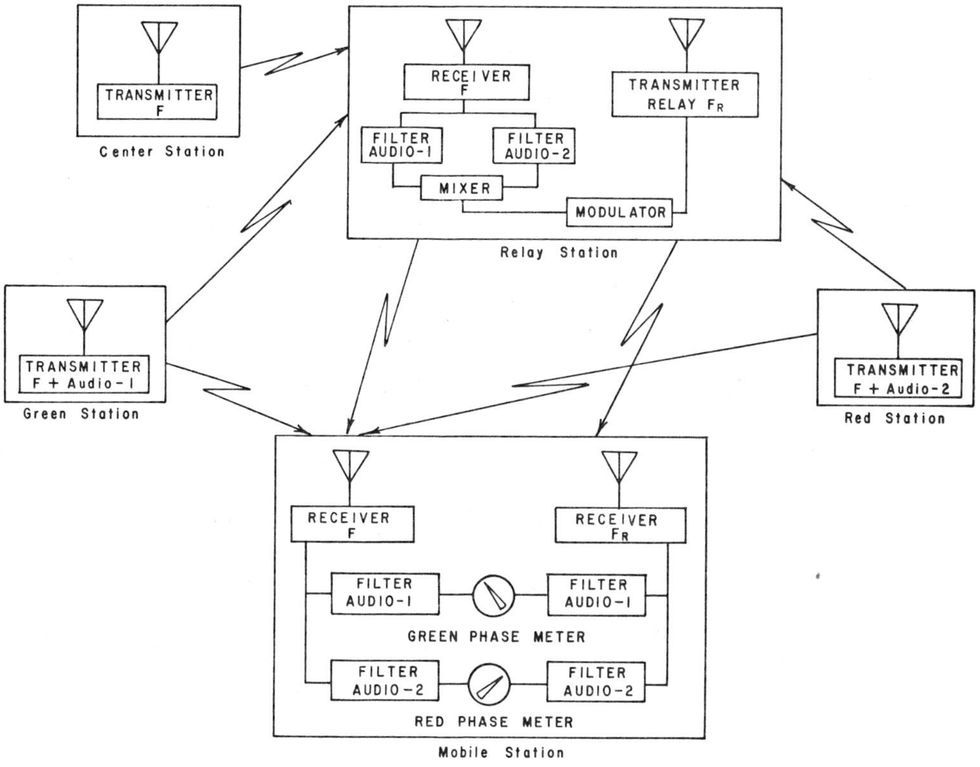

Fig. 114. **Lorac B system.**

of pulses from the master station and the slaves is determined, thus defining two hyperbolic lines of position which locate the mobile station. **4.** *Loran C* may also be operated in the "rho-rho mode" if the mobile station includes a stable atomic-frequency clock. The atomic clock is synchronized with the standard which initiates the master-station transmission so that the traveltime of the transmissions from the master and from a slave station can be determined, thus giving the range to these two stations. **5.** *Loran D* is similar to Loran C but uses shorter base-lines (distances between master and slave stations).

LOS: Line of sight.

losser: A circuit element which attenuates the gain upon proper instruction; used in AGC circuits.

loss tangent: A measure of dielectric loss defined by the relationship $\tan\delta = \sigma/\epsilon\omega$, where σ is the loss tangent, σ is the conductivity, ϵ is dielectric permittivity, and ω is angular frequency.

Love wave: A surface seismic wave associated with layering, characterized by horizontal motion perpendicular to the direction of propagation with no vertical motion. Designated Q-wave, L_Q-wave, G-wave, or SH-wave. Love waves may also be thought of as channel waves with the earth's surface the upper channel boundary. Total reflection occurs at the boundaries of the channel so that the waves represent energy traveling by multiple reflection. The energy builds up because of interference. Different interference modes are designated by the number of nodal planes within the layer but usually only zero mode is observed. The dispersion of Love waves can be used to calculate the thickness of the surface layer. Earthquake Love waves have velocities up to 4.5 km/sec, faster than Rayleigh waves. See Figure 65 **(earthquake seismology)**.

low: An area in which the beds are structurally lower than in neighboring areas; a syncline or structural depression.

low-cut filter: A filter that transmits frequencies above a given cutoff frequency and substantially attenuates lower frequencies. Same as high-pass filter.

low-level multiplexing: Multiplexing between the preamplifier and first amplifier stage.

low-pass filter: A filter that passes frequencies below some cutoff frequency while substantially attenuating higher frequencies. Same as high-cut filter.

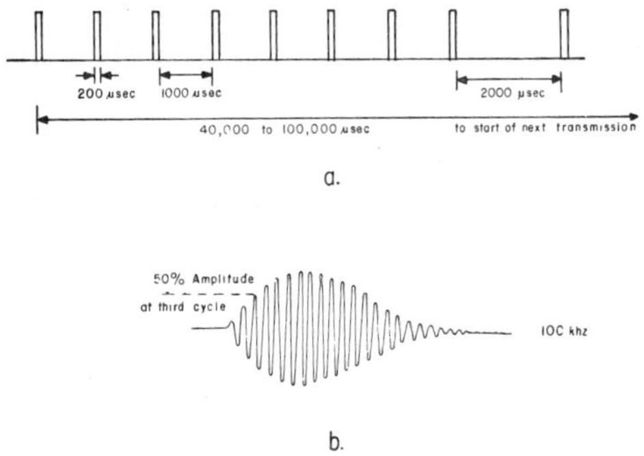

Fig. 115. **Loran C.** (a) Pulse transmission pattern; slaves do not transmit the 9th pulse. Phase coding of successive pulses identifies stations and helps in sky-wave identification. (b) Shape of one of the Loran C pulses.

low-velocity layer: 1. Weathering; a near-surface belt of very low-velocity material. **2.** A layer of velocity lower than that of shallower refractors. See **blind zone. 3.** The B-layer in the upper mantle (see Figure 63, **earth layering**) from 60 to 250 km deep, where velocities are about 6 percent lower than in the outermost mantle. **4.** The region just inside the earth's core.

loxodrome: Rhumb line.

LSB: Least significant bit.

L-spread: 1. A seismic spread in which the shotpoint is offset by an appreciable distance perpendicularly to the spread line, the shotpoint being opposite one of the end geophone groups. See Figure 171 **(spread)** and also **broadside. 2.** A spread often laid out for noise studies in which about half of the length is in-line with the shot and the remainder perpendicular.

lumped circuit: An electrical network composed of specific resistance, capacitance, and other elements, as opposed to distributed resistance and capacitance as in a transmission line or as opposed to a solid-state system.

LVL: Low-velocity layer. Usually implies the **weathering,** occasionally other situations.

L-waves: Long waves; seismic surface waves of long wavelength from earthquakes. L_Q denotes a Love wave, L_R a Rayleigh wave.

m-: Milli, a prefix indicating 10^{-3}.

ma: Subscript used with log terms to indicate the rock matrix.

machine language: A form of expressing a program so that it can be executed directly by a computing machine; an **object program.**

Maclaurin series: A special case of a **Taylor series** in which expansion is at the origin.

macroscopic anisotropy: The situation where resistivity perpendicular to the bedding differs from resistivity parallel to the bedding because of the inclusion of interbeds of markedly different resistivity; see **anisotropy.**

MAE: An analog time-domain process for reducing the effect of singing. See **Backus filter.**

mafic: Rich in magnesium rock-forming silicates and other dark minerals.

magnetic artifacts: See **artificial aeromagnetic anomalies.**

magnetic basement: The upper surface of extensive rocks having relatively large magnetic susceptibilities compared with those of sediments. Often but not necessarily coincident with the geologic **basement.** Generally excludes magnetic sediments and thin volcanic and other high-susceptibility rocks intruded into the sedimentary section, but thick volcanic rocks in the sedimentary section would be classed as magnetic basement where the magnetic effects of deeper bodies would not be resolvable.

magnetic core: A small toroidal ferrite ring used to store a bit of information in a computer's rapid-access memory.

magnetic dip: Inclination.

magnetic disc: A flat circular plate with a magnetic surface on which data can be stored for access by a computer.

magnetic disturbance: Magnetic storm.

magnetic equator: The line on the surface of the earth where the magnetic needle remains horizontal, that is, where the magnetic lines of force are horizontal. Also called the *aclinic line.*

magnetic field: The space through which influence on a magnet is exerted. The torque at any point in space which would tend to orient a current-carrying coil or magnet if it were located at that point. Also called the magnetic flux density or the magnetic induction and symbolized by **B. B** is a vector quantity. The unit of measure for **B** in the **SI** or mksa system is the tesla and in the cgs system the gauss or gamma (1 tesla = 1 weber/m^2 = 1 newton/amp-m = 10^4 gauss = 10^4 maxwell/cm^2 = 10^9 gamma). **B** is given in terms of the force *dF* produced on a small element of length *dl* which is carrying a current *I*:

$$dF = I\,dl \times B.$$

B is related to the magnetizing force **H** by a constant of the medium called the permeability μ:

$$B = \mu\mu_0 H.$$

The unit of measure for **H** in the **SI** or mks system is ampere turns/meter and in the cgs system, the oersted (1 amp turn/m = $4\pi\ 10^{-3}$ oersted = $4\pi\ 10^{-3}$ gilbert/cm). μ_0 is called the permeability of free space; $\mu_0 = 4\pi\ 10^{-7}$ weber/amp-m in **SI** units and 1 gauss/oersted in cgs units. **B** can also be expressed as

$$B = \mu_0 (H + M) \text{ in SI units}$$

$$B = \mu_0 (H + 4\pi I) \text{ in cgs units,}$$

where **M** and **I** are called the magnetization or intensity of magnetization.

A magnetic field of magnitude given by the inverse-square law surrounds a magnetic pole, and a magnetic field given by Ampere's law surrounds an electric current. **H** is sometimes called the magnetic field. See Table 6.

magnetic flux: The flux through a surface is the integral over the surface of the normal component of the **magnetic induction;** expressed in webers in mks units or in maxwells in cgs units (1 weber = 10^8 maxwells).

magnetic head: An electromagnet used for reading, recording, or erasing signals on a magnetic tape.

magnetic induction: See magnetic field.

magnetic intensity: Magnetic-field strength.

magnetic interpretation methods: The objective of magnetic data interpretation is locating the anomalous magnetic material: its depth, dimensions, and magnetization. Methods used include comparison with the fields over known areas and comparisons against the fields of model anomalies shown in a catalog of **master curves.** Two-dimensional convolution operations such as calculation of the **second-vertical-derivative map, downward continuation,** or **reduction to the pole** sometimes are used to help locate the anomalous body or bodies and determine its shape. Most petroleum search use of aeromagnetics involves determining the depth to the top of anomalous bodies (and thereby inferring the depth to magnetic basement), and the preparation of a contour map of the magnetic basement from the results. Various shape measurements are made on magnetic profiles or maps and used in conjunction with **depth rules.** The most common magnetic models used in interpretation are the dipping-dike and vertical prism for intrabasement bodies and the thin magnetic layer for structural features. Iterative methods involve calculating the observed field, modifying the model, and repeating the process until a satisfactory degree of fit between model field and observed field is achieved.

magnetic meridian: The direction of the horizontal component of the earth's magnetic field; the direction of magnetic north.

magnetic moment: That vector associated with a magnetized mass or coil carrying an electrical current, whose cross product with the magnetic-field intensity in which the mass or coil is immersed (ignoring the field distortion thereby produced) is a measure of the resulting torque. A magnetic moment of one ampere meter^2 feels a torque of one newton meter when perpendicu-

Table 6. Magnetic quantities and units

	SI (International System)		cgs System	
Term	**Symbol**	**Unit**	**Unit**	**Symbol**
Magnetic field strength Magnetic intensity Magnetic induction Magnetic flux density	$\mathbf{B} = \mu_0(\mathbf{H} + \mathbf{M})$ $= \phi/\text{area}$	1 tesla = 1 weber/m^2 = 1 newton/amp-m	$= 10^4$ gauss $= 10^9$ gamma $= 10^4$ maxwell/cm^2	$\mathbf{B} = \mathbf{H} + 4\pi\mathbf{I}$
Dipole moment	*md*	1 ampere-meter^2	$= 10^{10}$ pole cm	*md*
Magnetic pole	*m*	1 ampere-meter	$= 10^8$ unit poles	*m*
Magnetic flux Magnetic line of force	ϕ	1 weber	$= 10^8$ maxwell	ϕ
Magnetomotive force Magnetic potential	MMF	1 ampere turn	$= 0.4\pi$ gilbert $= 1.26$ gilbert	MMF
Magnetizing force	$\mathbf{H} = \mathbf{B}/\mu\mu_0$	1 ampere turn/m	$= 4\pi 10^{-3}$ oersted $= 0.0126$ oersted $= 4\pi 10^{-3}$ gilbert/cm	$\mathbf{H} = \mathbf{B}/\mu'$
Magnetization Magnetic dipole moment per unit volume Magnetic polarization Magnetization intensity	$\mathbf{M} = k\mathbf{H}$ $= md/V$	ampere-m^2/m^3 = ampere/m	$= 10^{-3}$ gauss/cm^3	$\mathbf{I} = k'\mathbf{H}$
Magnetic permeability			gauss/oersted	$\mu' = \|\mathbf{B}\|/\|\mathbf{H}\|$
Relative permeability	$\mu = \|\mathbf{B}\|/\mu_0\|\mathbf{H}\|$	dimensionless	dimensionless	$\mu = \|\mu'\|$
Magnetic susceptibility	$k = \|\mathbf{M}\|/\|\mathbf{H}\|$ $= \mu - 1$	dimensionless k_{SI}	dimensionless $= 4\pi k'$cgs	$k' = \|\mathbf{I}\|/\|\mathbf{H}\|$
Reluctance	$R = \text{MMF}/\phi$	1 ampere turn/ weber	$= 4\pi 10^{-7}$ gilbert 1 maxwell	$R = \text{MMF}/\phi$
Inductance	*L*	1 henry = 1 weber/amp		

An equal sign relates the magnitude of SI and cgs units.

μ_0 = permeability of free space $= 4\pi 10^{-7}$ weber/ampere meter
$= 12.57 \times 10^{-7}$ weber/ampere meter
$= 12.57 \times 10^{-7}$ henries/meter
$= 1$ gauss/oersted.

lar to a magnetic induction vector of one weber/meter[2].

magnetic permeability: The ratio of the magnetic induction **B** to the inducing field strength **H**; denoted by the symbol μ:

$$\mu = |\mathbf{B}|/\mu_0|\mathbf{H}|.$$

μ_0 is the permeability of free space $= 4\pi 10^{-7}$ weber/ampere meter (or henries/meter) in the **SI** system and 1 gauss/oersted in the cgs system, so that the permeability μ is dimensionless. The quantity $\mu\mu_0$ is sometimes considered the permeability (especially in the cgs system) in which case it has dimensions.

magnetic polarization: Magnetic moment per unit volume (occasionally per unit mass).

magnetic pole: 1. One of the two points near opposite ends of a magnet toward which the magnetic lines of force are oriented and concentrated. If the magnet is permitted to rotate about its center, the pole which points in the direction of the earth's north magnetic pole is the "north-seeking" or positive pole; the other pole is the "south-seeking" or negative pole. 2. The geomagnetic poles are the locations on the earth where the magnetic inclination is 90 degrees (dip poles) or, alternatively, the projections to the earth's surface of the axis of the geocentric magnetic dipole which best approximates the earth's magnetic field; these are not precisely equivalent.

magnetic resonance: Interaction between the magnetic moments (electron spin and/or nuclear spin) of atoms with an external magnetic field. Magnetic resonance is basic to the operation of the **proton magnetometer** and **optically pumped magnetometer.** See also **nuclear-magnetism log** and **Larmor frequency.**

magnetic scalar potential: The scalar function whose negative gradient equals the magnetic-field intensity and which represents the work required to bring a unit north pole from a point infinitely remote to the point at which the function is determined.

magnetic shield: High-permeability container which isolates its interior from external magnetizing forces.

magnetic signature: The shape of a magnetic anomaly.

magnetic storms: Rapid, irregular, transient fluctuations of the magnetic field which are greater in magnitude, more irregular, and more rapid than diurnal variations. These occur most commonly during unusual sunspot activity as a result of bombardment of the earth by high-energy particles from the sun. Magnetic storms commonly have amplitudes of 50 to 200 gammas, occasionally thousands of gammas. Magnetic prospecting usually has to be suspended during such periods.

magnetic survey: Measurements of the magnetic field or its component (such as the vertical component) at a series of different locations over an area of interest, usually with the objective of locating concentrations of magnetic materials or of determining depth to basement. Differences from the normal field are attributed to variations in the distribution of materials having different susceptibility and perhaps also remanent magnetization.

magnetic susceptibility: A measure of the degree to which a substance may be magnetized; the ratio k or k' of the magnetization **M** or **I** to the magnetizing force **H** that is responsible for it:

$$k = |\mathbf{M}|/|\mathbf{H}| \text{ in the mks system.}$$
$$k' = |\mathbf{I}|/|\mathbf{H}| \text{ in the cgs system.}$$

The susceptibility is dimensionless but of different magnitude in the two systems:

$$k = 4\pi k'.$$

The susceptibility is related to the magnetic permeability μ.

$$k = \mu - 1,$$
$$k' = (\mu - 1)/4\pi.$$

Susceptibility in cgs units is sometimes measured in units of 10^{-6}.

magnetic tape: A sheet or strip of plastic (such as Mylar) coated with a magnetically sensitive material on which information can be stored in the form of magnetization patterns. Half-inch tape containing nine tracks across is usual in seismic computers, though 7-track half-inch and 21-track one-inch tapes are also used. Analog seismic tapes occur in a variety of sizes and formats.

magnetization: Magnetic moment per unit volume (occasionally per unit mass), a vector quantity. Also called *magnetic polarization* or intensity of magnetization. Designated by symbols **M** or **I.** A measure of the effect of the medium on the magnetic field **B** when subject to a magnetizing force **H**:

$$\mathbf{B} = \mu_0(\mathbf{H} + \mathbf{M}) \quad \text{in SI system,}$$
$$\mathbf{B} = \mathbf{H} + 4\pi\mathbf{I} \quad \text{in cgs system,}$$

where μ_0 is the permeability of free space. The proportionality between magnetization and **H** is the magnetic susceptibility, k or k'.

magnetizing force: A measure of the influence of a magnet in the surrounding space. Symbol, **H.** See magnetic field and Table 6.

magnetohydrodynamics: Phenomena associated with the motion of an electrically conducting fluid, as a liquid metal or an ionized gas, through a magnetic field. Also called hydromagnetics.

magnetohydrodynamic theory: The theory that coupling between the mechanical and electrodynamic forces in the fluid core is responsible for the earth's main magnetic field.

magnetometer: An instrument for measuring **magnetic-field strength.** Ground magnetometers sometimes measure the vertical component of the magnetic field, sometimes a horizontal component, sometimes the total field. Most airborne magnetometers are of three types: **a. fluxgate, b. proton-resonance,** or **c. optically pumped;** all measure the total-field intensity. Vector and vertical component airborne magnetometers are also used occasionally.

magnetosphere: The space pervaded by the earth's magnetic field. Usually the space from the earth to more than 10 earth radii.

magnetostriction: Change in the strain of a magnetic material as a result of changes in magnetization. The dependence of magnetization (susceptibility or remanence) on applied stress is termed *inverse magnetostriction* or *piezomagnetism.* Magnetostrictive acoustical sources and hydrophones are extremely rugged. A magnetostrictive hydrophone might consist of a coil of wire wrapped around a cylinder of magnetostrictive material. A pressure wave acting radially induces hoop-stresses in the core which changes its permeability and thereby the flux linking the coil wrapped around it. The change of flux induces a voltage which is proportional to the derivative of the pressure-wave signal.

magnetotelluric method: A method in which orthogonal components of the horizontal electric and magnetic fields induced by natural primary sources are measured simultaneously as a function of frequency. Abbreviated MT. Apparent resistivity is defined as:

$$P_a = (E_i/H_j)^2.$$

To simplify the application of the method it is often assumed that the primary field is a uniform plane wave and that the earth consists of uniform horizontal layers. If these assumptions are valid, the data may be interpreted by curve-matching techniques. The predominant sources of energy for magnetotelluric measurements are micropulsations having frequencies of less than 1 hz. Sometimes magnetotelluric measurements are made at audio frequencies using energy from sferics; the method is then referred to as the audio-magnetotelluric method (AMT method). See **telluric current method.**

magnetotelluric noise: Unwanted voltage signals in the earth due to low-frequency earth currents whose source is electrical discharges in thunderstorms, power lines, ionospheric currents, or magnetospheric currents.

magnitude of an earthquake: A rating of an earthquake independent of the place of observation. The Richter scale of magnitude indicates the logarithm of the maximum amplitude observed (or which might have been observed) on an instrument of specified type 100 km from the epicenter. Empirical tables correct observations at other distances. Each step of one in magnitude means multiplying the amplitude by 10. Zero represents the smallest recorded earthquakes (when the scale was devised); the largest earthquakes are of magnitude about 9. Various relations are used to give the order of magnitude of the energy released in an earthquake from the magnitude *M.* Roughly, the energy *E* in ergs is given by $\log E = 10 + 2M$.

main lobe: The portion of a directivity graph which indicates the continuous band of directions (or apparent wavelengths) in which the greatest energy is radiated or which undergo least attenuation. See Figure 54 **(directivity graph).**

majority vote: A system used with **satellite navigation** whereby several observations of the satellite message are compared and if differences exist, the message is assumed to be that which was observed most often. Each bit position is majority-voted separately.

mantle: The part of the earth's interior between the core and the crust. The upper surface of the mantle is the **Moho,** characterized by a P-wave velocity of 8.2 ± 0.2 km/sec. The density of the mantle is about 3.3 – 3.4 gm/cm^3. The mantle includes the lower lithosphere and the asthenosphere. See Figure 63 **(earth layering).**

map: **1.** To transform information from one form to another or **2.** the product of such transformation. The transformation may involve the geographical distribution of observations or of calculations based on observations, as a *Bouguer anomaly map* or a *seismic reflection map.* The distribution may be with respect to variables other than geographic, as "to map from the time to the frequency domain." Transformations may involve a one point to one point correspondence or a one to many (involving multibranched surfaces). Examples of multibranched maps might be the map of a geological formation in the vicinity of a reverse fault where the same formation contact lies at two (or more) depths or a seismic reflection time map in a buried focus situation where the same reflector can be seen in several directions from the same surface point.

map convolution: A two-dimensional convolution often applied to potential field maps, whereby each point on the map grid is replaced by a weighted sum of the values at other grid points.

$$\psi_{x,y} = \sum_{\alpha}\sum_{\beta} f_{\alpha,\beta}\, \phi_{x-\alpha,\, y-\beta}.$$

ψ are the output values, ϕ the input values, and *f* is the weighting scheme, called the *template.* Simple residual maps are made by subtracting an average of values around the point from the value at the point. The values for different distances may be weighted and sometimes a bias is included so that the residual does not change sign very often. By weighting the points to give the horizontal derivatives and using Laplace's equation, second-vertical-derivative maps may be made. Other weighting schemes can be used for field **continuation,** wavelength filtering, etc. See **grid residual.**

map projection: A scheme for displaying the earth's curved surface on a plane surface. Some of the more common projections are shown in Figure 116. Distortions of one sort or another are inevitable. *Equivalence* is the projection property wherein the product of orthogonal scale factors is maintained constant so that areas are preserved. *Conformality* is the property wherein angles are preserved. *Standard lines* are great or small circles along which the scale is a uniform constant. *Tangent* projections are projections onto a surface (plane, cylinder, or cone) which is tangent to the earth; *secant* projections are those made onto a surface which intersects the earth. A *transverse* projection has its axis perpendicular to the earth's axis (sometimes merely at any angle to the earth's axis). An

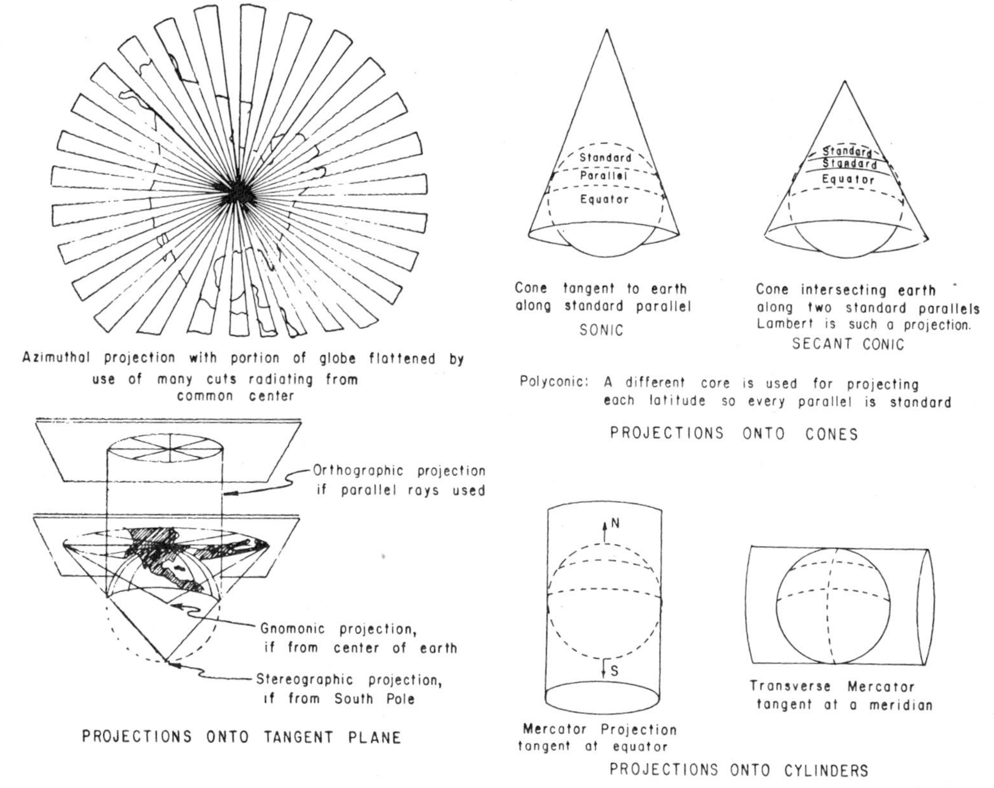

Fig. 116. **Map projections.**

azimuthal projection is onto a tangent plane; distortions increase as the distance from the point of tangency. A *stereographic* projection is both azimuthal and conformal. A *polyconic* has a straight central meridian and each parallel is the arc of a circle and is standard; the scale along meridians is therefore variable and the map is neither equivalent nor conformal. Transverse Mercator projections (onto a cylinder at right angles to the earth's axis) and **Lambert conic projections** form many official coordinate systems; the **Universal Transverse Mercator (UTM)** is one standard system in widespread use.

marker bed: 1. A bed or sequence of beds which yields characteristic reflections over a more-or-less extensive area. 2. A bed which accounts for a characteristic segment of a refraction time-distance curve and which can be followed over a reasonably extensive area. 3. A bed with distinctive magnetic or electrical effects.

marker velocity: The velocity with which head (refracted) waves are transmitted along a marker bed.

Markovian variable: A random variable which has a distribution of values which depends only on the preceding sample. Compare **stochastic.**

marsh buggy: A self-propelled vehicle designed to operate over marsh or extremely soft ground, characterized by very low ground pressure. Some have large wheels with very wide treads, tracks, or buoyant wheels or tanks so that they will float in water.

masked layer: Hidden layer.

masking effect: The effect whereby a highly conductive layer near the surface dominates resistivity measurements so as to make undetectable the effects of deeper resistivity variations.

massive sulfides: Rocks that are more than 20 perent sulfides by volume, as opposed to disseminated sulfides, which involve electrical phenomena that behave like massive metallic substances.

master curve: One of a set of theoretical curves calculated from models, against which an observed curve is matched in an effort to find a fit sufficiently close that the model parameters can be considered applicable to the actual situation. Also called *type curve.* Especially used in resistivity surveying, to a lesser extent in gravity and magnetic interpretation.

master station: 1. A radio transmitting station of a **positioning**-system net from which the synchronization of the other transmitters in the net (slave stations) are controlled. 2. The control station of a network even where synchronization is not maintained.

matched filter: A filter which yields a large peak output in response to a signal of particular shape. The elements of a matched filter are the elements of the signal in reverse order; hence filtering with a matched filter is equivalent to crosscorrelating with the signal. Used where the waveform of the signal is known, as in deconvolving Vibroseis data. Also called *crosscorrelation filter* and *correlator.* The frequency characteristics of a matched filter are the same amplitude-frequency response and the same phase-frequency response, except reversed in sign, as the waveform to which it is matched. A matched filter is the most powerful filter for identifying the presence of a given waveform in the presence of additive noise. See Anstey, 1964, Correlation Techniques: A Review: Geophysical Prospecting, v. 12, p. 355-382, and Treitel and Robinson, 1969, Optimum Digital Filters for Signal to Noise Ratio Enhancement: Geophysical Prospecting, v. 17, p. 248-293.

matrix: A rectangular array of numbers, called elements (which may be complex), which obeys certain rules. An $m \times n$ matrix $\mathbf{A} = \|a_{ij}\|$ has m rows and n columns:

$$\mathbf{A} = \|a_{ij}\| = \begin{Vmatrix} a_{11}, a_{12}, \cdots, a_{1n} \\ a_{21}, a_{22}, \cdots, a_{2n} \\ \cdots \qquad \cdots \\ a_{m1}, a_{m2}, \cdots, a_{mn} \end{Vmatrix}.$$

A *null matrix* **O** has all elements = 0, and is of any size. The *identity matrix* **I** is square, has ones on the principal diagonal, and zero's for all other elements, and is of any size. A *column vector* has only one column, a *row vector* only one row. A *square matrix* has same number of rows as columns. The "order" of the square matrix $m \times m$ is m. A *diagonal matrix* has zeros for all elements not on the principal diagonal (a_{ii}). The *transpose* $\mathbf{A}^T$ of a matrix has the rows and columns interchanged, and hence is of $n \times m$ size:

$$\mathbf{A}^T = \begin{Vmatrix} a_{11}, a_{21}, \cdots, a_{m1} \\ a_{12}, a_{22}, \cdots, a_{m2} \\ \cdots \qquad \cdots \\ a_{1n}, a_{2n}, \cdots, a_{mn} \end{Vmatrix}$$

A *symmetric matrix* equals its transpose $\mathbf{A} = \mathbf{A}^T$. A *skew-symmetric matrix* equals the negative of its transpose $\mathbf{A} = -\mathbf{A}^T$. The *cofactor* of the element a_{ij} of a square matrix is $(-1)^{i+j}$ times the determinant which is given by deleting the ith row and the jth column. The *adjoint* or *adjugate,* adj **A**, of a square matrix is the transpose of the matrix, where each element is replaced by its cofactor. The *determinant,* det **A**, of a square matrix is a single number and not itself a matrix.

$$\det \mathbf{A} = \sum_i a_{ik} A_{ik} = \sum_k a_{ik} A_{ik},$$

where A_{ik} is the cofactor of a_{ik}. The *inverse* of a matrix multiplied by the matrix yields the identity matrix

$$\mathbf{A}^{-1} \mathbf{A} = \mathbf{I}.$$

The inverse may be found by dividing the adjoint by the determinant:

$$\mathbf{A}^{-1} = \frac{\text{adj } \mathbf{A}}{\det \mathbf{A}}.$$

An *orthogonal matrix* has its inverse equal to its transpose: $\mathbf{A}^T = \mathbf{A}^{-1}$. Vectors $\mathbf{K}_i$ and $\mathbf{K}_j$ are "orthogonal" if $\mathbf{K}_i^T \mathbf{K}_j = 0$. A *Hermitian matrix* is a square matrix of complex numbers which equals the transpose of its complex conjugate: $\mathbf{H}^{*T} = \mathbf{H}$. A *minor* is a square determinant formed by deleting rows and/or columns. The *rank* of a matrix is the order of the largest nonvanishing minor. Matrices of the same size may be added:

$$\mathbf{C} = \mathbf{A} + \mathbf{B} \text{ and}$$

$$c_{ij} = a_{ij} + b_{ij}.$$

Matrices may be multiplied by a scalar:

$$\mathbf{D} = k\mathbf{A} \text{ and}$$

$$d_{ij} = ka_{ij}.$$

A matrix with m columns (**A**) may multiply from the left by a matrix with m rows (**B**):

$$\mathbf{E} = \mathbf{AB} \text{ and}$$

$$e_{ij} = \sum_{k=1}^{m} a_{ik} b_{kj}.$$

If **A** is of size $p \times m$ and **B** is $m \times n$, **E** is of size $p \times n$. Multiplication is (in general) not commutative.

$$(\mathbf{AB})^T = \mathbf{B}^T \mathbf{A}^T \text{ and}$$

$$(\mathbf{AB})^{-1} = \mathbf{B}^{-1} \mathbf{A}^{-1}.$$

If $\mathbf{X} = \mathbf{A}\,\mathbf{Y}$ is a set of linear equations, the unique solution is $\mathbf{Y} = \mathbf{A}^{-1}\mathbf{X}$, if det $\mathbf{A} \neq 0$. The *rank* of **A** is the number of such equations which are linearly independent. If the equations are inconsistent, the *least-squares* solution is

$$\mathbf{A}^T \mathbf{A} \mathbf{Y} = \mathbf{A}^T \mathbf{X}.$$

If **A** is square matrix of order n, the *eigenvalues* or *latent roots* λ are such that $\mathbf{A}\,\mathbf{q} = \lambda\,\mathbf{q}$. There are up to n such eigenvalues (some of which may be identical). The *eigenvector* $\mathbf{q}_i$ corresponding to λ_i consists of the n cofactors of a row of the matrix $\mathbf{A} - \lambda_i\,\mathbf{I}$. If **A** is symmetric, the matrix of the mutually orthogonal eigenvectors $\mathbf{q}_i$

$$\mathbf{\Lambda} = \mathbf{q}_1, \mathbf{q}_2, \cdots, \mathbf{q}_n.$$

has the property that $(\mathbf{\Lambda}^T \mathbf{A}\, \mathbf{\Lambda})$ is diagonal. Matrices are used in electromagnetic work, in wave propagation studies, in multichannel processing, and elsewhere.

maximum: An anomalous area in which measurements show larger values than in neighboring areas, as *gravity maximum.*

maximum convexity: The curvature on a seismic record section of a diffraction from a point in the plane of the section. This is the greatest convexity a coherent wave train can have except for reverse branches in buried-focus situations, diffracted reflections, and other special situations. See **diffraction curve.**

maximum-delay: Maximum-phase.

maximum-energy-sum filter: Output-energy filter.

maximum-phase: A two-term wavelet (a, b) is maximum-phase if $|a| < |b|$. Any wavelet is maximum-phase if the two-term wavelets into which its Z-transform can be factored are each maximum-phase. Same as maximum-delay. See **phase** and Figure 119 **(minimum-phase).**

Maxwell's equations: A set of differential equations to which electric and magnetic fields are subject at every ordinary point:

$$\nabla \times \mathbf{E} = \frac{-\partial \mathbf{B}}{\partial t}, \qquad \nabla \cdot \mathbf{D} = \rho, \text{ and}$$

$$\nabla \times \mathbf{H} = \mathbf{J} + \frac{\partial \mathbf{D}}{\partial t}, \qquad \nabla \cdot \mathbf{B} = 0,$$

where **E** and **H** are the electric and magnetic field intensities, respectively, **B** is the magnetic induction, **D** is the electric displacement, **J** is the current density, and ρ the charge density. In the cgs system, $1/c$ (where c = velocity of light in a vacuum) precedes the time derivatives and 4π precedes the **J** and ρ. An equivalent system of integral equations can be formulated.

md: Millidarcy or 1/1000 darcy.

M discontinuity: Mohorovčić discontinuity or **Moho.**

mean: Average. The *arithmetic mean* of n values x_i is $\frac{1}{n}\sum x_i$. The *geometric mean is* $\left[\frac{1}{n}\sum x_i^2\right]^{1/2}$. A "weighted average" is $\sum w_i x_i / \sum w_i$, where w_i = weights, and often $\sum w_i = 1$.

M-electrode: The potential drop is measured between the M- and N-electrodes in electrical-resistivity measurements.

membrane polarization: The induced-polarization effect primarily due to restrictions of ion mobility, as opposed to electrode polarization. See also **normal effect.**

memory: A part of a computing machine in which data can be stored and from which it can later be retrieved. Types of memory include core, magnetic drum, magnetic tape, registers, etc.

memory function: Impulse response or memory curve.

Mercalli scale: A descriptive scale which indicates the degree of shaking at a specific location as a result of an earthquake. An abridged version of the modified Mercalli scale is given in Table 7. Compare **magnitude** which is a measure of an earthquake independent of location.

Table 7. Modified Mercalli scale (abridged) of earthquake effect intensity

I.	Not felt
II.	Felt by persons at rest.
III.	Hanging objects swing; vibration like passing of light trucks.
IV.	Vibration like passing of heavy trucks.
V.	Felt outdoors; sleepers awakened; liquids disturbed; unstable objects displaced.
VI.	Felt by all; glassware broken; books off shelves.
VII.	Difficult to stand; noticed in motor cars; damage to some masonry; weak chimneys broken at roof line.
VIII.	Partial collapse of masonry; twisting, fall of chimneys; frame houses moved on foundations.
IX.	General panic; general damage to foundations; underground pipes broken; conspicuous cracks in ground.
X.	Most structures destroyed; large landslides; water thrown on banks.
XI.	Rails bent greatly; underground pipelines out of service.
XII.	Damage nearly total.

Mercator projection: A conformal cylindrical map projection developed on a cylinder tangent along the equator with the expansion of the meridians equal to that of the parallels.

mesosphere: The lower of three zones into which the outer part of the earth is divided, overlain by the asthenosphere and the lithosphere. The core is below the mesosphere. See Figure 63 **(earth layering).**

message: The desired information being sought. A satellite "message" consists of the ephemeral and Kepler data broadcast by the satellite which gives the location of the satellite; as opposed to the observation of the doppler shift of the carrier frequency which gives the position of an observer with respect to the satellite.

metal factor: A measure of the total change in conductivity or the capacitivity of a rock used in the interpretation of IP data; often written MF. Metal factor is the percent-frequency-effect (PFE) normalized by dividing by the measured resistivity (ρ in ohm-feet) and multiplied by a constant.

$$\text{MF} = 2\pi 10^3 (\text{PFE})/\rho.$$

1. Originally metal factor was defined as

$$\text{MF} = 2\pi 10^5 (\rho_{dc} - \rho_{ac})/\rho_{ac}\rho_{dc},$$

where ρ_{dc} is the low-frequency resistivity and ρ_{ac} is the high-frequency resistivity in ohm-feet. **2.** Most groups working in the frequency domain use the definition

$$\text{MF} = 2\pi 10^5 (\rho_{dc} - \rho_{ac})/\rho_{ac}^2.$$

3. In the time domain, the metal factor is

$$\text{MF} = 2000\, M/\rho_{dc},$$

where M is chargeability in millivolt-seconds per volt and ρ_{dc} is in ohm-meters. This unit is similar to Keller's parameter, **specific capacity** or *static capacity.* **4.** A constant times the ratio (sometimes called *electric susceptibility* or *capacitivity)* between induced-current dipole moment per unit volume (*P*) and electric field (*E*):

$$MF = -2 \times 10^6\ P/E.$$

The metal factor thus has units of capacitivity. It is also called *metallic-conduction factor.*

metallic-conduction factor: Metal factor.

Mev: Million electron volts.

mf: Subscript used with log terms to indicate values for the mud filtrate.

MF: Metal factor.

mgal: Milligal; 10^{-3} gal or 10^{-5} m/sec^2. A unit of acceleration used with gravity measurements.

mho: A unit of **conductance** or admittance, the reciprocal of **ohm.** Also called **siemens.**

mho per meter: A unit of conductivity; the conductivity for which a meter cube offers a resistance of one ohm between opposite faces. Reciprocal of ohm-meter.

mickey-mouse: Improvised; a short-cut method which may sacrifice rigor.

micro-: A prefix meaning 10^{-6}. Symbol: μ.

microhead: Very small **magnetic head** for use with **Decatrack** and similar equipment.

microinverse: See **microlog.**

microlaterolog: A microresistivity log of the **laterolog** type with a bucking electrode and two monitor electrodes arranged concentrically on a pad which is pressed against the formation so that the current flow is concentrated into a gradually flaring tube. Since the spacing is small, the log responds to the resistivity of a small volume of formation in front of the pad. Similar to the minifocused log, FoRxo log, or trumpet log. For the Proximity log the electrodes are mounted on a wider pad and the focusing is such that thicker mud cakes have less effect. See Figure 117 **(microlog).** Microlaterolog is a Schlumberger trademark.

microlog: A type of microresistivity log using three button electrodes spaced in a line one inch apart and located on a pad which is pressed against the borehole wall. The lower electrode is the "A" current electrode. The potential of the upper electrode with respect to a reference electrode on the surface gives a 2-inch *micronormal* and the difference between the two upper electrodes gives a 1 1/2-inch *microinverse* (lateral-type measurement). Because the mud cake usually has appreciably smaller resistivity than the formation, the microinverse will read less than the micronormal when mud cake is present. This difference (called *separation*) indicates a permeable formation. A caliper log is usually recorded at the same time. Similar to the contact log or Minilog. Microlog is a Schlumberger trademark. Figure 117.

micronormal: See **microlog.**

micro-omega: See **omega.**

microphonics: Electrical noise generated by mechanical vibration.

microprogram: 1. A hardware program which controls how a computer functions. It determines how a computer interprets an instruction in machine language. **2.** A software program constructed from the basic subcommands of a computer which hardware translates into machine subcommands. This provides a means of building various instruction combinations out of the subcommand structure of the computer in order to maximize the utility of the computer.

micropulsations: Small amplitude fluctuations in the earth's magnetic field, usually in the frequency range from 0.01 to 3 hz. Micropulsations having amplitudes up to tens of gammas result from interactions between plasma emitted from the sun (solar wind) and the earth's field. Micropulsations are classified as continuous (pc), irregular (pi), pearl (pp), etc. See also **bay** and **magnetic storm.**

microresistivity log: A well log designed to measure the resistivity of the flushed zone about a borehole, recorded with electrodes on a pad pressed against the borehole wall. See **microlog** and **microlaterolog.**

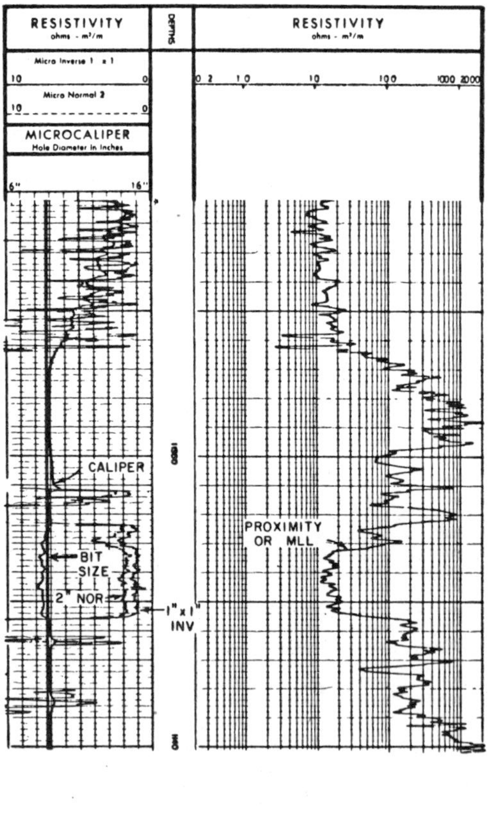

Fig. 117. **Microlog** (micronormal and microinverse) and microcaliper log (on left) and microlaterolog (on right). (Courtesy Schlumberger.)

Microseismogram log: Similar to the variable-density log or **three-D log**; see Figure 30 (**cement-bond log**). Microseismogram is a Welex trademark.

microseisms: Feeble earth tremors due to natural causes such as wind, water waves, etc.

microspread: A spread in which very short geophone group intervals (1 to 15 ft) are used, as in **noise analysis.**

migration: 1. Plotting of dipping reflections in their true spatial positions rather than directly beneath the point midway between the shotpoint and center of the geophone spread. Some automatic migration schemes utilize the complementary properties of wavefront curves and diffraction curves (curves of maximum convexity), shown in Figure 118 and described by Hagedoorn, 1954, A Process of Seismic Reflection Interpretation: Geophysical Prospecting, v. 2, p. 85-127. An unmigrated event is tangent to a diffraction curve (which defines the particular diffraction curve) and its arrival time defines the wavefront. The migrated event lies at the other intersection of these two curves and is tangent to the wavefront curve. Migration often is limited to the plane of the section, although, properly, three-dimensional positions should be considered. 2. Where the velocity varies laterally, data will migrate differently and ray tracing is used to determine the migrated positions. 3. Movement of ions in a solution because of electric-field gradient.

mil: 0.001 inch.

milli-: A prefix meaning 10^{-3}. Symbol m. Seismic events sometimes are timed in milliseconds (msec).

milligal: A unit of acceleration used with gravity measurements; 10^{-3} gal $= 10^{-5}$ m/sec^2. Abbreviated mgal.

milliradian: A unit of angle (or phase) measurement equal to 0.0573 degrees of arc. One degree equals 17.45 milliradians.

millisecond: 1. A thousandth of a second; abbreviated msec. 2. A unit of chargeability, the area under the decay curve of a pulsed (+, 0, −, 0) square wave. See **chargeability.**

Minilog: See **microlog**. Dresser Atlas tradename.

minimum: An anomalous area in which measurements show smaller values than in neighboring areas, as a *gravity minimum.*

minimum-delay: **Minimum-phase.**

minimum-phase: 1. A two-term wavelet or couplet *(a, b)* is minimum-phase if $|a| > |b|$. Same as *minimum-delay.* Any wavelet may be represented as the convolution of couplets; the wavelet is minimum-phase if all the couplets of which it is composed are minimum-phase. For example, the *z*-transform of a wavelet might be $6 + z - z^2$ which can be expressed as $(3 - z)(2 + z)$, each of which is minimum-phase and hence the wavelet is minimum-phase. Minimum-phase is sometimes expressed as having all roots outside the unit circle in the **z-plane** or as having no zeros in the right half of the Laplace transform *S*-plane. 2. A multichannel matrix of vectors is minimum-phase if its determinant (which can also be expressed as the product of couplets) is minimum-phase. A multichannel response produced by impulsive inputs might be:

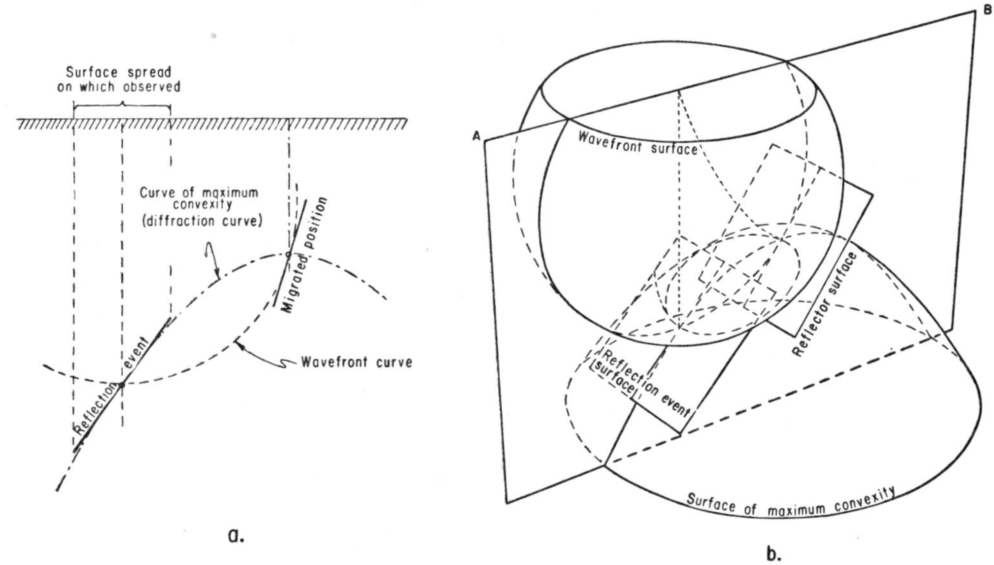

Fig. 118. **Migration.** (a) In two dimensions. (b) In three dimensions. (From Hagedoorn, Geophysical Prospecting, v. 2, no. 2, p. 96-97.)

	input channel 1	input channel 2
output channel 1	$(2 + z)$	(z)
output channel 2	(1)	$(6 + z)$,

which has the determinant $12 + 7z + z^2 = (3 + z)(4 + z)$ which is minimum-phase; hence the multichannel response is minimum-phase. The principal feature of minimum-phase is that the energy arrives in a burst which is not followed by a larger burst. **3.** A *minimum-phase filter* is that one of the set of possible filters with identical amplitude response, which delays the energy the least; it also is called the *minimum-delay filter.* If the input to a minimum-phase filter is itself minimum-phase, then the output will also be minimum-phase. Many of the filtering actions to which seismic signals are subjected are minimum-phase and much of the filtering done in digital processing is minimum-phase. A minimum-phase wavelet is sometimes called *front-loaded* because its energy is concentrated in the front end of the pulse. Maximum-phase or maximum-delay is the other extreme, and mixed-phase is intermediate. Figure 119.

minimum-time path: Least-time path.

Mintrop wave: Head wave or refraction wave.

mirror: Reflector.

misclosure: Mis-tie.

mise-a-la-masse method: An electrical-exploration method in which one current electrode is positioned in a conducting mineral either in outcrop or in a borehole. The other current electrode is a great distance away and the potential electrodes are moved about with the objective of mapping the mineral deposit. Also called *excitation-at-the-mass method.*

mis-tie: 1. The difference obtained on carrying a phan-

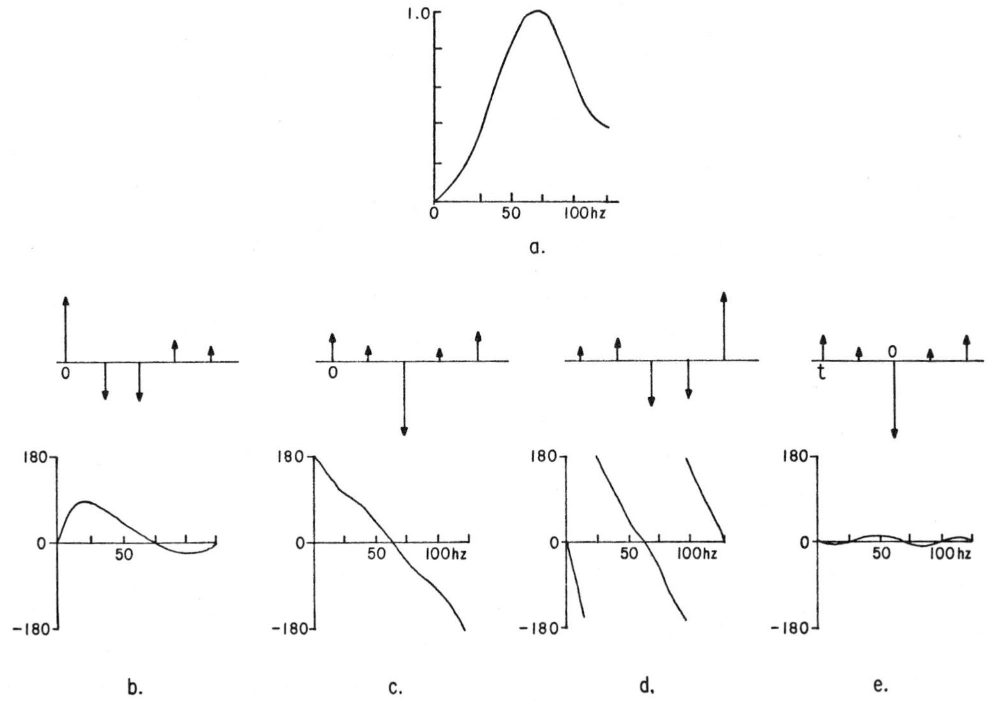

Fig. 119. **Minimum-phase,** linear phase, maximum phase, and zero phase.
(a) All wavelets have the same amplitude spectrum.
(b) Minimum-phase wavelet and phase spectrum:
$(1 - 0.8z)^2(1 + 0.5z)^2 = 1 - 0.6z - 0.71z^2 + 0.24z^3 + 0.16z^4$.
(c) Linear phase: $(1 - 0.8z)(0.8 - z)(1 + 0.5z)(0.5 + z) = 0.4 + 0.18z - 1.25z^2 + 0.18z^3 + 0.4z^4$.
(d) Maximum phase: $(0.8 - z)^2(0.5 + z)^2 = 0.16 + 0.24z - 0.71z^2 - 0.6z^3 + z^4$.
(e) Zero phase: $0.4z^{-2} + 0.18z^{-1} - 1.25 + 0.18z + 0.4z^2$; a zero-phase wavelet is anticipatory, i.e., begins before time zero. Phase curves depend upon the location of the time reference. Other mixed-phase wavelets can also be made from the same component doublets. (Courtesy Chevron Oil Company.)

tom or reflection or some other measured quantity around a loop. **2.** The difference of values at identical points on intersecting lines or of values determined by independent methods. **3.** The difference from zero of the algebraic sum of measured differences around a loop.

mixed-phase: A wavelet for which one or more of the component couplets into which it can be factored are **minimum-phase** and one or more is **maximum-phase.** Same as mixed-delay. See also **phase.**

mixing: Combining the energy of different channels, generally to cancel noise. Also called *compositing.* **1.** *Simple mixing* usually involves the combining of adjacent channels from the same record only. Mixed records frequently preserve two or more traces in unmixed form so that events can be picked on these traces without fear of distortion produced by the mixing. If mixing is done "toward the shot," the traces farthest from the shot are unmixed. The most common types of mixing are: **a.** simple 50 percent, in which the output traces (except the unmixed pair) contain equal contributions of energy from adjacent input channels; and **b.** *taper* or *resistive* mixes, in which input channels contribute to the output in inverse proportion to their proximity. For example, in a "50 percent taper mix," half of the energy in channel 6 would be added to the output of channels 5 and 7, a quarter to channels 4 and 8, an eighth to 3 and 9, except that the mix does not usually carry across the center of the record. **2.** Modern processing does mixing (though usually not so called) by various stacking operations, including **velocity filtering** and **coherency filtering,** by **ground mixing,** etc. See Table 10 **(stacking).**

mks: The meter-kilogram-second-ampere system of units. Also called **SI** (International System).
See **Appendix G** and Table 6 **(magnetic units).**

MM scale: Modified **Mercalli scale.**

mnemonic: An easily remembered code word such as designations of instructions to a computer.

mobility: The velocity of charge carriers per unit electric field. See **migration.**

mode: Manner of behaving. **1.** A form of behavior, as in a "Love wave of the first mode." **2.** A method of operation, as in "loran C in the rho-rho mode" or "doppler-sonar in the water-bottom mode."

mode conversion: Conversion of P-wave energy into S-wave energy or vice-versa by oblique incidence on an interface. See **Zoeppritz's equations.**

model: **1.** A concept from which one can deduce effects which can then be compared to observations; used in understanding the significance of the observations. The "model" may be conceptual, physical, or mathematical; Figure 120. For example, one might calculate the gravity, magnetic, or seismic effects for an assumed geological structure and compare these with actual data. Agreement between observations and effects derived from the model does not prove that the model represents the actual situation because of the lack of uniqueness in most geophysical interpretation problems. **2.** Some simple mathematical models are shown in Figure 121. The effects of more complicated models may be collected into a catalog of master curves or **type curves** for use in comparison with observed effects.

modem: A device which converts a digitized code into an audio signal which is more suitable for transmission over voice-grade telephone lines, or for converting the audio signal back into digitized code suitable for computer operations.

modulation: **1.** The process by which some characteristic of one signal is varied in accordance with another signal. Examples include frequency modulation (FM), amplitude modulation (AM), and pulse-width modulation; Figure 122. **2.** A measure of the intensity of magnetization impressed on a *direct recording* magnetic tape, often expressed as a percentage of that amount which will produce a certain degree of harmonic distortion. **3.** Sometimes used to imply the number of bits used to represent a maximum voltage in digital recording. See also **doublet modulation.**

modulus: The absolute magnitude of a complex number. If the complex number is $x + iy$, the modulus is $(x^2 + y^2)^{1/2}$.

modulus of elasticity: See **elastic constants.**

Moho: Mohorovčić discontinuity, the seismic discontinuity which separates the earth's crust and mantle. Situated 25-40 km below the continents, 5-8 km below the ocean floor, and 50-60 km where depressed by the roots of mountains. Characterized by a fairly abrupt increase of P-wave velocity from 6.5-7.2 to 7.8-8.5 km/sec, an increase of S-wave velocity from 3.7-3.8 to 4.8 km/sec, and an increase in density from approximately 2.9 to 3.2. The Moho is not the asthenosphere boundary, important in tectonics and gravity compensation. See Figure 63 **(earth layering).**

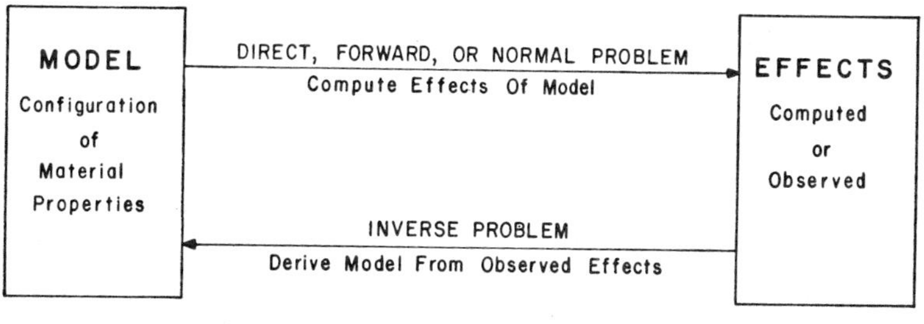

Fig. 120. The **model** and its effects.

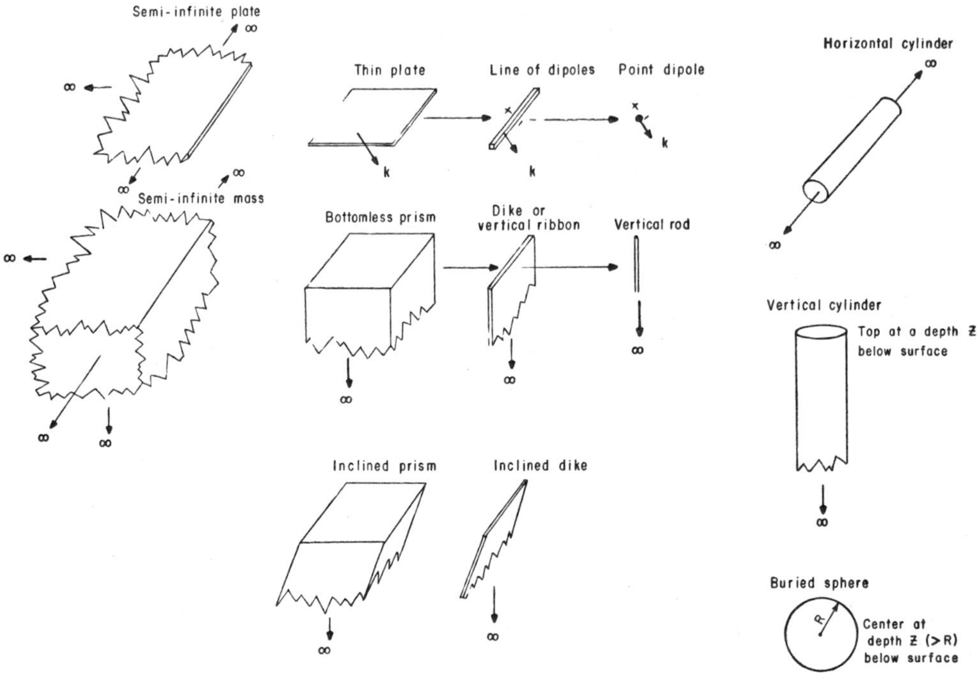

Fig. 121. **Models** of simple types.

monitor: 1. To inspect in order to verify that an operation is correct. 2. A piece of hardware or hard copy which permits verifying that an operation is correct.

monitor record: A record made as a check. 1. A record made in parallel with a magnetic-tape recording at the time of a shot. 2. A record of shipboard gravity and magnetic measurements made while the measurements are being obtained.

monochromatic: Containing only one frequency.

monocline: 1. A gentle structural flexure over which dip is everywhere in the same direction, though not of the same magnitude of dip. 2. A region of uniform dip, for which *homocline* is the preferred term.

Monte-Carlo method: A mathematical method whereby a calculation is repeated many times using random values. The result gives a statistical estimate for a result.

monument: An identifiable point on the ground to which surveys can be tied. May be an inscribed tablet on concrete, a steel fence picket with identification attached, etc. A bench mark.

moran: A positioning system similar to **shoran.**

motor rule: A rule for finding the direction in which a current will be deflected in a magnetic field. See Figure 101 **(induction).**

movable-oil plot: A well log calculated from other logs on which three **porosity** curves are plotted: "total" porosity ϕ_t such as is derived from the sonic log; "apparent water-filled" porosity ϕ_w, derived from a deep-investigation resistivity device such as the laterolog; and "apparent water-filled porosity of the flushed zone" ϕ_{xo}, derived from a shallow investigation resistivity device such as the microlaterolog. The separation between the first two curves indicates the volume fraction of hydrocarbons in the noninvaded zone and the separation between the last two curves indicates the volume fraction of movable oil. The remainder represents residual hydrocarbons left in the invaded zone. Abbreviated MOP. Figure 123.

moveout: Stepout. 1. The difference in arrival time at different geophone positions. Arrival times differ because of **a. normal moveout,** differences because of variable shotpoint-to-geophone distance along the reflection paths, **b.** dip moveout, differences because of reflector dip, and **c. statics,** differences because of elevation and weathering variations. See also **delta-t.** **2.** Dip moveout alone, especially across a single record. For corrected data from a single shot, often measured as the difference in arrival time Δt between an event at two geophone groups a distance Δx apart; hence, dip moveout $\Delta t/\Delta x$ might be measured in msec/1000 ft, for example. Where the difference in

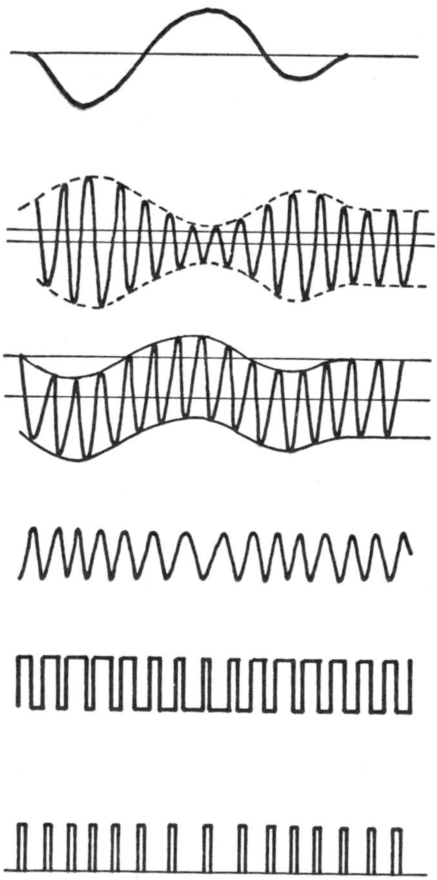

Fig. 122. **Modulation** types. (a) The message directly. (b) Amplitude modulation; the amplitude of the carrier wave varies as the message. The envelope of the carrier is the signal and its mirror image. (c) Ac-bias modulation; the bias varies as the message. (d) Frequency modulation; the frequency of the carrier wave varies as the message. (e) Pulse-width modulation; the timing of pulses varies as the message.

arrival time Δt is measured between the arrivals at shotpoints for two shotpoints a distance Δs apart, the dip moveout is $\Delta t/2\Delta s$, the factor of two being the ratio of spread length to subsurface coverage. The angle between the vertical and a reflection raypath at the shotpoint is $\sin^{-1}(V\Delta t/\Delta x)$, where V is the seismic velocity.

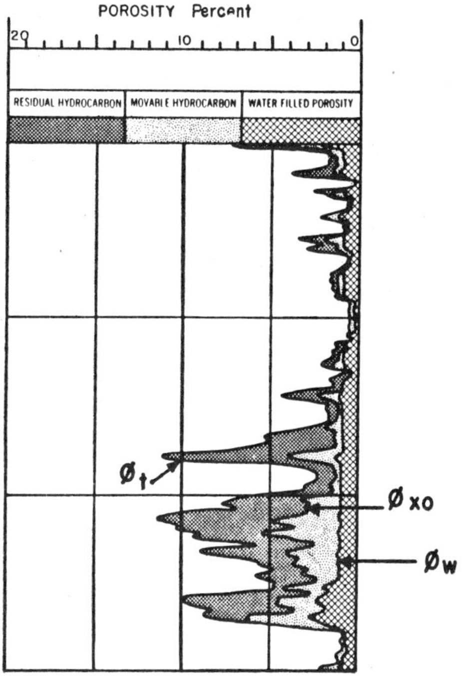

Fig. 123. **Movable-oil plot.** (Courtesy Schlumberger.)

moveout scan: Different amounts of normal moveout are introduced successively into a number of traces for different offset distances which are then stacked. Used to determine the optimum normal moveout for emphasizing certain events. See **velocity analysis.**

moving-coil geophone: A detector of seismic waves. See **geophone.**

moving-source method: A profiling method in which both source and receiver are moved about in a fixed configuration to explore an area. Usually applied to electromagnetic methods for which the free-space coupling between transmitter and receiver is fixed (unlike a **fixed-source method).**

MSB: Most significant bit.

msec: Millisecond or 10^{-3} second.

mud: An aqueous suspension used in rotary drilling. Mud is pumped down through the drill pipe and up through the annular space between it and the walls of the hole. The most common bases of drilling muds are bentonite, lime, and barite in a finely divided state. The mud helps remove cuttings, seal off porous zones, and hold back formation fluids.

mud cake: Filter cake, the residue deposited on the borehole wall as the mud loses filtrate into porous, permeable formations. The mud cake generally has very low permeability and hence tends to retard further loss of fluid to the formation. See **invaded zone.**

mud cup: A container used to measure mud resistivity.

mud filtrate: Fluid which enters permeable formations from the mud, leaving a mud cake on the borehole wall. See **invaded zone.**

mud log: A log made with a **microlog** sonde with the arms collapsed so that the measuring pad is not pressed against the borehole wall. Measures resistivity of the mud at in-hole conditions.

mud logging: Analysis of samples of the drilling fluid (or mud) and cuttings to detect signs of fluids which have entered the mud from the formations. Hydrocarbons may be evidenced by fluorescence, by chromatographic analysis, gas, and other ways. The mud is also monitored for salinity and viscosity to indicate water loss or "cut." Plots of such data often include a sample log and a drill-time curve.

multichannel filtering: 1. Filtering wherein the filter characteristics are partially based on the characteristics of other channels. **2.** A multichannel filter is often expressed as a matrix of output impulse responses on various channels to impulsive inputs on various channels:

	channel 1 input	channel 2 input	channel 3 input	···
channel 1 output	f_{11}	f_{12}	f_{13}	···
channel 2 output	f_{21}	f_{22}	f_{23}	
channel 3 output	f_{31}	f_{32}	f_{33}	
···	···			

multichannel processing: Data processing in which data from different input channels (different geophone groups, for example) are combined in some manner. Multichannel methods are involved in determining static trace-shift corrections, normal-moveout and stacking-velocity determinations, stacking, multiple attenuation, frequency filtering which is space and time variant, apparent-velocity (dip) filtering, and coherency filtering (including picking). See also **multichannel filtering.**

multiple: Seismic energy which has been reflected more than once. If this definition is strictly adhered to, virtually all seismic energy involves multiples, and primary energy which has been reflected only once is scarcely observable. The important distinction is between long-path and short-path multiples: a *long-path multiple* arrives as a distinct event whereas a *short-path multiple* arrives so soon after the primary that it merely adds tail to the primary. Usually it does not matter whether the reflections being mapped are primary but rather whether their attitude is indicative of the attitude of the appropriate portion of the section, so that the tailing produced by the many short-path *peg-leg* multiples does not detract seriously. The attitude of long-path multiples is not apt to be representative of the portion of the section associated with their arrival time. Usually the long-path multiples have traveled more in the slower (shallower) part of the section than primaries with the same arrival time, so that they ordinarily show more normal moveout and can be attenuated by CDP stacking. Figure 124.

multiple coverage: Seismic arrangement whereby the same portion of the subsurface is involved in several records, as with **CDP** shooting. The redundancy of measurements permits various types of noise to be attenuated in processing.

multiple geophones: A number of geophones (*group*) feeding a single channel; see **array.** Used **(a)** to attenuate ground roll and other undesirable energy which approaches the spread more or less horizontally (see Figure 54, **directivity graph**), and **(b)** to improve the signal-to-noise ratio by increasing the sampling and thereby randomizing planting factors, noncoherent energy, etc.

multiple regression: A mathematical procedure for finding the empirical equations which best fit a set of data in the least-square sense. See **factor analysis.**

multiple shotholes: Two or more shotholes which are shot simultaneously. The holes are usually spaced to minimize the effects of surface waves on the spread. Multiple shotholes are used in the same way as **multiple geophones.**

multiple-shot tool: A device run in a borehole to measure the direction of the borehole at several levels. See **directional survey.**

multiplex: 1. A process which permits transmitting several channels of information over a single channel without crossfeed. Usually different input channels are sampled in sequence at regular intervals and the samples are fed into a single output channel. Digital tapes are sometimes multiplexed in this way. Multiplexing can also be done by using different carrier frequencies for different information channels. **2.** A stereoscopic plotting instrument used in preparing topographic maps by stereophotogrammetry.

multiplexed format: A sequence of data in which the first sample of channel 1 is followed by the first sample of channel 2, then the first sample of channel 3, etc., until the first sample of all channels is given; then follows the second sample of channel 1, the second sample of channel 2, etc. As opposed to *trace-sequential format* in which the first sample of channel 1 is followed by the second sample of channel 1, etc., until all of channel 1 is given, followed by channel 2, etc. If such an array of data is thought of as a matrix,

	sample 1	sample 2	sample 3	···	sample n
channel 1	a_1	a_2	a_3	··	a_n
channel 2	b_1	b_2	b_3		b_n
···					
···	···			···	
···					
channel k	k_1	k_2	k_3	···	k_n

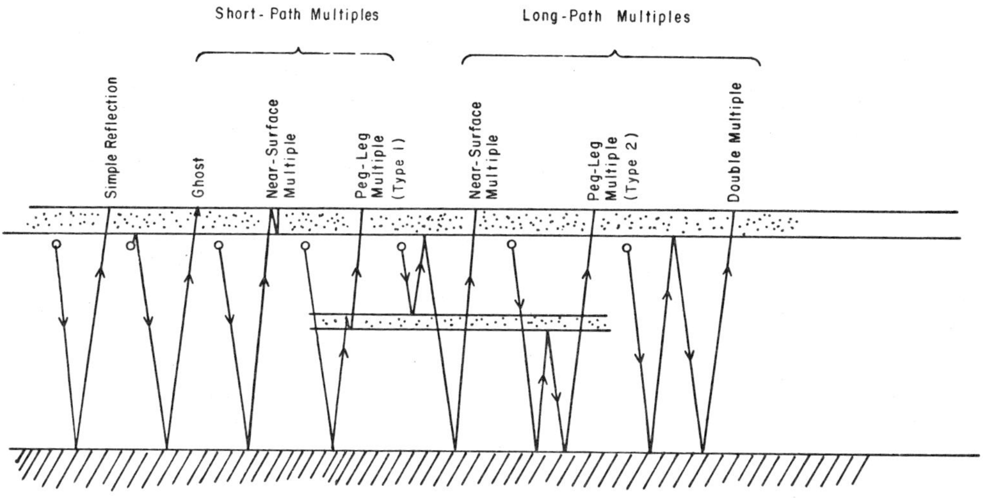

Fig. 124. **Multiple** types.

then multiplexed format may be thought of as reading by columns and trace-sequential format as reading by rows. The data output of digital recording equipment is in multiplexed format whereas most data processing is done in trace-sequential format. Converting from the one format to the other is one of the first steps in data processing and usually part of the **edit** routine. Such a format conversion is called *matrix transposition.*

multiplicity: **1**. Effort; the number of independent raypaths which add together to provide a single output trace. Given by the product of the number of source impulses times the number of geophones/group times the number of traces vertically stacked times the number of traces stacked in processing. **2.** The number of common-depth-point traces which sample essentially the same portion of reflector but with different offsets. For example, "12 fold" CDP recording involves recording each subsurface point 12 times, once from each of 12 different offset distances.

multivariant analysis: Factor analysis.

mute: To change the relative contribution of the components of a record stack with record time. In the early part of the record the long offset traces may be muted or excluded from the stack because they are dominated by refraction arrivals or because their frequency content after NMO correction is appreciably lower than other traces. The transition where they begin to contribute may be either abrupt or gradual. A large abrupt transition introduces frequencies which may not be otherwise present and may distort the design criteria for a deconvolution operator. Muting may be done over certain time intervals to keep ground roll, air waves, or noise bursts out of the stack. For this last function, compare **diversity stack**. Also called *fade.*

mutual: Relations between circuits, such as the mutual inductance, capacitance, or resistance (impedance) between the transmitter and receiver circuits of an IP survey system. See **coupling.**

nabla: Del, symbol ∇.
nadir: The point on the celestial sphere 180 degrees from **zenith.**
namelist: A semifree format for giving parameter values to a computer. A namelist might look something like this:

&LIST TIME = 3,NTRACE = 6,V = 5000,&END.

NAND gate: The negative of an AND gate. A circuit with multiple inputs which functions unless signal is present at all inputs. NAND for two inputs *A* and *B* is indicated $\overline{A.B}$; its truth table is:

A	B	$\overline{A.B}$
0	0	1
1	0	1
0	1	1
1	1	0

See Figure 88 **(gate).**
nano-: A prefix meaning 10^{-9}.
nanotesla: A unit of magnetic flux density. A nanotesla = 10^{-9} tesla = 1 gamma.
natural frequency: The oscillation or vibration frequency of a system in the absence of an oscillatory disturbing force.
Naudi method: Inflection-tangent-intersection method.
navigation: Directing a craft from one point to another; determining the position at a given moment and the direction and distance to a desired location. See **positioning.**
NCN: A nitrocarbonitrate shothole explosive; requires a primer to detonate.
near-dc: The commutated dc or low-frequency ac used in resistivity and IP surveying.
near field: The field near a source. Relationships near a source are apt to involve effects which attenuate so rapidly with distance that they are not involved when remote from the source. Hence, one sometimes distinguishes between near-field and far-field phenomena. If the distance from the source is *R* and the wavelength is λ, near-field implies $R < \lambda$. In the near-field zone of an antenna, fields vary predominantly as the inverse cube or square of the distance. Compare **far field.**
near surface: Often, the **weathering** or low-velocity layer.
near-trace gather: A record section which comprises only the data from the geophone group (or few groups) nearest the shot. Also called *short-offset section.*
Neel point: See **Curie point.**
negative area: An area subject to more-or-less continual subsidence.
negative frequency: The frequency of a sinusoidal wave train traveling in the negative direction.
negative IP effect: An IP decay voltage opposite in sign to that of the charging current, due to the geometric relationship of a shallow polarizable body and the measuring-electrode array.
negative screening: An electromagnetic phenomenon in which the presence of a conducting overburden enhances the anomaly due to a buried conductor. The effect involves current flow between the overburden and the conductor.
neper: A unit for measuring voltage ratios. *N* is the number of nepers if $N = \ln(V_1/V_2)$. It is thus similar to the decibel except using natural logarithms instead of logarithms to the base 10. One neper = 8.686 db.
neper/m: A unit of attenuation.
nest: **1.** An array of geophones, especially one containing many geophones close together; a patch. **2.** To embed a subroutine or block of data in a larger routine or set of data.
network: **1.** A set of points connected by communication channels. **2.** A system which converts an input function into an output function.
neutrally buoyant: Having the same buoyancy as the fluid in which it is immersed. Seismic streamers are usually nearly neutrally buoyant so that very little force is required to submerge or raise them.
neutral surface: The surface which separates compression and tension regions. When a layer is folded, the inside of the fold undergoes compression and the outside tension.
neutron-lifetime log: Similar to the thermal-decay-time log. A well log of the capture cross-section of thermal neutrons. A Van de Graaf neutron generator in the sonde periodically releases a burst of neutrons which enter the formation and begin to lose energy in collisions. At two discrete time intervals after a neutron burst, measurements may be made of the gamma rays which result from the capture of neutrons by nuclei (in the thermal-decay-time log), whereas measurements are made of the thermal neutrons in the neutron-lifetime log. The quantity plotted is (a) sometimes the reciprocal of the percentage which decay per unit of time, called the *thermal decay time* τ; (b) sometimes the time for the thermal neutron population to fall to half value, called the *neutron lifetime L*; and (c) sometimes the macroscopic **capture cross-section** Σ which is derivable from the foregoing ($\Sigma = 4.55/\tau = 3.15/L$). Thermal neutrons are captured mainly by the chlorine present and hence this log responds to the amount of salt in formation waters. Hydrocarbons result in longer decay times than salt water. Log readings are porosity-dependent and sensitive to clay content and permeability changes. This log is used in cased holes where resistivity logs cannot be run or to monitor reservoir changes to optimize production. It resembles a resistivity log with which it is generally correlatable. Dresser Atlas trademark. Figure 125.
neutron log: A porosity well log which measures mainly hydrogen density. Fast neutrons emitted by a source in the tool are slowed to thermal speed by collisions with (mainly) hydrogen atoms. The thermal neutrons are then captured by atomic nuclei of the surrounding material (mainly chlorine atoms) at which time a charactistic gamma ray of capture is given off. The neutron-log detector may record **(a)** the capture gamma rays ($n - \gamma$), **(b)** thermal neutrons ($n - n$), or **(c)** epithermal neutrons (those just above thermal speed). A low-hy-

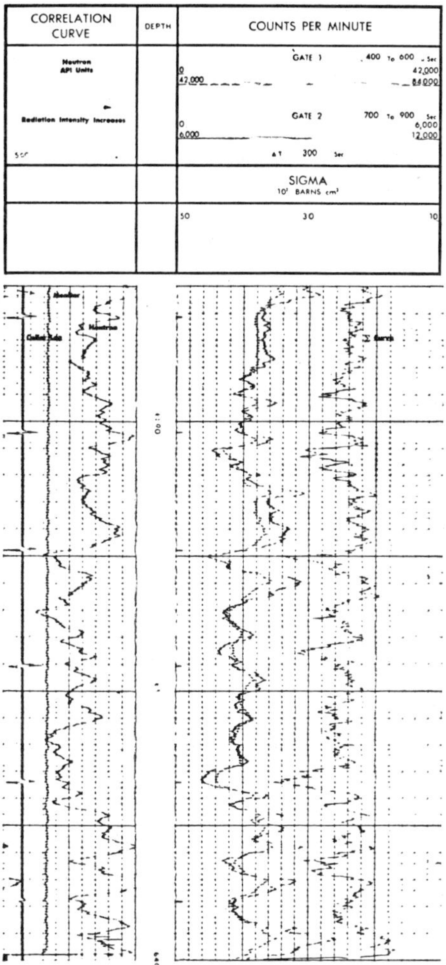

Fig. 125. **Neutron-lifetime log.** (Courtesy Dresser Atlas.)

drogen density indicates low liquid-filled porosity. Porosity calculated from the neutron log is affected somewhat by the formation matrix and by the presence of gas. Neutron logs are used in **crossplots** to detect gas and determine lithology. Neutron logs are sometimes scaled in API units, sometimes in porosity units assuming a limestone matrix. See also **sidewall neutron log.** The neutron log can be recorded in cased holes. Figure 126.

Newton-Raphson technique: An iterative method of finding a numerical solution of an equation, $f(z) = 0$. A value of z is tried, z_i ; the next trial value is $z_{i+1} = z_i - f(z_i)/f'(z_i)$, where f' is the first derivative with respect to z. Trials are repeated until successive values come sufficiently close together (where the process converges).

n-factor: The exponent by which the potential field caused by an anomalous mass falls off with distance. For example, n is less for a two-dimensional anomalous mass than for a three-dimensional one. For an actual anomaly, n may be obtained from use of a gradiometer arrangement or calculated from field measurements. Figure 127.

Nitramon: An ammonia-gelatin explosive, not cap sensitive, which requires a primer to detonate it.

NML: Nuclear-magnetism log.

NMO: Normal moveout.

NMR: Nuclear magnetic resonance. See **magnetic resonance.**

NNSS: Navy Navigation Satellite System; see **satellite navigation.**

nodal plane: The first motion from an earthquake will be either a push or pull, depending on the orientation of the station with respect to the epicenter and the direction of motion along the earthquake fault. A *nodal plane* separates the region where the first motion is a push and the region where it is a pull. There are two such planes for first energy P-waves, one of which is the fault plane.

node: A point where there is no wave motion as the result of interference between waves traveling in opposite directions.

noise: **1.** Any unwanted signal. **2.** A disturbance which does not represent any part of a message from a specified source. **3.** Sometimes restricted to energy which is random. To the extent that noise is random, it can be attenuated by a factor of $n^{1/2}$ by compositing n signals from independent measurements. **4.** Seismic energy which is not resolvable as primary reflections. In this sense noise includes microseisms, shot-generated noise, multiples (usually), tape-modulation noise, harmonic distortions, etc. Sometimes divided into *coherent noise* (including nonreflection coherent events) and *random noise* (including wind noise, instrument noise, and all other energy which is noncoherent). *Ambient seismic noise* refers to the background of random earth movements. Sometimes restricted to seismic energy not derived from the shot explosion. **5.** Disturbances in observed data due to more-or-less random inhomogeneities in surface and near-surface material. **6.** Noise in electrical or IP surveying can be due to atmospheric electrical discharges (sferics), interference from power lines, motor generator, or electronic components, or low-frequency magnetotelluric phenomena. **7.** Electrical circuit noise is caused by the randomness of conduction electrons **(Johnson noise)**, by the discreteness of magnetic transitions **(Barkhausen noise)**, by the discreteness of charge carriers in semiconductors **(shot noise)**, by modulation noise, and by other causes. **8.** *Geologic noise* is interference from unwanted geologic conditions.

noise analysis: A profile or set of profiles designed to gather data for an analysis of coherent noise trains. Usually consists of **microspreads** without any ground mixing so that low-velocity noise trains will be evident.

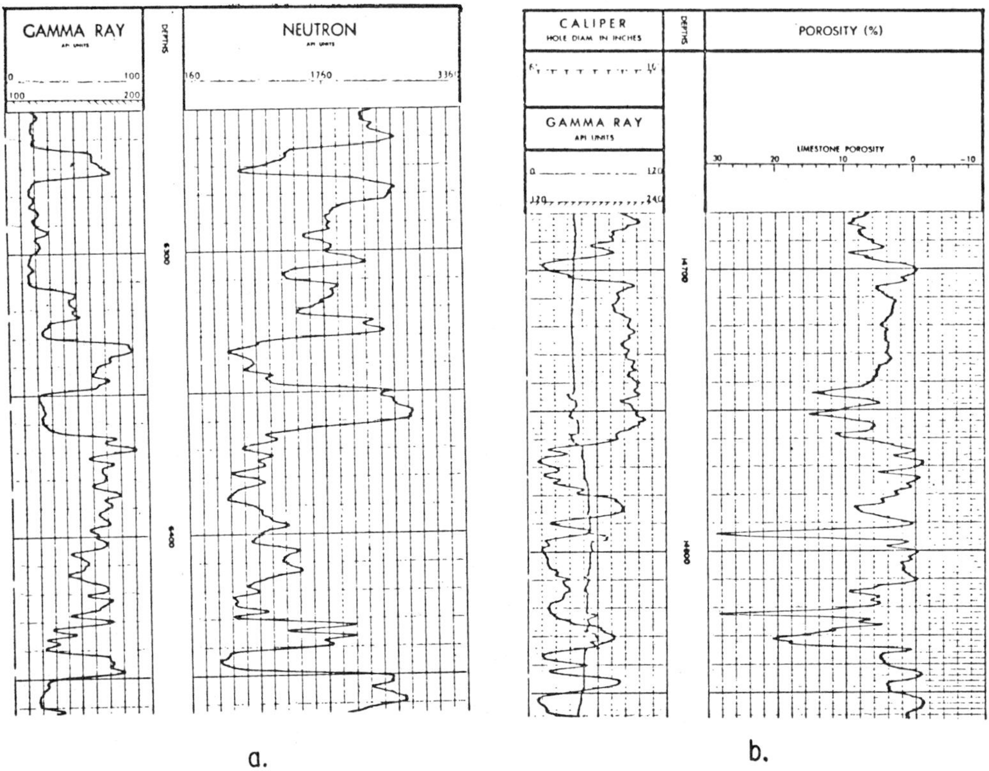

Fig. 126. **Neutron log** (a) and Sidewall Neutron log (b). (Courtesy Schlumberger.)

Results are often illustrated on a frequency versus wavenumber graph (see **f,k space**). Figure 128.

nonfaradaic path: The virtual passage of current near an electrode as a result of reorientation of the ionic layers of the double layer. The process is analogous to the charging of a capacitor and no charge carriers are transported across the interface.

nonnormal: Not **Gaussian.**

nonpolarizable electrodes: An electrode whose potential is not affected by the passage of current through

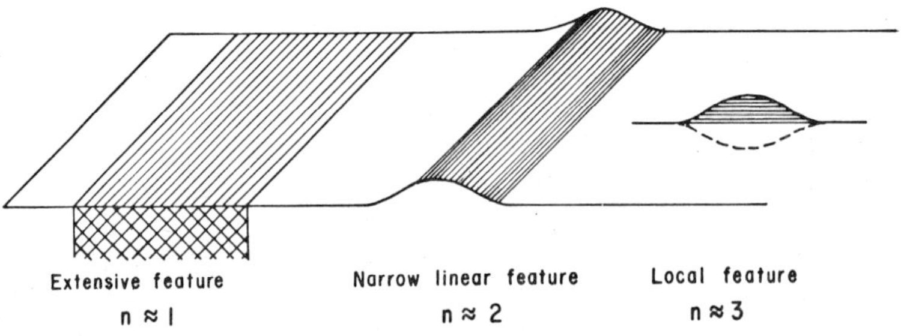

Fig. 127. **The n-factor.** The vertical gradient varies as $1/Z^n$, where Z is the distance from the mea surement elevation to the feature.

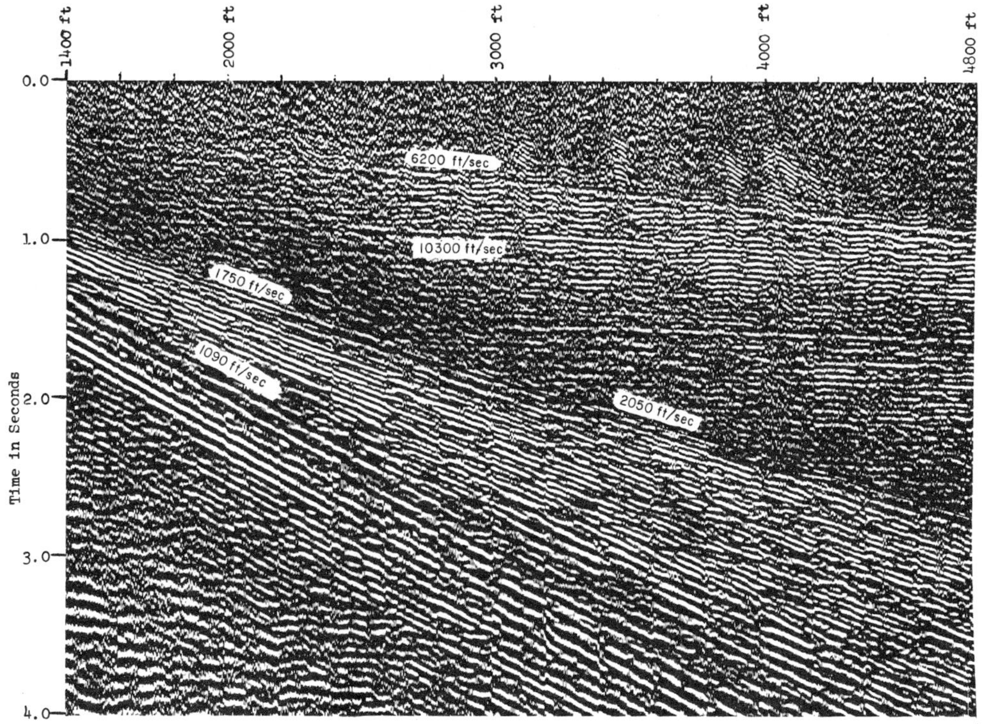

Fig. 128. **Noise analysis** or walkaway. Vibroseis source, geophones spaced 5 ft apart, offset to first phone is 1400 ft. The 6200 ft/sec arrival is a refraction from the base of the weathering; the 1750 and 2050 ft/sec arrivals are ground roll modes; the 1090 ft/sec arrival is an air wave. (Courtesy Chevron Oil Company.)

it. Electrodes which are free of potentials caused by electrochemical action between the electrode and the ground. See **porous pot** electrode.

NOR gate: The negative of an OR gate ("inclusive OR"). A circuit with multiple inputs which functions unless signal is present on any input. For inputs *A* and *B*, NOR is designated $\overline{A+B}$; its truth table is:

A	B	$\overline{A+B}$
0	0	1
0	1	0
1	0	0
1	1	0

See Figure 88 **(gate)**.

normal: **1.** Orthogonal; a line perpendicular to a surface or to another line. A condition of being perpendicular to a surface or line. **2.** A resistivity well log in which a constant current is passed between a current electrode in the sonde and a remote electrode (electrodes A and B) while the potential difference is measured between another electrode in the sonde and a reference electrode at the surface (electrodes M and N). The *spacing* is the distance between the A- and M-electrodes for the normal. A spacing of about 16 inches is used for the short normal and 64 inches for the medium or long normal. See Figure 129 and Figure 176 **(SP)**. **3.** In the absence of an anomaly, as in a *normal time-distance curve*. **4. Gaussian.**

normal correction: **1.** Subtracting the **normal magnetic field** from magnetic data. **2.** Normalizing the ratio of successive Turam readings by dividing by the ratio of the primary fields. Differs from the free-air Turam correction wherever the ground is conductive.

normal dispersion: A decrease of frequency with time

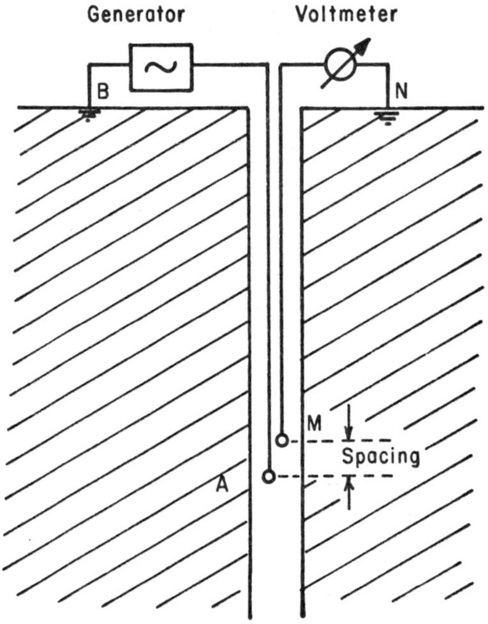

Fig. 129 **Normal** electric-log configuration.

for a seismic surface wave train. See **dispersion.**

normal distribution: See **Gaussian distribution.**

normal effect: An unwanted background IP effect due in part to membrane polarization, found to some extent in most rocks. See also **background polarization.**

normal equations: The set of linear simultaneous equations whose solution represents a least-squares fit (in particular, a **Wiener filter**):

$$\phi_{zx}(\tau) = \sum_t f_t\, \phi_{xx}(\tau - t).$$

A digitized input x_t passed through the filter f_t yields the actual output y_t which is optimum in a least-squares sense compared with the desired output z_t. $\phi_{xx}(\tau)$ is the autocorrelation of x_t and $\phi_{zx}(\tau)$ is the crosscorrelation of z_t with x_t. See **Levinson algorithm.**

normal fault: See **fault.**

normal gravity: The value of gravity at sea level according to a theoretical formula which assumes the earth to have a simple regular ellipsoidal shape. See **latitude correction.**

normal incidence: A wavefront striking an interface broadside; i.e., so that the angle between the wavefront and the interface (angle of incidence) is zero.

normalize: Forming a ratio of data with respect to a particular standard (the *normal*). A normalized value usually is dimensionless. Normalizing often consists of scaling such that "something equals one." The "something" may be the rms value, the maximum value, etc. For example, an array may be normalized by dividing each value by the rms (or energy) of the array. Hence for the array $X = (x_1, x_2, \ldots, x_n)$, the rms $= Y = [(x_1^2 + x_2^2 + \ldots + x_n^2)/n]^{1/2}$ and the normalized array is $(x_1/Y, x_2/Y, \ldots, x_n/Y)$. Autocorrelations are normalized by dividing by the value at zero time lag so that the maximum value of one indicates perfect correlation. *Type curves* (calculated effects for a model body) often are normalized so that the maximum effect is one.

normalized apparent resistivity: Apparent resistivity divided by the resistivity of the upper layer. In constructing type curves, normalized apparent resistivity is plotted against normalized electrode interval (electrode interval divided by depth to the second layer). See **apparent resistivity curve.**

normal log: See **normal.**

normal magnetic field: The smoothed value of a magnetic-field component as derived from a large-scale survey of continental scope or from satellite measurements. The normal field of the earth varies slowly with time and maps of it are as of a certain date.

normal moveout: The variation of reflection arrival time because of variation in the shotpoint-to-geophone distance (offset). Normal moveout depends on velocity and (to a lesser extent) dip as well as offset and decreases with reflection time. Many **velocity analysis** methods are based on normal-moveout measurements. To the extent that **long-path multiples** travel at a lower average velocity than primary reflections for the same arrival time (because velocity usually increases with depth), they will show greater normal moveout and can be attenuated by **common-depth-point stacking.** Offset sometimes is thought of as an independent variable and normal moveout is illustrated in **offset space.**

normal-moveout remover: Apparatus to remove normal moveout from a magnetic-tape record.

normal ratio: Ratio between the readings of two coils in electromagnetic surveying, in the absence of conductive material. See **primary ratio.**

normal traveltime curve: A time-distance curve (usually for first arrivals only) for a geologic section which does not contain geologic structures of the type sought. Departures from the normal may indicate structures. Used in **fan shooting.**

northing: Distance north of an east-west reference line. See **latitude.**

nose: 1. A plunging anticline with structural closure in three out of four directions. **2.** A gravity or magnetic anomaly for which the contours do not close; usually prefixed as *gravity nose* or *magnetics nose.*

notch filter: A **filter** which is designed to remove a single narrow band of frequencies. Often used to remove **high-line** effects.

NR: No reflection events. **1.** Denotes event not believed to be a reflection. **2.** Denotes absence of reflections at certain locations; used on maps and sections.

NRM: Natural **remanent magnetization.**

NRZ: Nonreturn to zero. A method of digital recording on magnetic tape in which magnetization in one direction indicates a zero and the opposite direction indicates a one. Compare **NRZI**.

NRZI: A method of digital recording on magnetic tape in which a change in the direction of magnetization indicates a one and no change indicates a zero. Compare **NRZ**.

NS: Not shot; designates a scheduled shotpoint location (on a map) which has not been shot.

NSC: Necessary and sufficient condition; a minimum complete set of conditions which are necessary for a solution to exist or for a situation to be true.

n-type semiconductor: A doped semiconductor with more electrons than holes available for carrying charges. Also called a *donor.*

nuclear cement log: A well log of scattered gamma rays, differing from the **density log** in that the gamma-ray source and detector are so spaced as to be sensitive to the density of material in the annulus. Used for distinguishing between cement and fluids behind casing. Can be run in empty hole.

nuclear-magnetism log: A well log which is dependent on the alignment of the magnetic moment of protons (hydrogen nuclei) with an impressed magnetic field. Abbreviated NML and also called free-fluid log. Protons tend to align themselves with the magnetic field and when it is removed they precess in the earth's magnetic field and gradually return to their original state. The proton precession produces a radio-frequency signal whose amplitude is measured as the free-fluid index (FFI). The rate of decay of the precession signal depends on interactions with neighboring atoms and hence on the nature of the molecule of which the proton is a part. The signal from the borehole fluid decays very rapidly because of disseminated iron (from steel worn from drill pipe and bits); by slightly delaying the time of measuring, the hole signal is minimized. Fluids bound to surfaces (such as the water in shales) do not give appreciable response. Thus, the FFI indicates the free fluid (the hydrogen in free fluid hydrocarbons and water). Gas gives a low reading because of its low hydrogen density. Sometimes thermal relaxation time, the rate of polarization buildup as a function of polarizing time, is measured to distinguish between water and oil.

nuclear-precession magnetometer: Magnetometer utilizing nuclear resonance; the resonance frequency is proportional to the absolute magnetic-field strength. See also **proton-resonance magnetometer** and **optically pumped magnetometer.**

null: Zero. A null instrument is one in which a balance is indicated by some quantity becoming zero, such as the current in one arm of a bridge circuit. The magnitude of the balancing "force" is then proportional to the quantity to be measured. Thus, spring tension in a gravimeter is adjusted until the spring tension balances out the gravitational force, the balance condition being indicated by null deflection.

number system: A method of coding numbers for digital-computer storage and manipulation. Systems include decimal (base 10), binary (base 2), quinary (base 5), octal (base 8), **biquinary, binary coded decimal,** (BCD), gray code, duodecimal (base 12), hexadecimal (base 16), one's complement, two's complement, etc. See Table 8.

Table 8. Number systems

decimal (base 10)	binary (base 2)	gray code	quinary (base 5)	octal (base 8)	BCD	BCDXS3	Biquinary
0	0	0	0	0	0000	0011	0 000
1	1	1	1	1	0001	0100	0 001
2	10	11	2	2	0010	0101	0 010
3	11	10	3	3	0011	0110	0 011
4	100	110	4	4	0100	0111	0 100
5	101	111	10	5	0101	1000	1 000
6	110	101	11	6	0110	1001	1 001
7	111	100	12	7	0111	1010	1 010
8	1000	1100	13	10	1000	1011	1 011
9	1001	1101	14	11	1001	1100	1 100
10	1010	1111	100	12	1 0000	100 0011	001 0 000
11	1011	1110	101	13	1 0001	100 0100	001 0 001
12	1100	1010	102	14	1 0010	100 0101	001 0 010

nutation: Motion of the axis of rotation of a body about its mean position.

Nyquist frequency: A frequency associated with sampling which is equal to half the sampling frequency. Also called *folding frequency.* Frequencies greater than the Nyquist frequency will **alias** as lower frequencies from which they will be indistinguishable.

Nyquist theorem: Sampling theorem.

object program: A computer program in machine language. Compare **source program.**

observer: **1.** The man in charge of the recording on a seismic crew. Sometimes the observer is also the field manager; sometimes he is principally an electronic technician. **2.** The man who reads the gravimeter on a gravity crew.

Occam's Razor: The simplest explanation of observations is the most probable. A dictum of scientific reasoning.

occultation: An eclipse of a star by the moon. The observation of an occultation from different points permits determining the geodetic locations of the observation points. The moon-position camera which photographs the moon against a background of stars is also used.

oceanic trench: A long narrow depression of the ocean floor. In **plate-tectonics** terms, often a discontinuity of the lithosphere where one plate plunges beneath another plate. Commonly associated with a volcanic island arc in the overriding plate displaced from the trench.

octave: Separation of two frequencies having a ratio of 2 (or 1/2). Filter rolloff is often given in decibels per octave.

odd function: A function which changes sign when its argument changes sign:

$$F(x) = -F(-x).$$

Also called *antisymmetric function.*

oersted: A unit of magnetic-field intensity (cgs-emu system) in free space; the field which would exert a force of 1 dyne on a unit magnetic pole. Equal to 1 ampere turn/($4\pi 10^3$m). See Table 6 **(magnetic quantities).**

off-line: An operation which is not continuous with that of the main system or an instrumental element which is not in the mainstream of data flow through a processing system. For example, an off-line plotter in a playback system is not directly connected to the main processing system.

offset: See Figure 130. **1.** The distance from the shotpoint to the center of the nearest geophone group. Often resolved into components: *perpendicular offset,* the distance at right angles to the spread line, and

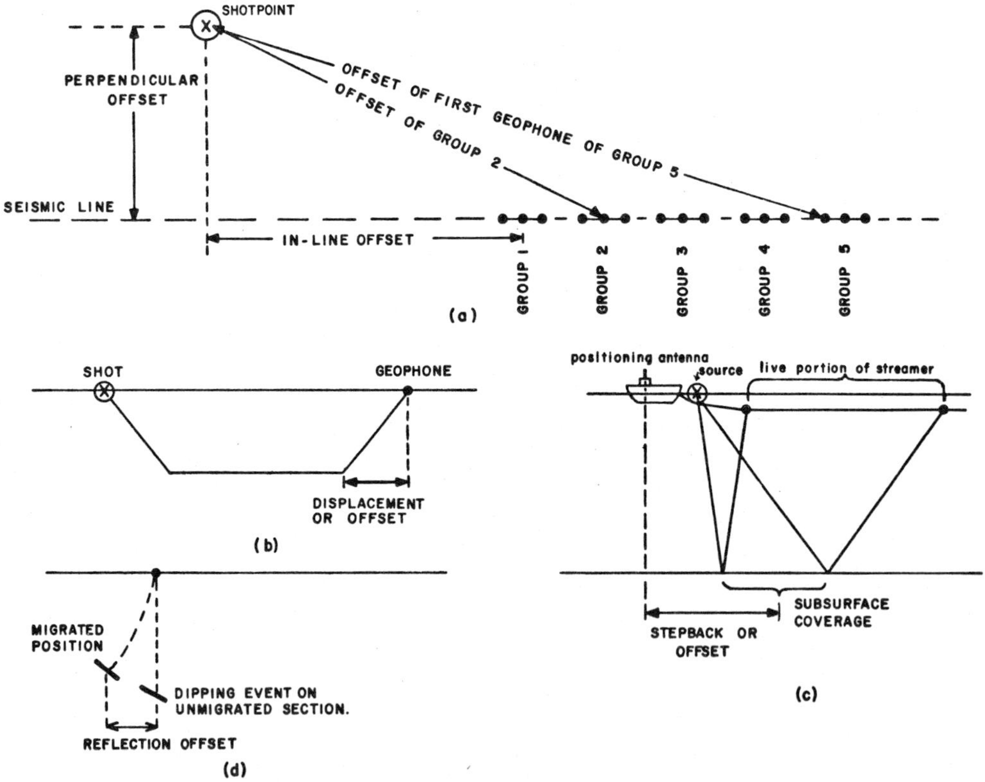

Fig. 130. **Offset** definitions.

inline offset, the distance from the projection of the shotpoint onto the line of the spread. **2.** The perpendicular offset only. **3.** The distance between the shotpoint and the center of any geophone group. **4.** The distance between the shopoint and any geophone. **5.** Sometimes (in refraction work) the **displacement**. **6.** In marine shooting, the distance from the recording boat where the radio-location equipment is usually located to the mid-point of the subsurface coverage; **stepback.** **7.** Displacement of the plot of a reflection to its proper position on a cross-section; **migration.** **8.** Displacement of a formerly contiguous body.

offset frequency: **1.** The difference between a frequency and a reference frequency. **2.** The difference between the frequency broadcast by a navigation satellite and 400 Mhz (or 150 Mhz for the low-frequency channel).

offset space: A way of thinking of seismic reflection data in which the independent variables are arrival time, location along the seismic line, and offset. Events curve in the offset direction because of **normal moveout,** which is involved in velocity analysis. See Figure 131.

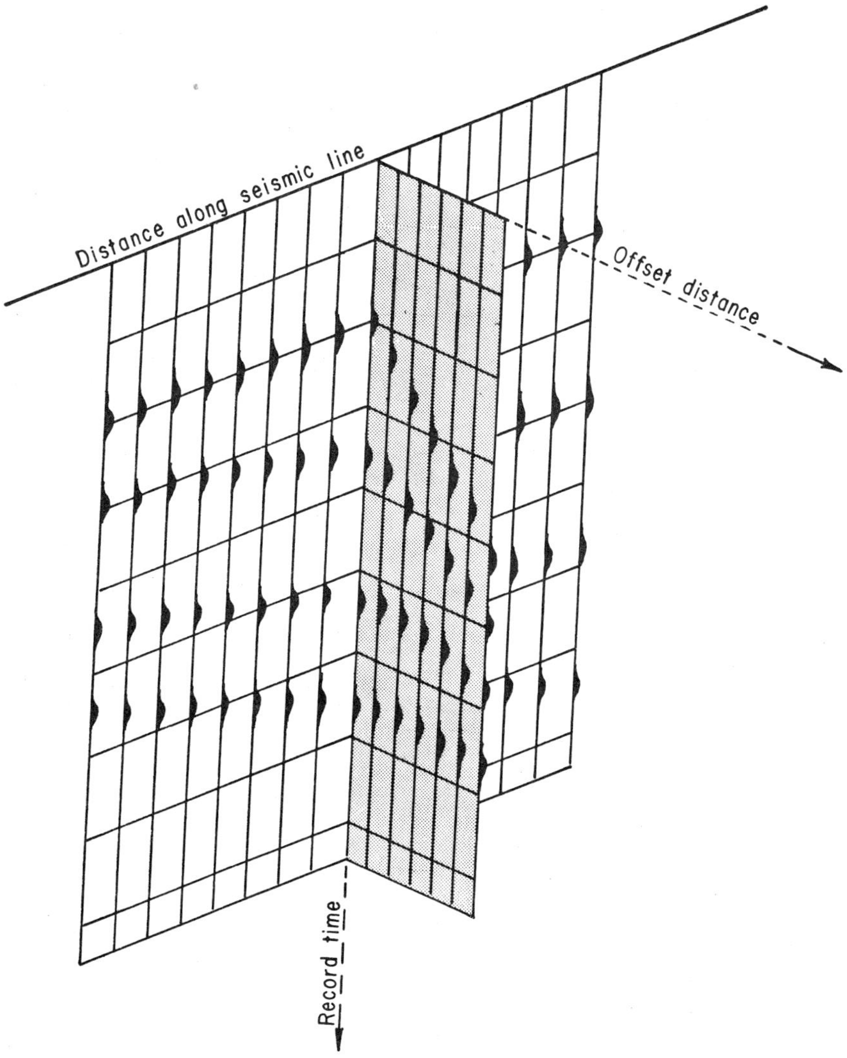

Fig. 131. **Offset space.** (Courtesy Compagnie Generale de Geophysique.)

off time: The time an IP pulse-type or an electromagnetic time-domain transmitter is off, during which the decay voltage is measured at the receiver.

ohm: A unit of electrical resistance or impedance; a resistance or impedance of one ohm has a potential drop across it of one volt per ampere of current.

ohmic: **1.** An electrical linear system, which obeys Ohm's law. **2.** The resistive component, as opposed to the reactive component, of an impedance.

ohm-meter: A unit of **resistivity**, also written ohm-meter^2/meter, being the resistivity of a meter cube which offers a resistance of one ohm to the flow of current between opposite faces. Reciprocal of mho per meter.

Ohm's law: The voltage across a linear element equals the current through it times its resistance. Earth materials are not necessarily linear and therefore do not always obey Ohm's law at high current densities.

Omega: A long-range very-low-frequency **positioning** system developed by the U.S. Navy, intended to provide all-weather positioning with an accuracy of about one mile (comparable with celestial navigation). Three frequencies (10.2, 13.6, and 11.333 khz) are transmitted in a sequential pattern synchronized in phase by atomic clocks (see Figure 132). The Omega receiver measures the difference in phase of signals from a pair of transmitters to define a hyperbolic line of position. Four stations presently (1973) in service (in Norway, Trinidad, New York, and Hawaii) cover the North Atlantic and eastern North Pacific oceans. The planned addition of four more stations will provide world-wide coverage. The differential-Omega technique (also called micro-Omega) uses information from a fixed receiver less than 100 miles from the mobile receiver to correct for **sky-wave** and other variable effects and thus improves accuracy.

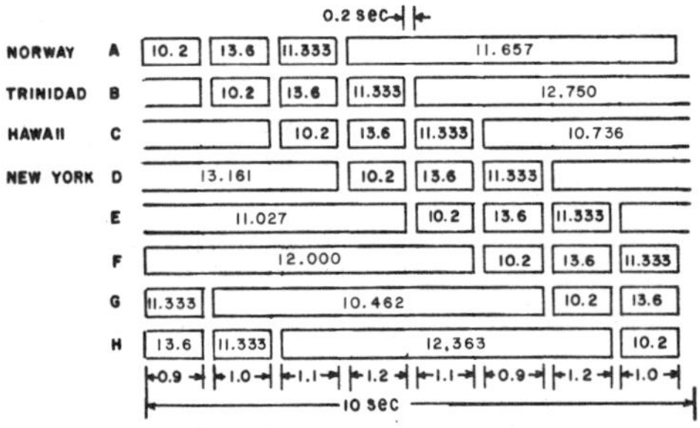

Fig. 132. **Omega** broadcast frequency sequence (in Mhz) showing the sequence for various stations.

Omnitape: A device for transcribing from one type of magnetic-tape recording to another.

one's complement: The radix-minus-one complement form for representing negative binary numbers. It can be found by replacing all the ones by zeros and all the zeros by ones. For example, the decimal number 27 might be represented as 0011011, and -27 as 1100100. When such a number is added to its negative all bit registers are full; e.g., $0011011 + 1100100 = 1111111$. This system contains two representations of zero; all ones or all zeros. Compare **two's compliment.**

one-sided function: **1.** A function which has the value of zero for all negative values of the argument; i.e., $F(x) = 0$ if $x < 0$ (or alternatively, zeros for all positive arguments). **2.** A function which is not defined for negative values of the argument.

one-way time: Half the corrected traveltime for a reflection arrival. One-way time multipled by average velocity gives reflector depth for a flat reflection and flat velocity layering.

on-line: **1.** A linear array of observation points, especially a seismic line. **2.** Equipment under the control of a central processing unit.

onset: The beginning of a wave train. See **break.**

on time: **1.** The time during which an IP transmitter is actually supplying current. **2.** The time during which the charging current or field from a pulse-type transmitter is received at the receiver.

open ended: The addition of new elements does not disturb the prearranged system.

open hole: A well bore which has not been cased, where measurements are made.

operating system: An integrated system of routines for supervising the operation of a computer. Also called *executive*. The OS system is stored in a core memory, the **DOS** system on a disc; **TSS** is a time-share system which permits work on several jobs simultaneously.

operation: A mathematical (or sometimes physical) process to be performed on data, usually indicated by a symbol. For example, a plus sign means the operation "add the number ahead of it to the number behind it." Differentiation, integration, convolution, Fourier transformation, crosscorrelation, etc., are likewise "operations." See also **operator.**

operational amplifier: A high-gain, high input-impedance amplifier requiring minimal current for operation. Ideally a voltage-controlled voltage source. External feedback components are used to obtain desired operations such as summing, integrating, differentiating, etc.

operator: **1.** The specific thing involved in a particular **operation.** Thus, a filter operator is a specific filter expression involved in filtering (convolution). See also **operator length.** **2.** **Observer.**

operator length: The time-domain length of the impulse response of a **convolution** operator. Often specified as a certain number of points; for example, a 56-point operator at 4-msec sample rate is 55 intervals (times $4 = 220$ msec) long.

optically pumped magnetometer: A magnetometer such as the **cesium** or **rubidium-vapor magnetometer.** Atoms in the vapor cell precess about the steady magnetic field which is to be measured, which is at an angle of 20-70 degrees (about 45 degrees in Figure 133) to the instrument axis. Monochromatic light, which is circularly polarized in the plane perpendicular to the instrument axis, has a component which can be absorbed

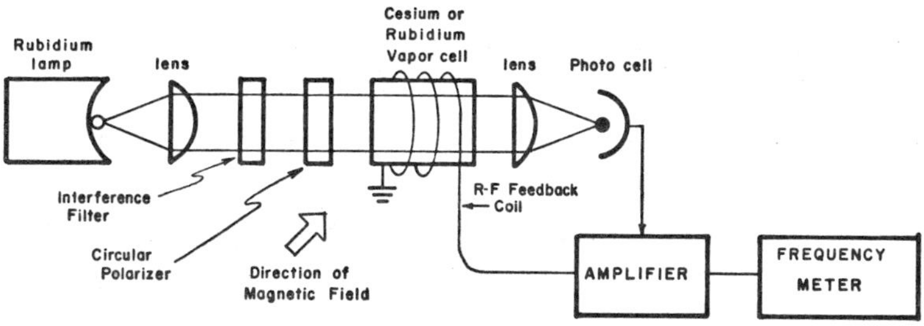

Fig. 133. **Optically pumped magnetometer.**

by the precessing atoms. Once this absorption is complete, no further absorption can occur and so the vapor cell becomes transparent, a condition which causes an increase in light to the photocell. The polarized precessing atoms will also have a component along the axis of a transverse RF field which will permit the atoms to undergo transition and hence become available to absorb more light. The precessing atoms thus become a transfer mechanism between the light and the transverse field when the field is at the Larmor frequency. The light is used to monitor the precession and automatically adjust the RF frequency to precisely the Larmor frequency, which can be measured to give the steady magnetic-field intensity. Such magnetometers can be made extremely sensitive. They measure absolutely the total magnetic field.

optimum: Best according to some criteria. A meaningless term unless the criteria is specified. Optimum in a least-squares sense means that the sum of the squares of all errors is minimized.

optimum filter: A filter designed to maximize or minimize a certain performance measure. Often means a **Wiener filter** in which the mean-square difference between actual and desired output is minimized.

optimum wide-band: A process of filtering and stacking which effects maximum cancellation (according to certain mathematical criteria) of one type of event (multiple in the case of a horizontal stack or ghost in the case of a vertical stack) regardless of frequency content, at the same time reinforcing another type of event (the primary). Application of the method requires precise knowledge as to the time differences between the events on the records to be stacked (i.e., precise differential-normal-moveout information for multiple cancellation, precise uphole data for ghost cancellation). For optimum wide-band horizontal stack, see Schneider, Prince, and Giles, 1965, A New Data Processing Technique for Multiple Attenuation Exploiting Differential Normal Moveout: Geophysics, v. 30, p. 348-362. For optimum wide-band uphole stack, see Schneider, Larner, Burg, and Backus, 1964, A New Data Processing Technique for the Elimination of Ghost Arrivals on Reflection Seismograms: Geophysics, v. 29, p. 783-805.

order of magnitude: The nearest integer to $|\log_{10}|X/S||$; where X and S are two quantities being compared. One order of magnitude indicates one quantity is of the order of ten (or a tenth) times the other value, two orders of magnitude 100 (or 1/100) times, etc. Used to give the error or uncertainty of measurement if the uncertainty is large or to make crude comparisons.

OR gate: A circuit with multiple inputs which functions when a signal is present at any input. Also called *inclusive OR* gate; for inputs *A* and B, signified by *A* ∪ *B* or *A*+*B*. The EXCEPT gate is called *Exclusive OR* but the inclusive OR is intended unless exclusive is specifically stated. Their truth tables are

A	B	OR or A∪B	EXCEPT
0	0	0	0
1	0	1	1
0	1	1	1
1	1	1	0

See Figure 88 (**gate**).

orogenic: Involving mountain-building by large-scale lateral forces; often involves thrust faulting and folding. See **tectonic types.**

orthogonal: Normal or at right angles. Linear combinations of functions can be made orthogonal if they are linearly independent, i.e., if they cannot be expressed as combinations of each other.

OS: Operating system.

oscillograph: Camera.

out-of-phase: The component of an electrical signal that has a 90-degree phase difference from the exciting or reference signal. Also called *quadrature.*

output: 1. The power, current, or voltage delivered by a circuit, system, or device. 2. The terminals where the power, current, or voltage may be delivered. 3. Data which have been processed.

output-energy filter: A filter which maximizes the energy of a signal while minimizing the energy of the filtered noise. See Treitel and Robinson, 1969, Optimum Digital Filters for Signal to Noise Ratio Enhancement: Geophysical Prospecting, v. 17, p. 248-294.

overburden: **1.** Material lying over an ore or valuable deposit. **2.** The section above a refractor or above a reflector.

overload point: The input-signal amplitude for which the ratio of output to input first differs by 3 db from the ratio within the linear operating range.

overvoltage: The extra potential (which in IP is proportional to impressed current density) due to an electrochemical and electrokinetic barrier set up at an electrode-to-electrode interface. *Activation overvoltage* is caused by current passage stimulating an electron-transfer reaction such that the electrode potential deviates from its reversible potential without appreciably changing the ion concentrations at the electrode surface. *Concentration overvoltage* is brought about by a depletion or accumulation of oxidized and reduced ion species at the electrode surface, causing a change in the reversible potential of the electrode. See **induced polarization.**

overvoltage method: **Induced-polarization method.**

package program: A standard set of computer programs which are used rather than tailoring the processing to the specific needs of the data. Applied especially to production processing of large volumes of data. Sometimes implies back-to-back processing without intermediate decisions between stages in the processing.

packing: 1. The number of bytes of information per unit length of magnetic tape, often measured in bytes per inch (BPI). 2. Increasing the density of stored data so that more data can be stored in the same space, as in placing more bits in a given length of magnetic tape. 3. Arrangement of particles in a matrix, as of grains in sandstone.

pad: Sidewall pad; a footing on the end of an arm which presses against the borehole wall.

paging: 1. The function in a printer routine which separates tabulations into separate pages. 2. The function in a plotter routine which adds a fixed amount to a coordinate so that a graph being plotted will not run off the paper. Magnetometer records and seismic-profiler records often are "paged" so that a large plotting scale can be used without requiring excessively large paper. 3. Division of data or program instructions into blocks. Some of the pages may be stored in some storage device other than the computer's rapid-access memory and brought into the rapid-access memory only as needed. In this way a program can be larger than the rapid-access memory. Such storage is called *virtual memory.*

paleomagnetism: Study of natural **remanent magnetization** of rocks and other natural materials in order to determine the intensity and direction of the earth's field at the time the materials were magnetized. It has as adjuncts *archeomagnetism* (study of the earth's magnetism during historical times) and *rock magnetism* (basic study of the magnetic properties of rocks and minerals).

paleosection: A cross-section showing attitude of bedding and structure as it is assumed to have been at some past time. May refer to a seismic section on which one horizon has been flattened, assuming that this horizon was laid down flat and that the flattened section therefore shows the attitude of deeper structure at the time of deposition of the flattened horizon; the compaction because of overburden and other changes subsequent to deposition are often ignored.

pantograph: 1. A device for copying a drawing at a different scale. 2. A device for mapping from one domain to another where there is a one-to-one correspondence between the domains. 3. A device for plotting seismic events in their migrated position.

paradox of anisotropy: The apparent transversal resistivity equals the true longitudinal resistivity. In a homogeneous but one-dimensionally anisotropic medium with resistivities perpendicular and parallel to the bedding ρ_T and ρ_L, respectively, the apparent resistivities measured in the transverse and longitudinal directions ρ_{aT} and ρ_{aL} are:

$$\rho_{aT} = \rho_L$$

$$\rho_{aL} = (\rho_T \rho_L)^{1/2}.$$

See also **coefficient of anisotropy.**

parallax: A change in the apparent position of an object (such as a meter needle) with respect to a reference (such as the meter dial) which is in a different plane, when viewed from a different position. Parallax error results when the observer is not correctly positioned for the reading.

parallel field: A uniform field in which current flow lines or equipotential surfaces are parallel.

parallel shot: A test record made with all the amplifiers connected in parallel and activated by a single geophone. Also called *bridle.* Used to check that the amplifier circuits perform similarly with respect to lead or lag, polarity, and phasing.

paramagnetic: Weakly magnetic with small positive susceptibility. The magnetic moments of individual atoms are uncoupled so that each atom behaves independently. Paramagnetism usually contributes only a few gammas to the magnetic field at the earth's surface. Compare **diamagnetic** and **ferromagnetic.**

parameter: 1. A variable which can be changed independently and (usually) arbitrarily between calculations but which remains constant during any calculation. 2. Quantities (each of which may represent a combination of quantities) which are sufficient to determine the response characteristics of a system.

parametric sounding: An electromagnetic depth sounding in which frequency is varied while holding the geometry constant, as opposed to *geometric sounding* where the frequency is held constant and the geometry is varied. Used to resolve resistivity layering, assuming horizontal layering. See **ellipse of polarization.**

parity bit: One of the bits in a **byte** which is dependent upon the other bits in such a way as to detect dropout anywhere in the byte. See **check.**

Parseval's theorem: For two aperiodic functions h_1 and h_2 with respective Fourier transforms H_1 and H_2,

$$\phi_{12}(0)=\int_{-\infty}^{\infty} h_1(t)h_2(t)dt=\int_{-\infty}^{\infty} H_1(f)H_2(f)df=P_{12}(0).$$

Thus the zero-lag value of the crosscorrelation $\phi_{12}(0)$ (the two values on the left) equals the integral of the product spectrum. Also the cross-power-spectral amplitude at zero frequency $P_{12}(0)$ equals the cross energy in the time domain.

partition gas chromatograph: A device for quantitative analysis of hydrocarbon constituents. A fixed quantity of sample is carried with a stream of sweep gas through a partition column packed with an inert solid coated with a nonvolatile organic liquid. The lighter fractions traverse the column faster than the heavier fractions so that the components appear separately at the column exit where their amounts can be measured.

party: The group of men working together to carry out

a geophysical field project. Also called **crew.** A land seismic party might be organized as shown in Figure 134.

party chief: The man in charge of a geophysical party. Often responsible for interpretation of the results.

party manager: The man working under the party chief who usually is responsible for the field work.

pass: 1. A complete cycle through a computer involving reading, processing, and writing; a machine run. 2. The passage of a satellite from rise to set over the horizon. Used for positioning in satellite navigation.

passband: The range of frequencies which can pass through a band-pass filter without significant attenuation.

passive: 1. Having no source of energy, as a "passive filter." A *passive beacon* is a radar reflector which merely reflects radar energy as opposed to an *active beacon* which transmits in response to a signal striking it. 2. Applied to a system which does not generate an output if there is no input. 3. A positioning system which does not involve the transmission of a signal but only the observation and measurement of angles, amplitudes, phases, times, etc., in a system which exists independently of the craft being positioned.

pass region: The band of frequencies which a system will pass with nearly the same gain while attenuating other frequencies.

patch: 1. A large geophone group which feeds a single channel. Patch arrays may be up to 600 by 1000 ft in size and may contain several hundred geophones. Used in **transposed** recording, especially with surface sources (e.g., Thumper and Vibroseis) in areas of poor record quality. See Figure 171 **(spread).** 2. A **jumper** or a temporary connection, especially one which can be changed easily, as a connection on a "patch panel." 3. A section of coding (or a subroutine) used to correct a mistake or alter a routine.

pattern: An **array** of shotholes or geophones. Shot arrays are used to cancel noise waves or to distribute the charge for improving shot efficiency. Geophone patterns are used to cancel unwanted noise trains and random noise. Array characteristics are illustrated by **directivity pattern** diagrams.

P-band: Radar frequencies between 225 and 390 Mhz; see Figure 68 **(electromagnetic spectrum).** Used in remote sensing because it penetrates vegetation and so shows a combination of vegetation and surface-soil effects.

PC: Continuous-type **micropulsations.**

PCM: Pulse-code **modulation.**

PDMI: Percent decrease in mutual impedance.

PDR: Potential-drop ratio.

peak: The maximum upward (positive) excursion of a seismic wavelet. Opposite of **trough.**

pearls: Continuous-type micropulsations "of the first kind" with periods from 0.2 to 5 sec and amplitudes from 0.05 to 0.1 gamma. Strip-chart records of pearls look like amplitude modulated sinusoidal waves resembling a pearl necklace.

peg-leg multiple: 1. A **multiple** reflection involving different interfaces so that its travel path is nonsymmetric. See Anstey, 1960, Attacking the Problems of the Synthetic Seismogram: Geophysical Prospecting, v. 8, p. 242-259. 2. A short-path multiple which is repeatedly reflected within thin formations, i.e., less than a half wavelength apart. See Trorey, 1962, Theoretical Seismograms with Frequency and Depth Dependent Absorption: Geophysics, v. 27, p. 766-785. See Figure 124 **(multiple).**

penetration: 1. The greatest depth from which seismic

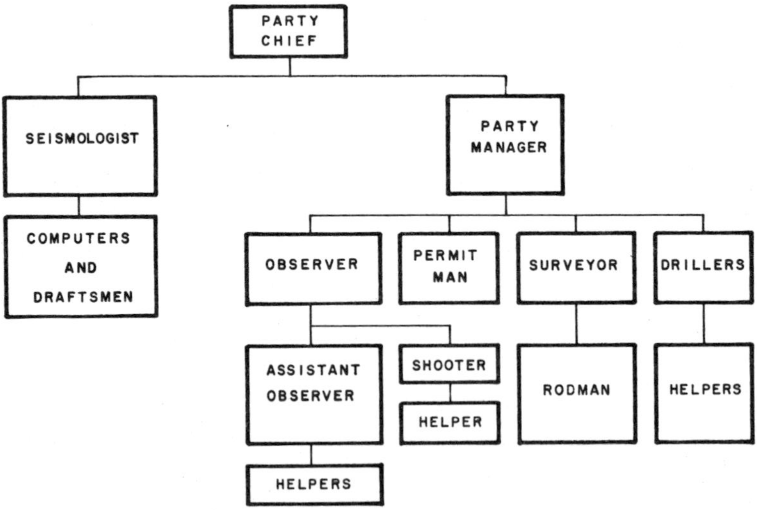

Fig. 134. **Party** organization for a land seismic party.

reflections can be picked with reasonable certainty. Dependent on the energy of the reflected wave, the presence of various types of noise, and the processing to which the data are subjected, as well as the characteristics of the recording system. **2.** The greatest depth at which the properties significantly affect measurements. **3.** See **skin effect.**

percent decrease in mutual impedance: The percent voltage change in a coupled circuit with respect to the low-frequency impedance. Used with reference to the amount of in-phase electromagnetic coupling.

percent frequency effect: **1.** The basic polarization parameter measured in the frequency domain, being the percent difference in resistivity measured at two frequencies,

$$\text{PFE} = 100\,(\rho_{dc} - \rho_{ac})/\rho_{ac},$$

where ρ_{dc} and ρ_{ac} are the low- and high-frequency resistivities. **2.** Decade-normalized PFE is multiplied by the log of the frequency ratio:

$$\text{PFE} = [100(\rho_{dc} - \rho_{ac})/\rho_{ac}]\,\log_{10}(f_{ac}/f_{dc}).$$

3. Keller suggests that PFE be defined as:

$$\text{PFE} = 100\,(\rho_2 - \rho_1)/(\rho_2\rho_1)^{1/2},$$

where ρ_1 and ρ_2 are resistivities measured at two frequencies measured a factor of 10 apart. **4.** PFE is closely related to chargeability if defined as suggested by Brant:

$$\text{PFE} = 100\,(\rho_{dc} - \rho_{ac})/\rho_{dc}.$$

percent mineralization: In IP surveying, the volume-percentage of metallic-luster minerals in a rock. This value is usually about half the metallic-luster mineral content by weight.

perforate: To open holes through casing and into a formation so that fluids can flow through the formation into the borehole or vice-versa.

perigee: Shortest distance from a satellite orbit to the earth's center; see Figure 69 **(ellipticity)**. The greatest distance is *apogee.*

period: The time T for one cycle. The time for a wavecrest to traverse a distance equal to one wavelength or the time for two successive wavecrests to pass a fixed point. For a monochromatic wavetrain,

$$T = 1/f = \lambda/V,$$

where f = frequency, λ = wavelength, and V = phase velocity.

permeability: **1.** The ratio of the magnetic induction **B** to the magnetic intensity **H**. See **paramagnetic, diamagnetic, ferromagnetic.** **2.** A measure of the ease with which a fluid can pass through the pore spaces of a formation. Measured in millidarcy (1/1000 **darcy**) units. The permeability constant k is expressed by Darcy's law as $\mu q/(dp/dx)$, where μ is fluid viscosity, q is linear rate of flow, and dp/dx is the hydraulic pressure gradient.

permeance: The reciprocal of **reluctance.**

permit: Permission from a landowner for a geophysical field party to work on his land.

permittivity: Electric inductive capacity, $\epsilon\mathbf{E} = \mathbf{D}$. The volumetric capacitance or **capacitivity** of a three-dimensional material, such as a dielectric. *Relative permittivity* is the dimensionless ratio of the permittivity of a material to that of free space; it is also called the *dielectric constant.*

Peters length: A measurement made on profiles across potential-field (especially magnetic) anomalies, the objective of which is to determine the depth to an anomalous mass which is a tabular body or dike model. Peters's rule gives the depth as 5/8 the horizontal distance between the points on the side of an anomaly where the slope is half of the maximum slope. See Peters, 1949, The Direct Approach to Magnetic Interpretation and its Practical Application: Geophysics, v. 14, p. 290-320.

PFE: **Percent frequency effect.**

PGC: Preset or **programmed gain control.**

phantom: A line on a seismic section drawn parallel to the dip of nearby reflection events. Phantoms are drawn and mapped where one cannot follow one event far enough to develop a map on that event alone.

phase: **1.** The argument of a wave. If the representation of a wave is $\psi(x - vt)$, the argument $(x - vt)$ is the phase. **2.** The angle of lag or lead (or the displacement) of a sine wave with respect to a reference; how far into the period rotation, oscillation, or variation has advanced, considered in relation to a reference or assumed instant of starting. Commonly expressed in angular measure. Phase information, being the measure with respect to the instant of starting, carries the timing information of a seismogram and hence proper phase preservation is of utmost importance. See also **phase response** and compare **phasing.**

phase angle: Tan^{-1} [quadrature component/in-phase component]. The phase angle is in quadrants 1 or 2 if the numerator is positive, in quadrants 1 or 4 if the denominator is positive. In **induced polarization,** phase angle is usually measured in milliradians.

phase comparison: A matching of the CW radio signals from two transmitters in order to determine a **line of position.**

phase control: The process of rapid on-off switching which connects an ac supply to a load for a controlled fraction of each cycle. An efficient use of silicon-controlled rectifiers in regulating current in an IP transmitter.

phase distortion: Change in waveshape which occurs because phase shift is not proportional to frequency. See also **phase response.**

phase inversion: A change of 180 degrees in phase angle. Mirror-imaging a trace about the zero-deflection position.

phase-lock: A technique in which a signal of almost constant frequency is generated within an instrument, which signal is brought to the same average phase as the external signal. Used as the reference signal in synchronous detectors to suppress noise.

phase response: A graph of phase-shift versus frequency, which illustrates the phase characteristics of a system. Filters with the same amplitude-frequency response but different phase characteristics will affect the shape of pulses put through them differently. See Figure 119 **(minimum-phase)**, and also the entries **minimum-phase, maximum-phase, linear-phase,** and **zero-phase.**

phase spectrum: Phase response.

phase velocity: 1. The velocity with which any given phase (such as a trough or a wave of single frequency) travels; may differ from **group velocity** because of **dispersion.** Sometimes called *trough velocity* or *peak velocity.* See Figure 55 **(dispersion). 2.** Apparent velocity.

phasing: A change in waveshape as a result of filtering or interference.

phasor diagram: A representation of a vector which rotates at some particular angular velocity. Characteristics can be interpreted by viewing the "frozen" position and spatial relationship of vectors. Master curves for the electromagnetic method are often presented as phasor diagrams.

photoclinometer: A well-logging device which photographically records the angle and azimuth of borehole deviation from the vertical. Compare **poteclinometer.**

photon log: A well log of scattered gamma rays, differing from a **density log** in that the sonde is not pressed against the borehole wall and hence the log is sensitive to changes in hole diameter and the density of the fluid in the borehole.

physically realizable: Satisfying two conditions: (1) Not existing (having values of zero) before some initial time and (2) containing finite energy (hence dying out toward infinity).

Pi: Irregular-type **micropulsations.**

pick: 1. To select an event on a seismic record, as to "pick" reflection events. **2.** An event or time on an event which has been selected. The evidences of an event on which picking must be based are amplitude and coherence (and some would add "character"). The arrival of a wave train signifying new energy should cause an increase in amplitude and should affect different channels in a systematic coherent way. Various statistical tests are used to make picking decisions, ranging from simple summing along possible coherent patterns to schemes like **semblance** criteria. Some criteria search only for phase coherence, others look at the amplitude buildup, integrate over several half cycles, equalize spectral variations, etc. **Grading** is intimately related to picking.

pickup: 1. Geophone. 2. Reception of a disturbance such as inductive or other input from an electric power line **(high line).**

pico-: A prefix meaning 10^{-12}.

pi diagram: Pole diagram.

Pie Slice: A fan filter or **velocity filter** process aimed at emphasizing a band of moveouts independent of frequency. Compare **butterfly filter.** Texas Instruments trademark.

piezoelectric: The property of a dielectric which generates a voltage across it in response to a stress, and vice-versa. In a **hydrophone** the stress is produced by the pressure and in an accelerometer the stress is produced by the inertia of the reaction mass. Piezoelectric transducers are commonly made of barium titanate or zirconate.

pigtail: The wire which connects a geophone to the seismic cable.

pigtail chart: See **dipmeter.**

pilot: A rough estimate for use as a basis of some analysis or process.

piloting: Positioning by determining location with respect to known geographical points. Many radio-location systems are sophisticated versions of piloting. Compare **dead reckoning.**

pilot trace: The seismic trace toward which other traces are adjusted in time shifting for static corrections or in cross-equalization processes. The pilot trace may be composited from the traces being adjusted.

pinger: 1. A transponder or device which emits an acoustic signal upon being activated by sensing a coded acoustic signal. Pingers placed on the sea bottom or in anchored buoys can be interrogated by a ship transmitting the coded acoustic (sonar) signal and the distance to the pinger determined by traveltime measurements. **2.** A shallow-penetration high-power transducer used in marine engineering studies in soft-bottom areas.

pitch: 1. Angle between the major axis of polarization and the horizontal. Also called dip or tilt. See **polarization ellipse. 2.** Rotational motion of a ship or aircraft about a horizontal axis perpendicular to the ship's course. Compare **roll, yaw,** and **trim.**

plane surveying: Surveying in which earth curvature is ignored.

plane table: A survey instrument consisting of a drawing board which can be leveled on a tripod. An object is sighted through an **alidade** (Figure 5) which rests on the table, allowing one to plot the line of survey directly from the observation by drawing a line along the ruler attached to the sighting telescope. Figure 135 shows the use of the plane table in traversing, intersection surveying, and resection.

plane wave: Having wavefronts which are planar (with no curvature), as might originate from a very remote source. A common assumption in seismic and electromagnetic wave analyses which is not strictly true in actual situations.

plane-wave field: An electric or electromagnetic field in which the phase is constant over a set of planes.

plant: 1. The manner in which a geophone is placed on or in the earth. **2.** The coupling to the ground. The nature and quality of the plant affects the overall system response. **3.** To place a geophone in its proper place on the ground.

planter: A device which pushes geophones (or hydrophones) into soft marsh, perhaps 8-10 ft deep.

plate tectonics: A concept which envisions the earth's

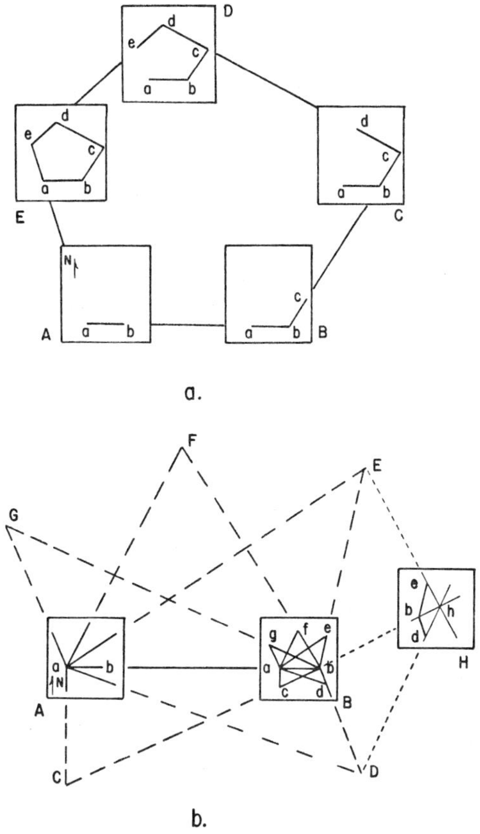

Fig. 135. **Plane table** methods. (a) *Plane table traversing.* The plane table is set at *A* and oriented north; the rod at *B* is sighted with alidade and line *ab* is drawn along the alidade edge, the line length depending on the stadia reading. The table is set up at *B,* oriented by backsighting on *A,* then the alidade is sighted on *C* and *bc* is drawn, and so on until the loop is closed at *E* by sighting on *A.* (b) *Plane table intersection* method. The plane table is set at *A* and oriented; the rod at *B* is sighted with the alidade and line *ab* is drawn along the alidade edge, of a length depending on the stadia reading. Points *C, D, E, F,* and *G* are sighted on and their directions indicated. The same points are then sighted on when the plane table is at *B.* (Also b)*Plane table resection.*The plane table is set at an unknown location *H* and the alidade is sighted on known points *e, b,* and *d.* Point *H* is determined by the line intersections.

crust divided into various plates which move slowly with respect to each other, being carried along by slow convection currents in the asthenosphere. Along major rifts (such as midocean ridges) the plates are separating and new crust is being created. Elsewhere plates are overriding one another (as at deep ocean trenches) or sliding by one another (as along the San Andreas fault). See also **transform faults** and **Benioff zone.** Figure 136.

playback: 1. To produce a new form of record from magnetic tapes (or other reproducible forms of recording). Seismic playback may include filtering, gain adjustment, time shifting, mixing, stacking, migrating, etc. A possible playback program is shown in Figure 141 **(processing). 2.** The result of such processing, as opposed to the original recording.

plot: 1. To draw lines representing seismic events on a **cross-section** or **plotted section. 2.** A graph or plotted section.

plot point: The location where a datum value is plotted relative to the measuring electrode array. For a symmetric array this is the midpoint of the array, but on an asymmetric array the convention may vary. See also **depth point.**

plotted section: Section on which seismic events are indicated by lines or sequences of points. The horizontal scale is usually the distance along the seismic line and the vertical scale is usually either depth or reflection time. Data may or may not be **migrated.** Figure 137.

plotter: 1. A device for making a seismic record or record-section display, frequently (but not necessarily) photographic. **2.** A device for graphing data, as an *X-Y* plotter. **3.** A person or device for drawing graphs, maps, or sections

plunge: 1. To set the horizontal crosswire of a **theodolite** in the direction of a grade. **2.** The direction of the axis of a fold with a downward component.

plus-minus method: A refraction interpretation method using reversed refraction profiles with shots *A* and *B* at opposite ends, also called *Hagedoorn method.* Let t_{AB} be the surface-to-surface time between *A* and *B* and let t_A and t_B be arrival times at various intermediate locations from shots *A* and *B,* respectively. "Minus" values $t_A - t_B - t_{AB}$ are calculated for each location and plotted to give the velocity of the refractor. "Plus" values $t_A + t_B - t_{AB}$ are calculated and plotted to give a picture of the refractor's depth. The refractor's depth under each location is *K* times the "plus" value where $K = 1/2\, V_1/(1 - V_1^2/V_2^2)^{1/2}$. The "minus" values are often plotted using an inclined grid so that the values do not extend beyond the graph paper on long profiles. The method can be extended for velocity variations above the refractor. See Hagedoorn, 1959, The Plus-Minus Method of Interpreting Seismic Refraction Sections: Geophysical Prospecting, v. 7, p. 158-182. Figure 138.

point: 1. Refers to operator length, as a "56-point filter." See **convolution. 2.** Shotpoint.

pointing error: Error in a ship's doppler sonar such as might be caused by misalignment of the sensors with

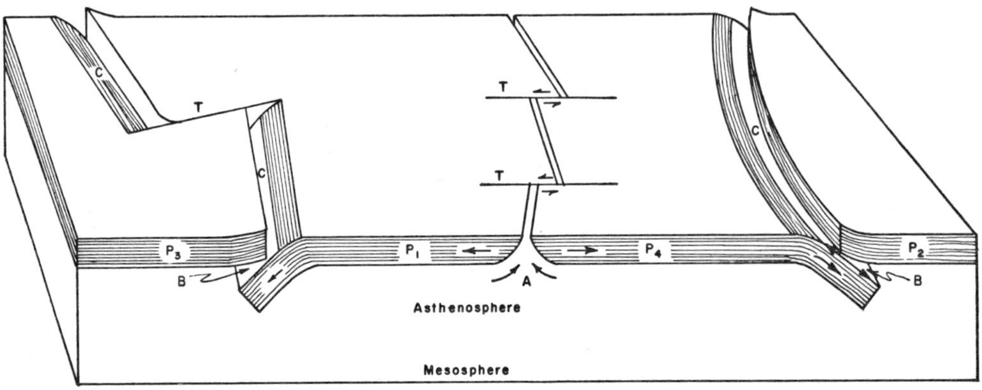

Fig. 136. **Plate-tectonic** model. *A* — Rift zone where plates P_1 and P_4 are moving apart, such as Mid-Atlantic Ridge where new crust is being created. *C* — Ocean trench resulting from overriding of plates P_3 over P_1 (or P_2 over P_4). *B* — Benioff zone of earthquakes dipping along contact of plates which are colliding. *T* — Transform faults where plates are sliding by each other. (From Conrath et al, Journal of Geophysical Research: v. 75, no. 30, p. 5855.)

the ship. Shows as a systematic fictitious cross-course velocity (or cross-course position error) even when in reliable bottom-lock mode.

point sort: **Gather.**

point source: **1.** A source whose actual size is unimportant as far as the effects being observed are concerned. **2.** A single current electrode whose companion is a great distance away, such as the current pole of the pole-pole or pole-dipole arrays.

poise: 100 **centipoise.**

Poisson's equation: In space where the source density is ρ, the Laplacian of a potential *U* is

$$\nabla^2 U = 4\pi \rho K,$$

where ∇ is the operator **del** and *K* is a constant (gravitational constant in case of mass and gravitational potential). The constant 4π is deleted in the mks system. In empty space where $\rho = 0$, this becomes Laplace's equation.

Poisson's ratio: The ratio of the transverse contraction to the longitudinal extension when a rod is stretched:

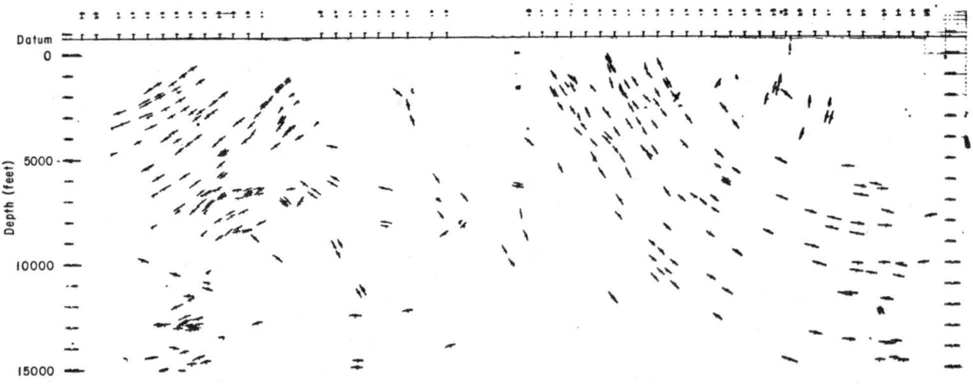

Fig. 137 **Plotted migrated seismic section.** Gaps in the coverage (indicated by the shotpoint distribution along the top) occasioned by permit difficulties affect the appearance. (Courtesy Chevron Oil Company.)

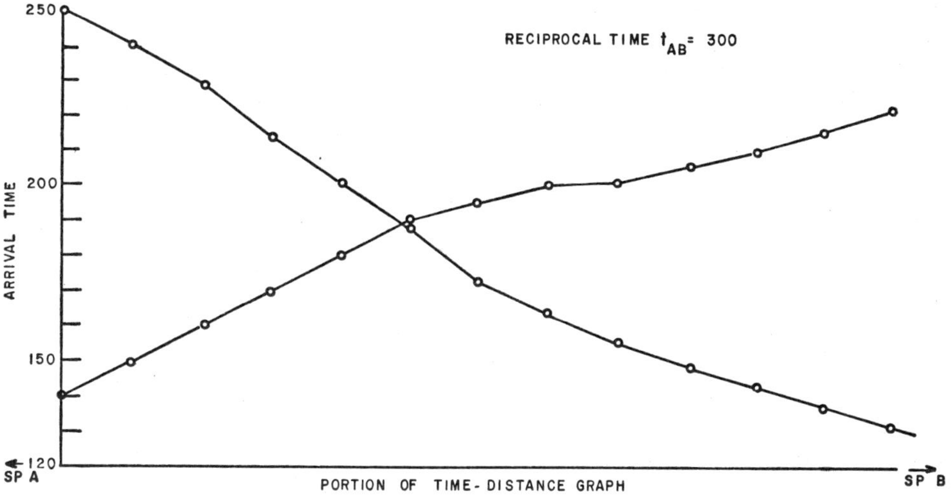

t_A	140	150	160	170	180	190	195	200	201	205	210	216	222
t_B	250	240	229	215	201	188	173	164	155	148	143	137	131
"plus"	90	90	89	85	81	78	68	64	56	53	53	53	53
"minus"	-410	-390	-369	-346	-321	-298	-278	-264	-254	-243	-233	-221	-209

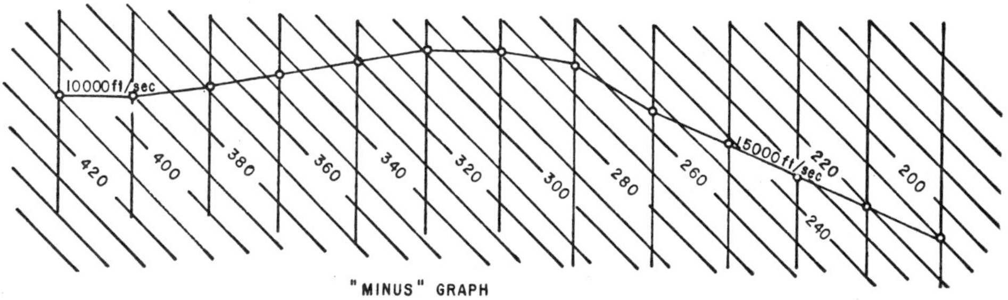

"MINUS" GRAPH

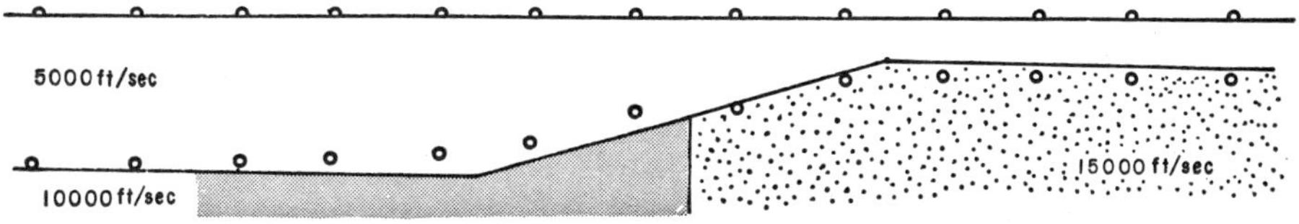

"PLUS" VALUES PLOTTED BENEATH GEOPHONES ASSUMING V_2 = 10000 ft/sec COMPARED WITH MODEL

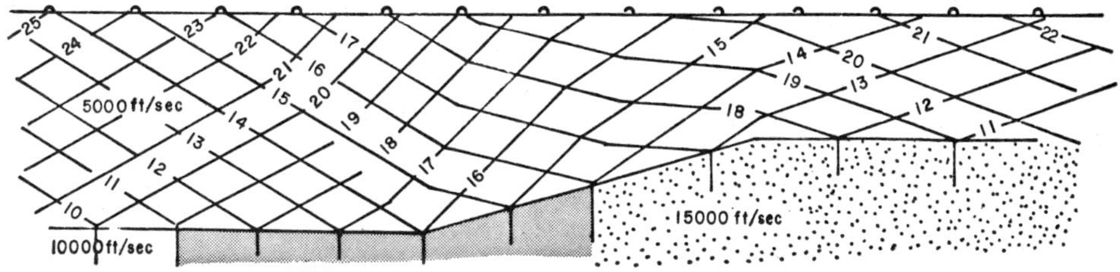

MODEL AND WAVEFRONTS

Fig. 138. Plus-minus refraction method.

$$\sigma = \frac{\Delta d/d}{\Delta L/L}$$

where Δd = change in diameter d and ΔL = the change in the length L. See also **elastic constant**. The ratio of the velocities of P- and S-waves, V_p and V_s can be expressed in terms of Poisson's ratio:

$$\frac{V_p}{V_s} = \left[\frac{2(1-\sigma)}{1-2\sigma}\right]^{1/2}.$$

For liquids, σ = 1/2 and V_s vanishes.

Poisson solid: An isotropic elastic material for which the Lamé **elastic constants** λ and μ are equal. A Poisson solid has Poisson's ratio σ = 1/4.

polarity: **1.** The condition of terminals being electrically positive or negative. If opposite terminals are connected, electrons flow from the negative to the positive in the external connector. **2.** The north or south character of a magnetic pole.

polarization: **1.** Dipole moment per unit volume. In induced polarization, current dipole moment per unit volume. **2.** The polarity or potential near an electrode. **3.** A preferential direction of wave motion, as the component of shear waves whose motion is confined to a horizontal plane (SH) or light with directional characteristics. **4.** Magnetic orientation concerning only the vector direction and not the magnitude.

polarization ellipse: The locus of points in space described by the superposition of two fields having different directions and whose variations in time are of the form $A\cos(\omega t)$ and $B\cos(\omega t + \theta)$. In general, in electromagnetic prospecting neither the direction nor the phase of the primary and secondary fields are the same so the superposition of the two fields results in a polarization ellipse. For the ellipse shown in Figure 139, the *modulus* of the magnetic wavetilt is given by $|H_r|/|H_z|$, the *tilt angle* or *pitch* by α, and the *ellipticity* by H_2/H_1.

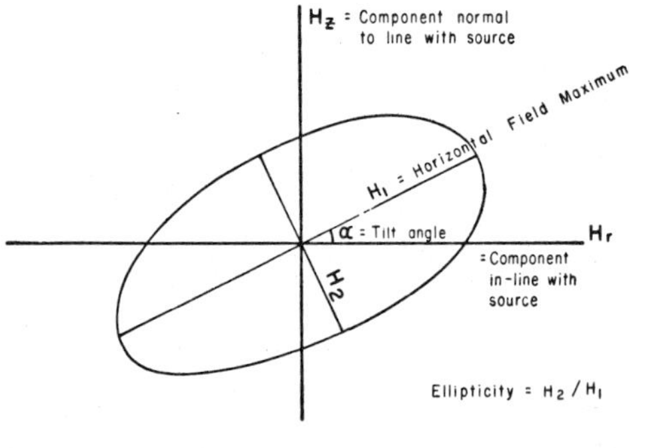

Fig. 139. **Polarization-ellipse** relations.

polarization filtering: A technique for enhancing one mode of wave propagation with respect to another by combining the outputs of three-component recordings. Phase relationships for P-, S-, Rayleigh waves, etc., are different at a free surface. See White, 1964, Motion Product Seismograms: Geophysics, v. 29, p. 288-299.

polarization resistance: The factor (RT/nFJ_o) employed in overvoltage theory, relating overvoltage η to current density J:

$$\eta = -(RT/nFJ_o)J.$$

R is the gas constant, T the absolute temperature, n the number of molar equivalents, F the Faraday, and J_o the exchange current density, all in cgs units. The units of polarization resistance are ohm-cm^2.

polarized: A connector is said to be polarized if the connector and its mate are so designed that they fit together in only one way, so as to prevent getting the wires connected incorrectly.

polarized electrode: See **ideal polarized electrode.**

pole: **1.** A singular point, where the value of a function becomes infinite. If a function has the factor $1/(x-a)^m$ and m = 1, then a is a *simple pole.* Otherwise, a is a pole of order m. **2.** A **magnetic pole. 3.** One electrode of a pair whose companion electrode (infinite electrode) is so far away that its location does not affect the measurements.

pole diagram: A **stereographic projection** on which the direction of lines is plotted. Also call *π diagram.* A plane is represented by the direction of the line perpendicular to it. Compare **cyclographic diagram.**

pole-dipole array: Electrode **array** used in induced-polarization surveying.

pole-pole array: Electrode **array** used in induced-polarization surveying.

polling: Calling stations to permit them to transmit information.

polyconic projection: See **map projection.**

poop shot: Weathering shot.

population: The aggregate of a set of observations whose subaggregates have the same statistical properties. Where subaggregates have different statistical properties, they may be said to be "of different populations." **Statistical measures** are used to characterize a population.

porosity: Pore volume per unit volume of formation. Often indicated by the symbol ϕ. Porosity is determined from cores or from sonic logs (see **Wyllie relationship**), from **density logs,** from neutron logs, or from resistivity logs (see **Archie's formula**). See also **movable oil plot.** *Primary porosity* refers to the porosity remaining after the sediments have been compacted but without considering changes because of subsequent chemical action or flow of waters through the sediments. *Secondary porosity* is the additional porosity created by subsequent chemical changes, especially fissures, fractures, solution vugs, and porosity created by dolomitization. *Effective porosity* is the porosity available to free fluids, excluding unconnected porosity and space occupied by bound water and disseminated shale.

porosity overlay: A plot of porosity values calculated

from different logs, plotted on top of each other. Compare **crossplot.**

porous pot: A nonpolarizable electrode which allows free ionic flow into the earth. A copper rod in a saturated copper-sulfate solution contained in a porous pot is such an electrode. Used in making voltage measurements where negligible current flows through it.

positioning: Determining the location of a survey ship or aircraft, usually with respect to geodetic coordinates but sometimes with respect to reference beacons which may not be at known locations. Positioning is sometimes divided into **(a)** *celestial navigation,* locating oneself by observing celestial bodies, which sometimes includes satellite navigation; **(b)** *piloting,* determining position with respect to geographical points, including many of the radio-navigation methods; and **(c)** *dead reckoning,* positioning by the extrapolation of track and direction from a previously known point of departure, including inertial positioning, doppler-sonar, and doppler-radar methods. Some modern positioning systems measure traveltime or differences in traveltime from reference stations, some measure the phase in standing-wave patterns set up by pairs of stations, some measure doppler frequency shifts, some measure the direction of strongest signal. See Figure 113 **(line of position).** Features of some systems are listed in Table 9. See also **Decca, doppler radar, doppler sonar, Lambda, Lorac, loran, Omega, Raydist, shoran, Toran.**

positive: An anomalous area in which values are larger than in neighboring areas, as a **gravity positive.**

positive area: An area characterized by uplift. Positive often is used in a relative sense and sometimes is used for an area which is subsiding less rapidly than surrounding areas.

postplot: Computation of positioning locations which have been previously occupied, based on the best reconciling of all available data.

pot: 1. To fire a small charge in a hole to create space for loading a large charge. **2.** A **potentiometer. 3.** A **porous pot.**

poteclinometer: A device for measuring the angle and direction of borehole deviation continuously during a log run. A pendulum moves a variable-resistor arm so that the resistance is a measure of the angle with the vertical and a compass needle moves another arm so that another resistance is a measure of the azimuth. Often run with dipmeters. Compare **photoclinometer.**

potential: 1. Electrical voltage with respect to a reference point. **2.** The amount of work required to position a unit charge, unit pole, or unit mass at a given position, usually with respect to infinity. The electrical, magnetic, and gravitational fields are scalar potential fields. The gradient of a potential field is called the *field strength* or *flux density;* see **Gauss law.**

potential-drop ratio: An electrical-survey method which compares ratios of voltages between two adjacent, aligned pairs of potential electrodes.

potential electrode: The contact of the IP and/or resistivity receiver circuit with the ground, usually a **porous-pot** electrode.

potentiometer: An electrical instrument for measuring low-level dc voltages without drawing current from the measured circuit, by using the unknown voltage as an arm in a direct-current **bridge** circuit.

pot resistance: The electrical resistance at a porous-pot potential electrode due to the resistance of adjacent material. Too high a pot resistance will "load" the voltmeter's input impedance, reducing the reading and making it susceptible to electrical noise.

Potsdam ellipsoid: The geodetic reference ellipsoid used in most of Europe, Africa, the Middle East, and the Soviet Union. See Table 5 **(geodetic systems).**

Potsdam gravity: See **gravity standard.**

Poulter method: See **air shooting.**

powder: Explosive.

power series: An expression of the form $y = a + bx + cx^2 + \dots$ See **Taylor series.**

power spectrum: 1. The power density versus frequency relationship. The power spectrum is the square of the amplitude-frequency response or the Fourier (cosine) transform of the autocorrelation function. **2.** Occasionally implies cumulative power,

$$\int_0^f P(f)\,df.$$

power-transfer function: The function of frequency which represents the ratio of output-power density to input-power density.

ppm: Parts per million.

Pratt hypothesis: A model of compensation for isostasy in which topographically high areas are compensated by lower-density crustal material. The densities are so related to the surface elevations that pressures are everywhere equal at a uniform depth of compensation. See Figure 105 **(isostasy)** and compare **Airy hypothesis.**

preamplifier: An amplifier which precedes the main amplifier. Located near the signal source to improve signal-to-noise ratio. Often has a high input impedance to prevent loading and to give maximum signal transfer.

precession: The tendency of a gyroscope to turn when under the influence of a torque which tries to change the direction of its axis of spin. See Figure 140.

precision: The repeatability of an instrument, different from **accuracy,** measured by the mean deviation of a set of measurements from the average value.

precision index: See **error function.**

predictive deconvolution: Use of information from the earlier part of a seismic trace to predict and deconvolve the latter part of that trace. Some types of systematic noise, such as reverberations, multiples, etc., can be predicted. The difference between the predicted value and the actual value is called the *prediction error.* The prediction error is sensitive to new information such as primary reflections. Predictive deconvolution may also be used in a multitrace sense, where one tries to predict a trace from neighboring traces in an ensemble of data.

preemphasis: 1. Frequency filtering before processing

Table 9. Positioning and navigation systems

Type	System	Frequency	Range (mf)	Accuracy (ft)	Principle	Remarks
Absolute reference	Celestial		unlimited	15000	Sighting on sun or stars	
	Satellite	150 Mhz 400 Mhz	unlimited	150	Doppler shift	U.S. Navy operation Fix every 1-2 hr with 4-5 satellites
Piloting	Sightings		near land		Sightings	Identify landmarks, buoys
	Bathymetry				Fathometer	Identify features in water bottom
	Pingers				Sonar transponders	Used in relocating a site such as a well head
Low-freq. Radio	Loran A	1900 khz	600-750	25000	pulse timing	Government maintained; hyperbolic
	Loran C	100 khz	1500	100-1500	pulse+phase timing	Government maintained; hyperbolic
	Rho-rho mode		1500	100-500	timing	Circular; atomic clock reference
	Omega	10-14 khz	6000	1500-10000	pulse phase	Goverment maintained; hyperbolic
Medium frequency radio	Decca	70-130 khz	200-300	25-500	CW phase	Hyperbolic; 4 frequencies
	Hi-Fix	1.6-2 Mhz	150-200	25-300	CW phase	Circular or hyperbolic
	Lambda	100-200 khz	150-400	50-500	CW phase	Circular or hyperbolic; 4 freq.
	Lorac	1.6-2.5 Mhz	150-250	25-300	CW phase	Hyperbolic; 2 frequencies
	Raydist DM	1.6-3.3 Mhz	50-200	25-200	CW phase	Circular; 4 frequencies
	DR	1.5-5 Mhz	50-200	25-200	CW phase	Circular; 2 frequencies
	N	1.5-3 Mhz	150-250	25-300	CW phase	Hyperbolic; 2 frequencies
	Toran	1.6-3.8 Mhz	150-250	25-300	CW phase	Hyperbolic or Circular; 2 freq.
	Toran O	100 khz	600	50-300	Phase timing	Circular; atomic clock reference
High frequency	Shoran	210-310 Mhz	70	30-100	pulse timing	Circular; 3 frequencies
	XR Shoran	210-310 Mhz	150-200	50-150	pulse timing	Uses tropospheric scattered waves
	RPS	9300-9500 Mhz	60	10-50	pulse timing	Circular; uses transponders
	Autotape	2900-3100 Mhz	60	10-20	pulse phase	Circular; 3 frequencies
	Hydrodist	2800-3200 Mhz	30	10-100	CW phase	Circular; 4 frequencies
	Trisponder	9500 Mhz	15-50	10-20	pulse timing	Circular; uses transponders
	Tellurometer	3000 Mhz	15	5	pulse timing	Circular; uses transponders
Dead Reckoning	Doppler Radar	8800 Mhz			Doppler shift	Used with aircraft
	Doppler Sonar	300 khz		600/hr	Doppler shift	
	Inertial				Accelerometers	Accumulates error with time

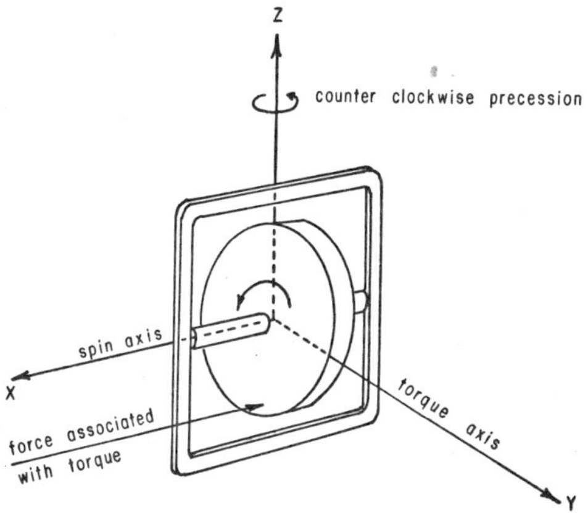

Fig. 140. **Precession** rule: The spin axis precesses toward the torque axis.

(as in field recording) so as to emphasize certain frequencies compared to others. **2.** Recording so as to emphasize higher frequencies.

preplot: **1.** The location programmed for shotpoints (or boat position) before the points are actually occupied. **2.** A list of programmed points in radio-navigation-system coordinates. **3.** To calculate the navigation-system coordinates for programmed points.

preprocessor: A computer which operates on data prior to the main processing. Often located apart from the main computer. Seismic preprocessing sometimes includes vertical stacking, reformating, adding headers, editing, resampling, demultiplexing, etc.

pressure detector: **Hydrophone.**

pressure wave: **P-wave** or longitudinal wave.

presuppression: **Initial suppression.**

prewhitening: Preemphasis designed to make the spectral density more nearly constant.

prills: Pellets of ammonium nitrate used as a shothole explosive.

primacord: An explosive rope which can be used to connect charges. A detonation at one end starts an explosion wave traveling down the cord, detonating other explosives attached to the cord. Primacord is an Ensign Bickford Co. trademark. See also **broomstick** charge.

primary field: The electromagnetic field which would be generated by a source if the source were placed in free space.

primary radar: Radar which relies on reflected energy to indicate targets, as opposed to using active transponders as targets.

primary ratio: The ratio of electromagnetic field readings at two locations in the absence of conducting material. Also called *normal ratio.*

primary reflection: Energy which has been reflected only once and hence is not a multiple. Strictly speaking, primary reflections are virtually undectable because they contain so little energy and short-path multiples usually are included.

primary voltage: In IP surveying, the peak asymptotic charging voltage observed at a time-domain receiver. This occurs during the later time that current is being put into the ground at the IP transmitter. Used to determine apparent resistivity and chargeability.

primary wave: **P-wave** or longitudinal wave.

prime: To prepare an explosive for firing, as to insert a cap in a stick of dynamite.

primer: An intermediate explosive which is set off by a cap and whose function is to detonate another explosive which is not cap-sensitive.

principal alias: A frequency between the Nyquist frequency and twice the Nyquist frequency.

printed circuit: A thin laminated board on which electrical circuits are drawn and components mounted. Usually easily removed for testing and replacement.

private line: A channel dedicated to some exclusive use.

probable error: The range within which half of a series of readings of the same quantity probably lie. For a Guassian distribution, the probable error is 0.674 times the **standard deviation.** The probable error is not one which is more probable than other errors.

probing: **Sounding.**

processing: Changing the form of data. The objective of processing usually is to improve the signal-to-noise ratio so as to facilitate interpretation of the data. Processing operations include applying corrections for known perturbing causes, rearranging the data, filtering it according to some criteria, combining data elements, transforming, etc. A possible sequence of seismic processing operations is shown in Figure 141.

producibility-index log: A calculated well log showing effective porosity and the percent (**q**) of the total porosity occupied by clay. Low clay content and high fluid-filled porosity suggest good permeability.

production log: A well log run inside tubing. Small-diameter sondes are used so that they can be lowered through 2-inch ID tubing. Devices include continuous flowmeter, packer flowmeter, gradiomanometer, manometer, densimeter, watercutmeter, thermometer, radioactive tracer tools, through-tubing caliper, casing-collar locator, and fluid sampler.

profile: **1.** The series of measurements made from a shotpoint location into a recording spread. Precise definition is sometimes required because of the number of profiles is used as a production statistic of seismic work. Additional shots from the same general shotpoint location into the same spread are considered part of the same profile even though different shotholes may be used. However, if the same spread is shot into from a different shotpoint location, it is a different profile, or if the same shotpoint is shot into a different spread, it is also a different profile. **2.** A **refraction profile** denotes the ensemble of individual profiles (as defined above) shot from the same shotpoint. Use of "profile" for both the component records and for the ensemble sometimes produces confusion. *Refraction set* is also

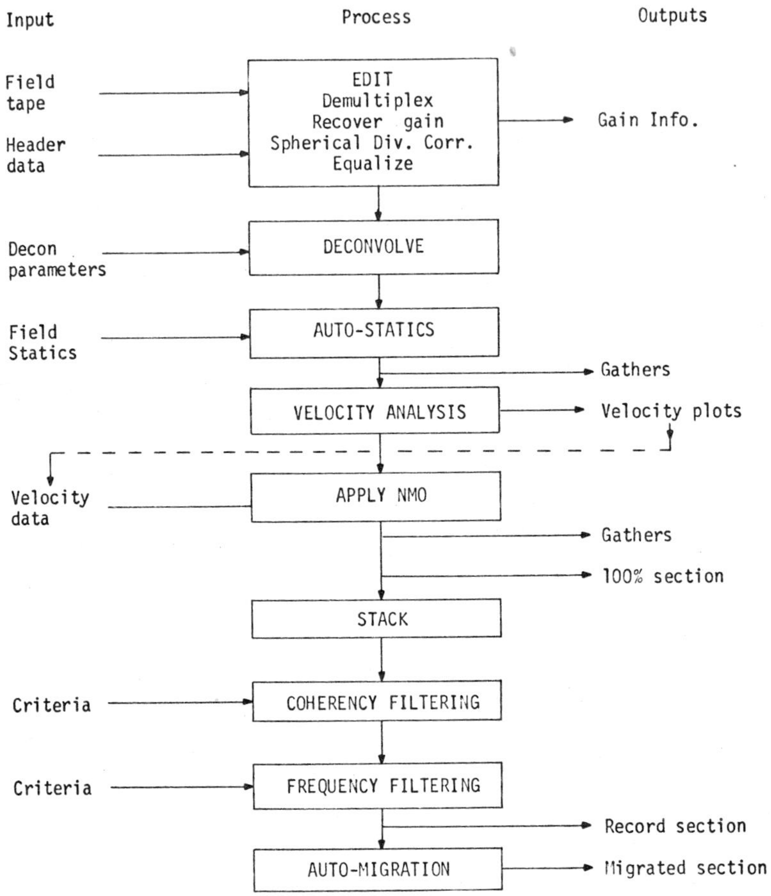

Fig. 141. **Processing** flow chart for seismic data.

used for the ensemble. **3.** A drawing showing a vertical section of the ground along a line. **4.** A graph of a measured quantity against horizontal distance, as in a gravity profile.

profile line: The line along which measurements are made.

profiler: A marine seismic reflection system usually involving a low-energy source such as the **sparker** and only one or two hydrophone groups. Sometimes the only record (Figure 142) is produced by a single-channel plotter on electrosensitive paper, sometimes the data are recorded in reproducible form (usually on magnetic tape) so that subsequently they can be processed. Often the high-frequency, low-energy source results in limited penetration and the use of only one channel prevents the use of offset to discriminate against water-bottom multiples. The recording of deep data is often paged back on top of the recording of the shallow data. Usually this presents no problem, such as in deep water where there are no shallow data (corresponding to the lack of reflectors within the water layer). The profiler is used especially in oceanographic work, as a reconnaissance tool in conjunction with gravity and other type surveys, in engineering studies, etc.

profiling: A resistivity, IP, or electromagnetic field method whereby an array with a fixed electrode or antenna array is moved progressively along a traverse to create a horizontal profile of the apparent resistivity. Occasionally refers to vertical profiling or **sounding**.

program: **1.** The work schedule for a geophysical party; e.g., the lines to be shot on a seismic prospect. **2.** The instructions for processing data, as through a computer. **3.** The plan for gain control variations, as in **programmed gain control.**

programmed gain control: **Gain control** of a seismic amplifier in which the function describing amplifier gain with respect to time after the shot is set previous to the

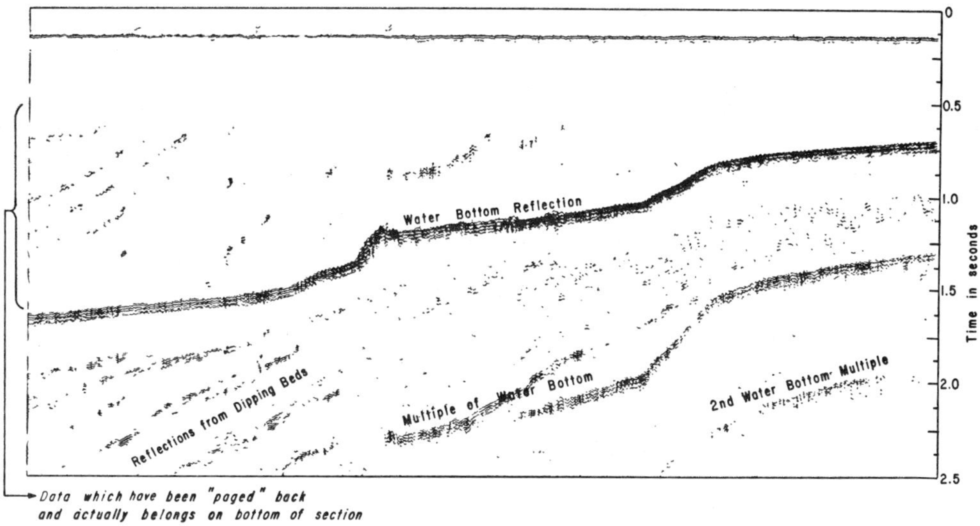

Fig. 142. **Profiler** record. (Courtesy Teledyne Exploration.)

shot. The gain does not depend on output level. Abbreviated PGC.

propagation constant: In electromagnetic theory, the propagation constant k is given by:

$$k^2 = \mu\epsilon\omega^2 + i\mu\sigma\omega$$

for the time factor $e^{-i\omega t}$, where μ is magnetic permeability, ϵ is dielectric permittivity, ω is angular frequency, and σ is electrical conductivity. Sometimes $\gamma = ik$ is called the propagation constant. Also called wavenumber.

proper crossover: If the transmitter is west of the receiver and if a counterclockwise rotation is recorded as a south dip angle, a "proper crossover" occurs where electromagnetic dip angles change from south on the south side of a point on the traverse to north on the north side; the axis of current concentration lies underneath. Opposite of *backwards crossover.*

prospecting: Exploration of an area with the objective of locating oil, minerals, etc.

protection ratio: The ratio of transmission at a desired frequency to the transmission at an undesired alias of that frequency.

proton-resonance magnetometer: A nuclear-precession magnetometer which measures the proton-resonance frequency. The proton (hydrogen) nucleus has a magnetic moment because of its spin. The spin axes normally are oriented randomly and precess about the earth's magnetic-field direction. A polarizing field normal to the earth's field is impressed on the nuclei for a short period to polarize the nuclei and is then removed. The nuclei, now oriented together, precess about the earth's magnetic field H at a frequency f_L (Larmor frequency) which is directly proportional to the strength of the earth's field. In so doing, a voltage at the precession frequency is induced in a measuring coil. As the protons gradually relax into random orientation, the induced-field strength drops to zero. The induced frequency is measured by a counting arrangement (Figure 143) to determine the value of the earth's magnetic field. The Larmor frequency is given by:

$$f_L = \gamma H/2\pi ,$$

where γ = gyromagnetic ratio ($2\pi/23.4868$ hz/gamma for protons). For the normal earth field of about 50,000 gammas, $f_L = 2100$ hz. Compare **optically pumped magnetometer.**

proximity log: A microresistivity log similar to the **microlaterolog** but less sensitive to mud-cake thicknesses. "Proximity" is a Schlumberger trademark.

proximity survey: A survey to determine how far a well is from a structural feature; for example, shooting at the surface into a geophone in a deep well with the objective of determining the position of the flank of a salt dome with respect to the well.

prune effect: A surface with many local highs. Some methods of finding the maximum of such a surface may find one of the local subsidiary maxima instead if they begin with a bad initial guess.

pseudoanisotropy: A difference in values because of the direction of measurement other than anisotropy such as crystals possess. Hence most geophysical **ani-**

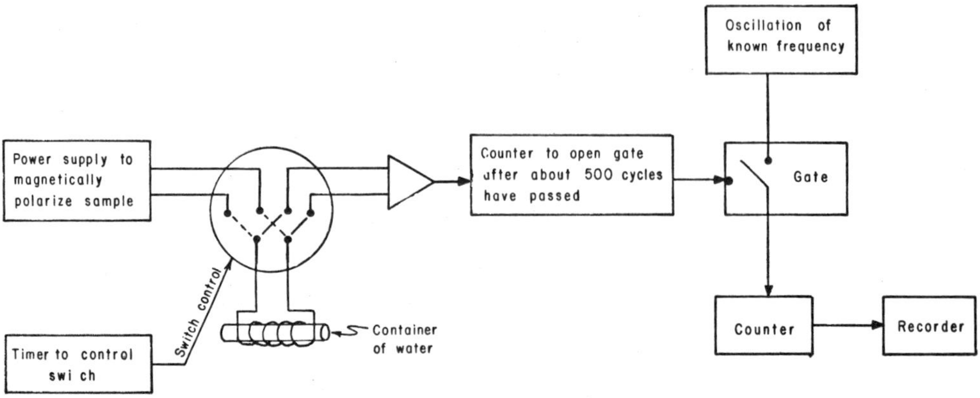

Fig. 143. **Proton-resonance magnetometer.** The free (Larmor) precession frequency controls the length of time the gate is open; this time is measured by counting the cycles of a standard frequency.

sotropy concerns pseudoanisotropy. Also called *apparent* or *effective anisotropy.*

pseudogeometric factor: A coefficient used for estimating the response of a resistivity measurement R_a at different invasion depths:

$$R_a = R_{xo}J + R_t(1 - J),$$

where R_{xo} = flushed-zone resistivity, R_t = uncontaminated-zone resistivity, and J = pseudogeometric factor which is a function of invasion depth.

pseudosection: A plot of apparent-resistivity measurements made along a line at various electrode separations. Also called *quasi-section.* The data are plotted (Figure 144) beneath the point midway between electrodes at a depth equal to half the electrode separation. The pseudosection thus conveys the variation of apparent resistivity with location and with penetration depth but cannot be simply and directly converted into resistivity distribution. Sometimes resistivity and IP pseudosections are plotted as mirror images of one another, the upper section being inverted.

pseudostatic SP: The "static SP" of a shaly sand. Strict-

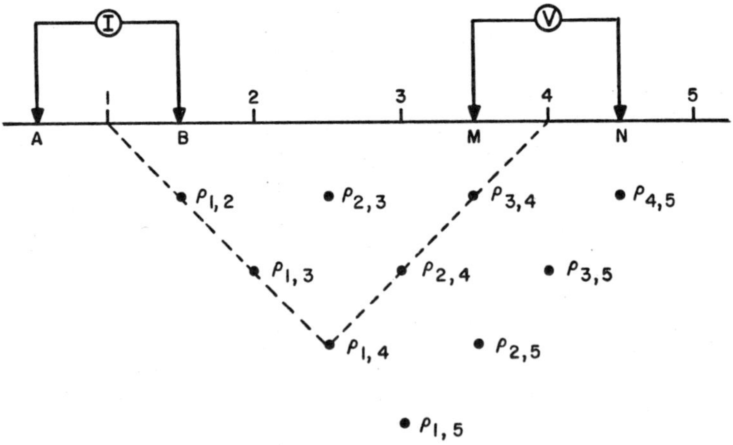

Fig. 144. **Pseudosection.** The apparent resistivity $\rho_{1,4}$ (when the current electrodes *A* and *B* are centered at 1 and the potential electrodes *M* and *N* are centered at 4) is plotted 45 degrees below 1 and 4. Data for other pairs of measurement are plotted similarly and apparent resistivities are contoured to yield a "quasi-section" or psueudosection.

ly, **static SP** refers to a clean sand. Abbreviated PSP.

psi: Pounds per square inch, a unit of pressure.

p-type semiconductor: A doped semiconductor with more **holes** than electrons available for carrying charge; an acceptor.

pu: Porosity unit or 1 percent porosity.

pull boat: A boat or raft on pontoons; it has a power winch and moves by reeling in a line previously anchored ahead. Used in travel through swamps containing heavy timber.

pulse: A waveform whose duration is short compared to the time scale of interest and whose initial and final values are the same (usually zero). A seismic disturbance which travels like a wave but does not have the cyclic characteristics of a wave train; a *wavelet.*

pulse curve: Beta curve.

pulse method: Observation of the voltage decay after cessation of a transmitted current pulse. Also called the time-domain IP method, the pulse-potential method, and dc-pulse method.

pulse-position modulation: Similar to frequency modulation but uses pulses instead of full wave. See Figure 122 **(modulation).**

pulse shaping: To change the shape of a pulse into a more desired shape. Used especially t sharpen-up the onset of energy, as to make square waves or to shape the time-break pulse so as to make the time break more definite.

pulse-width modulation: A type of magnetic-tape recording using a square carrier wave the width of whose pulses are proportional to the amplitude of the modulating wave. See Figure 122 **(modulation).** Pulse-width modulation is also used in communication.

push-pull wave: P-wave.

P-wave: An elastic body wave in which particle motion is in the direction of propagation. The type of seismic wave assumed in conventional seismic exploration. Also called *compressional wave, longitudinal wave, primary wave, pressure wave, dilatational wave,* and *irrotational wave.* In an isotropic homogeneous solid, the P-wave velocity V_P can be expressed in terms of the elastic constants and the density:

$$V_p = [(\lambda + 2\mu)/\rho]^{1/2}$$

$$= [E(1 - \sigma)/\rho(1 - 2\sigma)(1 + \sigma)]^{1/2},$$

where λ and μ are Lame's constants, E is Young's modulus, k is the bulk modulus, and σ is Poisson's ratio. See also **wave notation.**

Q

Q: 1. The ratio of 2π times the peak energy in a cycle to the energy dissipated; the ratio of 2π times the power stored to the power dissipated. The Q of rocks is of the order of 50 to 300. Q is related to other measures of absorption:

$$\frac{1}{Q} = \frac{\alpha V}{\pi f} = \frac{\alpha \lambda}{\pi} = \frac{\delta}{\pi} = \frac{2\Delta f}{f_r},$$

where V, f, and λ are, respectively, velocity, frequency, and wavelength. The absorption coefficient α is the term for the exponential decrease of amplitude with distance because of absorption; the amplitude of plane harmonic waves is often written as $A\, e^{-\alpha x} \sin 2\pi f(t - x/V)$, where x is the distance traveled. The logarithmic decrement δ is the natural log of the ratio of the amplitudes of two successive cycles. The last expression relates Q to the sharpness of a resonance condition; f_r is the resonance frequency and Δf is the change in frequency which reduces the amplitude by $1/\sqrt{2}$. **2.** The ratio of the reactance of a circuit to the resistance. **3.** A term to describe the sharpness of a filter; the ratio of the midpoint frequency to the band-pass width (often at 3 db). **4.** A designation for Love waves.

q: The fraction of total porosity occupied by dispersed shale. See **producibility-index log.**

Q-band: Radar frequencies between 36 and 46 Ghz; see Figure 68 **(electromagnetic spectrum).**

QC: Quality control.

Q-factor: Koenigsberger ratio.

Q-type section: A **three-layer resistivity model** in which the resistivities of the three layers decrease with depth.

quadrant: Surveying angles are often measured in degrees clockwise from north; the northeast quadrant is called the first quadrant, the southeastern quadrant the second, the southwestern quadrant the third, and the northwestern the fourth. This is a different notation from that usually employed in mathematics. Figure 145.

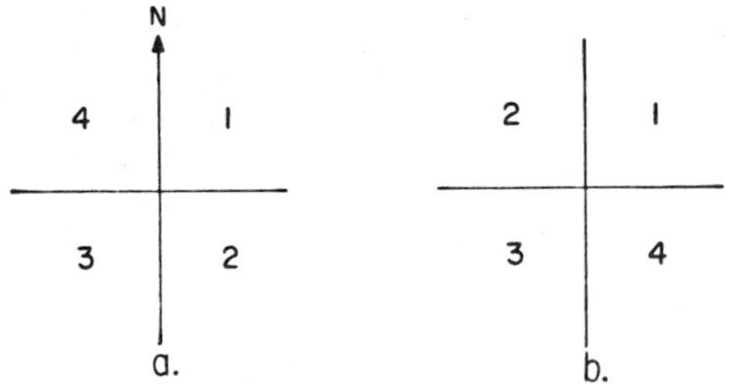

Fig. 145. **Quadrant** numbering conventions. (a) As used in surveying. (b) As used in mathematics.

quadratic spline: An interpolating operator whose slope ϕ at $(X_1 + \Delta X)$ is a linear combination of the slope at nearby points X_1 and X_2.

$$\phi = \phi_1 + (\phi_2 - \phi_1)\, \Delta X/(X_2 - X_1).$$

quadrature: Out of phase. The quadrature component of a signal is the out-of-phase component; the part of an induced signal which is out-of-phase with the generating signal.

quadrature filtering: Polarization filtering.

quarterboat: A boat or barge used by a geophysical crew as living quarters and/or base of operations.

quarterline: The lines parallel to the boundaries of a section of land (640 acres or one square mile) which bisect the section. The cross of quarter lines thus divides the section into quarter sections of 160 acres.

quasi-polynomials: Whereas polynomials are one-sided (i.e., involve increasing powers), quasi-polynomials are two-sided (i.e., involve both increasing positive and negative powers). Thus $(a_o + a_1x + a_2x^2 + \ldots)$ is a polynomial and $(\ldots + a_{-1}x^{-1} + a_o + a_1x + \ldots)$ is a quasi-polynomial.

quasi-section: A plot of apparent-resistivity measurements made along a line with various electrode separations, the data being plotted beneath the point midway between electrodes at a depth equal to half the electrode separation. See Figure 144 **(pseudosection).**

quasi-static: Varying very slowly, so that a solution which assumes constant values is an adequate approximation. In quasi-static–electromagnetic problems, displacement currents are neglected. The quasi-static electromagnetic solution becomes poor where the loss tangent becomes less than 1, such as very resistive situations (on glaciers or deserts in the absence of groundwater) where the displacement currents dominate over conduction currents.

quaternary gain: A **gain control** system in which amplification is changed only in discrete steps by factors of 4. Compared with binary gain, fewer gain jumps are required for a recording and an additional bit is required in the gain word, but one bit is saved in the amplitude recording to still have the same range of gain.

quefrency domain: Fourier transform of the frequency domain. Sometimes implies a transform back to the time-domain after nonlinear processing has been done in the frequency domain.

queue: A backlog of jobs awaiting action.

queuing: A system for handling random arrivals with minimum interference and delay. Rules for selection of items in the queue involve priorities and the demands of the arrivals on the various capabilities of the system.

Q-wave: Love wave.

R: Earthquake designation of a **Rayleigh wave.** See **wave notation.**

radar: A system in which short electromagnetic waves are transmitted and the energy which is scattered back by reflecting objects is detected. Most ships use radar to help "see" other ships, buoys, shorelines, etc. **Beacons** are sometimes used to provide distinctive targets. Radar is used in aircraft navigation (see **doppler radar**), in **positioning,** and in **remote sensing.** The radar spectrum is sometimes subdivided: P-band, 225-390 Mhz; L-band, 390-1550 Mhz; S-band, 1550 to 5200 Mhz; X-band, 5200-11000 Mhz; and K-band, 11-36 Ghz; Q-band, 36-46 Ghz; V-band, 46-56 Ghz. See Figure 68 **(electromagnetic spectrum).**

radar doppler: **Doppler radar.**

radar imagery: Mapping from an aircraft using short electromagnetic waves. A narrow radar beam which is transmitted perpendicularly to the aircraft flight line reflects from the ground to a receiver on the aircraft. The data displayed on an oscilloscope screen are photographed and give the appearance of an aerial photo.

radial array: An arrangement of electrodes in electromagnetic dipole-dipole surveying. See **array.**

radial refraction: **1.** A pattern of shooting comparable with **fan shooting.** **2.** Use of a detector deep in a borehole to receive and record seismic waves from sources located near the ground surface at different distances and azimuths. Used in determining salt-dome boundaries. Travel paths which are partly in salt will show a **lead** which will depend on the amount of salt travel. See Figure 11 **(aplanatic surface).**

radial survey: See **azimuthal survey.**

radian: A unit of angular measure such that the arc equals the radius. Abbreviated rad. One radian = 57.2958 degrees; one degree = 0.017453 radian.

radioactive-tracer log: A log involving the detection of radioactive materials dissolved in water or oil to determine the movement of fluids. A quantity of radioactive material ("slug") might be injected into the fluid and movement of the slug monitored to detect casing leaks or points of fluid entry or exit; or points where fluid enters formations might be shown by residual radioactivity at those levels.

radioactivity log: A well log of natural or induced radiation. Usually refers to a **gamma-ray log,** but sometimes also to a **density log, neutron log, neutron-lifetime log,** or other types of logs.

radioactivity survey: Measurements of variations in gamma radiation with the objective of mapping the distribution of radioactive elements (usually K, U, and Th). Observations are also affected by radionuclides, nuclear fallout, radon in the air, and cosmic radiation. See **gamma-ray surveying.**

radio-altimeter: An aircraft instrument in which the round-trip traveltime of radio waves which are beamed at the earth and reflected is measured to determine the height above ground.

radio earth: A model of the earth whose radius is 4/3 that of the earth. The index of refraction for radio waves in the atmosphere decreases with altitude in a nearly linear manner, resulting in a bending of radio waves toward the earth. By assuming that the earth's radius is 4/3 its actual radius, this refraction is compensated for and radio raypaths can be drawn as straight lines.

radio frequency: A frequency above 3 khz. Radio frequencies are subdivided into **bands.** See Figure 68 **(electromagnetic spectrum).**

radiometer: A device which measures radiation, such as the infrared radiation used in thermal imaging.

radix: **Number system** base; the number of different digits used to express a number.

raised kernel function: The **kernel function** plus 1/2. Used in electrical exploration. See Koefoed, 1968, Application of the Kernel Function in Interpreting Geoelectrical Resistivity Measurements, in Geoexploration Monographs Series 1, No. 2: Berlin, Gebruder Borntraeger.

ramp: **1.** To change from one set of parameters to another, usually implying in a linear manner; e.g., the change between filter cutoffs selected for the early portion of a seismic record to those selected for the later portion may be "ramped" or distributed uniformly over the intervening transition portion. Or the amplitude of a Vibroseis sweep may increase gradually from zero to the amplitude which is held during most of the sweep. As opposed to an abrupt step. **2.** A method of terminating a data window. See Table 12 **(window).** **3.** Motion between horizontal and vertical used to measure cross-coupling in testing shipboard gravimeters. **4.** The integral of a unit step function.

random: A relationship between two or more quantities A and B where knowledge of one quantity does not help determine the other. A random process is called *stochastic.*

random access: Equal facility of access to any of the locations in a computer's memory.

random noise: Energy which does not exhibit correlation between distinct receiving channels; energy which is random with respect to the source parameters. By adding together n elements, random noise can be attenuated by the factor $n^{1/2}$ with respect to coherent signal. Random seismic noise is attenuated by the use of many geophones per group or the stacking of several records. In contrast to coherent noise such as unwanted modes of wave travel like ground roll, air wave, etc.

range: **1.** Source-to-detector distance in refraction work; offset. **2.** The distance to a **positioning** station, especially when measured directly. **3.** The extreme distance at which useful signals can be detected. **4.** The row of townships between successive meridian lines six miles apart. American townships are specified in relation to standard reference meridians (which often are not stated explicitly). R-2-E (Figure 182, **township**) designates the "second row east of the principal meridian." Canadian townships are specified by counting west from a principal meridian. For example, 36-39-10-W1 designates section 36 of township 39 of range 10 west of principal meridian 1.

range finder: An instrument for finding the distance to

other points at which no instruments are placed. It employs the principle of triangulation, using a short base line included in the instrument. Because the angle of intersection of the two lines of sight is small, distance measurements are not of high precision.

range line: The north-south boundary between **townships.**

range pole: A long pole which is sighted on when making long shots with a transit. Range poles are temporary survey markers and often have some distinguishing arrangement (perhaps survey flagging) on top of them to aid in locating and identifying them.

ranging: Making a distance measurement as by sighting on a survey rod or with an electronic instrument such as the **tellurometer.** See also **positioning.**

rate: One of two radiopositioning range measurements necessary to determine a position. The other measurement is sometimes called **drift.**

ratiometer: An instrument for determining the ratio of two quantities. Ratiometers or compensators are used extensively in electromagnetic exploration equipment for measuring the ratio of two phasor voltages or currents. Bridge circuits are often used in ratiometers.

ray: See **raypath.**

Raydist: A medium-range radio positioning system in which three or more transmitters emit continuous waves which differ by an audiofrequency. The position of a mobile station is determined by measuring the phase difference of the audiofrequency pattern which results from the interference of the waves emanating from transmitter pairs. The phase of the beats from heterodyning the received waves are measured. Uses 1.6 to 3.3 Mhz frequencies and has a range of around 150 miles. A number of station configurations are used to give various patterns of lines of position. See Figure 113 (**line of position**). **1.** *Type-N Raydist* (similar in principle to **Lorac B,** Figure 114) consists of a fixed center station and two fixed slave stations, "red" and "green". The center station transmits a *CW* frequency f, one slave the frequency $f+\Delta f_1$, the other slave the frequency $f+\Delta f_2$. The frequency f is of the order of 1.7 Mhz, Δf_1 is 210 hz, Δf_2 is 370 hz. These frequencies are received at a fixed relay station where they are used to modulate an unrelated reference frequency broadcast from that station. The waves from the two slave stations and the center station set up standing-wave patterns in which mobile stations can locate themselves by measuring phase differences. Lines of constant phase difference are hyperbolas for which the pairs of stations are foci. Additional frequencies often are used to provide lane identification. **2.** In *type-DR Raydist* (Figure 146a) the mobile station transmits a *CW* signal of frequency f, the red fixed station the *CW* frequency $(f+\Delta f_R)/2$ and the green fixed station $(f+\Delta f_G)/2$, where Δf_R and Δf_G are different audio frequencies ($\sim$ 400 hz). Comparisons of the phase being transmitted and the signal from the mobile station are made at each fixed station and the information radioed by a "link transmitter" to the mobile station, where the information is compared with the mobile transmission. The phase comparisons at the mobile station are thus rendered insensitive to small frequency drifts. The lane counts give the range from each fixed station. **3.** In *type-DM Raydist* (Figure 146b) the mobile station transmits the frequency $2f+\Delta f$ and one fixed station (red) transmits the frequency f. A fixed "green" station and the red station each compare the phase of the signal from the mobile and red stations and relay the comparison information to the mobile station by link transmitters. The lane counts give the range to each fixed station. Similar to **Lorac, Toran,** and **Decca.** Hastings-Raydist, Inc. trademark.

Rayleigh wave: 1. A type of seismic wave propagated along a free surface; the dominant wave involved in ground roll. Symbolized *R*-wave or L_R-wave. Ground roll such as encountered in seismic exploration sometimes involves modes other than pure Rayleigh wave and is sometimes called a "pseudoRayleigh" wave. Particle motion is elliptical and retrograde (i.e., the particle moves opposite to the direction of propagation at the top of its elliptical path) in the vertical plane containing the direction of propagation; see Figure 147. Its amplitude decreases exponentially with depth, and the elastic properties to a depth of about one wavelength determine its velocity. But because the elastic constants change with depth, long wavelengths depend on the elastic properties of different material than the short wavelengths and hence different wavelengths travel at different velocities. This dispersion can be used to calculate the thickness of surface layers. For a Poisson ratio of $\sigma = 1/4$, the Rayleigh-wave velocity is 0.9194 times the S-wave velocity. A Rayleigh wave along a free surface can be thought of as a special case of a Stoneley wave (wave travel along an interface). **2.** A surface wave in a borehole is sometimes called a Rayleigh wave; see **acoustic wave.**

Rayleigh-Willis relation: The oscillation period T for the bubble effect varies as the cube root of the energy **Q** (in joules) and inversely as the 5/6th power of the pressure. The relation is

$$T = 0.0450\, Q^{1/3}/(D+33)^{5/6},$$

where D is the depth in feet. Experimental data for various sources is compared with this relationship (from Kramer, Peterson, and Walter, 1968, Seismic Energy Sources Handbook, Bendix-United Geophysical Co.) in Figure 148.

ray parameter: The quantity $p = (\sin\theta_i)/V_i$ which is constant everywhere along a raypath in the case of horizontal velocity layering. θ_i is the angle with the vertical and V_i is the instantaneous velocity.

raypath: A line everywhere perpendicular to wavefronts (in isotropic media). A raypath is characterized by its direction at the surface, often expressed as stepout. In the situation of horizontal velocity layering, $(\sin\theta_i)/V_i$ is constant for any ray (where θ_i = angle with vertical and V_i = instantaneous velocity). While seismic energy does not travel only along raypaths (i.e., seismic energy would reach the point by diffraction even if the raypath were blocked), raypaths constitute a useful

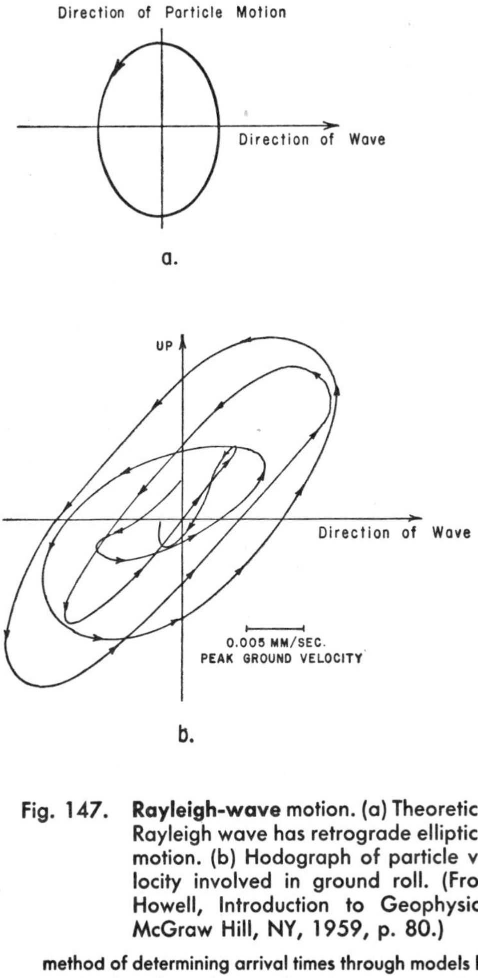

Fig. 147. **Rayleigh-wave** motion. (a) Theoretical Rayleigh wave has retrograde elliptical motion. (b) Hodograph of particle velocity involved in ground roll. (From Howell, Introduction to Geophysics, McGraw Hill, NY, 1959, p. 80.)

method of determining arrival times through models by **ray tracing**. Raypaths are often shown on wavefront charts; see Figure 193 **(wavefront chart)**.

ray tracing: Determining the arrival time at detector locations by following raypaths which obey Snell's law through a model for which the velocity distribution is known.

reactance: The opposition to alternating current flow offered by inductance or capacitance. The quadrature component of **impedance**.

read: To input data into a computer.

readability: The least discernible change in a readout device which can be readily estimated. Compare **sensitivity** and **resolution**.

real: 1. The component of a complex number (vector) in the direction of the real axis; as opposed to the imaginary component. **2.** A component which is in-phase, as a component of an electromagnetic field which is in-phase with some reference field or signal (such as the input voltage). Compare **quadrature component**.

realizable: See **physically realizable.**

real time: 1. Having the same time scale as actual time. Processing of seismic data at the same rate as that at which they were recorded. **2.** Processing data at the time of detecting and recording them.

receiver: As used in IP surveys, a sensitive, filterable ac or dc voltmeter with SP buckout controls. Generally a frequency-domain receiver is ac coupled and a time-domain voltmeter is dc coupled.

reciprocal sonde: A sonde with the current and measuring electrodes interchanged. See **lateral.**

reciprocal time: The traveltime between common points on reversed refraction profiles. Surface-to-surface time from a shotpoint at *A* to a geophone at *B* must equal that from a shotpoint at *B* to a geophone at *A*.

reciprocity principle: 1. The potential at a point *M* with respect to a current source at *A* is the same as if the points of measurement were reversed. Applied in electrical exploration. **2.** The seismic traveltime from a source at *A* to a geophone at *B* is the same as from a source at *B* to a geophone at *A*.

reconnaissance: 1. A general examination of a region to determine its main features, usually preliminary to a more detailed survey. **2.** A survey whose objective is **(a)** to ascertain regional geological structures, **(b)** to determine whether economically prospective features exist, or **(c)** to locate prospective features. As opposed to "detail" surveys which have the objective of mapping individual structures.

reconstitute: 1. To convert digitized data to a finer sample period, as to input 4 msec samples and output 2 msec samples. Opposite of **decimate. 2.** To convert sampled data to continuous data, as with a D/A converter.

record: 1. A recording of the energy from one shot (or other type of energy release) picked up by a spread of geophones. May be on photographic or other paper or on magnetic tape. Figure 149. **2.** A collection of data, such as data to be input to a computer. A record might be a single magnetic tape containing several seismic records. **3.** A group of data handled by a computer as a single block of data. **4.** To make a record.

recording truck: A vehicle which contains the amplifiers and recording gear for making a seismic record in the field. The cab where the record is made is often called the "doghouse" and is often mounted on small boats, pontoons, etc. for work where a truck cannot go.

record section: Display of seismic traces side-by-side to show the continuity of events. Record sections were originally made by splicing together individual seismic records (and hence the name) but the entity of individual records now has been largely lost.

record time: 1. Time after the shot instant. **2.** Time after a certain reference.

recovery: 1. The amount of core recovered compared to the amount cut. **2.** The amount of fluid in the drillpipe on a drill-stem test which did not produce enough fluid to fill the pipe and flow at the surface. **3.** A visit to a preestablished survey location, as for use in tying a new survey to an older survey.

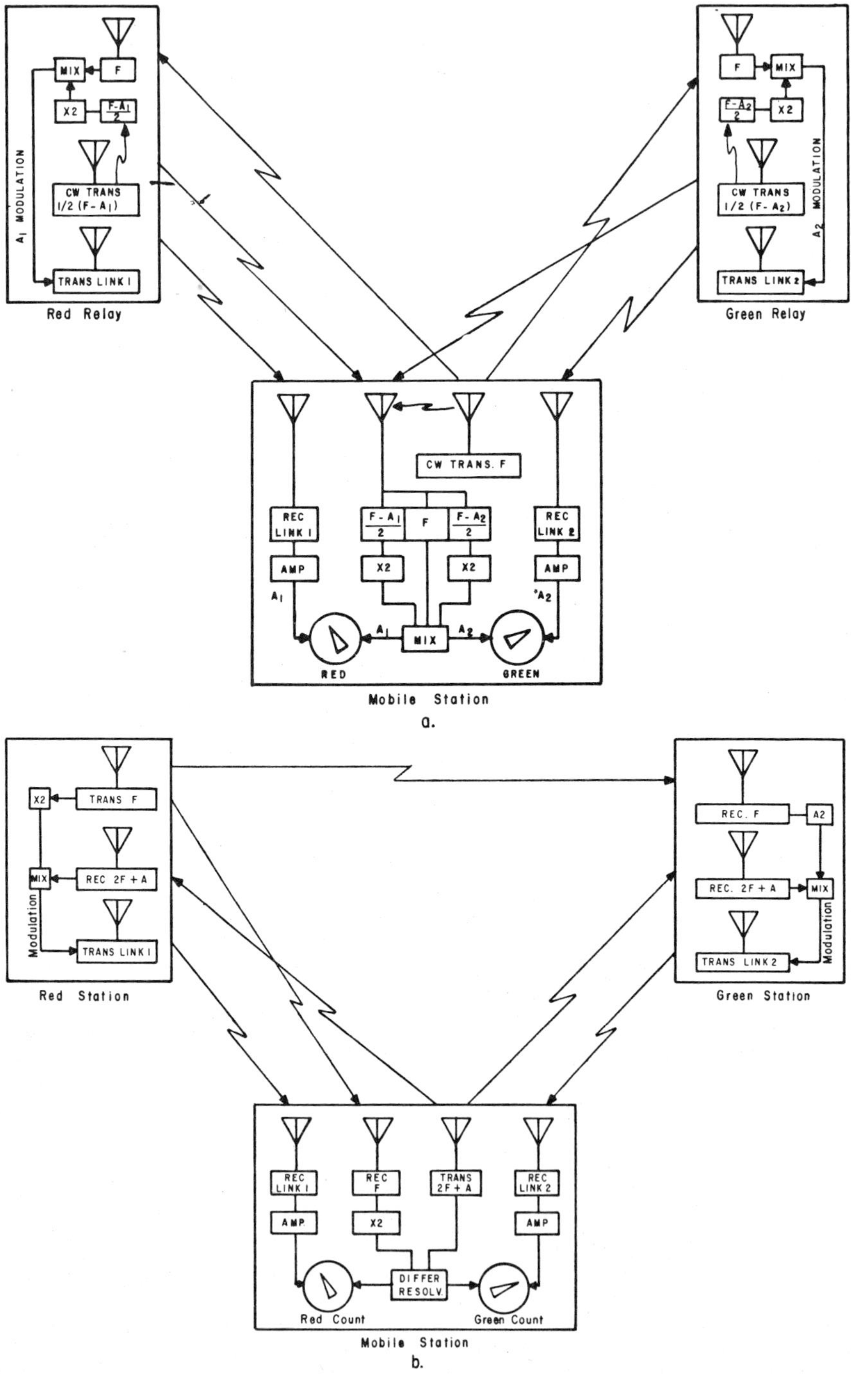

Fig. 146. **Raydist** systems. (a) Type DR Raydist. (b) Type DM Raydist.

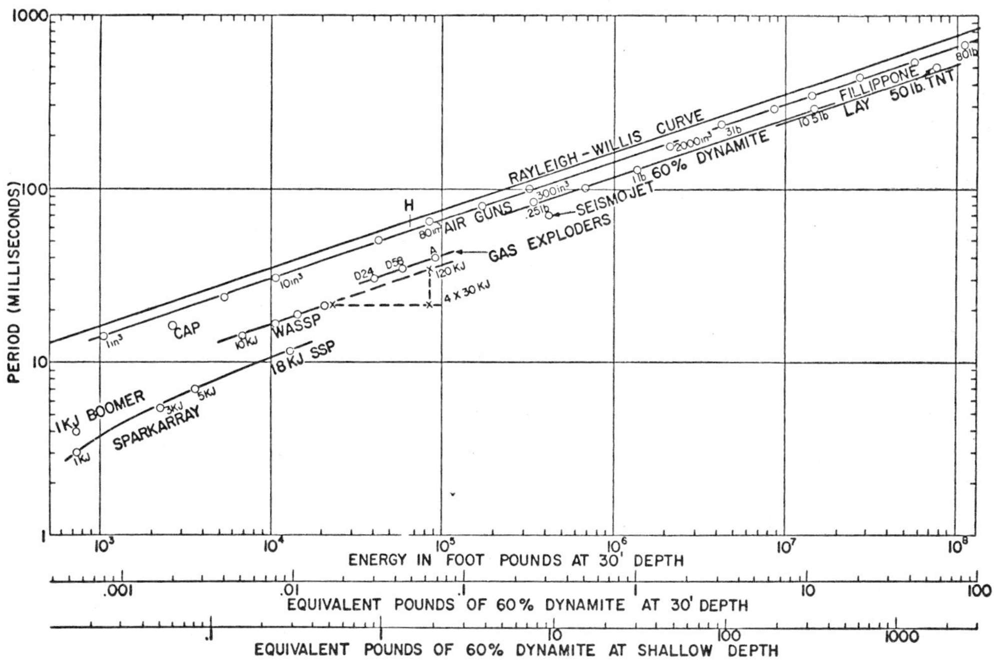

Fig. 148. **Rayleigh-Willis relation** for representative energy-sources. Source depth, 30 ft. (Courtesy Bendix-United Geophysical Company.)

rectify: **1.** To allow current to pass in one direction only. See Figure 25 **(bridge rectifier)**. **2.** To adjust a borehole log for true vertical depth. Logs in slant holes are usually recorded in distance measured along the hole, which makes them difficult to correlate with other logs which intersect the formations differently. Also sometimes used for adjusting a log to what would be seen without formation dip.

rectilinear: A system of straight lines at right angles to each other, as with ordinary *x-y* coordinate paper.

recursive filter: A filter for which the output depends on previous outputs as well as the input and the filter response. Part of the output is delayed and added to the input. Some types of filtering which require a long operator can be accomplished with a short operator in this way. Also called *feedback filter.*

reduced ratio: The ratio of electromagnetic field strength at two locations after being corrected for the *normal ratio* or *free air correction* (the electromagnetic-field gradient in the absence of perturbing conductive bodies). Used with the Turam method.

reduced traveltime: **1.** The difference between the observed refraction traveltime and the traveltime which would have been observed if only one velocity material had been present. The result yields the total **delay time** (shot delay time + geophone delay time) if the assumed velocity is the refractor velocity. An assumed velocity of 6 km/sec is often used as the reduction velocity in crustal studies. **2.** Refraction records sometimes are plotted with the time reference advanced by an amount proportional to the offset so that arrivals with the refractor velocity appear flat. Thus relief on the refractor produces apparent relief on the event. Such a plot also facilitates picking weak refraction arrivals (especially when secondary events are picked). **3. Intercept time.**

reduction: The process of substituting for observed values ones which are derived from the observed values but which have been corrected for known sources of error.

reduction to the pole: A method of removing the dependance on the angle of magnetic inclination. "Reduction to the pole" converts data which have been recorded in the inclined earth's magnetic field to what the data would have looked like if the magnetic field had been vertical.

redundancy: A repetition of information, such as the same measurements made several times (perhaps in several ways). Redundancy permits the attentuation of some effects involved in the measurement. The successive measurements may differ from each other in that data components other than the objective components are different. Hence, the objective data will be recorded the same each time and thus may be distinguished from data components which vary for the different measurements. For example, six-fold CDP

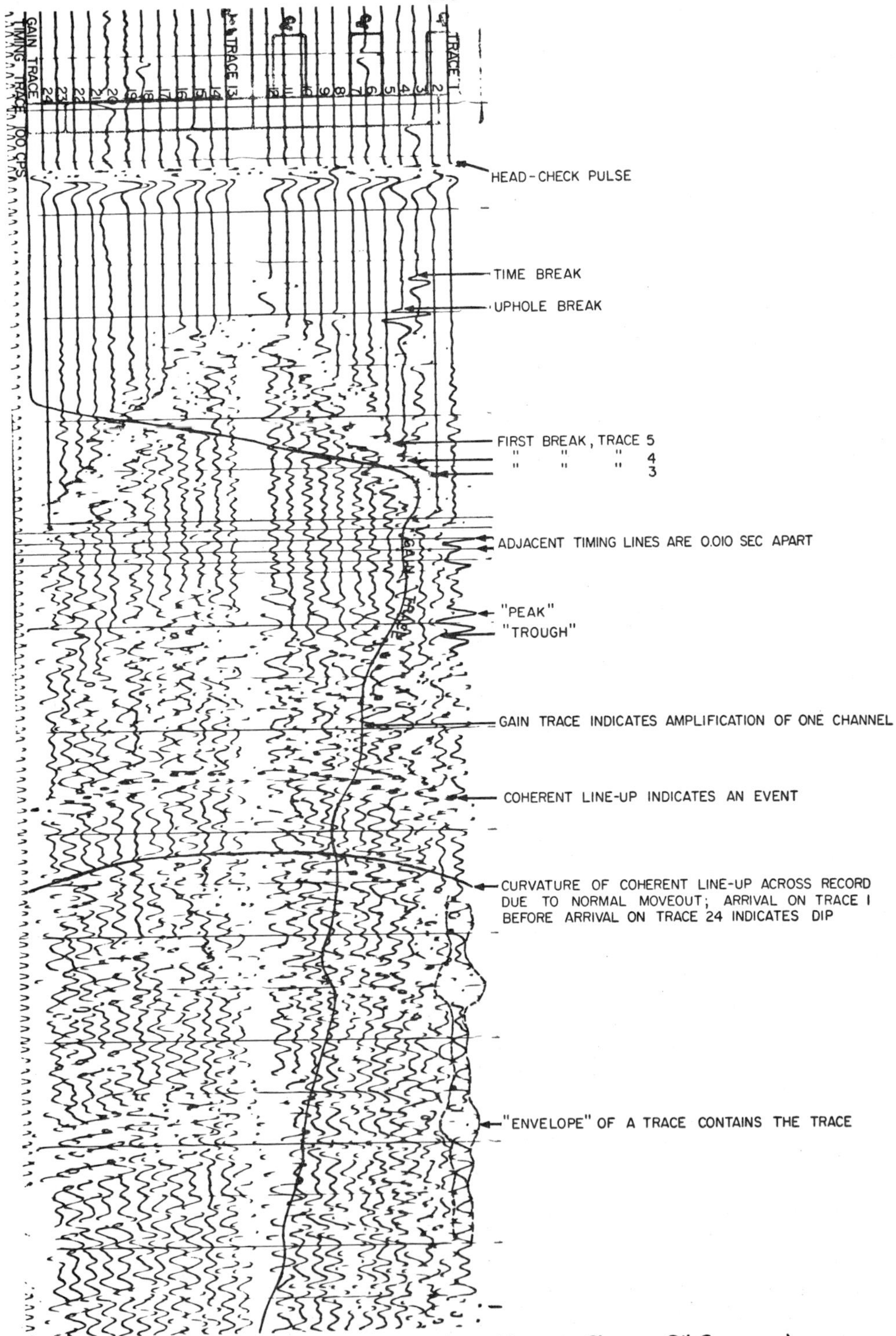

Fig. 149. **Record** in wiggle-trace form. (Courtesy Chevron Oil Company.)

shooting involves measuring the reflected energy from a given portion of the subsurface six times and hence has a redundancy of 6.

reel truck: A vehicle for transporting seismic cables and geophones.

reference plane: Datum.

reference seismometer: A geophone which records successive shots to verify that they are similar. Used in **well shooting** for velocity determination to eliminate the possibility that errors are introduced by shot variability.

reflectance: Reflection coefficient.

reflected refraction: 1. An event which results from refracted energy being reflected or diffracted back from a discontinuity in the refractor, such as by a fault. Characterized by the apparent velocity of the refractor. 2. A refraction multiple, in which the energy is multiply reflected between reflectors (in marine shooting, for example, this might be the surface and the bottom of the water layer) above the refractor. The extra travel may occur before or any time during the travel along the refractor. The effect of this type of refraction multiple is often to add cycles to the refraction arrival. 3. Any of a number of travel paths which involve reflection at an interface as well as travel through a high-velocity refractor. Figure 150.

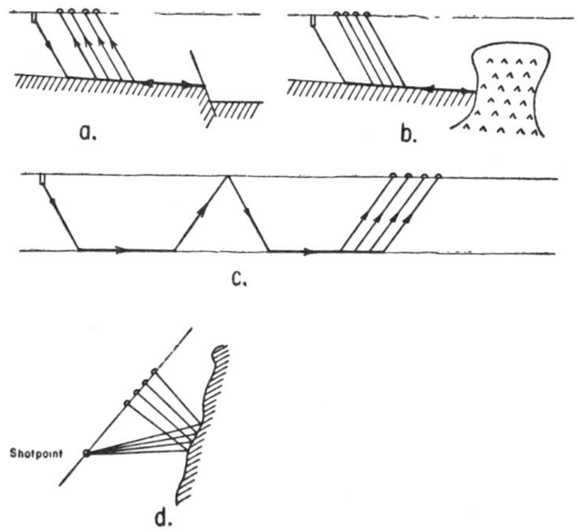

Fig. 150. **Reflected refractions.** (a) Refraction reflected at a fault. (b) Refraction reflected by a salt dome. (c) Multiply reflected refractions. (d) Plan view of refractions from base of weathering being reflected from offline feature.

reflection: The energy or wave from a shot or other seismic source which has been reflected (returned) from an acoustic-impedance contrast *(reflector)* or series of contrasts within the earth. The objective of most reflection-seismic work is to determine the location and attitude of reflectors from measurements of the arrival time of **primary reflections** and to infer from the reflectors the geologic structure and stratigraphy. See **reflection coefficient.**

reflection coefficient: 1. The ratio of the amplitude of the displacement of a reflected wave to that of the incident wave. The relationship is obtained by assuming continuity of displacement and continuity of stress at an interface. The ratio of the reflected energy to the incident energy is the reflection coefficient squared. 2. For normal incidence on an interface which separates media of densities ρ_1 and ρ_2 and velocities V_1 and V_2, the reflection coefficient for a plane wave incident from medium 1 is

$$\frac{\rho_2 V_2 - \rho_1 V_1}{\rho_1 V_1 + \rho_2 V_2}$$

This implies that displacement is measured with respect to the direction of wave travel. Thus a compression is reflected as a compression when the reflection coefficient is positive. A negative reflection coefficient implies phase inversion, that a compression is reflected as a rarefaction. Where displacement is measured with respect to a space-fixed coordinate system (e.g., positive means downward displacement), the reflection coefficient equation is the negative of that given above. Where the density does not vary appreciably, i.e., where $\rho_1 \approx \rho_2$, the reflection coefficients are approximately given by the difference in velocities divided by the sum of the velocities. 3. In the more general case of plane-wave incident at an angle, both reflected P- and S-waves and transmitted P- and S-waves will be generated. The amplitude of each of these waves may be found from Knott's equations or **Zoeppritz's equations.** 4. The square of the quantities described in 1 to 3 above sometimes is called the reflection coefficient. It denotes the ratio of the reflected energy density to the incident energy density. 5. A ratio of resistivities as derived from the method of **images.**

$$k = (\rho_2 - \rho_1)/(\rho_2 + \rho_1).$$

Used in describing **type curves.** Also called *resistivity-contrast factor* or *reflection factor.*

reflection peak: 1. The **peak** of an event identified as a reflection. 2. In electrical logging, an increase in resistivity reading as the upper (A) electrode of a lateral sonde (see Figure 111, **lateral**) passes a thin high-resistivity formation.

reflection survey: A program to map geologic structure employing the seismic-reflection method. Measurements are made of the arrival time of events attributed to seismic waves which have been reflected from interfaces where the acoustic impedance changes. The objective usually is to map variations in the depth and attitude of the interfaces, which usually are parallel to the bedding. A second objective is to define stratigraphic variations from normal-moveout measurements or from the amplitude and character of reflection events.

reflectivity: Reflection coefficient.

reflector: 1. A contrast in acoustic impedance, which

gives rise to a seismic **reflection**. **2.** A contrast in electrical properties which gives rise to a electromagnetic reflection.

reformat: To change the form of data from one **format** to another.

refraction: 1. The change in direction of a seismic ray upon passing into a medium with a different velocity. See **Snell's law** and **refraction wave. 2.** Involving head waves, i.e., involving a travel path in a high-velocity medium parallel to the bedding. See Figure 95 **(head wave).**

refraction count: The difference between doppler frequency counts from a navigation satellite for 400 Mhz and 150 Mhz radio waves. This difference is usually within ±25 hz. Ionospheric refraction (which is inversely proportional to frequency) can be determined from the difference, allowing correction for refraction above elevation angles of about 7.5 degrees. See Figure 151.

refraction marker: Refractor.

refraction survey: 1. A program to map geologic structure by using **head waves.** Head waves involve energy which enters a high-velocity medium (refractor) near the critical angle and which travels in the high-velocity medium nearly parallel to the refractor surface. Head wave arrivals are identified in terms of time after the shot and of distance from the shot. The objective is to determine the arrival times of the head waves in order to map the depth to the refractors in which they traveled. Among refraction interpretation methods are **Barthelmes method, Baumgarte ray-stretching method, Gardner method, Hale method, plus-minus method, Slotnik method, Thornburgh method. 2.** Refraction surveying also includes the detection and mapping of high velocity masses such as salt domes (see **fan shooting)** and the delineation of such masses by shooting into seismometers located in deep wells (see Figure 11, **aplanatic surface).**

refraction test: Depth probe.

refraction wave: A wave which travels obliquely downward from a source to a high-velocity formation (or *marker*), then within the formation, and, finally, obliquely upward to detectors. The angles of incidence and of emergence at the marker are critical angles. Typically, refracted waves from successively deeper

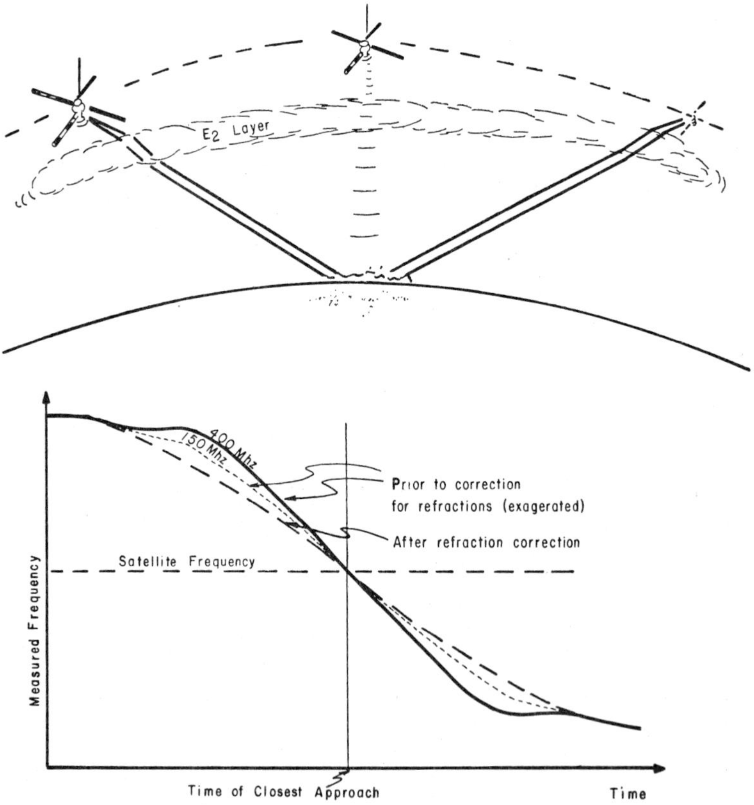

Fig. 151. **Refraction correction** to satellite observations (schematic). The shift because of ionospheric refraction is proportional to frequency.

and faster markers appear as first arrivals with increasing offset (increasing shot to detector distance). Also called *head wave, Mintrop wave,* and *conical wave.* See Figure 181 (**time-distance curve**).

refractor: A layer of higher velocity than overlying layers, through which a head wave (refraction wave) travels. In order to be useful for mapping, refractors must be sufficiently thick (~ 1/10 of wavelength) so that the refraction wave can carry energy over an applicable distance, sufficiently extensive that the same refractor is mappable over an appreciable area, sufficiently distinctive in its velocity that the refracted wave can be distinguished from the waves carried in other layers, and not be "hidden" by a shallower refractor.

regional: The general attitude or configuration disregarding features smaller than a given size. Regional dip is the general dip attitude ignoring local structure. The *regional gravity* is the gravity field produced by large-scale variations ignoring anomalies of smaller size. See **residualize.**

regression analysis: A method of finding the statistical dependence of one quantity on other quantities. Often implies the same as **factor analysis** (or multivariate analysis).

reject region: The range which a filter attenuates to a considerable extent. May refer to those frequencies which frequency filters will not pass, to those dips or apparent velocities which velocity filters attenuate, etc.

relative apparent resistivity: The ratio of apparent resistivity to the true resistivity of a portion of a model; for example, apparent resistivity divided by the resistivity of the upper layer in a simple two-layer case. Such dimensionless ratios are used in resistivity **type curves.**

relative bearing: Azimuth of hole deviation with respect to the reference on a **dipmeter** sonde (such as the No. 1 electrode).

relative ellipse area: In telluric surveying, the electric-field vectors are observed at a station (E_u and E_v) and also simultaneously at a base (E_x and E_y). The vectors are related to each other in linear combinations:

$$E_u = aE_x + bE_y$$
$$E_v = cE_x + dE_y.$$

The determinant of this transformation $J = ad - bc$, called the *relative ellipse area,* is a measure of the electrical properties at the station relative to the base.

relative permeability: **1.** The ratio of the permeability for a given phase compared with the permeability when other phases are not present; the ability of a porous medium to permit fluid flow through it when two or three phases are present in the pore space. The relative permeability depends on the fraction of the pore space occupied by that phase. Since the various phases inhibit the flow of each other, the sum of the relative permeabilities of all phases present is always less than unity. **2.** Relative **magnetic permeability.**

relative permittivity: Dielectric constant, normalized by dividing by the permittivity of free space (8.85×10^{-12} farad per meter) so as to give a dimensionless quantity.

relative thickness: The ratio of thickness of a layer to the electrode interval. A dimensionless ratio used in drawing **apparent-resistivity curves.**

relaxation time: See **time constant.**

relay: A switching device, usually controlled by a separate electrical circuit. Mercury relays are sometimes used to switch current in IP transmitters.

release time: See **AGC time constant.**

relief: The range of values over an anomaly or within an area. The difference between the highest and lowest elevations in an area. While elevation differences are usually meant, one also speaks of the "gravity relief" for the magnitude of a gravity anomaly, etc.

reluctance: Magnetic flux per unit magnetomotive force.

remanence: See **remanent magnetization.**

remanent magnetization: Remanence; the magnetization remaining in the absence of an applied magnetic field. **a.** *Natural remanent magnetization* is the residual magnetization possessed by rocks and other materials in situ; unless otherwise qualified, this is the meaning which remanent magnetization usually implies. **b.** *Thermoremanent magnetization (TRM)* remains after a sample has been cooled from a temperature above the Curie point in a magnetic field. **c.** *Chemical remanent magnetization (CRM)* is acquired when a magnetic substance is chemically formed or crystallized in a magnetic field at temperature below the Curie point. **d.** *Depositional (or detrital) remanent magnetization (DRM)* is acquired in sediments when magnetic mineral particles are preferentially aligned during deposition (usually by settling through water) in response to the effect of the ambient magnetic field. **e.** *Isothermal remanent magnetization (IRM)* is remanent magnetization in the ordinary sense, i.e., the magnetization after application and subsequent removal of a magnetic field; it is not involved in paleomagnetism. **f.** *Pressure (or piezo) remanent magnetization (PRM)* is remanence acquired as a result of the application of stress; the effects generally become more pronounced as the strain proceeds from elastic to plastic deformation. See also **viscous magnetization.**

remote electrode: Infinite electrode.

remote sensing: Measurements made from large distances, as from high flying aircraft or from earth satellites. Especially refers to measurements of either natural radiation or radiation from a source in the sensor which has been reflected back from the earth. Any of several portions of the electromagnetic spectrum (ultraviolet, visible, infrared, microwave, or radar) may be used. Both passive methods (measuring natural radiation) and active methods (beaming from a source on the aircraft and measuring the reflected energy) are used. Displays sometimes use imaging (so as to produce a picture of the radiation, as in a photograph) and sometimes nonimaging (so as to produce a profile of the variation of radiation along the flight path). May also include measurements of the magnetic field or other nonradiation measurements.

remote triggering: A method of controlling "on" and "off" voltage switching of IP receiving equipment to record the decay signal. A synchronous-detection method which uses the ground signal for a timing channel.

repeatability: The maximum deviation from the average of corresponding data points taken from repeated tests under ideal conditions.

replicating function: Comb or shah.

reproducible recording: Recording in a form which readily permits conversion into electrical signals, such as on magnetic tape, but excluding an ordinary print on paper.

resample: To change the sampling frequency (or the interval between adjacent samples). To decrease the number of samples is to **decimate;** for example, a data set at 2 msec intervals can be "decimated" to a data set a 4 msec intervals by dropping every other sample. To increase the number of samples is to **reconstitute;** thus values can be interpolated between the values of a data set a 2-msec intervals to yield a data set at 1-msec intervals.

resection: Determination of the horizontal location of a survey station by the intersection of lines indicating the direction from other stations. See Figure 135 **(plane table).**

resident programs: Programs stored within a computer's memory and which hence do not need to be specifically loaded in order to use them.

residual: 1. What is left after the regional has been removed. See **residualize. 2.** What is not accounted for. **3. Salt residual.**

residual disturbance: Disturbances that persist during quiet days following a magnetic storm.

residualize: 1. The process of separating a curve or a surface into its low-frequency parts (called the *regional*) and its high-frequency parts (called the *residual*). Residualizing is an attempt to sort out of the total field those anomalies which result from local structure; that is, to find local anomalies by subtracting gross (regional) effects. This separation cannot be done uniquely and some subjectivity is often desirable. Thus one might want certain types of effects classed as "regional" for one type of study and as "residual" for another study. The regional may be drawn graphically by smoothing contours on a map or by drawing smooth curves through profiles or networks of profiles. See Figure 152. Or the regional may be determined analytically by **surface fitting,** by **grid-residual** methods (including **second-derivative map** methods), by Fourier analysis and filtering, and by other methods, most of which can be thought of as two-dimensional **convolution** operations. Some of these methods produce **halo effects** about local anomalies but these can be reduced by biasing. Sometimes residualizing is done in steps, a very smooth regional being removed from the residual of the first operation. **2.** The process of determining what is not accounted for by a particular model. The effects of the model are calculated and subtracted from the observed field, the residual being those effects still unaccounted for by the model. See **salt residual.**

residual normal moveout: A small amount of normal moveout which remains because of incomplete normal-moveout removal. Compare **differential normal moveout.**

resistance: Opposition to the flow of a direct current. Compare **impedance** and **reactance** and see **Ohm's law.**

resistivity: The property of a material which resists the flow of electrical current. Also called *specific resistance.* The ratio of electric-field intensity to current density. The reciprocal of resistivity is *conductivity.* In nonisotropic material the resistivity is a **tensor.** See **ohm meter** and **apparent resistivity.**

resistivity-contrast factor: The ratio, $(\rho_2 - \rho_1)/(\rho_2 + \rho_1)$, which appears in the analysis of resistivity relationships between materials of resistivity ρ_1 and ρ_2. Also called *reflection coefficient.*

resistivity index: The ratio of the resistivity of a formation bearing hydrocarbons to the resistivity if 100 percent saturated with formation water. See **Archie's formulas.**

resistivity logs: 1. Well logs which depend on electrical resistivity: **normal, lateral, laterolog,** and **induction log.** Most resistivity logs derive their readings from 10 to 100 ft^3 of material about the sonde. **Microresistivity logs** on the other hand derive their readings from a few cubic inches of material near the borehole wall. **2.** Records of surface **resistivity methods.**

resistivity method: 1. Observation of electric fields caused by current introduced into the ground as a means for studying earth resistivity in geophysical exploration. The term is normally restricted to include only those methods in which a very low frequency or direct current is used to measure the **apparent resistivity.** Includes **electrical profiling** and **electrical sounding.** Various **array** types are used. See also **Gish-Rooney method** and **Lee partitioning method. 2.** Sometimes includes **induced-polarization** and **electromagnetic-survey** methods also.

resistivity spectrum: The resistivity of a polarizable material, measured at successive frequencies and plotted against frequency. This spectrum can be transformed into an IP decay curve and vice versa. The IP resistivity spectrum of a polarizable material sometimes appears to be distinctive of the nature of the substance (type of mineralization).

resistor: An electrical circuit element possessing a certain resistance. Resistor color codes are shown in Figure 153.

resolution: 1. The ability to separate two features which are very close together. The minimum separation of two bodies before their separate identity is lost on the resultant map. **2.** The smallest change in input that will produce a detectable change in output **3.** The ability to localize an event seen through a window; usually taken as the half width of the major lobe.

resolve: To separate into parts. **1.** To determine the component orthogonal vectors which add together to form a given vector. The direction of the component vectors has to be prescribed. Usually resolution implies finding the components in the directions of an orthogonal coordinate system. **2.** To show two features as

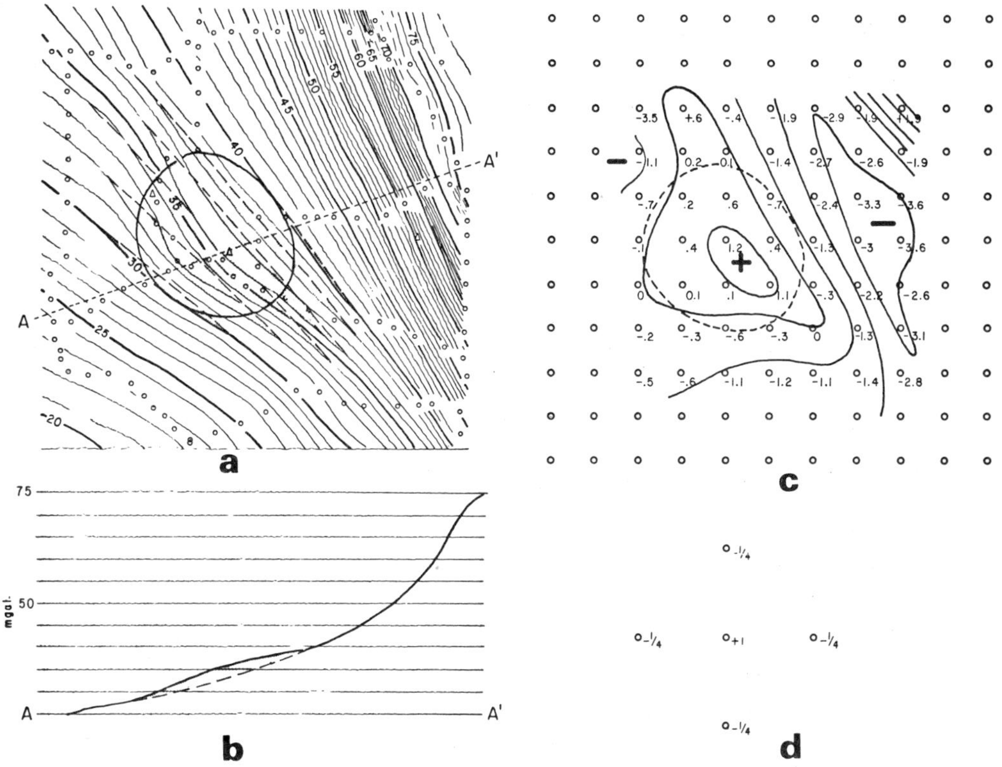

Fig. 152. **Residualizing** methods. (a) Graphical smoothing of map contours. The difference between smoothed and actual contours localizes an anomaly. (b) Graphical smoothing of profile. The profile is drawn where the contours are controlled by the data. (c) Reading values on a regular grid and convolving with a template. Margins are lost in the process so the residual covers a smaller area. Template and weighting can be varied to yield second derivative and other types of maps. (d) Template used to make map shown in (c) above.

seperate rather than blended together.

resolved bands: The *number of resolved bands* is the ratio of the Nyquist frequency to the resolution.

response time: The time elapsed between the generation of an inquiry and the receipt of a response.

restored section: A cross-section showing reflection events or geologic beds as they presumably would have been at some time in the past. A seismic section in which one event is made flat, the event being attributed to a horizon which was presumably laid down flat. Also called paleosection.

retrocorrelation: Convolving a function with itself; autoconvolution. Retrocorrelation of a seismic trace yields a *retrocorrelogram,* which contains no primary events but only multiples which involve the surface. See Anstey and Newman, 1968, The Sectional Autocorrelogram and the Sectional Retrocorrelogram: Geophysical Prospecting, v. 14, p. 389-426.

retrograde: Rotational motion opposite to the usual direction. Rayleigh waves are sometimes called *retrograde waves* because particles travel in elliptical orbits in the direction such that the particle is traveling opposite to the direction of the wave while at the top of the ellipse. See Figure 147 **(Rayleigh wave)**.

reverberation: Multiple reflection in a layer, usually the water layer in marine work; *singing.* Sometimes the term is reserved for the case where the successive multiples blend together into a more or less steady oscillation, thus excluding the situation where the water is so deep that the successive multiples are discrete. Reverberations sometimes occur on land records also. Removing reverberation effects is the objective of much **deconvolution** in digital processing. See **Backus filter.**

reversal: A local change of approximately 180 degress in the direction of the dip along a line. Often refers specifically to a dip reversal which denotes a structural lead.

reverse branch: A seismic event where right-to-left orientation on the event corresponds to the reverse orientation (left-to-right) on the reflector. The seismic

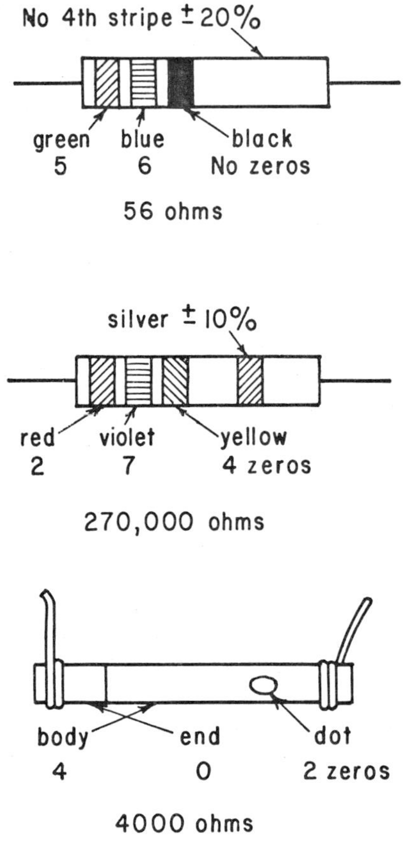

Fig. 153. **Resistor** identification code. The color code is: brown = 1, red = 2, orange = 3, yellow = 4, green = 5, blue = 6, violet = 7, gray = 8, white = 9, black = 0, gold = 0.1 or 5 percent if tolerance, silver = 0.01 or 10 percent if tolerance.

reflection energy of a reverse branch observed at a sequence of points along a seismic line corresponds to the backward sequence of reflection points on the reflecting interface. For zero offset, this occurs where the radius of curvature for the concave upward reflector is less than the reflector depth. See Figure 28 **(buried focus)**.

reverse control: Profiles shot from opposite directions; two-way control. **1.** The portions of a refractor for which there is overlap of relative-delay-time data resulting from shooting in opposite directions (i.e., the subsurface coverage from which arrivals are seen in opposite directions). **2.** Sometimes merely the portion of a refraction profile which has been shot from opposite directions. **3.** Reflection shooting where dip can be verified by data shot in opposite directions.

reversed: A spread or refraction **profile** shot from opposite directions.

reverse migration: Determining where the reflection event from a given portion of reflector would be observed at normal incidence. Figure 154. Dipmeter data from a borehole may be reverse migrated to tie them to events seen on a reflection seismic section or a model of an assumed structure may be reverse migrated to see what seismic data shot across it would show.

reverse SP: Where the mud is more saline than the formation water, SP characteristics are reversed with respect to the usual situation.

reverse symmetric: A **matrix** for which $A_{xx}(t) = A^{T}_{xx}(-t)$.

reverse-wound geophone: See **humbucking.**

reversible process: A physical or chemical change which can be caused to proceed in either direction by an infinitesimally small alteration in one of the controlling equilibrium conditions such as concentration, temperature, or pressure. If a small change in current shifts the equilibrium position, a change back to the original value restores the equilibrium to its original position. See **equilibrium.**

review: To reinterpret data or to rework.

Reynolds number: The ratio of inertial force to viscous force for viscous fluid flow. In the bubble effect, $R = 2ua\rho/\mu$, where u = radial water velocity, a = bubble radius, ρ = water density, and μ = water viscosity.

RF: Radio frequency.

rho-rho determination: Positioning in which the distance from two known points is used to determine the location. Other ways of determining location fixes include *rho-theta* determinations, difference in distance measurements, azimuthal systems, dead reckoning, etc.

rho-theta determination: Positioning in which the distance and direction from a fixed point are used to determine the location.

rhumb line: A line of constant azimuth (constant direction with respect to geographic north); loxodrome. A straight line on a Mercator projection.

ribbon: A very thin sheet used as a magnetic model; a very thin dike model.

Richter scale: A scale for measuring the **magnitude** of an earthquake.

Ricker wavelet: A particular seismic waveform which results from the transmission of a seismic impulse through an ideal viscoelastic material with attenuation proportional to the square of the frequency. See Ricker, 1953, The Form and Laws of Propagation of Seismic Wavelets: Geophysics, v. 18, p. 10-40.

Rieberize: To time-shift adjacent traces and sum, thus emphasizing energy which comes from a particular direction and deemphasizing energy from other directions. By varying the amount of time shift, various directions can be searched to find a maximum for a particular arrival, thus ascertaining the direction of the raypaths associated with its arrival. Interfering events which arrive from different directions can be sorted out in this way. The display of sums as a function of arrival

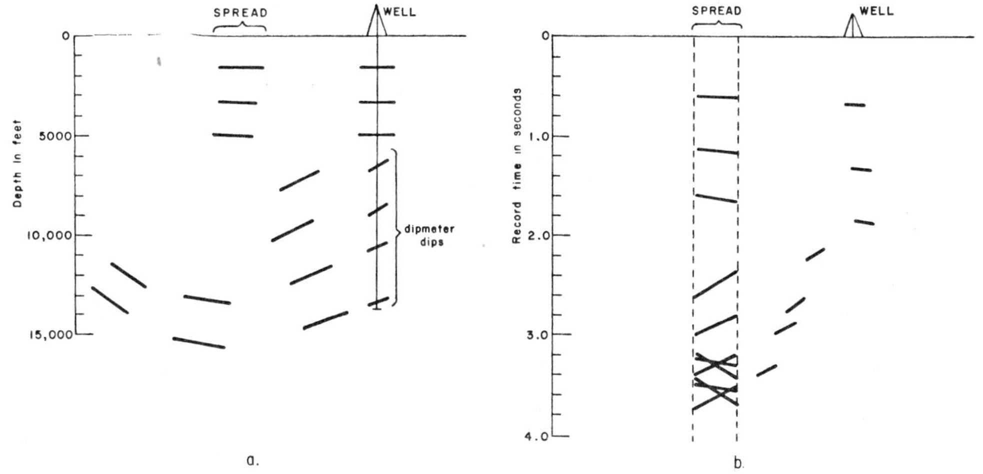

Fig. 154. **Reverse migration.** (a) Depth section showing migrated seismic events. (b) Time section showing reverse-migrated dipmeter dips. Structural models may be reverse migrated to find where their seismic evidences should be sought.

time for various amounts of trace shift is called a sonograph or sonogram. Rieber, 1936, Visual Presentation of Elastic Wave Patterns: Geophysics, v. 1, p. 196-218.

rift: 1. A **fault**; especially a long strike-slip fault such as the San Andreas. 2. A long graben such as the rift valleys of Africa. 3. A long graben associated with the upwelling of the mantle and the consequent tension and injection of igneous materials.

rig: **Drill.**

right ascension: 1. The angular distance east of the vernal equinox line (first line of Aries) on the celestial sphere to a celestial body. 2. The angle between the vernal equinox line and the intersection of the plane of the orbit of a satellite with the plane of the celestial equator. See Figure 107 (**Kepler coordinates**).

right-hand rule: A rule which gives the direction of the force on a current in a magnetic field. See Figure 101 (**induction**).

right lateral fault: See Table 2 (**fault types**).

rigidity modulus: See **elastic constants.**

ringing: 1. **Reverberation** or singing; the oscillatory effect on seismic waveforms produced by short-path multiples in a shallow water layer. 2. The oscillatory effect produced by passage through a narrow-band filter.

ringy: Looking like a modulated sine wave; oscillatory.

rise-time error: A delay in measuring the onset of an event because of the time required to reach a detectable amplitude level.

rms: **Root mean square.**

rms error: The square root of the average of the squares of the differences between a series of measurements and their mean value; the standard deviation. For a normal distribution, 0.683 of the population lies within the limits. See **statistical measures.**

rms positional error: The circle whose radius is the square root of the mean of the sum of the squares of the distances of measured positions from a point. If the uncertainties in *x* and *y* directions are equal, the probability of a measurement lying within the rms positional-error circle is 41 percent. Called *one-sigma circle* because this radius is equal to the standard deviation. Compare **CEP.**

rms velocity: Root-mean-square velocity. The velocity measured in a **velocity analysis** based upon normal-moveout measurements where the assumption of parallel velocity layering is valid. For a series of parallel layers of velocity v_i, where the traveltime for seismic energy perpendicularly through each is t_i, the rms velocity is

$$V_{rms} = \left[\frac{\sum v_i^2\, t_i}{\sum t_i}\right]^{1/2}.$$

The average velocity, on the other hand, is

$$V_{av} = \left[\frac{\sum v_i t_i}{\sum t_i}\right].$$

rockbit: **Roller bit.** See Figure 23 (**bit**).

rod: A graduated pole used as a target in surveying. See **stadia.**

rodman: A surveyor's assistant, one of whose jobs is holding the **stadia** rod.

roll: Rotational motion of a ship about the axis of principal direction of motion. As opposed to average tilt to starboard or port, which is called list. Compare **pitch** and **yaw.**

roll along: The field method for recording for a **common-depth-point stack.**

roll-along switch: A switch which permits connecting different geophone groups to the recording instruments, used with common-depth-point recording.

roller bit: A type of drill bit with conical-shaped, hard-surfaced, rolling cutting edges, which rotate and dig at the bottom of a shothole, used to drill in hard rock. Also called *rock bit.* See Figure 23 **(bit).**

roll off: 1. The frequency beyond which a **filter** produces significant attenuation. **2.** Attenuation produced by a filter, often given in decibels per octave.

roll over: Reversal of dip direction.

root: 1. Relatively light crustal material projecting down into the mantle and thus balancing topographically high areas (mountains). See Figure 105 **(isostasy). 2.** The values of x which satisfy the equation $F(x) = 0$.

root mean square: The square root of the average of the squares of a series of measurements. Abbreviated rms. This is also the autocorrelation value (without normalizing) for zero lag. For a sine wave, it is $(1/2)\sqrt{2}$ times the peak amplitude. Often used because it is related to power.

rose: A polar diagram in which radial length indicates the relative frequency of observation at a certain angle. Also called *rosette.* Used to illustrate the direction probabilities of wind, dip directions from dipmeter measurements, lineations, faults, etc. Figure 155.

rotary drill: See **drill.**

rotary-field electromagnetic method: A method using circularly polarized electromagnetic energy. See Sinha, 1970, Theoretical Studies of the Rotary Field Electromagnetic Method of Prospecting: Geophysics, v. 35, p. 310-330.

rotary table: See **drill.**

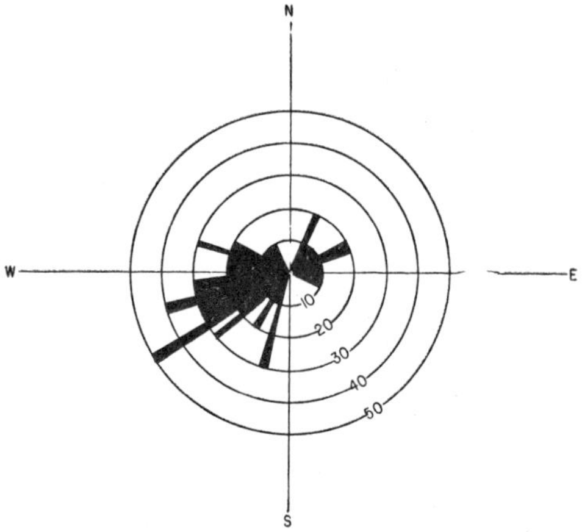

Fig. 155. **Rose** showing frequency of observations in various directions.

R-wave: Rayleigh wave.

rubidium-vapor magnetometer: A type of **optically pumped magnetometer.**

rugosity: Irregularity of a borehole wall. The readings of logging tools which have to be held in contact with the hole wall (such as density and microresistivity logs) are affected by rugosity.

RZ: Return to zero; a method of digital recording on magnetic tape in which a one is indicated my magnetization in one direction whereas a zero is indicated by magnetization in the opposite direction. Compare **NRZ** and **NRZI.**

S: Admittance; see **S-rule.**

salinity: The amount of total salts in solution (sodium, potassium, chloride, sulfate, etc.). Normal seawater has around 35,000 ppm (parts per million) salinity; less than 2,000 ppm often is regarded as fresh.

salt residual: A gravity map from which the calculated effects of a salt-dome model has been subtracted, thus showing the gravity effects for which the model does not account.

sample interval: The interval between readings, such as the time between successive samples of a digital seismic trace or the distance between gridded gravity values. Also called *sample period.*

sampling function: An infinite sequence of impulses occurring at equal time intervals. The *comb* or *shah,* symbolized III(t). Multiplying a waveform by the sampling function produces the set of sample values which represents the waveform. See Figure 34 **(comb).**

sampling skew: A small systematic channel-to-channel time displacement in seismic digital recording, produced because the multiplexer samples adjacent channels successively in order to present them to a single channel A/D digitizer.

sampling theorem: Band-limited functions can be reconstructed from equi-spaced data if there are two or more points per cycle for the highest frequency present. Also called *cardinal theorem.* See Figure 4 **(alias).**

sand count: **1.** The total effective thickness of a permeable section excluding shale streaks or other impermeable zones. Often determined from electrical logs (SP or microlog). **2.** The number of separate permeable sands separated by impermeable zones.

sand line: **1.** A line that can be drawn through the maximum deflections for thick, clean sands on an SP log in a section where the formation water is of constant salinity. See Figure 170 **(SP).** **2.** A wire line on a drilling rig often used to run or recover tools inside the drillpipe.

satellite navigation: A determination of location based on observations of a navigation satellite. Satellites in polar orbits transmit ultra-stable carrier frequencies of 150 and 400 Mhz (accurate to one part in 10^{11}). Each successive two-minute transmission begins precisely on every even minute, thus providing timing. Information about the satellite's position is carried by phase modulation (doublet modulation) of the carriers; see Figure 107 **(Kepler coordinates).** This position information is updated by ground tracking stations and reinjected to the satellite every twelve hours. The distance to the nearest point of the satellite's orbit can be determined from frequency measurements; higher frequency is seen as the satellite approaches because of the doppler effect and lower frequency after the satellite has passed (Figure 156). In the long-doppler-count mode, the number of cycles in a two-minute observation period are counted; eight such counts are possible on a satellite pass. In the short-doppler-count mode, 23-sec observation periods are used. Three or more long counts (including one before and one after closest approach) or 15 short counts are adequate for a fix. With four or more satellites, a satisfactory fix usually can be obtained every couple of hours. The accuracy of satellite fixes depends on the accuracy with which the velocity of the observing station is known (both speed and direction) but can be around a hundred feet under good conditions. The accuracy can be improved by the **translocation method.** Satellite fixes are commonly combined with other navigation tools such as **doppler sonar** or **loran C** in the rho-rho mode to give locations between satellite fixes and to determine the velocity of the observing station. See also **refraction count.**

satellite pass: The transit of a navigation satellite which can be used for a determination of position.

satnav: Satellite navigation.

saturable system: A radio-positioning system which can be used by only one (or a few) users at a time.

saturation: **1.** Apparent nonlinear resistivity or IP behavior due to large constrasts in electrical property and extreme values of resistivity contrast. These conditions make it difficult to evaluate the true resistivity and IP effect of a body but easier to find its depth. **2.** In induced-polarization measurements, the IP response sometimes varies nonlinearly with charging current, probably due to exceeding the current-density limit of the polarizable body. **3.** The limiting value of a nonlinear variable. **4.** The maximum magnetization as the magnetic field is increased.

saturation exponent: See **Archie's formula.**

saturation prospecting: The blanket use of several exploration methods over an area, in contrast to a more selective choice of exploration methods.

sausage powder: A long plastic tube of low-velocity explosive, used to provide a **directional charge.**

sawtooth SP: A jagged SP curve. When a very permeable salt-water sand containing a shaly streak is invaded by fresh-mud filtrate, the filtrate tends to accumulate just below the shaly streak, setting up an electrochemical cell which causes the SP to develop a sawtooth appearance.

S-band: Radar frequencies between 1550 and 5200 Mhz; see Figure 68 **(electromagentic spectrum).**

scaling: Changing the amplitude by multiplying by a constant. See **linear system.**

scan: One complete sequence of events, such as sampling all record channels or sonogramming for all dips (stepouts).

scatter diagram: A diagram on which points are plotted against two (or more) variables to demonstrate the relationship between the variables.

scattering: The irregular and diffuse dispersion of energy caused by inhomogeneities in the medium through which the energy is traveling.

schlieren: Streaks of density and index-of-refraction differences in a turbulent fluid. Such streaks permit photographic detection of the density variations.

Schlumberger electrode array: Electrode arrangement used in resistivity surveying consisting of four colinear electrodes, with the outer two serving as current sources and the inner two (which are closely spaced about the midpoint of the outer pair) serving as

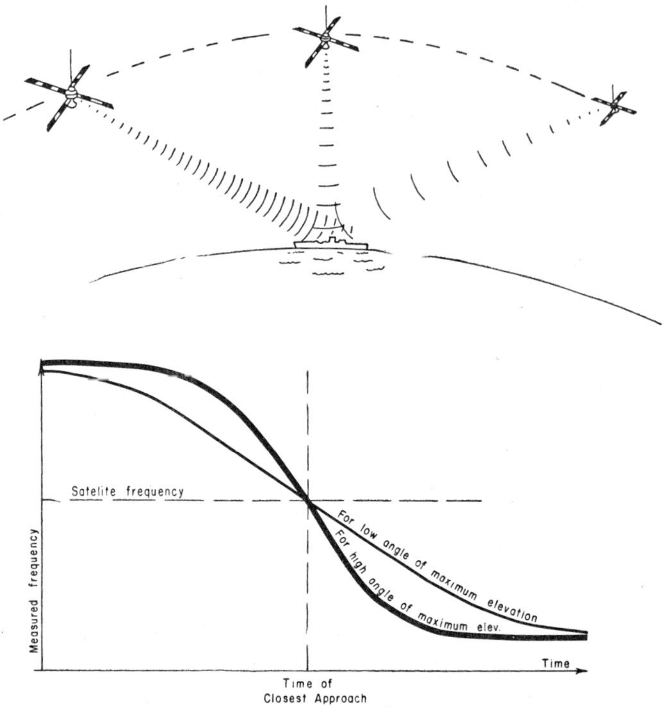

Fig. 156. **Satellite** pass. For satellites in polar orbits, the time of closest approach gives the latitude and the rate of change of frequency gives the longitude, both with respect to the satellite.

potential-measuring points. See **array**.

Schmidt diagram: A polar plot where the angle indicates dip or drift direction and the distance from the origin indicates the dip or drift magnitude. In the *modified Schmidt diagram* used for plotting low dips, zero dip is on the outside and dips become larger toward the diagram center. Figure 157.

Schmidt field balance: A magnetometer which consists of a permanent magnet pivoted on a horizontal knife-edge. The torque of the magnet trying to align with the earth's field is balanced by a gravitational torque because the magnetometer is not pivoted at its center of gravity.

Schmidt net: An equal-area plot of latitude and longitude, used in plotting geologic data such as the direction of structural features. The Schmidt net is the same as Lambert's azimuthal equal-area projection (different from Lambert's conic projection). See Figure 173 (**stereographic projection**).

Schuler period: The time equal to $2\pi(R/g)^{1/2}$, where R is the earth's radius and g is the acceleration of gravity, equal to 84 minutes. The Schuler period is a natural precession rate for gyrocompasses.

Schumann resonance: An electromagnetic waveguide phenomenon between the earth and the ionosphere; the space between acts as a cavity resonator. Frequencies of the lowest-order modes are about 8, 14, 20, and 26 hz.

scintillometer: An instrument for measuring radioactive radiation, especially from gamma rays. Gamma radiation impinging on a sensitive phosphor causes it to emit light (scintillations) which is measured by a photo-multiplier tube.

sea chest: A fitting in a ship's hull below the water line, such as used to mount sonar transducers. See Figure 158.

sea state: A scale of the height of sea waves. The *Douglas sea state* scale is shown in Table 1 (**Beaufort number**).

secondary: **1.** An energy train, especially a refraction event, which is not the first arrival. **2.** A **ghost** or simple **multiple**.

secondary field: **1.** The electromagnetic fields which result from the induction of currents in a medium by a primary field. **2.** In methods where galvanic energization of the earth is used, the secondary field is that resulting from the difference between the actual current system in the earth and the system which would exist if the earth were homogeneous.

secondary lobe: A pass region other than the principal.

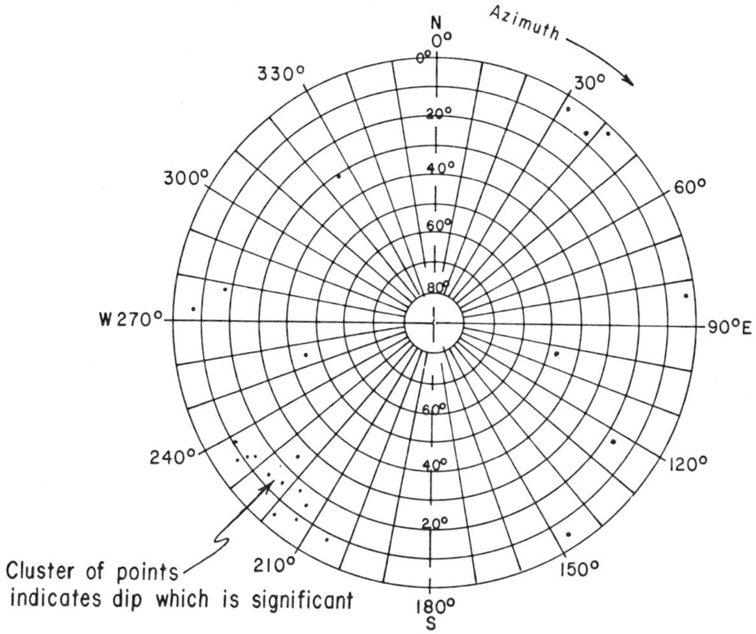

Fig. 157. **Schmidt diagram** (modified) used for plotting dipmeter data.

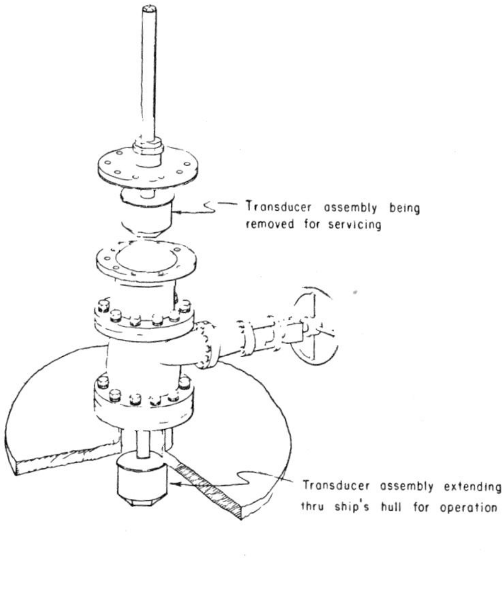

Fig. 158. **Sea-chest** opening in ship's hull for doppler-sonar transducer.

Applies especially to the **directivity patterns** of arrays, the mixing of traces in velocity filtering, etc.

secondary patterns: The use of the sum and difference frequencies *(red + green* and *red − green)* to yield "coarse" networks which can be used to remove lane ambiguities. Used with medium-frequency radio-positioning systems such as **Decca.**

secondary porosity: **Porosity** resulting from the alteration of the formation such as by fractures, vugs, solution channels, dolomitization, etc.

secondary-porosity index: A measure of the secondary porosity, calculated from sonic-log values in conjunction with either density-log or neutron-log values. Symbol: SPI. If ϕ_D is the porosity calculated from a density (or neutron) log and the ϕ_{sonic} is the porosity calculated from a sonic log, SPI is sometimes defined as $(\phi_D - \phi_{sonic})$, usually as $(1 - \phi_{sonic}/\phi_D)$.

secondary voltage: In IP surveying, the polarization voltage observed at a time-domain receiver immediately after the primary current is turned off. Sometimes called *initial transient* or *initial decay voltage.*

secondary wave: An **S-wave,** transverse wave, or shear wave.

second-derivative map: A map of the second vertical derivative of a potential field, such as gravity. Such maps tend to emphasize local anomalies and isolate them from a regional background. Often made using **Laplace's equation** relating the second vertical derivative to the second horizontal derivatives, which can be determined from differences in the values near the point. See also **grid redisual.**

second-order correction: 1. A correction which is significantly smaller than first-order corrections. 2. Frequently refers to refinements in normal moveout or static corrections.

section: 1. What might be seen by slicing through a solid object, such as a slice through the earth. 2. A chart showing the geologic formations which would be exposed in a vertical cut or some physical property of what would be so exposed. 3. A plot of seismic events, as in a **record section.** 4. Geologic formations, as in "the section in this area is Mesozoic." 5. A square mile = 640 acres. Sections are numbered in a systematic way in the township-range system, as shown in Figure 182 (**township**). An American section might be designated Sec. 10, T-2-N, R-1-E. A Canadian section might be designated 36-39-10-W1 for Section 36, Township 39 (N), Range 10 west of principal meridian 1 (which is near Winnipeg).

sectional correlogram: The autocorrelation of successive traces, displayed like a record section.

section gauge: Caliper log.

secular variation: Time variations whose periods are measured in decades, as "the secular variation of the earth's magnetic field" with periods of 30 to 300 years.

SEG A, SEG B, or SEG C: An SEG standard magnetic tape **format.** See Northwood et al, 1967, Recommended Standards for Digital Tape Formats: Geophysics, v. 32, p. 1074-1084; and Meiners et al, 1972, Recommended Standards for Digital Tape Formats: Geophysics, v. 37, p. 36-44.

seis: Seismometer or **geophone**; a device for detecting seismic energy and transforming it into electrical voltage (or other measurable quantity). Sometimes also used for a hydrophone.

seismic: 1. Having to do with elastic waves. Energy may be transmitted through the body of an elastic solid by **P-waves** (compressional waves) or **S-waves** (shear waves) or along boundaries between media of different elastic properties by **surface waves.** Often equated with **acoustic, elastic,** and **sonic.** 2. Having to do with natural **earthquakes.**

seismic discontinuity: 1. A discontinuity in elastic properties and/or density at which seismic velocity and/or acoustic impedance change abruptly. 2. Especially the **Moho** discontinuity between the earth's crust and mantle, the Gutenberg discontinuity between the mantle and the core, and the gradational change between the outer and inner core.

seismic-electric effect: A voltage between two electrodes inserted in the ground caused by a seismic wave passing through the region between the electrodes.

seismicity: 1. The likelihood of an area being subject to natural earthquakes. 2. The relative frequency, intensity, or magnitude, and kind of natural earthquakes.

seismic survey: A program for mapping geologic structure by creating seismic waves and observing the arrival time of the waves reflected from acoustic-impedance contrasts or refracted through high-velocity members. A **reflection survey** is usually implied unless **refraction survey** is specifically mentioned. The energy source for creating the waves is usually impulsive (i.e., energy is delivered to the earth for a very short period of time) although energy is introduced for considerable time with the Vibroseis method. The energy is detected by arrays of geophones or hydrophones connected to amplifiers, and the information is amplified and recorded for interpretation. The data often are **processed** to enhance the wanted information (**signal**) and displayed in record-section form. Signal is recognized as a coherent **event,** although noise often is coherent also. Events considered to be reflections from acoustic-impedance contrasts (reflectors) are used to locate the reflectors, it being assumed that their attitudes are that of the geologic structure. Events attributed to be head waves are used to locate the refractors of which they are characteristic, it being assumed that the attitudes of these refractors are those of the geologic structure. **Velocity analysis** is also done on reflection data where the offset varies.

seismogram: A seismic record.

seismograph: 1. A seismic recording system. 2. A **geophone.** See **seismic.**

seismologist: 1. One versed in seismic principles such as required in oil exploration or earthquake analysis. 2. An individual occupying a position on a seismic field party which requires knowledge of seismic principles, often in charge of the office.

seismology: The study of **seismic** waves; a branch of **geophysics.** Especially refers to studies of **earthquakes** or to the conduct of **seismic surveys** in exploration for oil, gas, minerals, engineering information, etc.

seismometer: Geophone.

Seisviewer: Borehole televiewer. Birdwell trademark.

self-potential: Spontaneous potential or **SP.**

self-potential method: Observation of the static natural voltage existing between sets of points on the ground, sometimes caused by the electrochemical effects of ore bodies. Used in mining exploration, especially for sulfide bodies shallower than about 1000 ft. Compare **telluric method** and see also **SP.**

semblance: A measure of multichannel coherence. The energy of a sum trace divided by the mean energy of the components of the sum. The normalized output/input energy ratio resulting from stacking several channels according to some scheme for time-shifting the channels before stacking. If $f_{i,j}$ is the jth sample value of the ith trace, then the semblance coefficient S_c is

$$S_c = \frac{\sum_{j=k-\frac{1}{2}N}^{k+\frac{1}{2}N} \left[\sum_{i=1}^{M} f_{i,j} \right]^2}{\frac{1}{M} \sum_{j=k-\frac{1}{2}N}^{k+\frac{1}{2}N} \sum_{i=1}^{M} (f_{i,j})^2},$$

where M channels are summed; the coefficient is evaluated for a window of width N centered at k. It is therefore basically the energy of the stack normalized by the mean energy of the components of the stack. This is equivalent to the zero-lag value of the autocorrelation of the sum trace divided by the mean of the zero-lag values of the autocorrelations of the component traces.

semiconductor: A substance such as germanium and silicon whose electrical conductivity at normal temperature is usually intermediate between that of a metal and an insulator, and whose conductivity is anisotropic. Its concentration of charge carriers increases with temperatures over a given range. Current flow may be by the transfer of positive holes or missing electrons (p type) as well as by movement of electrons (n type). Many common metallic sulfides and oxides are actually semiconductors. See **transistors.**

semiinfinite: Extending so far that some of the boundaries have no effect. Thus a semiinfinite slab is a horizontal unit of finite thickness bounded by a vertical plane on one side but extending so far in other directions that the boundaries in those directions create no measurable effects. A semiinfinite prism is a vertical prism with a bottom so remote as not to affect measurements.

sender: An IP waveform-generator control device, also called a **transmitter.**

sensitivity: The least change in a quantity which a detector is able to perceive. An instrument can have excellent sensitivity and yet poor **accuracy**. Compare **readability.**

separation: 1. A difference in values between two measurements, especially well-log-based measurements. **2.** The difference in resistivity readings from two tools which have different depths of investigation (see Figures 117 and 102, **microlog** and **induction log**). Low-resistivity mud cake (which becomes thicker where formations are more permeable) causes the apparent resistivity of a shallow measurement to be lower than that of a deeper measurement, a situation called "positive sepatation." See also **movable-oil plot. 3.** Apparent displacement on a fault.

setup: A particular arrangement of cables, geophones, shotholes, etc., for making a recording in the field. Repeated records may be taken at one setup.

sextant: A doubly reflecting instrument for measuring angles, particularly the altitude angle of a celestial body above the horizon.

sferics: Natural "atmospheric" fluctuations of the electromagnetic field, generally at frequencies from 1 to 10^5 hz, caused by lightning discharges. See **Schumann resonance.**

SH: 1. An S-wave which has a horizontal component of motion only. **2.** The horizontal component of any S-wave. **3.** A surface **Love wave.**

shadow zone: 1. A portion of the subsurface from which reflections are unobservable because their raypaths do not reach the surface. The overlying beds may have such dips and velocity contrasts that raypaths to or from reflectors within the shadow zone become refracted or undergo total reflection. **2.** A portion of the subsurface which does not evidence itself on refraction profiles, such as beds whose velocity is lower than an overlying refractor. See also **blind zone, hidden layer, channel wave.**

shah: An infinite sequence of impulses occuring at unit time intervals. Also called *comb* and sampling function and symbolized $\text{III}(t)$.

shake table: A test instrument on which geophones can be mounted and driven with known frequencies and amplitudes so that the geophone characteristics can be determined.

shale base-line: 1. A line drawn through the minimum deflections characteristic of impermeable shales on an SP log (see Figure 170, **SP**), which is used as the reference in making measurements to determine the characteristics of sands and their formation waters. See also **base-line shift. 2.** A line drawn through the characteristic of thick shales on the gamma-ray log.

shale potential: A part of the electrochemical **SP.**

shaliness: The content of shale (or clay) in a dominantly nonshale formation; the degree to which ion-exchange processes contribute to resistivity measurements. Electrical conduction in shales is an ion-exchange process whereby electrons move between exchange sites on the surface of clay particles. See **dirty.**

shallow-water survey: A geophysical survey in waters where conventional marine survey ships cannot operate easily because of shallow water, reefs, or other obstructions, etc.

shear modulus: See **elastic constants.**

shear wave: A body wave in which the particle motion is perpendicular to the direction of propagation. Also called **S-wave** or **transverse wave.** See also **acoustic wave.**

shielding: The enclosure of electrical wires or components (such as in an IP receiver) in order to reduce the effects of noise and electrostatic, magnetic, or electromagnetic coupling.

shingling: Forming an echelon pattern. **1.** For refraction recordings, a successive loss of visibility of early cycles with increase in range (Figure 159). A shift of energy to successively later cycles as distance increases. **2.** Incorrect migration of segments of a continuous reflection event so that the migrated segments do not join to form a continuous event.

shipboard gravimeter: A system for measuring the acceleration of gravity from a ship in motion. Complex arrangements are used to insulate the meter from the many accelerations to which the ship is subject, a computer is necessary to make corrections to the data, and the data are heavily filtered to separate out the gravity effects. The velocity of the ship has to be accurately known to make the **Eötvös correction** to the data.

shock wave: A wave motion which propagates at greater than seismic (sonic) velocity. In contrast to an **elastic wave.**

shoot: 1. To fire an explosive. **2.** To carry out a seismic survey, as "to shoot a prospect."

shootback method: An electromagnetic-surveying method which employs two tilted coils, each of which serves both as a transmitter and a receiver. At every station, readings are taken with each coil serving as transmitter for one and as receiver for the other. By

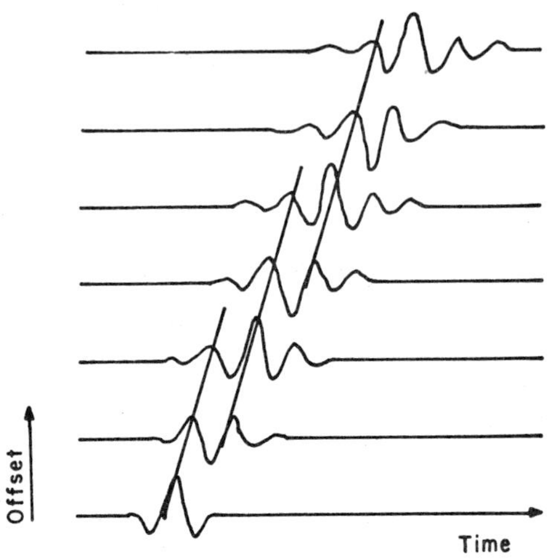

Fig. 159. **Shingling** resulting from shift of energy in a wavetrain.

averaging the readings, errors due to misalignment are essentially eliminated.

shooter: The man on a seismic party in charge of detonating the explosive.

shooting poles: Loading poles.

shooting under: Shooting from a shothole on one side of a property into a spread on the opposite side in order to obtain subsurface coverage under the property itself. Used when the surface of the property is inaccessible, such as on a line crossing a navigable river.

shoot on paper: Thinking through a field program before starting it; includes calculating the results which may be expected, trying to anticipate problems which may occur and ambiguities which are likely to arise, and, in fact, deciding whether the desired objectives are likely to be achieved.

shoran: Short-range navigation; a positioning system where the distances from reference transponders are determined by measuring the traveltime of pulsed radio waves to and from the stations. See Figure 160. A 210-Mhz pulse is transmitted to one fixed station (the *drift* station) and alternately a 260-Mhz pulse to the other fixed station (the *rate* station). Each fixed station rebroadcasts on 320 Mhz upon receiving the pulse. (Frequencies may differ somewhat.) Shoran's range is line-of-sight limited; the range is roughly

$$k(h_t^{1/2} + h_r^{1/2}),$$

where h_t and h_r are the heights of the transmitting and receiving antennas and k depends on antennas gain, transmitted power, receiver sensitivity, and other factors; k is apt to be 1.5 to 2.5 for the range in miles when the antennas heights are given in feet. See also **extended-range shoran.**

short doppler count: A satellite-navigation measurement in which cycles are counted for less than the whole 2-minute transmission cycle. Short counts are apt to be for 23 sec (four counts after the beginning of a 2-minute cycle) plus one count for 28 sec.

short normal: A **normal** resistivity log made with the *A* and *M* electrodes in the sonde about 16 inches apart.

short-path multiples: A multiple reflection in which energy is reflected back and forth over a small portion of the section so as to blend with the primary pulse, changing its waveshape and adding a tail. See Figure 124 **(multiple).**

short shot: Weathering shot.

short-trace section: A record section composed of short-offset traces only. Usually implies a 100 percent section extracted from data shot for CDP stacking.

shot: 1. The detonation of an explosive. **2.** Any impulsive source of seismic energy. **3.** Any source of seismic energy. **4.** A measurement through an alidade or transit sighting on a stadia rod.

shot bounce: Noise on a seismic record caused by mechanical motion of the recording truck.

shot break: Time break. 1. The instant of the explosion. **2.** Initiation of a seismic wave by a nonexplosive source.

shot depth: The distance down the hole from the surface to the top (usually) of the explosive charge, often measured with loading poles which come in 10 ft lengths. With small charges the shot depth may be recorded to the center or bottom of the charge, but with large charges the distances to both the top and bottom of the column of explosives usually are given.

shot instant: Time break, the time at which a shot is detonated.

shot-moment line: An electric line wrapped around an explosive charge which is ruptured when the charge is detonated. A primitive way to record the **time break.**

shot noise: 1. Hole noise; noise caused by the motion of the waste gases from an explosion and the material filling the shothole. Includes both the rumbling in the shothole following an explosion and the ejection of mud and debris from the hole and its subsequent impact back on the earth. **2.** Noise in a semiconductor because of the discreteness of current carriers. Such rms current noise is proportional to the square root of the product of current and bandwidth.

shotpoint: Abbreviated SP. **1.** The location where an explosive charge is detonated in one hole or in a pattern of holes to generate seismic energy. **2.** The location of other sources of seismic energy such as Thumper drops, Dinoseis pops, etc. **3.** The area surrounding the shotholes.

shotpoint gap: A greater distance between geophone groups on opposite sides of the shotpoint in an otherwise uniform spread. Used so that the geophone groups nearest the shotpoint will be far enough from it that the hole noise will have less adverse effects.

shotpoint seis: Uphole geophone.

shoulder-bed effect: Effect of adjacent beds on a log reading. Also called the *adjacent-bed effect.* For example, high-resistivity beds adjacent to a low-resistivity bed may result in more current flowing in the low-

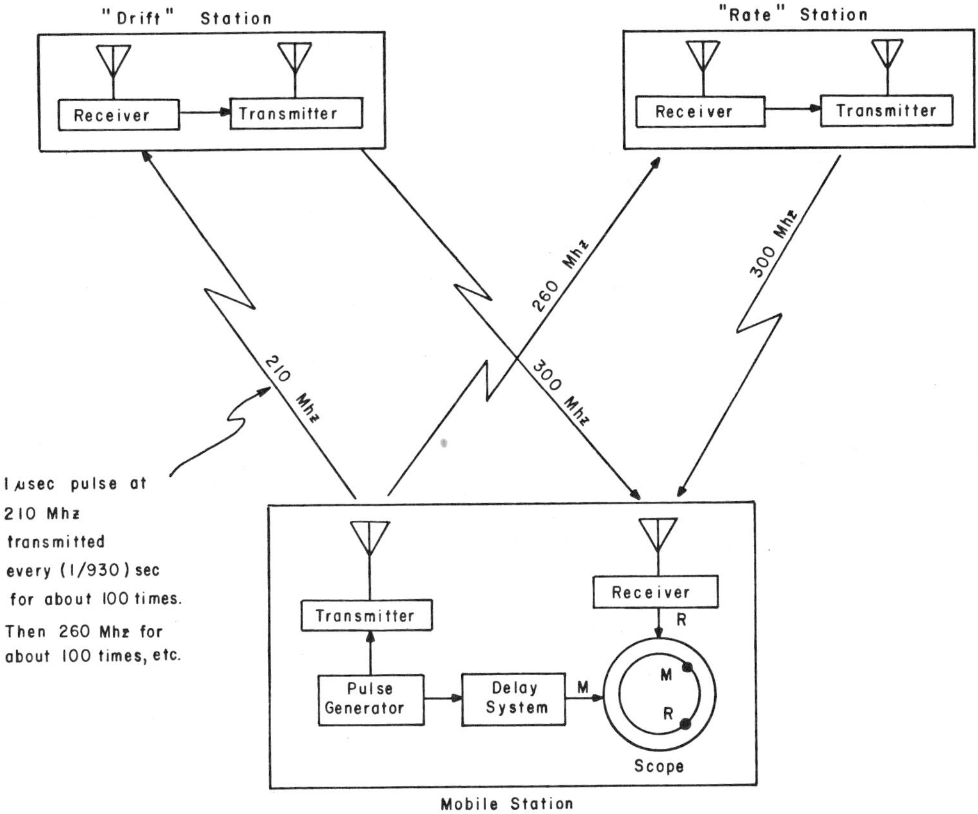

Fig. 160. **Shoran** system.

resistivity bed than it the high-resistivity bed were not present, thus changing the apparent resistivity of the low-resistivity bed.

shuttle: The moving part of an airgun; compressed air is confined in the chamber when the shuttle is closed but released into the water when the gun is fired by opening the shuttle. See Figure 3 **(airgun).**

SI: International System of units, also called the mks or mksa system. See **Appendix G** and Table 6 **(magnetic quantities).**

sial: Granitic, gabbroic, acidic, or continental crust, plus the overlying sediments. Of dominantly silica-alumina composition and specific gravity about 2.7. Compare **sima.**

side lobe: 1. Alias passbands. **2.** Subsidiary passbands on either side of the main passband on filter curves, array directivity patterns, multichannel mixing arrangements, velocity filters, etc. See Figure 54 **(directivity graph).**

side-scan sonar: A method of locating irregularities on the ocean bottom. A pulse of sonar energy (typically 120 khz) is emitted from a **fish** which is towed 50 to 500 ft above the bottom, depending on the range and resolution sought. The sonar beam is narrow in the direction of traverse because the source consists of a line array of elements. Bottom irregularities (rock outcrops, pipelines, shipwrecks, boulders) and variations in bottom sediments produce changes in the amount of energy return; see Figure 161. The arrival time measures the distance from the fish to the reflecting object.

side shot: A reading or measurement from a survey station to locate a point which is not intended as a base for an extension of the survey. Usually made to determine the position of some object to be shown on the map.

sidewall core: A formation sample obtained with a wire-line tool from which a hollow cylindrical bullet is fired into the formation and retrieved by a cable attached to the bullet. The type of bullet and size of charge is varied to optimize recovery in different formations.

sidewall neutron log: An epithermal **neutron log** made with a skid which is pressed against the borehole wall and may cut into the mud cake to minimize borehole effects. Abbreviated SNP. SNP is a Schlumberger trademark.

sidewall pad: A measuring device which is pressed

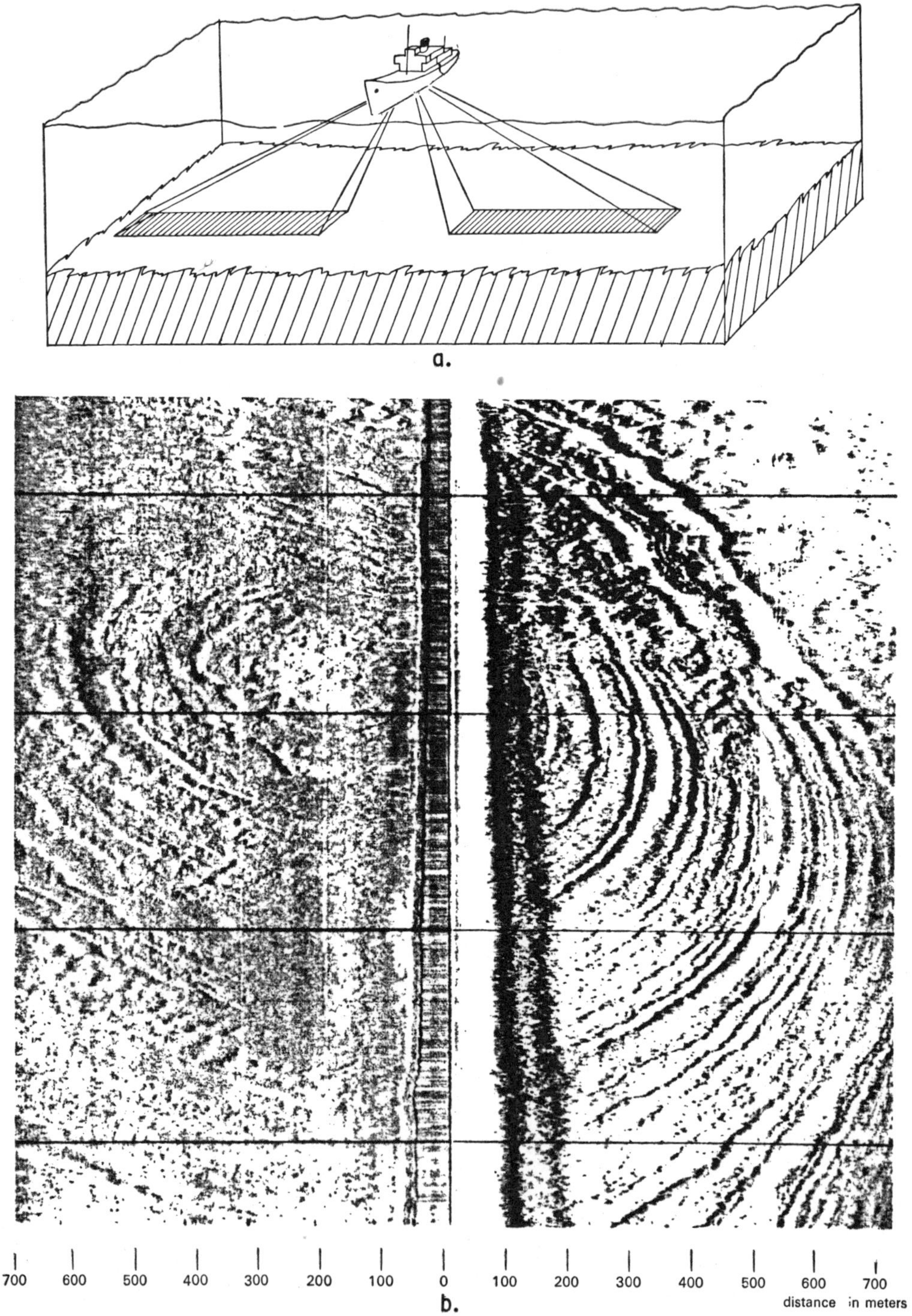

Fig. 161. **Side-scan sonar.** (a) Schematic diagram. (b) Record appearance. (Courtesy Compagnie Général de Géophysique.)

against the side of a borehole, such as used with micro-resistivity logs, density logs, many radioactivity logs, etc.

sidewall sampler: A wireline device for taking **side-wall cores.**

siemens: A unit of electrical conductivity; the reciprocal of ohm. Also called mho.

sigma: 1. Standard deviation. 2. Poisson's ratio.

sigma unit: A unit of measure of **capture cross-section.** Also called **capture unit.**

sight: 1. A bearing or angle measured with a compass, transit, or alidade. **2.** Any established point on a survey. **3.** An opening through which an object can be seen, used to determine the direction and/or distance to the object.

signal: That which is sought, which carries desired information. As opposed to **noise.** See **signal-to-noise ratio.** Sometimes "message" is used for the desired information, and signal is used to include both message and noise.

signal correction: A correction for time differences between reflection times resulting from changes in the outgoing signal from shot to shot. See **hole fatigue.**

signal restoration: 1. Correction for instrument response. **2. Inverse filtering.**

signal-to-noise ratio: The energy of desired events divided by all remaining energy **(noise)** at that time. Often abbreviated *S/N.* Sometimes the energy of the desired events is measured with respect to the total energy at that time, S/(S + N). This ratio is difficult to determine in practice because of the difficulty in separating out the signal which constitutes the desired events; one must presuppose some characteristic of the signal to effect the separation. This distinction is sometimes made on the basis of moveout, it being assumed that all coherent energy within a certain band of moveouts is signal and all else is noise. Signal-to-noise is sometimes represented in **f,k plots** on which the signal is indicated as a specified area. Amplitude ratios are sometimes used instead of energy ratios.

signature: Waveshape which is characteristic of a particular source, transmission path, or reflecting sequence. **Character.**

sign bit: The **bit** which indicates the algebraic sign of the number; plus or minus.

significance: The ratio of the smallest change which can be detected to the magnitude of a quantity at that time, such as the number of significant bits in a quantity. Compare **dynamic range.**

significance level: The complement of probability; e. g., 10 percent significance = 90 percent probability.

sima: Basaltic, peridotitic, basic, or oceanic crust, of specific gravity 3.0 to 3.3. Name derives from silica-magnesium composition. Compare **sial.**

similar fold: See **folding.**

simple multiple: A long-path **multiple** which has undergone only three reflections (that is, twice reflected from a deep interface and once from the shallow interface at the base of the weathering or at the surface). Most multiples are more complex than this.

simplex: Half duplex.

sinc x: $(\sin \pi x)/\pi x$, also called the **diffraction function:** This function is the Fourier transform of a unit **boxcar** and is extensively used in seismic data processing. Sometimes sinc $x = (\sin x)/x$.

sines, law of: In any plane triangle (Figure 162), the ratio of the lengths of any two sides is as the ratio of the sines of the angles opposite the sides:

$$\frac{a}{\sin \mathrm{A}} = \frac{b}{\sin \mathrm{B}} = \frac{c}{\sin \mathrm{C}}.$$

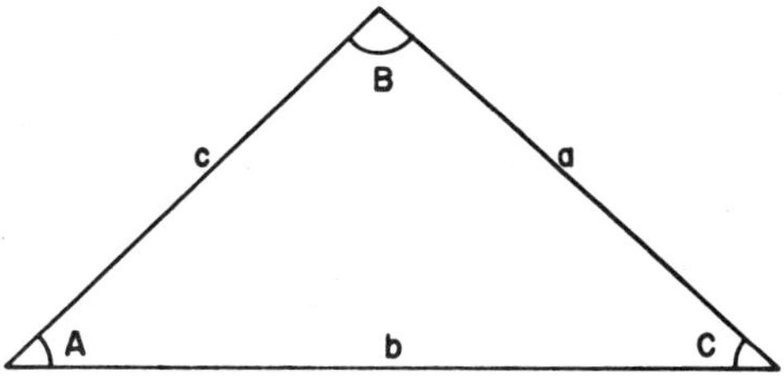

Fig. 162. **Sine** and cosine law. The law of sines is: $a/\sin A = b/\sin B = c/\sin C$. The law of cosines is: $C^2 = a^2 + b^2 - 2ab \cos C$.

sing around: A system where the echo from one transmitted pulse triggers the next pulse; used in **velocimeters.** The sing-around frequency is related to the path length and the propagation velocity.

singing: Short-path **multiples** in a water layer which produce **ringing; reverberation.**

single-ended spread: A reflection profile which is shot from one end. Also called **end-on** spread. See Figure 171 **(spread).**

single-shot tool: A device to obtain one measurement of the direction of a borehole at a particular depth. See **directional survey.**

singularity: A value of a variable for which a function becomes infinite. Also called a *pole.*

singy: Having an oscillatory character; **ringy.**

sinistral: Rotational to the left or counter-clockwise. A sinistral wrench fault is also called *left lateral.* Opposite is **dextral.** See Table 2 **(fault).**

sixfold: Having a redundancy of six. "Sixfold CDP shooting" involves recording from the same subsurface at six different offset distances.

skew: A condition that occurs when a magnetic tape is not mounted properly with respect to the magnetic heads. The tracks may deviate from their proper position and produce crossfeed, time displacement between channels, parity errors, etc.

skewness: Asymmetry in a distribution. A set of values X_i which has the mean value $\overline{X}$, and the standard deviation σ has skewness γ, where

$$\gamma = \frac{1}{n}\sum_{i=1}^{h}\left[\frac{X_i - \overline{X}}{\sigma}\right]^3.$$

See **statistical measures.**

skid: **1.** A mounting for a borehole sonde which cuts into the mud cake and presses the sonde against the borehole wall to minimize borehole effects. **2.** A sled on pontoons on which geophysical equipment is carried and which is dragged from one location to another. Used in marsh work.

skidded shots: A shotpoint which has been moved from its normal location, usually because of access difficulties.

skin depth: The *effective depth* of penetration of electromagnetic energy in a conducting medium when displacement currents can be neglected. The depth at which the amplitude of a plane wave has been attenuated to $1/e$ (or 37 percent):

$$\text{skin depth} = (2/\sigma\mu\omega)^{1/2} \text{ meters},$$

where σ = conductivity in mhos/meter, μ = permeability in henries/meter, and ω = angular frequency in radians/second.

skin effect: **1.** The tendency of alternating currents to flow near the surface of a conductor. **2.** A reduction in apparent conductivity. The propagation of an electromagnetic wave through a conductive formation results in a phase shift. In induction logging in high-conductivity formations, a correction is made for this effect. See **skin depth.**

skip: See **cycle skip.**

skip distance: The minimum distance from a transmitting antenna at which a **sky wave** can normally be received.

skip mixing: The mixing of data from nonadjacent channels. Adjacent output channels do not contain any common input information.

sky wave: Electromagnetic (radio) waves which reflect from ionized layers in the ionosphere. Involved in **sky-wave interference** and in making radio waves receivable beyond the line-of-sight horizon. Compare **tropospheric scatter.**

sky-wave interference: Interference between the direct (or ground) radio wave and waves reflected from ionized layers in the ionosphere. The ionization results from sunlight and hence the ionized layers vary around sunrise and sunset and sky-wave interference is variable at these times. Sky-wave interference degrades the accuracy of radio-positioning systems.

slave station: A transmitting station which is controlled by another station (the "master" station). Used to retransmit **positioning** signals in positioning systems which involve standing-wave patterns such as the phase-measurement systems. "Slave" is also used for subsidiary stations in radio-positioning networks even where not under the control of the master station. Lorac, Raydist, and Decca involve several transmitters—one master station and two or more slave stations.

sleeper: An explosive charge loaded into the hole an appreciable time (up to several days) before it is to be used.

sleeve exploder: A seismic marine energy source in which a gas (propane or butane) is exploded in a thick rubber bag (the sleeve), and from which the waste gases are vented directly to the air rather than into the water, thus reducing the bubble effect. Also called *Aquapulse.* Figure 163. Esso Production Research patent.

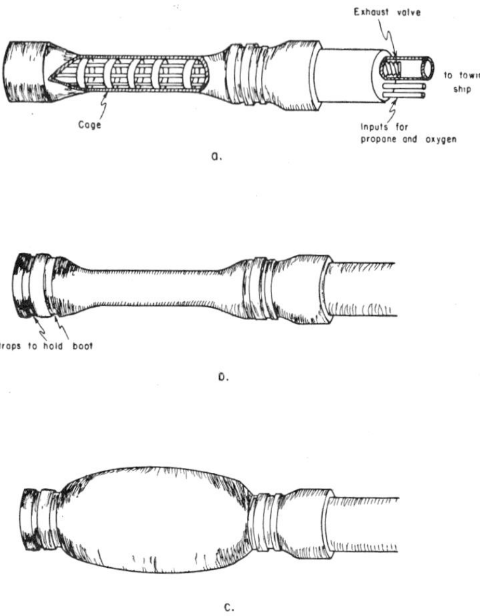

Fig. 163. **Sleeve exploder.** (a) Metal framework over which boot is placed. (b) Thick rubber boot fits tightly over cage prior to explosion. (c) Boot expands because of pressure of exploding gases. (Courtesy Esso Production Research.)

S-line: See **S-rule.**

slingram: A moving-source electromagnetic profiling method in which the mutual coupling between two loops is measured. Sometimes called **horizontal-loop** method.

slip-vector analysis: The first motion from earthquakes appears as a push or pull depending on the location of the observing station with respect to the epicenter and the direction of motion involved in the earthquake. Objective of analysis is determining the fault motion involved. See **nodal plane.**

Slotnick method: A graphical refraction-interpretation method applicable for plane multilayer refractors. For a single interface the solution is: (1) Calculate emergent angle of V_{2u}, $\alpha = sin^{-1}(V_1/V_{2u})$; emergent angle of V_{2d}, $\alpha' = sin^{-1}(V_1/V_{2d})$; critical angle $\theta = (\alpha + \alpha')/2$; dip angle $\phi = (\alpha - \alpha')/2$; reverse time T; $V_2 = V_1/(sin\ \theta)$. (2) Draw the lines OQ and $O'Q$ at the angles α and α' and OP at the angle ϕ (Figure 164). Measure the distance OP and $O'P$. (3) Calculate $OA = V_2T - OP)/2(V_2/V_1 - V_1/V_2)$. See Slotnick, 1950, A Graphical Method for the Interpretation of Refraction Profile Data: Geophysics, v. 15, p. 163-180.

slowness: The reciprocal of **velocity.** Transit time over an interval as measured by the sonic log is "slowness."

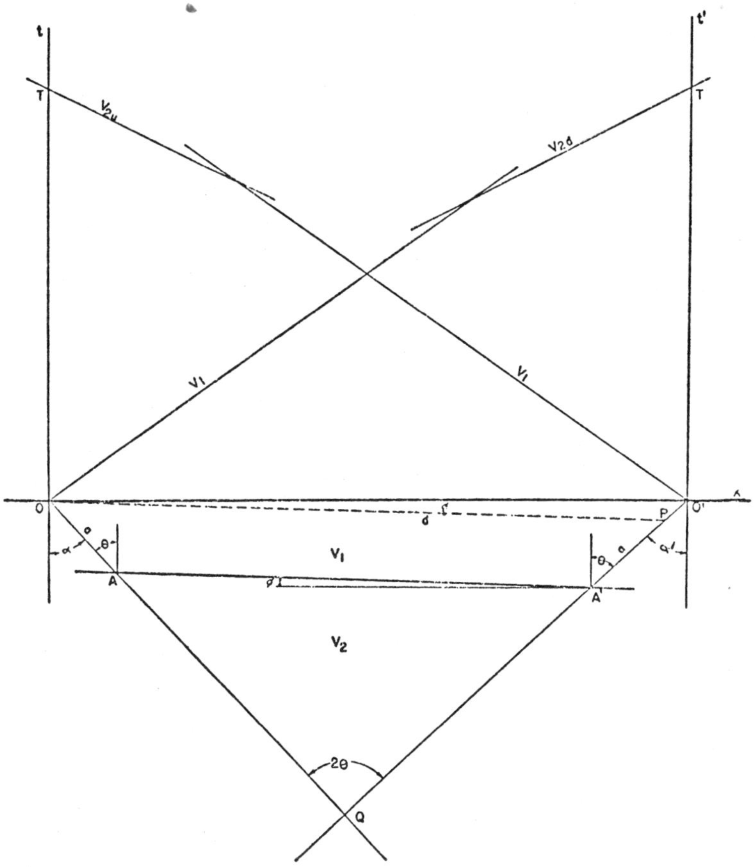

Fig. 164. **Slotnick refraction method.** (From Slotnick, Geophysics, v. 15, no. 2, p.165.)

SLR: Side-looking radar. A remote-sensing method which involves sweeping the earth to either side of an aircraft with a radar beam and recording the reflected signals, which are displayed so as to give the appearance of an aerial photograph. An aircraft at 20,000 ft can map a strip about 12 miles wide.

slue: To turn about its own axis. The slueing characteristics of a gyrocompass enable it to follow faithfully variations of the ship's heading. Also called slew.

slurry explosive: A bulk-type explosive which can be poured into boreholes. Not cap sensitive, requires a primer charge to detonate.

slush pit: The pit used in rotary drilling for storage of water or mud for circulation through the hole. Muds can be mixed in the pit. Sometimes a pit is dug in the ground, sometimes a portable sheet-iron pit is used.

small slam: A log combination and computation procedure. Involves calculating the resistivity of the invaded zone, the true resistivity, the diameter of invasion, and the water saturation based on a dual-induction, laterolog, and a porosity log of some type, often a sonic log.

smear: **1.** To mix data not in complete register. **2.** To average seismic data originating from sources at different locations or recorded by geophones at different locations or both. The amount of smear is the distance over which the sources are placed plus the distance over which the geophones feeding one channel are planted. The data from several shots or other source impulses may be combined without adjustment by **vertical stacking,** a method extensively used with surface and nonexplosive sources. See also **mixing** and **ground mixing.**

smoothing: Averaging adjacent values according to some scheme; filtering out higher frequencies. Convolving with a boxcar is one form of smoothing; convolving twice with a boxcar is a form of tapered smoothing (triangular smoother).

S/N: Signal-to-noise ratio.

Snell's law: When a wave crosses a boundary between two isotropic media, the wave changes direction such that the sine of the angle of incidence (angle between the wavefront and a tangent to the boundary) divided by the velocity in the first medium equals the sine of the angle of refraction divided by the velocity in the second medium. Snell's law applies to each of the P- and S-waves which result from an incident P- or S-wave (Figure 165):

$$\frac{\sin i}{v_i} = \frac{\sin \theta_{p1}}{v_{p1}} = \frac{\sin \theta_{s1}}{v_{s1}} = \frac{\sin \theta_{p2}}{v_{p2}} = \frac{\sin \theta_{s2}}{v_{s2}},$$

where i is the incident wave with a velocity $v_i = v_{p1}$ if a P-wave, or $v_i = v_{s1}$ if an S-wave; θ_{p1} and θ_{s1} are the angles of reflection of the P- or S-waves in medium 1, which have velocities v_{p1} and v_{s1}, respectively; θ_{p2} and θ_{s2} are the angles of refraction of the P- and S-waves in medium 2 which have velocities v_{p2} and v_{s2}, respectively. If $\sin \theta_{p2}$ or the $\sin \theta_{s2}$ exceed 1 as given by this equation, total reflection occurs. See also **Zoeppritz's equations.** The relationships become complicated if anisotropy is present; then additional shear wavefronts are generated and the wavefronts are not necessarily orthogonal to the direction of wave propagation.

snorkel: A tube to the surface from an underwater energy source through which waste gases escape. Used to attenuate **bubble pulses.**

SNP: Sidewall neutron (porosity) **log.**

soak: To allow a system to come to equilibrium with its surroundings. For example, a borehole gravity meter may have to rest at the bottom of a deep well bore for a period of time to allow it to adjust to the temperature and pressure regimen before meaningful readings can be obtained from it.

SOFAR: Seismic velocity in the oceans decreases with depth to a value of about 4890 ft/sec at about 750 fathoms (the exact depth varies because of near-surface temperature variations) and then increases with further depth. Seismic energy which attempts to escape from this velocity-minimum channel tends to be refracted back toward the velocity minimum. Consequently energy within this channel attenuates slowly and sonar signals generated in the channel can be detected in the channel great distances away. This is the SOFAR principle and the channel is called the SOFAR channel. See Figure 166 and **channel wave.**

soft formation: A poorly consolidated sand-shale sequence.

soft magnetism: That component or portion of rema-

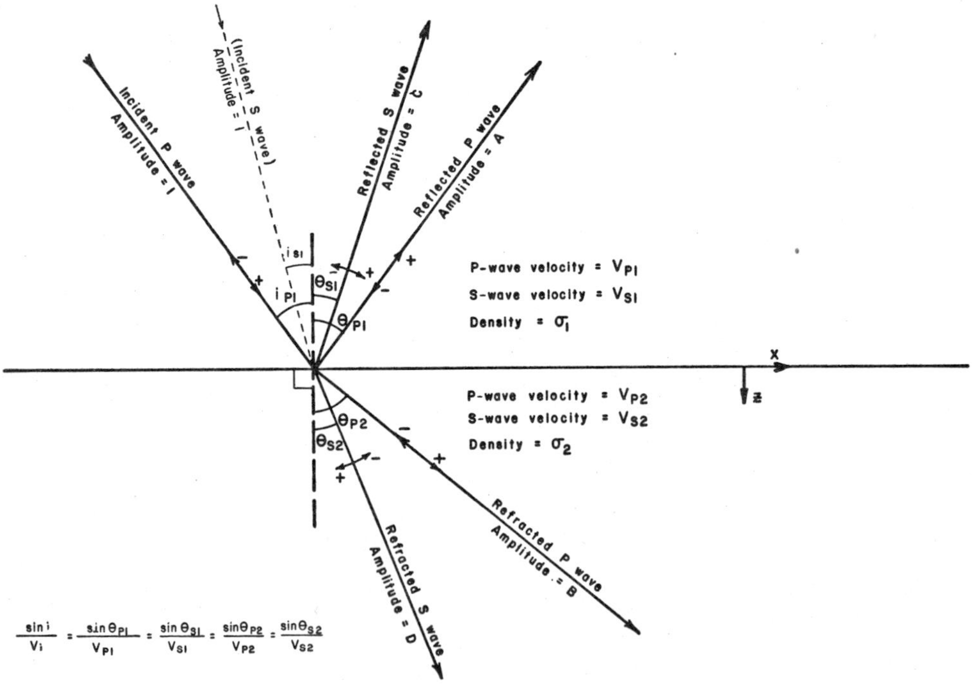

Fig. 165. **Snell's law** relations for either incident P-wave or S-wave.

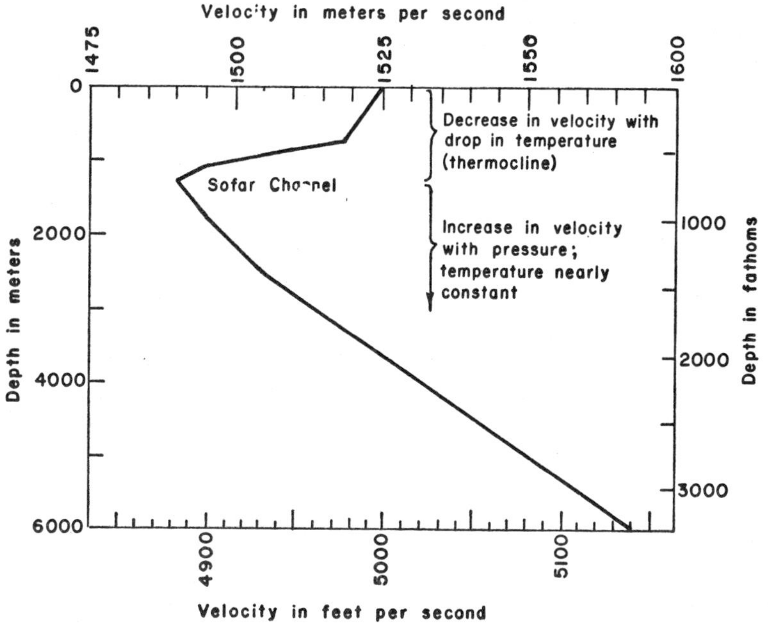

Fig. 166. **SOFAR channel** resulting from velocity minimum. Sonic velocity versus depth in the ocean. (From Ewing, Worzel, and Pekeris, 1948, GSA Memoir 27.)

nent magnetization which has relatively lower coercive force. In paleomagnetism it is this softer magnetization which is preferentially removed (say by alternating-field demagnetization) in order to isolate the harder and more diagnostic remanence.

soft mantle: Upper mantle which has lower-than-normal velocity which occurs under plate boundaries.

soft rock: Sedimentary rock. Used to distinguish between mining (hardrock) and petroleum (soft rock) objectives.

soft spring: A spring with very low natural frequency which effectively insulates from high-frequency noise.

software: Programs for data processing, including those that control the internal operation of the processing system itself.

solar wind: Ionized particles flowing radially outward from the sun. Transient magnetic disturbances (see **K-index**) may be correlated with variations in the solar wind.

sole: The lowest movement surface of a fault, especially where a gravity listric fault or a thrust fault becomes a bedding-plane fault.

solid-state circuitry: The use of semiconductor elements such as **transistors,** integrated circuits, etc., which do not require much space or power.

sonar: Involving the use of sonic (seismic) waves in the water for navigation and positioning and sometimes communications. See **doppler sonar** and **side-scan sonar.**

sonde: A logging tool such as is lowered into a borehole to record resistivity, sonic, radioactivity, or other types of logs.

sonic: Pertaining to longitudinal (P-) waves; *acoustic.* Often includes other wave modes also and hence becomes synonomous with *seismic* and *elastic.*

sonic log: A well log of the traveltime (transit time) for **acoustic waves** over a unit distance, and hence the reciprocal of the longitudinal wave (P-wave) velocity. Also called *acoustic-velocity* log and *continuous-velocity* log. Usually measured in microseconds per foot. Especially used for porosity determination by the **Wyllie relationship.** The interval transit time is integrated down the borehole to give the total traveltime. For the compensated sonic log, two transmitters are pulsed alternately; averaging the measurements tends to cancel errors due to sonde tilt or changes in hole size. See also **cycle skip, three-D log, character log, cement bond log,** and **fracture log.** Figure 167.

sonobuoy: 1. A device used in marine refraction surveys for detecting energy from a distant shot and radioing the information to the recording ship; see Figure 168. The sonobuoy is a free-floating buoy which usually is simply thrown off the shooting ship. Once the sonobuoy is in the water, seawater activates the buoy's batteries, one or more hydrophones are dropped, and a radio antenna is extended upward into the air. As the ship travels away from the buoy, firing charges (or other energy source) as it goes, the head waves are detected by the hydrophones and transmitted to the ship where they are recorded and timed. The distance

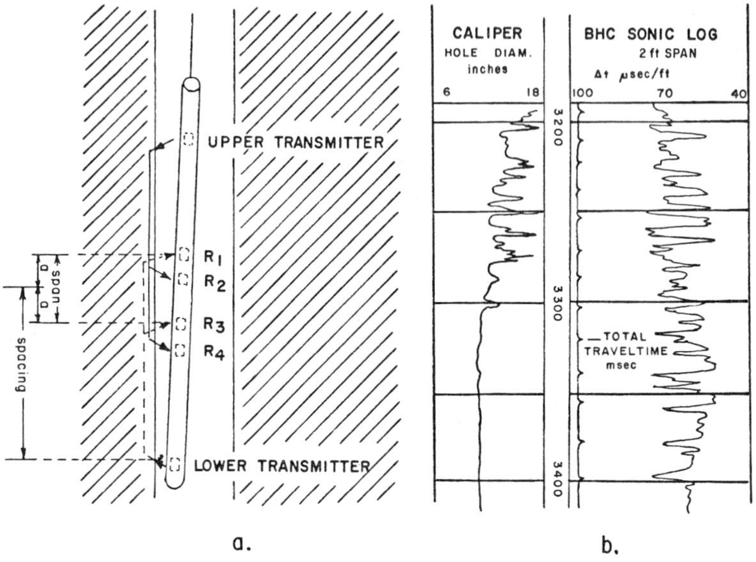

Fig. 167. **Sonic log.** (a) Schematic compensated sonic logging sonde. (b) Sonic log. (Courtesy Schlumberger.)

from the energy source to the sonobuoy can be determined by the arrival time of the wave which travels directly through the water. The buoy is expendable and sinks itself after a certain time; the cost of the buoy usually is less than the cost of retrieving it. **2.** A buoy which automatically transmits a radio signal when triggered by a water-borne sonic signal; used in positioning.

sonogram: sonograph.

sonograph: A display of seismic information as a function of dip or the apparent velocity of events. An ensemble of input traces such as the component traces of a reflection profile are time shifted proportional to their position, so that events with a certain apparent velocity line up. The ensemble is stacked to produce a single trace on the sonograph so that large energy is present where a coherent event occurs, whereas random noise tends to cancel. The ensemble of input traces is then shifted by a different amount and the next sonograph trace is formed. The complete sonograph thus is the transform of the data from the space-time domain to the moveout-time domain. Picking of events can be done on the sonograph. Figure 169. Sonographing is also called *Rieberizing* and *sonogramming.*

Sonoprobe: A type of marine echo sounder that generates sound waves and records their reflections. Usually has more power and more penetration than a fathometer but less than a sparker or gas exploder. Mobil Oil trademark.

sophisticated: Complex or intricate. Often refers to methods which were not feasible before the use of high-speed digital computers. "Sophisticated processing" implies that methods are being used which are more complete and correct than were formerly used. The dictionary gives another definition which is sometimes appropriate: deceptive or misleading; the newest technique which one wants to sell.

sorption: The binding of one substance to another by mechanisms such as adsorption (holding on the surface) or absorption (taking in completely).

sound channel: See **SOFAR.**

sounding: A *depth probe* or *expander.* Especially a series of electrical resistivity readings with successively greater electrode spacing made while maintaining one point in the array fixed, thus giving resistivity versus depth information (assuming horizontal layering). Also used with electromagnetic, magnetotelluric, and other types of geophysical surveying. See also **geometric sounding** and **parametric sounding.**

source program: A computer program written by a human, prior to machine decoding; e.g., a program in some symbolic language such as Fortran.

source-receiver product: The number of separate raypaths mixed together to produce the final display. Also called *effort.* Effort is equal to the product of the number of holes per shot (or impulses per record), the number of geophones per group, and the number of records stacked.

southing: See **latitude.**

SP: *Spontaneous potential* or *self potential.* **1.** A well log of the difference between the potential of a movable electrode in the borehole and a fixed reference electrode at the surface. The SP results from **electrochemical SP** and **electrokinetic potentials** which are

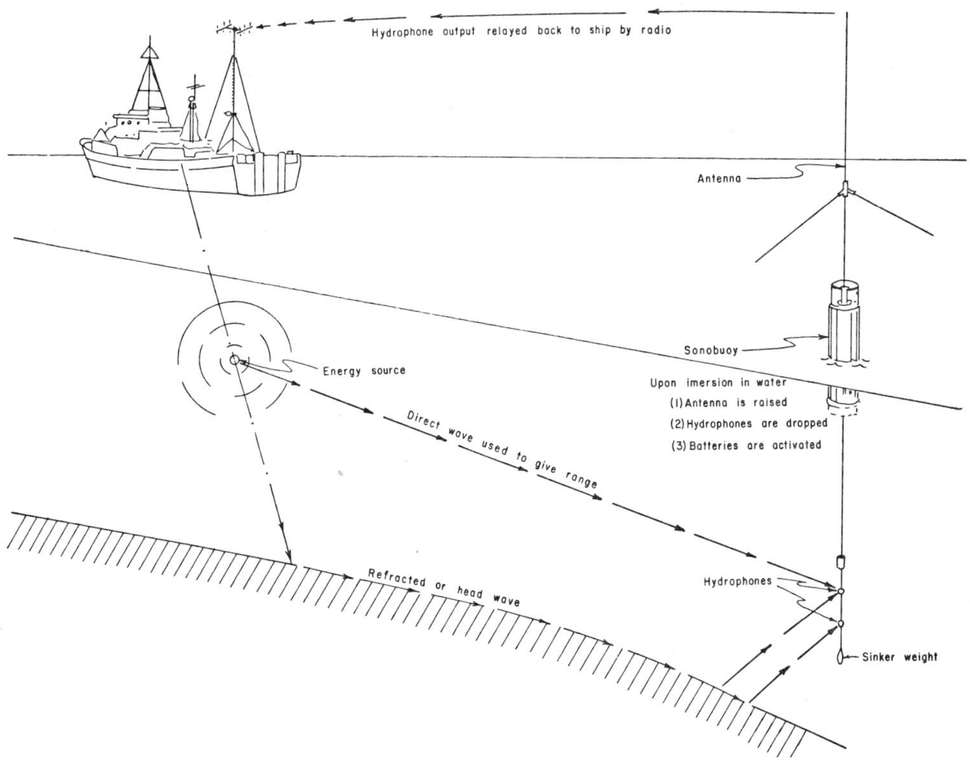

Fig. 168. **Sonobuoy** operation. (Courtesy Aquatronics)

present at the interface between permeable beds adjacent to shale. In impermeable shales, the SP is fairly constant at the *shale base-line* value (see Figure 170). In permeable formations the deflection depends on the contrast between the ion content of the formation water and the drilling fluid, the clay content, the bed thickness, invasion, and bed-boundary effects, etc. In thick, permeable, clean nonshale formations, the SP has the fairly constant *sand line* value, which will change if the salinity of the formation water changes. In sands containing disseminated clay (shale), the SP will not reach the sand line and a **pseudostatic SP** value will be recorded. The SP is positive with respect to the shale base-line in sands filled with fluids fresher than the borehole fluid. See also **SSP. 2.** The dc or slowly varying natural ground voltage observed between nearby nonpolarizing electrodes in field surveying. In many mineralized areas this is caused by electrochemical reaction at an electrically conducting sulfide body.

space-time filter: A **velocity filter** or wavenumber filter. Also called *beam pointing.*

spacing: The separation of certain electrodes or sensors on logging sondes. In nuclear devices usually the distance from the source to the detector. See **normal** and **density-log** figures. See also **span.**

span: The separation of certain sensors on logging sondes. On the sonic sonde, span is the distance between two receivers of a pair whereas *spacing* is the distance from transmitter to the midpoint of the corresponding receiver pair. See Figure 167 **(sonic log).**

span adjustment: Calculation of a log which would have resulted from the use of a **span** different from the one actually used.

sparker: A seismic source in which an electrical discharge in water is the energy source. See also **exploding wire.**

spatial frequency: Wavenumber, symbolized **k.** The number of wave cycles per unit of distance in a given direction (direction of the spread). Specifically, the reciprocal of the apparent wavelength λ_{app} along the spread direction.

$$k = \frac{1}{\lambda_{app}} = \frac{f}{V_{app}},$$

where f = frequency and V_{app} is apparent velocity. If a wavefront makes the angle θ with the given direction,

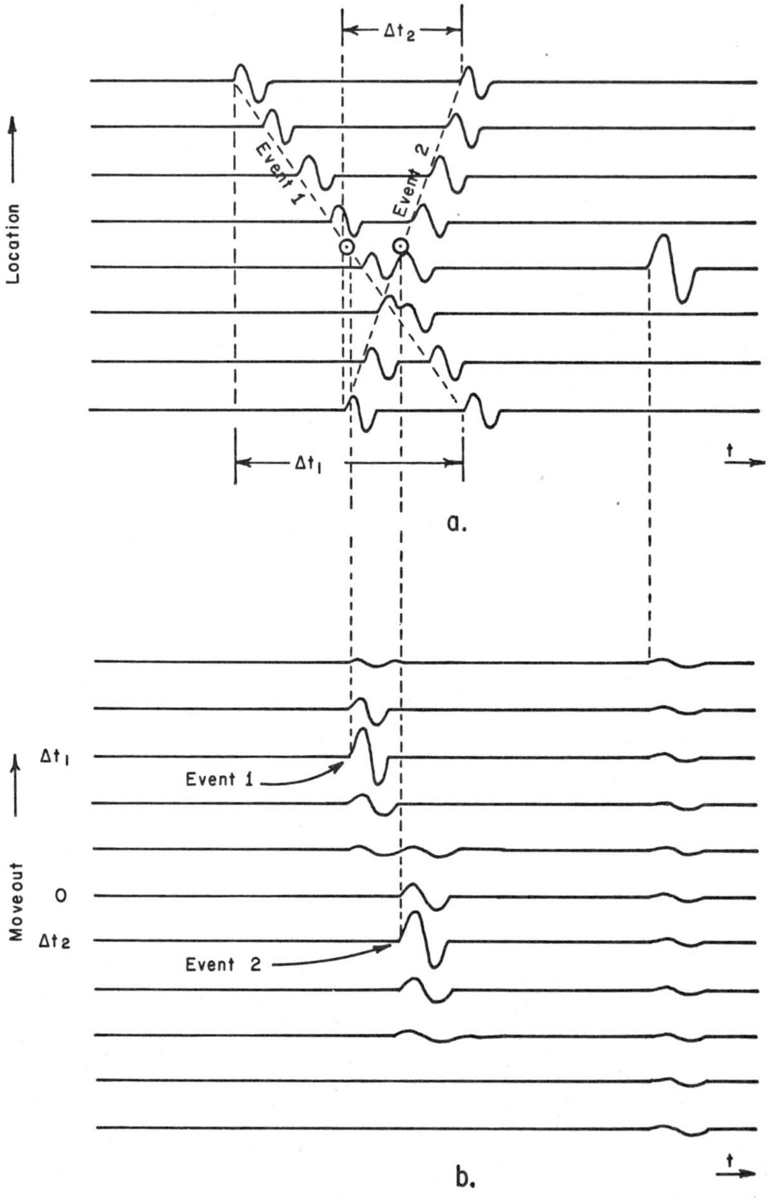

Fig. 169. **Sonogram** or sonograph. (a) Hypothetical seismogram showing two events and a noise burst on one channel. (b) Sonogram of these events. (From Trorey, Geophysics: v. 26, no. 6, p.761.)

$$k = \frac{f \sin \theta}{V},$$

where V is the actual velocity of the wavefront. Hence a wavenumber of zero indicates a wavefront striking the line of detectors simultaneously. See **f-k plot.**

SP buckout: A variable voltage–compensation circuit in series with the input terminals of an IP, resistivity, or SP receiver. Used to match the input voltage level of the voltmeter-receiver to that of the ground. The buckout voltage is equal and opposite to the dc self-potential. Also called *self-potential compensation.*

specific capacity: A polarization parameter in the time domain similar to the metal factor in the frequency domain. The long-time **chargeability** (area under decay curve) divided by resistivity. Also called *static capacity.* Dimensions are farad per meter.

specific factor: See **factor analysis.**

specific impedance: See **resistivity.** "Specific" refers to the normalized dimensional or volume properties of a material.

spectral log: A gamma-ray log which records the energy of gamma rays as well as the intensity of gamma rays.

spectrum: 1. The amplitude and phase characteristics as a function of frequency for the components of a wavetrain or wavelet. See **Fourier analysis. 2.** Filter response characteristics. See **transfer function. 3.** Other "quantities" displayed so as to show the relative content of various components. Thus a *velocity spectrum* (or *normal-moveout spectrum*) shows the amount of coherent energy which appears to have various amounts of normal moveout as a function of arrival time. A *dip spectrum* (or *sonograph*) shows the amount of coherent energy which appears to have various amounts of dip moveout or apparent velocity.

speed: Used synonomously with **velocity** in seismic work.

spherical divergence: The decrease in wave strength (energy per unit area of wavefront) with distance as a result of geometric spreading. A spherical wave traveling through the body of a medium continually spreads out so that the energy density decreases. The spherical divergence from a point source causes the energy density to decrease inversely as the square of the distance the wave has traveled. For energy which travels along a surface, the analogous term is *cylindrical divergence,* which varies inversely as the distance. Other mechanisms by which a seismic wave loses energy involve absorption and loss at interfaces by reflection (including diffraction, mode conversion, and scattering).

spherical excess: The amount by which the sum of the angles of a triangle on the surface of a sphere exceeds 180 degrees. For a sphere this excess is the area of the triangle. For a spherical triangle on an ellipsoid such as the earth approximates, the spherical excess ϵ is approximately:

$$\epsilon = mbc \sin \alpha,$$

where b and c are two adjacent sides of the triangle which intersect at the angle α, m = latitude function = $\rho/2RN$, ρ = number of seconds of arc/radian = 206264.8, R = radius of curvature in the meridian, and N = radius of curvature in the prime vertical.

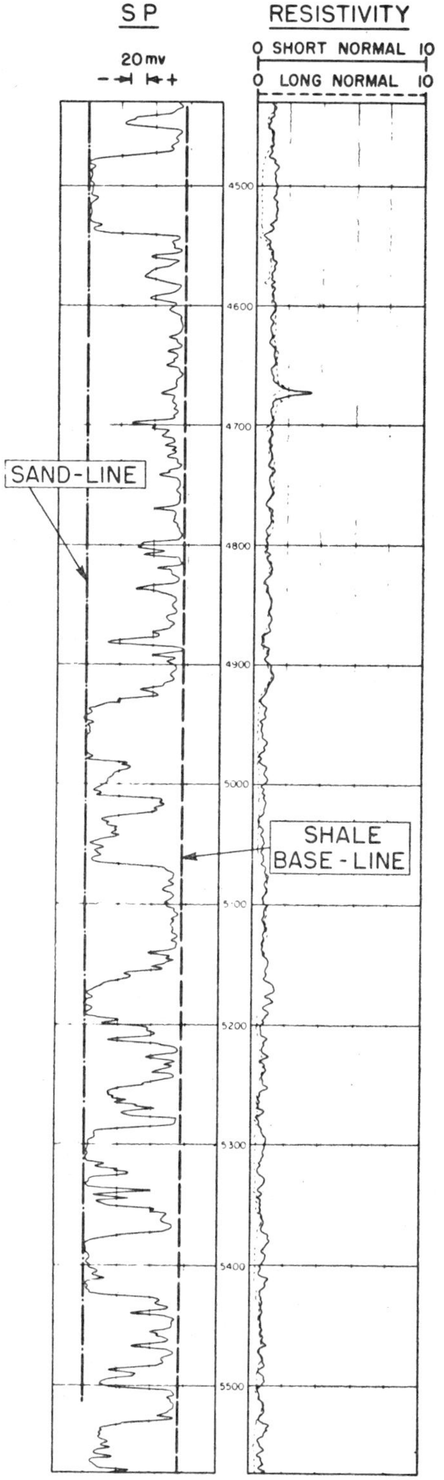

Fig. 170. **SP log** in sand-shale series with fresh mud in the borehole. (Courtesy Schlumberger.)

Values of *m* for various ellipsoids are obtained from tables; the value for the center of the triangle to the nearest half degree is usually used. For greater precision a correction factor is often applied in iterative fashion.

spherical triangle: The triangle formed on the surface of a sphere by the intersections of three arcs of great circles.

spheroid: The oblate ellipsoid of revolution used to approximate the earth's shape. See Table 5 (**geodetic systems**).

SPI: Secondary-porosity index.

spike: 1. To work away from the crew's normal area; **hot shot.** 2. An **impulse.**

spinner survey: A log of the rate of flow of fluid in the wellbore, casing, or tubing at specified levels in the wellbore. See **flowmeter.**

S-P interval: In earthquake seismology, the time interval between the first arrivals of longitudinal and transverse waves, one measure of the distance from the earthquake hypocenter.

spit-out: 1. **Gather.** 2. To print out the detailed data. Sometimes, to dump.

S-plane: The **Laplace transform** domain σ versus ω, where $s = \sigma + j\omega$.

spline: 1. A long flexible strip used in drawing a smooth curve. 2. A curve fitting scheme which assures a desired degree of smoothness to the fitted curve. An *n*th order spline has all derivatives up to $n-1$ continuous. Thus a quadratic spline has a continuous first derivative and a cubic spline has both first and second derivatives continuous. 3. Both the long flexible strip and analytic splines are sometimes used in residualizing, the smooth curve representing the regional and the difference between the smooth curve and the gravity profile representing the residual. By extension, a smooth surface used to represent a regional gravity field.

split: Split spread.

split-dip shooting: See **split spread.**

split spread: A method of reflection shooting in which the shotpoint is at (or perpendicularly offset from) the center of the geophone spread. Commonly used in **continuous profiling** and in **dip shooting.** See Figure 171 (**spread**).

spontaneous potential: Self-potential or **SP.**

spot correlation: Correlation of reflections on nonadjacent seismic records based on character or intervals between events.

spread: 1. The layout of geophone groups from which data from a single shot are recorded simultaneously. Reflection spreads usually consist of 24 groups with 50- to 400-ft intervals between adjacent group centers. Arrangements include broadside, cross, end-on, **fan shooting,** in-line, offset, interlocking, L spread, reversed, split spread, and T spread; see Figure 171. Spreads are *interlocking* if the geophone group location and the shotpoint for one profile are located at the shotpoint and geophone group location (respectively) for another profile (for example, shot at A into geophone 24 and shot at C into geophone at 13, in

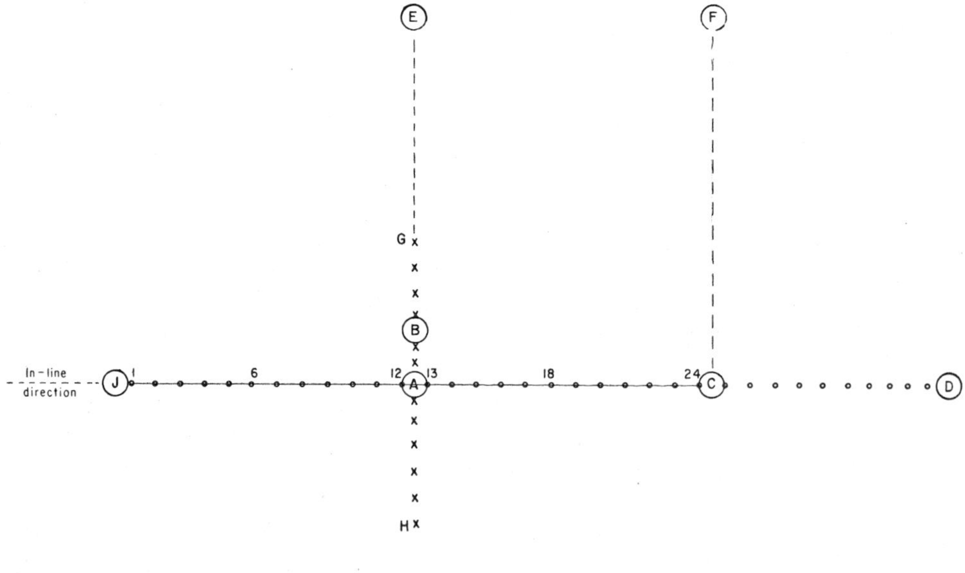

Fig. 171. **Spread** arrangements for 24 in-line geophone groups. *Split dip* if the shotpoint is at *A, offset* split dip if the shotpoint is at *B, end-on* if the shotpoint is at *C, in-line offset* if the shotpoint is at *D, broadside T* if the shotpoint is at *E, broadside L* is the shotpoint is at *F,* and *cross* if some geophone groups are at right angle (*G* - *H*).

Figure 150). Spreads are *reversed* if the same array of geophones is shot into from shotpoints in opposite directions in-line (for example, shot into spread from 1 to 24 from both shotpoints at C and at J). A *microspread* has very small geophone group interval (~2-15 ft). **2.** The layout of electrodes or antennas in resistivity or electromagnetic surveying. See **array.**

spread correction: **1. Normal moveout. 2.** Correction applied to refraction data to make **reduced traveltime** curves.

spreading: Divergence; loss of amplitude because of **spherical divergence** (for body waves) or **cylindrical divergence** (for surface waves).

SP reduction factor: The ratio of actual **SP** to **SSP.**

SPS: Shotpoint seismometer or **uphole geophone.**

spudder: A drill used for making holes in hard rock. The bit is raised and dropped and the resulting cuttings are removed by a bail or a pipe with a flap valve at the bottom. Also used to pound casing into gravel or formations with boulders such as glacial drift.

square wave: A *full square wave* is a current (or voltage) waveform consisting of equal magnitude "positive on" and "negative on" portions. A *half square wave* is switched "on" and "off." A *pulsed square wave* has portions which are "positive on, off, negative on, and off."

squelch circuit: A control used in uphole geophone amplifier circuitry which permits the uphole geophone signal to be recorded on one of the ordinary geophone channels prior to the first breaks, but then blocks or disconnects the uphole geophone signal after the uphole break has been recorded, so that the uphole geophone will not interfere with the subsequent record.

squiggle: A **wiggle trace** or trace of galvanometer deflection versus time. See Figure 56 **(display).**

S-rule: A rule in resistivity sounding and an application of the equivalence principle. Sounding graphs over a series of strata with resistivities ρ_i and thicknesses h_i above a highly resistive substratum possess the same asymptote if the sum of longitudinal conductance (h_i/ρ_i) is constant, i.e., if:

$$S = \sum \frac{h_i}{\rho_i} = \text{const.}$$

S is called the *admittance.*

S/(S + N) filter: See **Wiener filter.**

SSP: Static self potential; the maximum **SP** that would be recorded when the borehole logging sonde passes from a position well inside a very thick, porous, permeable, clean sand to a point well within a thick shale. The electrochemical SSP (the electrokinetic SP is often neglected) is approximately given by:

$$SSP = -K \log_{10} (a_w/a_{mf}),$$

where a_w is the **activity** of the formation water and a_{mf} that of the mud filtrate. Because of the inverse relationship between activity and equivalent resistivity, this equation can be written:

$$SSP = -K \log_{10} (R_{mfe}/R_{we}).$$

For NaCl muds that are not too saline, $R_{mfe} = R_{mf}$, for other muds an activity correction should be made. In these equations $K = 60 + 0.133\ T$, where T = Fahrenheit temperature.

stabilized platform: A platform on which instruments (such as gravimeters) can be mounted where they will remain nearly level despite tilt of the platform support. The platform, mounted on gimbals, is controlled by a gyroscope coupled to an accelerometer-controlled servo system on each gimbal axis. Used in measuring gravity on a ship in motion and in mounting inertial navigation sensors.

stable filter: A filter is stable if the energy of its impulse response is finite. Stable minimum-phase filters have stable inverses. Maximum-phase wavelets do not have stable inverse memory functions but inverse filtering can be accomplished by stable anticipation functions. The inverse of a mixed-delay wavelet requires both a stable memory function and a stable anticipation function.

stack: A composite record made by mixing traces from different records. Some of the types of stacking are listed in Table 10. See **common-depth-point stack, diversity stack, uphole stack,** and **vertical stack.**

Table 10. Stacking

Ground mixing — Combines outputs of several geophones and/or sources located over a limited area of the ground to attenuate data of certain apparent velocities.

Instrument mixing — Combines traces without time shifting or normal moveout removal; traces so combined are usually adjacent and effect is same as ground mixing.

Vertical stacking — Combines traces which have nearly identical source-group positions. May or may not involve static shifting before the stacking.

Uphole stacking — Combines data from different shot depths after static shifting based on uphole-time measurements.

Common-depth-point (CDP) stacking — Combines data which have a common midpoint between source and receiver locations after correcting for normal moveout.

Common-offset stacking — Combines data from a limited area for which the offset distance is the same.

Velocity filtering — Combines data which have the same apparent dip.

Coherency filtering/Auto-picking — Combines data which satisfy certain trace-to-trace coherence criteria.

Diversity stacking — Combines data elements which have been weighted inversely according to amplitude before the stacking. Various weighting schemes are used, often based on rms amplitude over a few cycles. One scheme sets the value to zero when amplitudes exceed a threshold.

Automigrating — Combines weighted data elements along curve of maximum convexity (or equivalent operation by other methods).

stacking chart: A diagram showing the interrelationships among the traces from common-depth-point shooting. Used to determine the proper traces for stacking, and for determining parameters for shifting traces. Components along various alignments represent common depth point, common shot, common geophone, or common offset. Figure 172.

stacking velocity: Velocity calculated from normal-moveout measurements. Used to maximize events in common-depth-point stacking. Sometimes called *rms velocity* (not always correctly). See **velocity analysis.**

stadia: **1.** An instrument for measuring distances, consisting of a telescope through which a vertical graduated rod can be seen overlain by horizontal parallel crosshairs (located in the focal plane of the telescope eyepiece); the amount of rod between the crosshairs allows one to determine the distance to the rod. **2.** The rod alone.

stadia tables: Tables giving values of $(\sin 2\alpha)/2$, $\cos^2\alpha$, and/or $\sin^2\alpha$ as a function of α. Used in calculating horizontal *(H)* and vertical *(V)* distances from a transit station to a stadia rod. If α is the angle which the line of sight makes with the horizontal, *F* is the stadia interval factor (often 100), and *X* is the distance on the rod between the cross hairs, then

$$H = FX \cos^2\alpha = FX - FX \sin^2\alpha, \text{ and}$$

$$V = FX(\sin 2\alpha)/2.$$

stake: **1.** A marker used by field parties to locate gravity stations, shotpoints, geophone locations, survey locations, etc. Usually indicates a temporary location as opposed to a **monument,** which is a permanent location. **2.** To locate the site for a well. **3.** An electrode, such as might be used to ground a seismic truck or be used as an electrode in electrical exploration. **4.** To mark the boundaries of a mineral claim.

stake resistance: The electrical resistance between a current electrode and the ground.

stand alone: Complete in itself rather than as a part of a larger system.

standard curve: Master curve.

standard deviation: The standard deviation σ of n measurements of a quantity X_i with respect to the mean $\bar{X}$, is

$$\sigma = \left[\sum (X_i - \bar{X})^2/n\right]^{1/2}.$$

With a normal or Gaussian distribution of data, 68.3 percent of the data fall within a standard deviation about the mean. The square of the standard deviation is the *variance.* See **Statistical measures.** For two degrees of freedom, measurments (X_i, Y_i) with respect to the mean $(\bar{X},\bar{Y})$, σ is

$$\sigma = \left\{\sum[(X_i - \bar{X})^2 + (Y_i - \bar{Y})^2]/n\right\}^{1/2}.$$

For a Rayleigh distribution of data, 40.5 percent fall within a circle of radius σ (called *one sigma).*

standard error: Standard deviation.

standard meridan: See **map projection.**

standard parallel: See **map projection.**

standard section: A diagram showing all the stratigraphic units in an area in their sequence of deposition; used as a standard for correlation. May show the maximum thickness of units.

standoff: **1.** The distance a sonde is from the wall of the borehole. **2.** A device for keeping the sonde from lying against the borehole wall.

standout: The amount by which the amplitude of an event exceeds the mean amplitude.

state variable: One of the sets of variables which completely describes a system at any time. A state variable may represent a derivative of a quantity which is itself

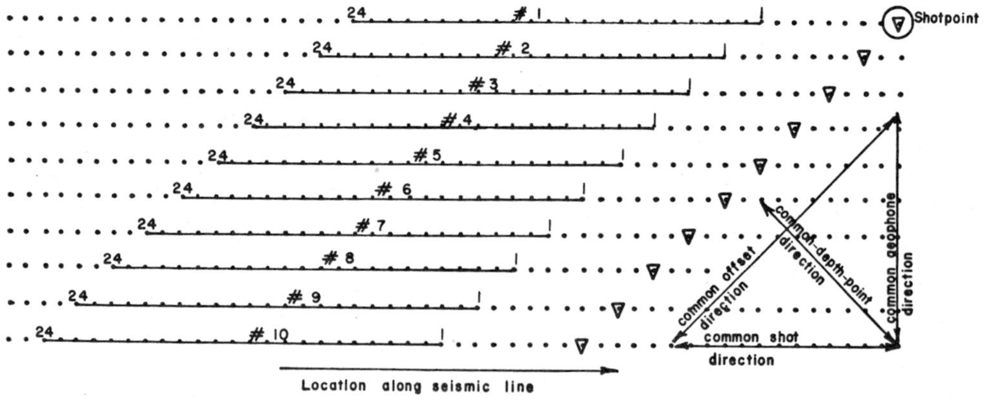

Fig. 172. **Stacking chart.** Each horizontal line shows the location of geophone groups and shotpoint for a single record. In the right corner are shown the directions for finding all traces with various elements in common.

a state variable, allowing differential equations to be expressed as a set of linear simultaneous equations. For example, the voltage drop around an electrical circuit which includes capacitance, inductance, and resistance may be expressed by the differential equation:

$$E(t) = C\int I\,dt + RI + L\,dI/dt.$$

Using state variables of I, $Q = \int I\,dt$, and $P = dI/dt$ permits this to be written as a set of three simultaneous equations:

$$\begin{aligned} E(t) &= CQ + RI + LP, \\ dQ/dt &= I\text{, and} \\ dI/dt &= P. \end{aligned}$$

See also **parameter.**

static capacity: Specific capacity.

statics: Corrections applied to seismic data to eliminate the effects of variations in elevation, weathering thickness, or weathering velocity. The objective is to determine the reflection arrival times which would have been observed if all measurements had been made on a (usually) flat plane with no **weathering** or low-velocity material present. The information on which these corrections are based derives from uphole-time data, refraction first breaks, and/or data-smoothing considerations. **a. Uphole shooting,** the direct measurement of near-surface traveltimes. This is so expensive that one cannot sample near-surface conditions everywhere along a spread; when surveying with a surface source one may not have any uphole information at all. **b.** "First-break" **ABC method** and variations for more complex assumptions are used for this determination. **c.** Data-smoothing methods assume that patterns of irregularity which most events have in common result from near-surface variations and hence the static corrections are those trace shifts which minimize such irregularities. Most automatic statics-determination programs work with this assumption and employ statistical methods to achieve the minimization. **d.** Underlying the very concept of static corrections is the assumption that a simple time shift of an entire seismic trace will yield the seismic record which would have been observed if the geophones had been displaced vertically downward to the reference datum. This assumption, of course, is not strictly true and a complete solution of the problem requires consideration of other factors, including the interplay of dynamic and static corrections and lateral as well as vertical velocity variations.

static SP: SSP or static spontaneous potential.

station: 1. A ground position at which a geophysical instrument (gravity meter, geophone, etc.) is set up for an observation. **2.** An input and/or output point on a communications system.

stationary: Having statistical properties which do not change with time and/or position. The statistics are the same if the time origin is changed.

stationary filter: A filter which is not time variant.

statistical measures: Many "measures" of a distribution of data are used. The most common for a set of n values X_i are

$$\text{mean} = \overline{X} = \frac{1}{n}\sum_{i=1}^{n} X_i;$$

$$\text{mean deviation} = \frac{1}{n}\sum_{i=1}^{n} |X_i - \overline{X}|;$$

$$\text{standard deviation} = \sigma = \left[\frac{1}{n}\sum_{i=1}^{n}(X_i - \overline{X})^2\right]^{1/2};$$

$$\text{variance} = \sigma^2;$$

$$\text{coefficient of variation} = \sigma/\overline{X}\text{ ; and}$$

$$\text{skewness} = \gamma = \frac{1}{n}\sum_{i=1}^{n}\left[\frac{(X_i - \overline{X})}{\sigma}\right]^3.$$

If X_i is a Gaussian distribution, then σ^2 is a chi-square distribution, and γ is approximately Gaussian.

steady state: Equilibrium conditions observed when short time variations are absent.

steam gun: Vaporchoc.

steepest descent: 1. An iterative method of approaching a minimum by taking an increment along the steepest gradient to arrive at the next approximation, the step length often being proportional to the magnitude of the gradient. Provision can be made to prevent oscillation about the minimum and to speed-up convergence on the minimum. It is assumed in the steepest descent that the function is continuous and that the initial estimate is in the correct minimum, in the event that the function has more than one minimum. Sometimes called *steepest ascent* when it is desired to approach a maximum. **2.** A method to compute the asymptotic behavior of an integral. Also called the *saddle-point method.* See Morse and Feshbach, 1967, Methods of Theoretical Physics, Part 1: New York, McGraw-Hill Book Co., Inc., p. 437.

steer: To introduce a time shift into an ensemble of traces so that energy approaching from a given direction appears at the same time on all traces. Used in sonogramming, in studying earthquakes with large arrays, etc. Variations include weighting the components and/or filtering before the summation.

Stefanesco function: Kernel function.

stepback: The correction applied to a location (such as the location of a seismic ship determined by radio methods) to yield the mid-point of the subsurface coverage for a seismic record, allowing for the position of the cable and the shot with respect to the navigation antenna. See Figure 130 **(offset).**

step function: An abrupt increase or decrease from one constant value to another (often zero to one or vice versa). The first derivative of a step function is an **impulse.** Also called *Heaviside function.*

step-function response: Output of a system when the input consists of a **step function.** For a linear system it is the integral of the impulse response. Has fallen into disuse compared to **impulse response.**

stepout: Differences in arrival time because of dip; **moveout.** "Normal moveout" is used for differences because of offset.

stepping method: A surveying method for determining the vertical displacement of the stadia rod with respect to the transit. A point level with the transit is sighted through the telescope which is then tilted so that the bottom cross hair is aligned with the point. Another point in the line of the top cross hair is sighted and the procedure is repeated until the stadia rod comes into the view of the telescope. The number of steps or times of retilting the telescope is counted and converted to Beamans, the full intercept between the stadia wires being one **Beaman.**

steradian: A unit of measure of solid angle. A sphere $= 4\pi$ steradians.

stereographic projection: **1.** A representation of directional information used in three-dimensional structural problems. A *stereonet* or *Wulff net* is used if angular relations are to be preserved and a *Schmidt* or *Lambert net* if areas are to be preserved. Lines are represented as points indicating their direction, planes as either great circles *(cyclographic projection)* or by points for the axial lines perpendicular to the planes *(polar projection).* Figure 173. **2.** A projection used to map the earth; see Figure 116 **(map projection).**

stick: A time-domain representation of energy at a particular time. An impulse which has been scaled. See **stickogram** and compare **stick plot.**

stickogram: **1.** Graph of reflection coefficients as a function of depth, usually made from a sonic log as an intermediate step in synthetic-seismogram preparation; see Figure 176 **(synthetic seismogram).** Stickograms occasionally include sticks which represent multiples also. **2.** A time-domain diagram of a sampled wavelet or filter showing the successive quantized values of the wavelet or of the **impulse response** of the filter.

stick plot: A presentation of dipmeter results where the well bore is represented by a line according to the projection of the well onto a vertical plane, and the components of dip in this plane are indicated by short line segments. See Figure 154 **(reverse migration).**

stochastic: Random; value determined from a specified distribution by chance. Compare **Markovian variable.** Opposite of **deterministic.**

Stoneley wave: **1.** A type of seismic wave propagated along an interface. **2.** A surface wave in a borehole. See **acoustic wave.**

stone-slab correction: Bouguer correction.

storage: A computer **memory** system; a device where data can be stored and from which it can be retrieved.

storm: Magnetic storm.

straddle spread: Split spread.

straight: Not **mixed.**

strain: The change of shape associated with a stress. Strain is usually expressed in dimensionless units such as change of length per unit of length, angle of twist, change of volume per unit of volume. Rotation or translation without change of shape is not strain. See **elastic**

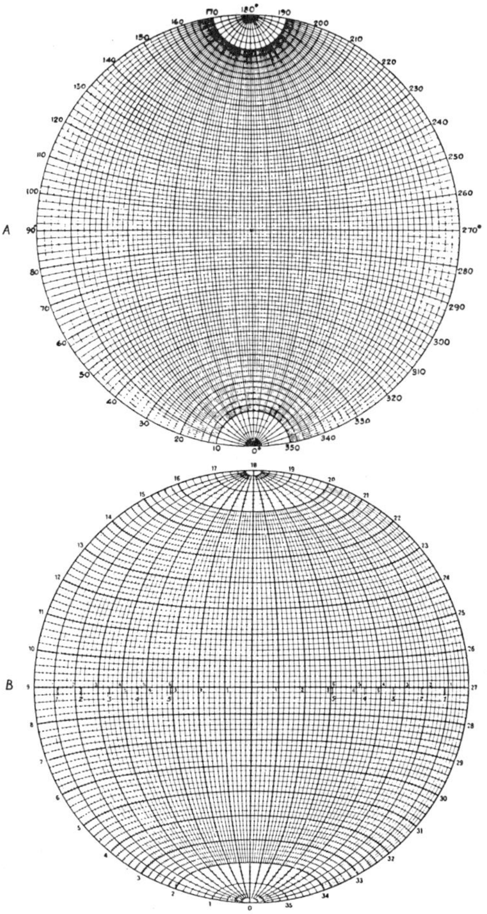

FIG. VI–5
A. Stereographic or Wulff net.
B. Lambert's equal area plat, or Schmidt net.

Fig. 173. **Stereographic** or Wulff net (a) and Lambert's equal area or Schmidt net (b). (From Hills, 1963, Elements of Structural Geology, New York, Wiley & Sons, p. 145.)

constants.

strain ellipsoid: A representation of strain by showing the ellipse into which a circle would be distorted if subject to the same strain. Figure 174.

streamer: A marine cable incorporating pressure hydrophones internally, designed for continuous towing through the water. A marine streamer (Figure 175) is typically made up of 24 "active" or "live" sections which contain hydrophone arrays separated by "spacer" or "dead" sections. Usually the streamer is nearly neutrally buoyant and "depressors" or "depth controllers" are attached to depress the streamer to the proper towing depth. The entire streamer may be 5000 to 9000 ft in length.

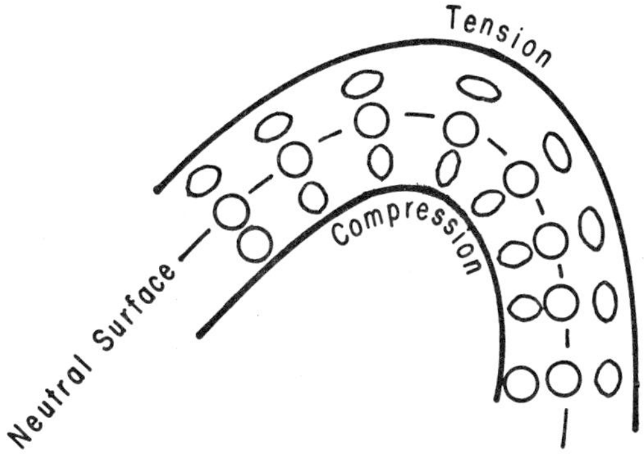

Fig. 174. **Strain ellipsoids** show the distortion which circles would undergo in folding.

streaming potential: See **electrokinetic potential.**

strength: The limiting stress before failure.

stress: The intensity of force acting on a body, in terms of force per unit area.

stretch modulus: Young's modulus; see **elastic constant.**

strike: **1.** The direction of the intersection of a surface and a horizontal plane; hence the horizontal direction at right angles to the dip. As in the "strike of a bed." **2.** The projection on the horizontal of the major axis of the ellipse of polarization.

strike-slip fault: See Table 2 **(fault).**

string: Several geophones which are permanently connected together; a *flyer.* Typically a string might consist of 6 geophones separated by 15 ft, with a polar connector on one end for attaching to the cable.

stringer: A thin high-speed layer (whose presence may be erratic) within low-speed formations. A stringer is too thin or not continuous enough to carry refracted energy very far.

strip log: A record (often in colors and symbols) of the lithology penetrated by a borehole, indicating also shows and tests of oil, gas, etc.

strobe: To read or measure at discrete time intervals.

structure: **1.** Deformation of rocks. **2.** A subsurface area characterized by folding and/or faulting. **3.** Seismic anomaly, usually a closed high.

strum: See **cable strum.**

stunt box: A device which controls the nonprinting functions of an output device upon receiving orders.

su: Sigma unit or **capture unit;** a unit for measuring capture cross-section.

subcarrier: A carrier which is applied as a modulating wave to another carrier.

subduction zone: Benioff zone.

subroutine: A computer program called for as part of a larger program.

subshot: A seismic shot or other energy release, the record from which will be stacked vertically to make one profile for input to a processing system. Marine seismic work often employs 2 or 4 subshots per shotpoint.

subsurface coverage: **1.** The position under which reflection points from plane horizontal reflectors lie. (Actual reflection points will not lie under the subsurface-coverage position if the reflectors dip.) The position (Figure 47, **depth point**) which extends from the midpoint between the source and the nearest geophone group to the midpoint between the source and the most distant geophone group, therefore of length equal to half the spread length. **2.** Sometimes the position of the actual reflection points, allowing for dip.

subsurface trace: The locations on a reflector which are perpendicular to raypaths from locations on a seismic line; hence, the locus of subsurface reflecting points as the seismic line is traversed, making allowance for migration perpendicular to the line as well as along the line. Reflection data properly should be plotted along the subsurface trace rather than along the seismic line. Apt to be different for each reflector.

subweathering velocity: Velocity immediately below

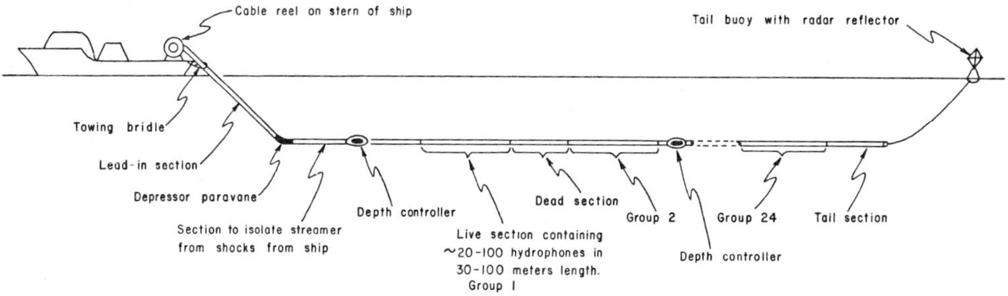

Fig. 175. **Marine seismic streamer.**

the base of the **weathering**. Sometimes taken as the velocity of a refraction at the base of the weathering.

summation method: A method of calculating a **weathering correction** to seismic arrival times. For adjacent interlocked split-dip profiles with shots just below the low-velocity layer, the correction for each group is half the sum of the first-arrival times at that group from the two interlocking records minus the average high-velocity time between the shotpoints (obtained by subtracting the uphole time from the first-arrival time of the group at one of the shotpoints when the shot is fired at the other shotpoint).

superimposed mode: A display in which two presentation modes are superimposed on each other. Usually refers to wiggle trace superimposed on variable density or variable area. See Figure 56 **(display).**

supernormal pressure: Abnormal pressure.

superposition: The situation where the same end result is obtained by an operation on a whole input as is obtained from the sum of the results of operations on the components of the input. If h, α, and β are operations to be performed on $a(t)$ and $b(t)$, then superposition dictates that

$$h[\alpha a(t) + \beta b(t)] = \alpha h[a(t)] + \beta h[b(t)].$$

Superposition allows a problem to be broken into a number of component problems which may be easier to solve than the entire problem, and yields the solution for the larger problem. Superposition is a necessary and sufficient condition for linearity; see **linear system.**

supervisor: 1. The man overseeing the work of two or more geophysical parties. Immediate supervisor of the party chief. **2.** The function of a computer which keeps tasks in order; *executive.*

suppressed layer: A layer whose resistivity is intermediate between the resistivities of the enclosing layers and which may not be "seen" unless it is very thick.

suppression: See **initial suppression.**

surface anomaly: An anomaly caused by variations at or near the surface of the earth.

surface conductivity: Conduction along the surfaces of certain minerals due to excess ions in the **diffuse layer.**

surface density: 1. Mass per unit area. **2.** The density used in calculating the **Bouguer correction. 3.** Occasionally refers to the *Green's equivalent layer,* a surface density distribution which has the same potential field outside the closed surface as does a mass distribution over the volume contained by the closed surface. See **Gauss's Law.**

surface fitting: A method of approximating a surface by a mathematical surface, usually of low order. Used in **residualizing;** the mathematical surface represents the regional and the departures of the original surface from the mathematical surface represent the residual.

surface-ship gravimeter: Shipboard gravimeter.

surface SH-waves: Love waves.

surface wave: Energy which travels along or near the surface; **ground roll.** Includes Rayleigh, Love, hydrodynamic waves, etc. Also called *interface wave.*

survey: To determine the form, extent, position, subsurface characteristics, etc., of an area or prospect by topographical, geologic, or geophysical measurements.

surveying: 1. Determining surface location. Conventional plane surveying is done using **level** and **chain, transit, plane table,** or other methods. Location networks may be resolved by **traversing, triangulation,** or **trilateration.** Surveys over larger areas require corrections for earth curvature; see **spherical excess** and **Legendre theorem.** Various radio, sonic, and satellite **positioning** methods are used in marine and airborne work. **2.** Carrying out other types of measurements as a function of location.

susceptibility: See **magnetic susceptibility.**

SV: Shear- (S-) wave energy polarized so that motion is in the vertical plane which also contains the direction of wave propagation. Converted waves are (mostly) SV.

S-wave: Shear wave or transverse wave; a body wave in which the particle motion is perpendicular to the direction of propagation. Shear waves are generated by the incidence of P-waves on interfaces at other than normal incidence, whereupon they are sometimes called *converted waves.* The velocity of shear waves V_s depends on the elastic properties of the medium:

$$V_s = (\mu/\rho)^{1/2} = \left\{E/[2\rho(1+\sigma)]\right\}^{1/2},$$

where μ is the shear modulus, ρ is the density, E is Young's modulus, and σ is Poisson's ratio. See **SH** and **SV.**

sweep: 1. To explore the effect of varying a parameter. *Velocity sweeping* consists of trying various normal moveouts on a set of CDP data to see which stacking velocity emphasizes desired events. *Dip sweeping* is done in the **sonograph** process. **2.** The continuously varied frequency of the Vibroseis source. The frequency is varied continuously during a "sweep" period of 7 sec or longer. See Figure 190 **(Vibroseis). 3.** The steady movement of the electron beam across a cathode-ray screen. Both linear and circular sweeps are used. **4.** A cycle of operations such as a radar antenna making one rotation or a side-scan sonar cycle.

synchronous detection: A method of enhancing signal and suppressing noise by synchronizing the detection period of the voltmeter receiver with the "on" cycle of the current transmitter or reference signal.

synergetic: A combination of data elements such that more information is apparent from the combination than from the elements treated independently. Synergetic Log is a Schlumberger trademark.

synthetic seismogram: An artificial seismic reflection record manufactured from velocity-log data by convolving the reflectivity function (a stickogram) with a waveform which includes the effects of filtering by the earth and recording system (Figure 176). Used to compare with an actual seismogram to aid in identifying events or in predicting how stratigraphic variation might affect a seismic record. Often constructed from

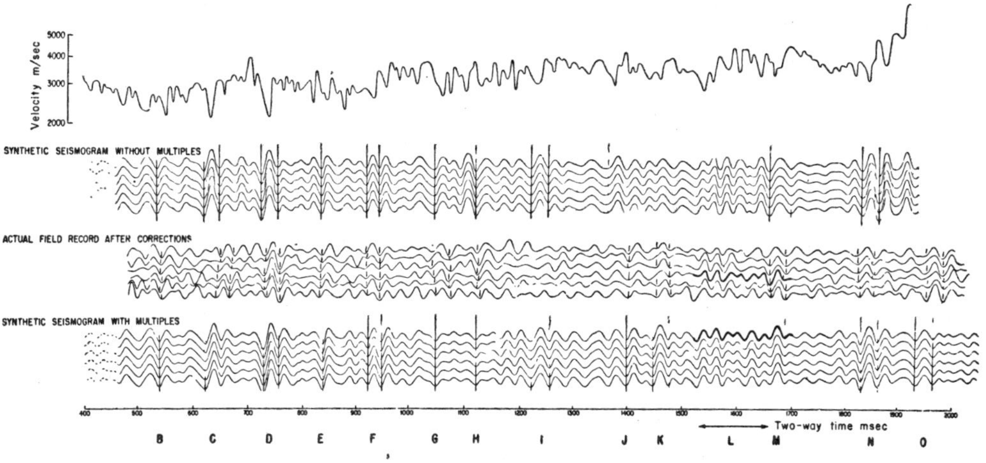

Fig. 176. **Synthetic seismogram.** Comparison with an actual field record shows that events at *B, D, F,* and N correlate well as primary reflections, events K, M, and possibly *O* correlate with multiples, and correlation of events at *C, E,* and *J* is poor. The region *G - H - I* shows little correlation of waveform whereas near *L* correlation is good. Agreement should be better if density variations had been included. Multiples involving near-surface formations do not show because the logs did not give shallow data. (Courtesy Compagnie Générale de Géophysique.)

sonic-log data alone although density data may also be incorporated. Generally assumes plane interfaces and plane waves, sometimes a point source. Synthetic seismograms sometimes show primary events only, primaries plus selected multiples, or primaries plus all multiples; they may be constructed by analog, digital, or manual methods. Also called **theogram.**

systematic error: An error not attributable to chance alone. Such errors do not average out merely by including more measurements. Systematic errors may be caused by bias produced by the instruments (for example, incorrect scale factor or incorrect zero), by the observer or measuring procedures (for example, not selecting representative samples or altering samples before the measurements), or by the action of factors or physical laws which are not suspected. Much geophysical noise is systematic.

T. Period; the time between adjacent corresponding points on a periodic wave; the reciprocal of **frequency**.

tabular body: A body of finite thickness with one edge horizontal but other edges infinitely remote; an *infinite dike*. One type of model used in potential-field calculations. Implies that a body's width is more than 50 times its thickness.

tactical characteristics: The characteristics of a ship which determine its maneuverability under various sea conditions.

tadpole plot: A type of plot of dipmeter or drift results; an *arrow plot*. The position of a dot gives the dip angle versus depth and a line segment pointing from the dot gives the direction of dip, using the usual map convention of North being up. See Figure 53 **(dipmeter)**.

Tafel's law: A relationship between overvoltage η and current density J at an anode or cathode:

$$\eta = a - b \log_{10} J ,$$

where a and b are experimentally determined constants. This law applies over a greater current-density range than is used in IP field measurements.

Tagg method: A method of interpreting resistivity-sounding data obtained with the Wenner array over a two-layered earth.

tailing: Lengthening of a waveform as by adding extra cycles.

takeout: A pair of leads from a multiconductor cable to which geophones or geophone flyers can be connected. Such takeouts often are polarized to reduce the likelihood of making the connection backwards.

tamp: To pack material about an explosive in a shothole. Objectives of tamping are to effect better coupling of the explosive energy with the earth and to retard the expulsion of the gaseous and other products of the explosion, thereby improving the conversion of explosive energy to seismic energy and delaying hole-blow effects. Water or mud is usually used, sometimes sand and earth.

tandem survey: Electromagnetic-survey method in which both transmitting and receiving coils are moved simultaneously, maintaining a constant separation between them. Equivalent to *moving-source method.*

tangential wave: S-wave.

tape: 1. A magnetic tape. 2. A chain.

tape-guide pins: Pins of nonmagnetic material which help position magnetic tape and prevent tape **skew**.

tapered array: A source or geophone array in which the various elements contribute unequally. Tapered source arrays may be achieved by loading different amounts of explosives in the different shotholes of a pattern, by varying the number of source impulses (pops, weight drops, etc.) or their spacing, or by varying the weighting while vertical stacking. Tapered geophone or hydrophone arrays may be achieved by varying the output of the different elements or (more commonly) by varying the spacing of the elements of the group. Array tapering is used to remove nulls or otherwise change the shape of array **directivity patterns** but at the price of widening the major lobe (for the same number of elements and the same overall array size).

tapered sweep: Vibroseis sweep where source varies as $A(t) \cos [(w_o \pm bt)t]$. Generally $A(t)$ becomes smaller as the limiting frequencies are approached.

tapered window: A window or gate with gradual edges. A tapered window compared to an abrupt boxcar window lessens the dependence on high frequencies in order to represent it adequately, simplifying many types of processing. Some tapers in common use are shown in Table 12 **(window)**. See also **Gibb's phenomena**.

taper mix: See **mixing.**

taphrogenic: A term referring to regional block-faulting tectonics. See **tectonic types.**

tap test: A recording made as a geophone is tapped lightly, showing which channel that geophone feeds. Used to check that the spread is properly connected and oriented and also that the geophone is live.

TAR: True-amplitude recovery; a process for removing the effects of variable gain in the field recording and adjusting the amplitude to compensate for spherical divergence and other time-dependent attenuation.

tare: A discontinuity in data indicating an error in measurement or computation rather than an actual jump in the quantity being measured. Sometimes spelled "tear."

target: A sliding marker on a stadia rod on which one sights while leveling.

Tarrant method: A graphical refraction interpretation method applicable where refractor shape varies; see Figure 177. For a refractor of velocity V_2 covered by an overburden of velocity V_1, the refractor is tangent to an ellipse for which the geophone is at one focus. Tarrant approximates this ellipse by the arc of a circle. The center of the circle is at the distance $(V_2 T \tan^2 C)$ toward the shotpoint, where C is the critical angle and T is the delay time attributable to the geophone ($T = t - t_{sp} - X/V_2$, where t = arrival time, t_{sp} = delay time attributable to the shotpoint, and X = offset distance). The line AB is drawn making the angle C with the vertical and the center of the circle is found by completing the right triangle. The refractor surface is the common tangent to the arcs from the various geophones. See Tarrant, 1956, Rapid Method of Determining the Form of a Refractor from Line Profile Results: Geophysical Prospecting, v. 4, p. 131-139.

Taylor series: A function $f(x)$ can be expressed in terms of the value of the function and its derivatives (f', f'', f''', etc.) at any point $x = b$. In one variable this is

$$f(x) = f(b) + \frac{f'(b)}{1!}(x-b) + \frac{f''(b)}{2!}(x-b)^2 + \frac{f'''(b)}{3!}(x-b)^3 + \cdots +,$$

where $f^n(b) = (dx^n)/(d^n f)$, when $x = b$ and ! denotes factorial (e.g., 3! = 3·2·1 = 6). This series converges if

$$\lim_{n\to\infty} \frac{f^n(b)\,(x-b)}{n f^{n-1}(b)} < 1.$$

The Maclaurin series is the special case where $b = 0$.

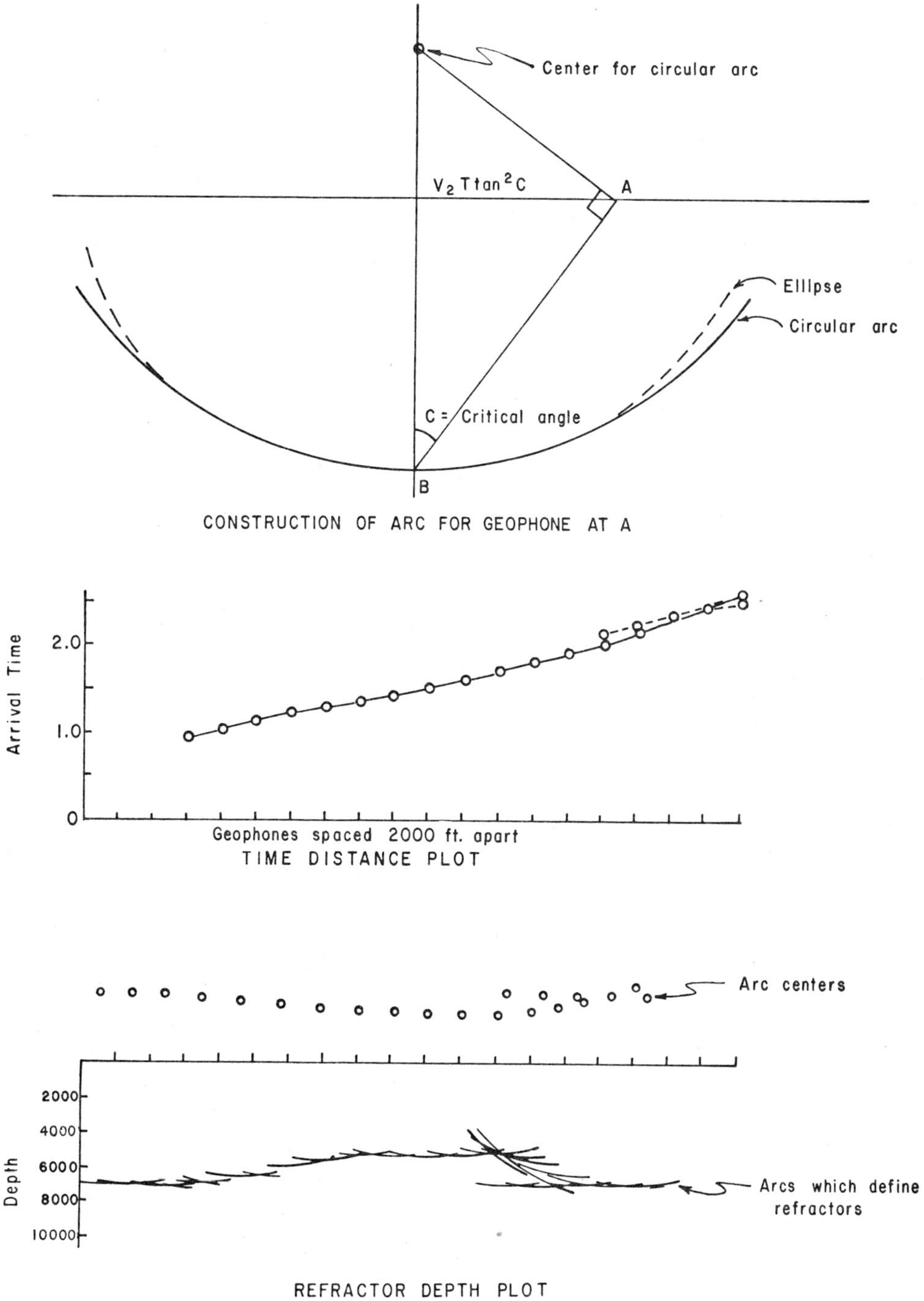

Fig. 177. **Tarrant refraction method.** (From Tarrant, Geophysical Prospecting: v.4, no. 2, p.133-135.)

TB: **Time break.**

Tchebyscheff array: **Chebychev array.**

TD: Total depth. The maximum depth reached by a well.

T-D curve: **1.** Time-distance curve or *T-X* plot. **2.** Time-depth plot.

T^2-D^2: A method of velocity analysis from seismic surface data. More often written X^2-T^2.

T-ΔT analysis: Normal moveout *(ΔT)* as a function of offset *(X)* and total time *(T)* can be used to yield **stacking velocity** *($\overline{V}$)* and information about multiples which show abnormal velocities for their arrival times. The formula for velocity is

$$\overline{V} \approx X/(2T\,\Delta T)^{1/2}.$$

In the case of nearly flat parallel bedding, this yields the rms velocity.

tear: See **tare.**

Techno: The name of an analog magnetic recorder and magnetic tape.

tectonic map: A map showing major structural features produced by uplift, downwarp, compression, or faulting, with the more significant lineations associated with such features. The term is usually applied to maps covering large areas, while maps of smaller areas showing the same features are called structural maps.

tectonic types: Four types of broad structural deformations are **a.** *Orogenic* or mountain-building which involves lateral forces, folding, and thrusting, **b.** *Epeirogenic* or relatively gentle warping, **c.** *Taphrogenic* which involves mainly vertical forces and block faulting, and **d.** *Lineagenic* which involves strike-slip faulting.

telemetering: The transmission of data from a point of observation to a recording point, often by radio. Telemetered time signals can be used for synchronous detection of resistivity and IP signals.

teleprocessing: A data-processing and communications system so that input/output devices may be remote from the processing devices.

teleseism: An earthquake whose epicenter is over 1000 km away. Earthquakes originating nearer are "local" earthquakes.

televiewer: See **borehole televiewer.**

telluric: Of the earth. Often refers specifically to **telluric currents.**

telluric current: A natural electrical earth current of very low frequency which extends over large regions and may vary cyclically in that direction. Telluric currents are widespread, originating in variations of the earth's magnetic field.

telluric-current method: A method in which orthogonal components of the horizontal electric field associated with currents induced in the earth by natural sources of energy are measured simultaneously at two or more stations. The measurements from one station, which serves as a base, are used to normalize the measurements from the other stations. The normalized measurements, plotted as vectors at each station, may outline an ellipse if signals from several different azimuths are recorded. The relative area of the ellipse at each station is (ideally) inversely proportional to the conductance of the sedimentary section above the basement. The orientation of the ellipse yields information about the direction of current flow. Other methods of processing or recording the data such as the triangle or the vectogram method can be used to obtain the relative ellipse areas.

tellurometer: An electronic survey instrument for measuring distances with great accuracy. A high-frequency (3×10^9 hz) radio pulse is transmitted to the "rod" transponder, where it is retransmitted back to the master transmitter which measures the time that has elapsed since the original transmission. Accuracy of a few inches in several miles can be achieved with proper corrections (mainly for moisture content of the air). Line-of-sight limited. Trade name of Tellurometer, Ltd.

temperature log: A well log of temperature, often made with a resistance thermometer (thermistor). Used for locating cement behind the casing (because the setting of cement is exothermic and hence raises temperature), intervals which are producing gas (because the expansion of gas as it enters the borehole lowers the temperature), and fluid flows (particularly behind the casing). The differential-temperature log records the difference between two thermometers which are usually about 6 ft apart; this log is especially sensitive to very small changes in temperature gradient.

temperature surveying: Measurements of temperature to locate thermal-energy sources or to investigate ground-water problems, thermal springs, karst cavities, sulfide deposits, dikes, structure (faults) affecting groundwater flow, etc. See **geothermal prospecting.**

template: **1.** A pattern. **2.** A sheet of transparent plastic marked with the seismic or normal-moveout pattern as a guide in picking seismic events. **3.** A transparent overlay for calculating gravity or magnetic effects such as a **dot chart.** Templates have been used for **terrain correction, isostatic correction,** or **residualizing.** See also **graticule** and **zone chart.**

template scan: Filtering with a **matched filter.**

temporal frequency: Ordinary **frequency,** as opposed to *spatial frequency* (or *wavenumber).*

tensor: A set of functions that relates different vector fields, such as might involve a change in coordinate system. For example, the vector electric field **E** and the vector current-density field **J** are related by resistivity ρ in which $\mathbf{E} = \rho\mathbf{J}$; in an anisotropic material ρ is a tensor, usually given as a matrix. Similarly, the elastic "constants" are components of tensors for anisotropic media, relating a stress vector to a strain vector.

tent poling: Sudden deflections of a sonic log resulting from **cycle skip** or tool sticking.

terminal: An input and/or output device.

terrain correction: **1.** A correction to gravity data required because the surroundings are not all at the same elevation as the meter. Relief in the immediate vicinity of the station may require special surveying, called *terrain surveying,* whereas corrections for relief more remote from the station often are made from a topographic map using a terrain-correction template or **zone chart. 2.** A correction to seismic data because of **topographic loading. 3.** A correction to magnetic or electrical data because of terrain effects.

Terrapak: A seismic energy source in which a plate (hammer) is driven against the ground by compressed air. General Dynamics trademark.

tesla: A unit of magnetic field strength **B.** 1 tesla = 1 weber/m^2 = 1 newton/amp-m = 10^4 gauss = 10^9 gamma. 1 nanotesla = 1 gamma.

theodolite: A precision instrument for measuring horizontal and vertical angles (Figure 178). It consists of a telescope mounted so as to swivel vertically and secured to a revolvable table carrying a vernier for reading horizontal angles. A graduated circle for measuring vertical angles and a compass are included. See also **transit.**

theogram: Synthetic seismogram.

theoretical seismogram: Synthetic seismogram.

thermal-decay-time log: See **neutron-lifetime log.** $\tau D \tau$ is a Schlumberger trademark.

thermal imaging: Mapping with infrared radiation. See **thermal-infrared.**

thermal-infrared: A remote-sensing method in which an infrared beam outside the natural thermal range is swept back and forth across the earth from an aircraft and the reflected energy recorded. The display often simulates an aerial photograph.

thermal neutron: A neutron whose motion energy corresponds to ambient temperatures; neutrons with mean energies of the order of 0.025 electron volts.

thermal noise: Noise resulting from random thermal energy; *Johnson* noise. The mean-square voltage because of thermal noise in an electrical circuit varies as the absolute temperature, the bandwidth, and the resistance.

thermal-relaxation time: See **nuclear-magnetism log.**

thermistor: A device whose electrical resistance varies with temperature. Used with doppler-sonar units to measure the temperature and hence the velocity of sea water with an assumed salinity.

thermocline: The decrease in water temperature with depth in the ocean. See Figure 166 **(SOFAR channel).**

thermocouple: Two dissimilar conductors welded together at one end. When the junction is heated, a voltage develops across it which is proportional to the temperature difference between the junction and the open ends.

thermoremanence: See **remanent magnetism.**

thixotropic: A property of gels which allows them to become liquid when agitated. Drilling muds are often thixotropic.

Thompson-Haskell method: A frequency-domain method for plane waves in a multilayered half-space, used for both body-wave propagation and surface-wave dispersion problems. Haskell, 1953, The Dispersion of Surface Waves on Multi-Layered Media: Bull.

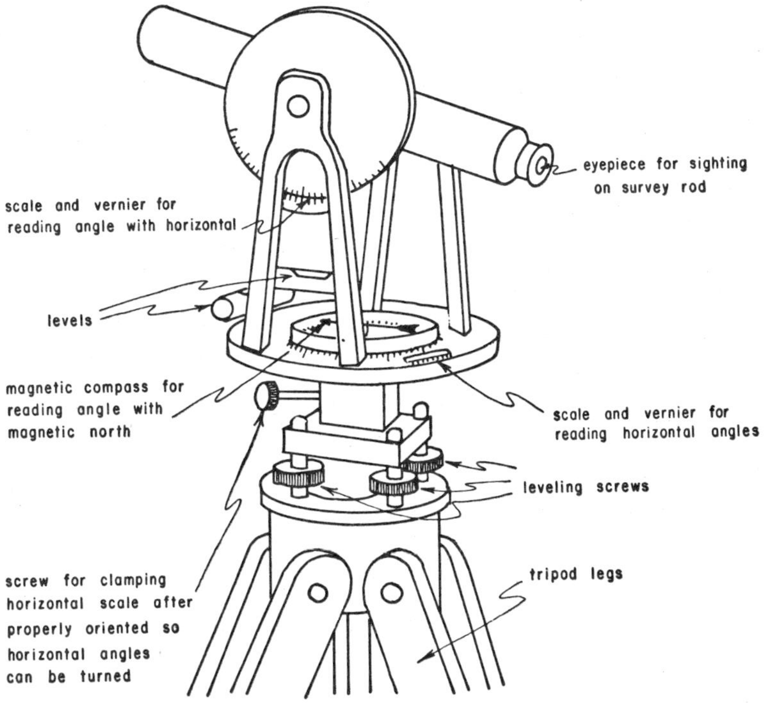

Fig. 178. **Transit-theodolite.**

Seism. Soc. Am., v. 43, p. 17.

Thornburgh's method: A refraction interpretation method which involves using Huygen's principle to construct wavefronts from reciprocal shotpoints by working backwards from the observed arrival times at the surface. The velocities above the refractor must be known for the construction. See **wavefront methods** and Thornburgh, 1930, Wave Front Diagrams in Seismic Interpretation: Bull. Am. Assoc. Petrol. Geol. v. 14, no. 2, p. 185.

three-array: A special case of the pole-dipole **array** in which the three electrodes are equally spaced.

three-D log: Three-dimensional velocity log; a display of the seismic or **acoustic wavetrain** received a short distance (3 to 12 ft) from a sonic wave transmitter. Also called *microseismogram* log and *variable density* log. See also **character log, cement-bond log,** and **sonic log.** Three-D log is a Birdwell tradename. Figure 179.

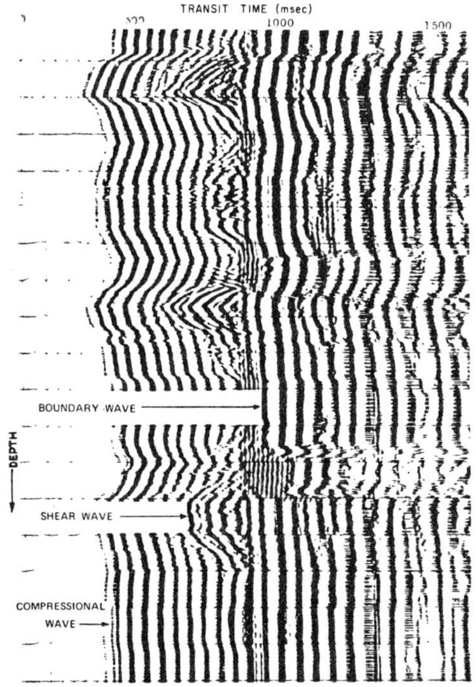

Fig. 179. **Three-D log.** (Courtesy Birdwell Division of Seismograph Service Corporation.)

three-layer resistivity models: Models of horizontal electrical layers are given "type" names (Figure 180). The curves for such models are called **type curves** and used in the interpretation of electrical-resistivity observations.

three-point method: 1. A method for locating a station by taking backsights on three previously located stations. **2.** Determining geographic position by the intersection of bearing lines from three stations.

three-point operator: A **Backus filter.**

threshold: The lower limit that will produce a phenomenon. For example, the IP saturation threshold is the current density above which the IP phenomenon becomes nonlinear.

throw: The vertical component of separation of a bed by a **fault.**

thrust: A reverse **fault,** especially a low-angle reverse fault where the dip of the fault plane is less than 45 degrees.

Thumper: Device for dropping a weight to provide seismic energy. Typically a three-ton weight is dropped 10 ft. Trademark Mandrel Industries.

Tiburg rule: A **depth rule** used in magnetic-data interpretation: the depth to a pole is 2/3 the horizontal distance at half the maximum amplitude.

tidal effect: Variations in gravity observations resulting from the attraction of the moon and sun and the distortion of the earth so produced. Tidal corrections to gravity observations are made by means of tables or are included with drift corrections.

tie: Observations connecting two previously observed points, with the objective of establishing or checking the relationship of the two, or an independent measurement at a previously observed point with the same objective.

tie-in: To relate a new station to previously established stations.

tie-line: A survey line which connects other survey lines. Especially such a line which closes a traverse loop.

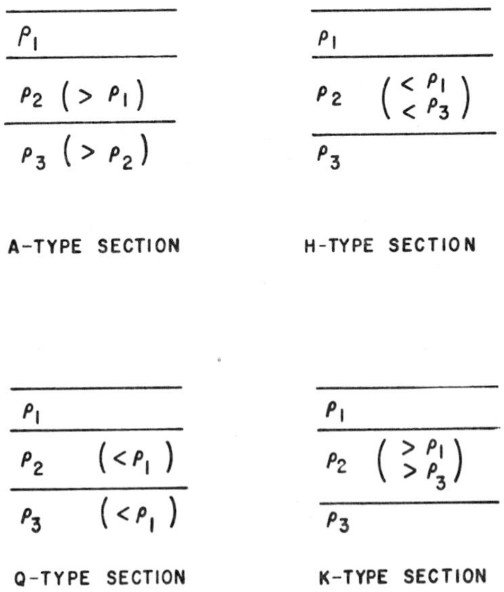

Fig. 180. **Three-layer resistivity** types.

tight: **1.** Having very low permeability. **2.** Data held confidential.

tilt: **Pitch.**

tilt angle: See **polarization ellipse.**

tilt table: A device on which a meter can be mounted and then tilted by precisely known amounts. Used to test geophones and to calibrate certain kinds of gravity meters.

time: See **record time.**

time-average equation: The traveltime Δt through a rock with matrix velocity V_m and porosity ϕ which is filled with fluid of velocity V_f is

$$\Delta t = (1 - \phi)/V_m + \phi/V_f.$$

time branch: One of several reflection events produced by the same curved reflector. At a given location coherent reflections may be seen from different portions of the same reflector where synclinal curvature produces buried-focus effects. See Figure 28 **(buried focus).** Three time branches normally occur as a result of excessive two-dimensional synclinal curvature, more than three because of more complicated curvature or excessive curvature in three dimensions. Time branches due to two-dimensional curvature in the plane of the seismic line will migrate into a syncline if the proper velocities are used in the migration, but if the curvature is outside the plane of the line, they will not.

time break: The mark on a seismic record which indicates the shot instant or the time at which the seismic wave was generated. See Figure 149 **(record).**

time constant: **1.** The time taken for the current in a circuit having a steady emf to reach a definite fraction of its final value after the circ ı is closed. The fraction is $(1 - 1/e) = 0.632$. **2.** The time taken for the current to decay to $1/e = 0.368$ of its value after the emf is removed. Also called *decay constant* or *relaxation time.* **3. AGC time constant.** **4.** A time over which readings are averaged to remove statistical fluctuations, as with nuclear-log readings.

time delay: See **delay time, filter correction,** and **El-cord.**

time-depth chart: A graph or table of reflection time (or sometimes one-way time) against reflector depth for vertically traveling energy. Often written *T-D chart.* Specific for a particular velocity distribution. Used in converting times to corresponding depths. Compare **time-distance curve.**

time-distance curve: A plot of the arrival time against the shotpoint-to-geophone distance. Also called a *T-X curve.* Used in **noise analysis** and in interpreting refracted events (head waves). The slopes of segments of the curve give the reciprocals of the apparent velocities for various refractor beds. See also **normal-traveltime curve** and **reduced traveltime.** Figure 181.

time domain: **1.** Expression of a variable as a function of time, as opposed to its expression as a function of frequency (frequency domain). Processing can be done using time as the variable, i.e., "in the time domain." For example, convolving involves picking up values at successive time intervals, multiplying by appropriate constants and recombining; this is equivalent to filtering by passing a signal through frequency-selective circuitry. It is also equivalent to Fourier transforming, multiplying the amplitude spectra and adding the phase spectra ("in the frequency domain"), and then Fourier transforming again. **2.** Time-domain induced polarization is called the **pulse method.**

time lag: A delay of a wavelet.

time lead: See **lead.**

time of closest approach: The time when a navigation satellite is closest and reaches its maximum elevation angle during a pass. Abbreviated TCA.

time series: The series of values of a function sampled at regular time intervals. Sometimes represented as a set of values, as a stickogram, in **Z-transform** form, etc.

time sharing: **1.** Simultaneous use of a computing machine on two or more different programs. The operations on the different programs involve different parts of the computer and the computer's central processing unit exercises the control to keep the results separate. Thus, a computer might be simultaneously reading, searching its memory for certain data, calculating, and writing results as parts of different programs. **2.** Especially such simultaneous use of a computer by remote users communicating with the computer using phone lines and a teleprinter console.

time tie: Identifying events on different records by their arrival time where they possess common raypaths.

time-variable filtering: Varying the frequency band pass with record time. Time variable deconvolution is sometimes used to compensate for the shift of reflection energy to lower frequencies at late record times.

time variant: Describing an operation in which the parameters vary with record time, as in "time-variant filtering."

timing lines: Marks or lines at precise intervals of time such as used on seismic records (usually 0.01-sec intervals) to help measure the arrival times of seismic events. See Figure 149 **(record).** The timing mechanism in older recording systems commonly was an accurate tuning fork, and in newer systems a crystal-controlled oscillator.

timing word: A word at the head of a block of data which gives the elapsed time since the shot instant.

tin hat: **1.** A hard hat worn as protection to the head. **2.** A hole plug.

Toeplitz property: Property when all the elements on a given diagonal of a matrix are identical.

tool: **Sonde.**

topographic correction: **Terrain correction.**

topographic-loading effect: The effect of variable overburden on seismic velocity in areas of large surface relief.

Toran: **1.** A medium-range electronic phase-comparison **positioning** system similar to **Raydist,** Lorac, and Decca. Toran in the *Z-mode* consists of pairs of fixed free-running transmitters whose frequency differs by an audio frequency. The beat between the signals from these is observed at a fixed reference station and the information is rebroadcast on another frequency. The mobile station compares the beat between the two

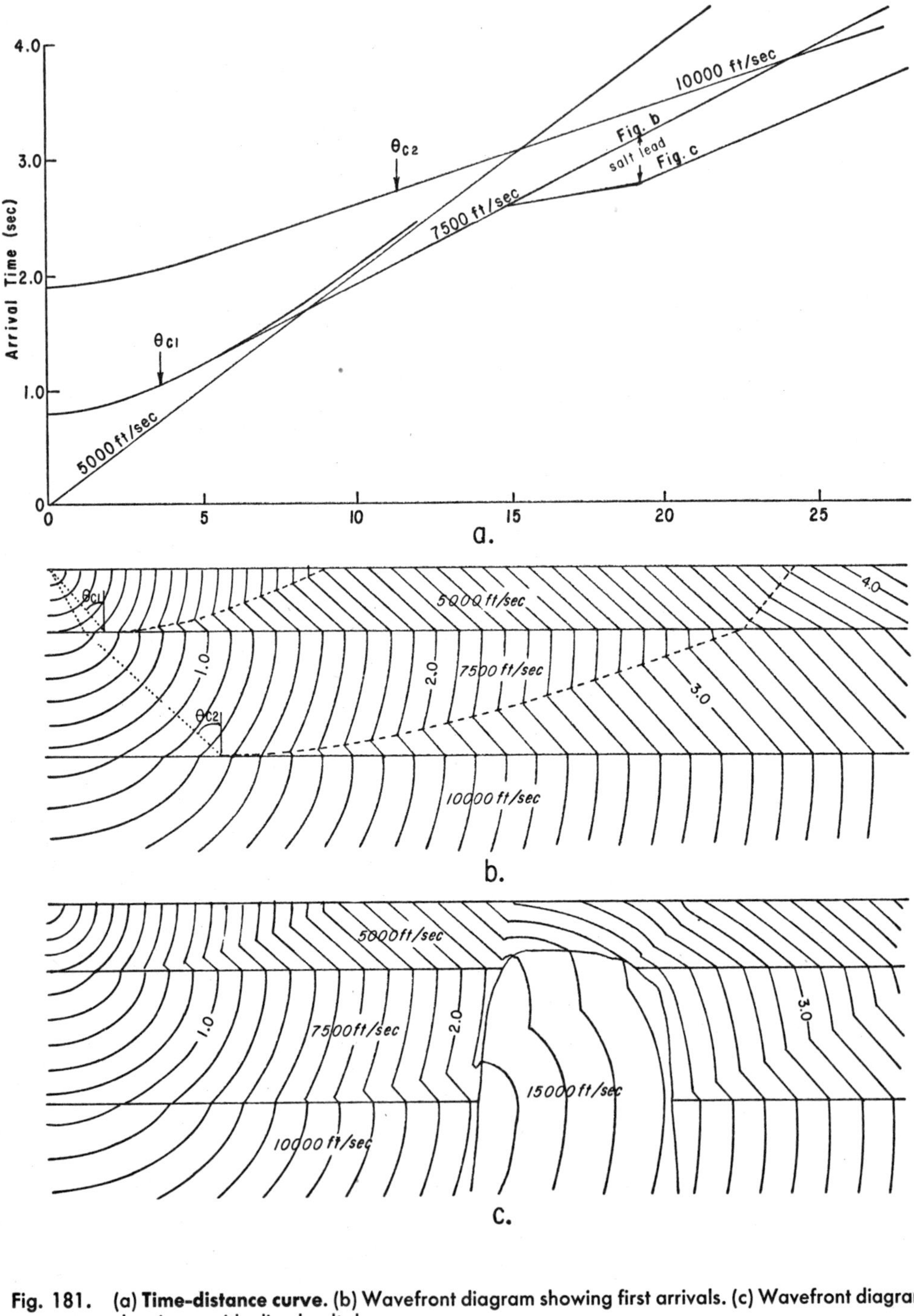

Fig. 181. (a) **Time-distance curve.** (b) Wavefront diagram showing first arrivals. (c) Wavefront diagram showing an idealized salt-dome.

free-running transmitters with the information from the reference transmitter to locate itself on a hyperbolic line of position. The *X-mode* is similar except that the reference receiver is used to modulate one of free-running stations, thus dispensing with the reference transmitter. In the *Y-mode* one of the free-running transmitters is aboard the mobile station. **2.** *Toran O* is a direct ranging system which utilizes highly stable frequency standards at both fixed and mobile stations so that transmission times and, hence, ranges can be measured.

torr: A unit of pressure; the pressure required to support a column of mercury one mm high at standard gravity. Equals 1.333 mbar.

torsion balance: **1.** An instrument for measuring second derivatives of the gravitational potential. In a non-uniform field different forces act on equal masses at opposite ends of a horizontal beam suspended by a very fine torsion wire, producing a torque. The resulting measured gradients can be integrated to make a gravity map. The curvatures have characteristic patterns associated with gravity anomalies. **2.** A device for measuring other sorts of force fields, such as magnetic or electrical.

tortuosity: The length of the path of a fluid passing through a unit length of rock.

total intensity: Usually refers to the total magnetic intensity as opposed to the components of the intensity in the vertical or horizontal directions.

total reflection: Reflection where the angle of incidence exceeds the **critical angle.** At and beyond the critical angle the energy is either reflected or converted.

tour: A shift in drilling operations. Pronounced "tower."

tower: Tour.

towing bridle: Assembly by which the towing ship holds a marine streamer in the proper position amidship and keeps it from rubbing the stern.

township: **1.** A survey unit of area, nominally six miles on a side. It is subdivided into 36 sections, numbered as shown in Figure 182. **2.** A designation of east-west row of townships. American townships are specified in relation to standard reference parallels which often are not stated explicitly. T-3-N indicates a township in the third row north of a reference parallel. Canadian townships are counted northward from the 49th parallel. See also **range.** Township is also abbreviated Twp.

T.P.: Turning point.

trace: **1.** A record of one seismic channel, electromagnetic channel, etc. See Figure 149 **(record).** **2.** A line on one plane representing the intersection of another plane with the first one, such as a *fault trace.* **3.** The sum of the diagonal elements of a matrix; spur.

trace analysis: Determining and plotting the corrected arrival time of events for every trace.

trace equalization: Adjusting a seismic channel so that the amplitudes of adjacent traces are comparable in the sense of having the same rms value over some specified interval or some other criterion.

trace gather: See **gather.**

trace integration: A form of mixing. Assume 20 input traces are mixed to form one output trace, and the last 15 of these input traces plus the next 5 are mixed to form the next output trace; then the trace integration is 20/5. Especially used with the weight-drop method.

trace sequential: A format on magnetic tape in which one channel (trace) is recorded without interruption followed sequentially by other channels. As opposed to *multiplexed format* in which the data for one record time is recorded for all channels followed by the data for the next record time, etc.

track: **1.** A **trace.** **2.** The data positions which can be read by a single magnetic head. **3.** To follow the movements of an object.

track plotter: A device which continuously displays the position of a ship, operated from radio-**positioning** system signals. Often displays position in terms of the positioning-system coordinates.

train: A series of successive events, as a "train of waves."

transceiver: Device which is both a transmitter and a receiver, such as used in sonar. See **transponder.**

transcord: Transcribe.

transcribe: To transform information from one form to another, as to make a magnetic tape from a paper seismic record. See also **reformat.**

transcurrent fault: A strike-slip fault or wrench fault. See Table 2 **(fault).**

transducer: **1.** A device which converts one form of energy into another. Many types of transducers are reversible, for example converting electrical energy into acoustical energy and vice versa. The electrodynamic **geophone** is a reversible transducer; it converts mechanical motion to electrical voltage or passing a current through the coil causes the coil to move with respect to the case. Other reversible transducers are electrostatic, variable reluctance, magnetostrictive, piezoelectric, etc. Piezoelectric transducers of barium or lead zirconate or titanate are used in many **hydrophones** and sonar transducers. **2.** Especially a piezoelectric transducer.

transduction factor: Ratio of output to input for a transducer. For digital-grade velocity geophones, it is of the order of 1v/cm/sec. For a hydrophone, of the order of 20 v/bar.

transfer characteristics: Transfer function.

transfer function: The ratio of output to input as a function of frequency. The frequency-domain characteristics of a system (e.g., a **filter**). The complex function which changes cosinusoidal inputs into outputs. Multiplying the frequency-domain transform of the input by the transfer function yields the frequency-domain transform of the output. The transfer function usually is represented by the amplitude-versus-frequency and phase-angle-versus-frequency curves; these contain the same information as the impulse response in the time domain and are convertible into the inpulse response through the Fourier transform.

transfer impedance: The complex ratio of a potential difference at one pair of terminals or electrodes to the current at the other pair.

transform: To convert information from one form into

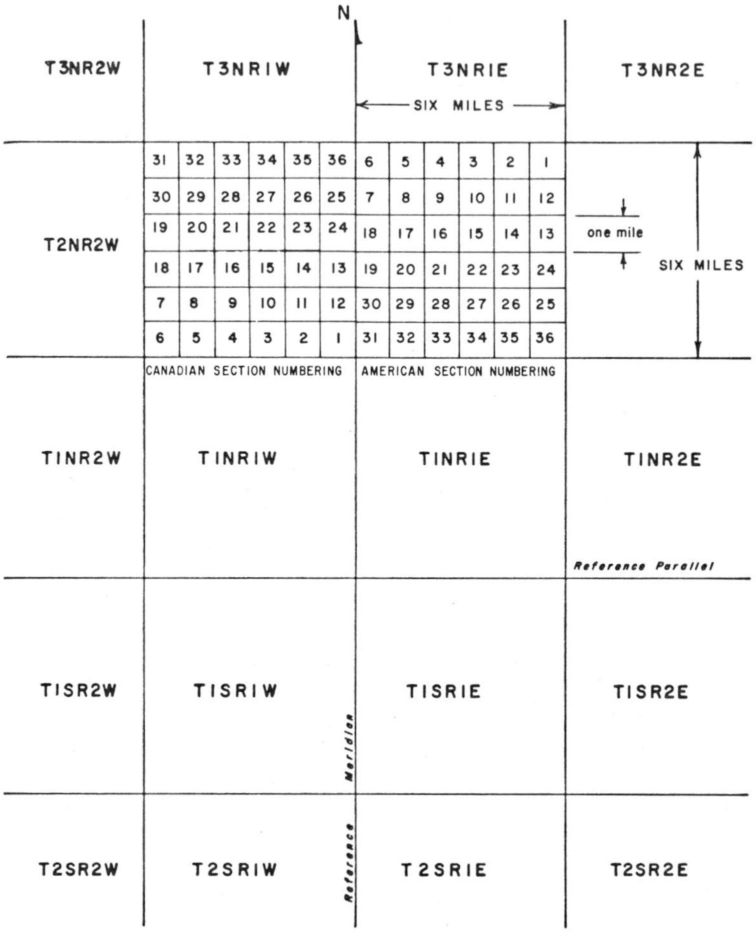

Fig. 182. **Township-range** location system showing the system for numbering sections.

another, as with the **Fourier transform** or **Laplace transform.**

transformed wave: Converted wave.

transform fault: A fault which displaces the rift zone associated with the generation of new crust or displaces the zone where plates collide. See Table 2 **(fault)** and Figure 136 **(plate tectonics).**

transform pair: A waveform and its frequency-domain equivalent, or a time-domain operation and its frequency-domain equivalent. See Table 4 **(Fourier pairs).** Used also for transforms other than Fourier.

transient: A nonrepetitive pulse of short duration, such as a voltage pulse or seismic pulse.

transient electromagnetic method: An electromagnetic method in which the waveform of the transmitted signal is a pulse, step function, ramp, or other form which can be considered to be nonperiodic and in which measurements are made after the primary field has stopped changing. Some methods, which may be called "quasi-transient" or "quasi-time domain methods," use a train of primary pulses with measurements being made during the off-time between pulses. Principal advantages of transient methods over continuous-wave methods are that the primary field is not present during the measurement of the secondary field and that measurements of the secondary field as a function of time is equivalent to continuous-wave measurements over a wide frequency range. Transient methods are used primarily for depth sounding but quasi-transient airborne methods are used for continuous profiling. See **Input.**

transient IP method: See **pulse method.**

transistor: An electrical device with three or more terminals using a semiconductor for controlling the flow of current between two terminals by means of current flow between one of these terminals and a third terminal. It functions similarly to a vacuum tube without requiring current to heat a cathode. Ideally it is a current-controlled current source which operates in only one direction. See Figure 183.

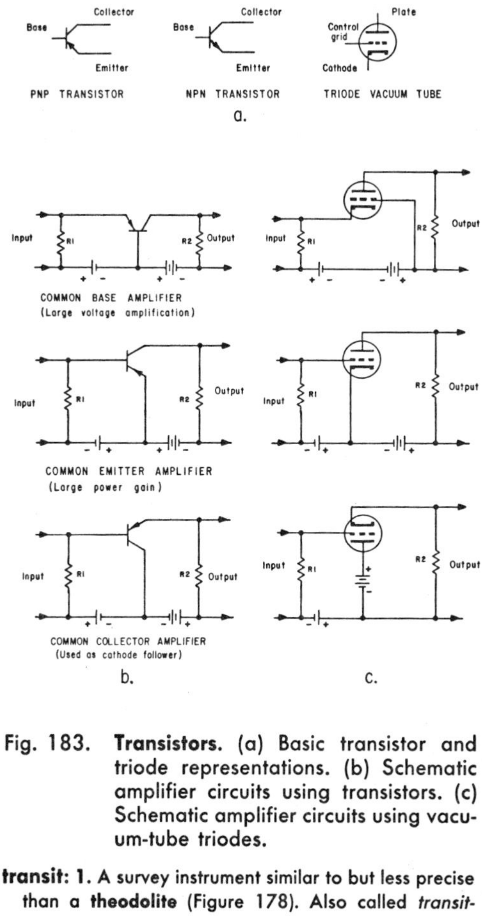

Fig. 183. **Transistors.** (a) Basic transistor and triode representations. (b) Schematic amplifier circuits using transistors. (c) Schematic amplifier circuits using vacuum-tube triodes.

transit: 1. A survey instrument similar to but less precise than a **theodolite** (Figure 178). Also called *transit-theodolite.* An accuracy of about 10 sec of arc can be achieved, though 30-sec accuracy is more likely. 2. The passage of a celestial body across a celestial meridian. 3. The passage of a satellite involved in **satellite navigation.** "Transit" is the system name of the Navy Navigation Satellite System.

transit time: The traveltime of a compressional acoustic wave over one foot of distance. See **sonic log.**

translocation method: A positioning technique which uses a nearby fixed station to improve the accuracy of a mobile station. 1. Observation of a satellite transit from a nearby fixed station in order to correct for minor variations in satellite orbit. 2. Observation of radio-positioning signals at a fixed station for information on sky-wave variations.

transmission coefficient: 1. The ratio of the amplitude of the displacement of a wave transmitted through an interface to that of the wave incident upon it. 2. A measure of the amplitude of a wave passing through an interface at normal incidence. 3. Because we are usually interested in energy which has passes through an interface twice (once going down and once returning upward), a *two-way transmission coefficient T* for normal incidence is often used:

$$T = 1 - R^2 = \frac{4\rho_1 V_1 \rho_2 V_2}{(\rho_1 V_1 + \rho_2 V_2)^2},$$

where R is the reflection coefficient, ρ_1, ρ_2, V_1, and V_2 are the densities and velocities in the upper and lower media. 4. The ratio of the energy density of a wave transmitted through an interface to the energy density of the wave incident on the interface.

transmitter: In IP surveying, a current waveform generator. Also called a *sender.*

transponder: A device which transmits a signal upon receiving another signal. When the receiver in the device detects the "interrogating signal" it triggers the transmitter which replies with a coded pulse or sequence of pulses. Transponders are used with both electromagnetic waves and sonar acoustic waves. A radar transponder is also called a radar *beacon,* a sonar transponder a *pinger.*

transpose: The transpose A^T of a matrix A is the matrix whose rows are identical to the columns of A. Thus if matrix A is of size $(m \times n)$, matrix A^T is of size $(n \times m)$. For conformable matrices, $(AB)^T = B^T A^T$. For any matrix, $(A^T)^T = A$.

transposed method: A method of seismic field recording in which a single large geophone group (or a small number of such groups) records energy generated at a succession of positions along the spread line, as opposed to "conventional" shooting in which single or simultaneous shots (or other energy sources) are recorded by a number of geophone groups.

transverse filter: A normal convolution as opposed to a recursive filter. The output is the input convolved with the filter's impulse response.

transversely isotropic: Having elastic properties which are the same in two orthogonal directions but different in the third orthogonal direction.

transverse Mercator projection: See **map projection.**

transverse resistance: Resistivity times thickness.

transverse wave: Shear wave or **S-wave.**

trap: 1. A waveguide phenomena; see **channel wave.** 2. A configuration of rocks which is able to confine fluids (such as oil) which float on other fluids (water). A closed structure in porous formations may be a trap if it has an impermeable cap; an unclosed structure may also be a trap if permeability variations block off the escape route of fluids. Compare **closure.**

traveltime: The time between time break and the recording of a seismic event.

traverse: 1. A survey line or a series of connected survey lines. 2. A sequence of connected profiles, as a seismic line. 3. A series of measured distances at measured angles; see Figure 135 **(plane table)**. Compare **triangulation** and **trilateration.**

trend: The systematic smooth component of a function.

trend analysis: 1. The fitting of an analytic surface (the *regional*) to data points as a representation of the "order" in the data as opposed to the erratic element (the *residual*). The objective may be to analyze trends in the data or to interpolate between data points. The underlying assumption is that the data may be decomposed into a relatively low-order, smooth surface plus more-or-less random noise. The number of independent parameters in the analytic surface should be appreciable less than (usually less than 10 percent of) the number of data points. Also called *surface fitting.* **2.** The two-dimensional Fourier analysis of a surface into spatial frequency components (or wavelengths) so that one can determine preferred orientations and wavelengths.

triad: Three stations (one master and two slaves) which constitute a **positioning** system, such as loran, raydist, etc. A master and three slaves constitute a *star.*

triangulation: Establishing locations by a system of overlapping triangles where the angles are directly measured but only a few of the sides are directly measured. Sometimes called "method of intersection." *First-order triangulation* has an accuracy from one part in 25,000 to one part in 100,000; *second-order,* an accuracy of one part in 10,000; *third-order* one in 5000. Triangulation with a plane table is shown in Figure 135 **(plane table)**. Compare **traverse** and **trilateration.**

trigonal: A grid of equilateral triangles (or regular hexagons) produced by three series of equally spaced lines positioned at 60 degrees to each other, used in making **grid-residual** maps. Figure 184.

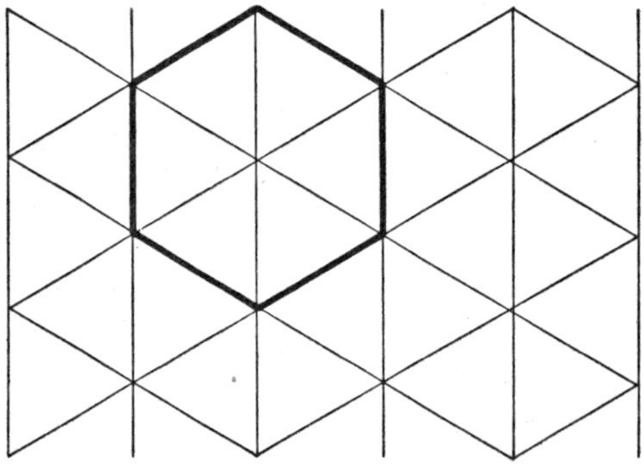

Fig. 184. **Triogonal grid** of equilateral triangles (or interlocking hexagons).

trilateration: Establishing locations by a system of overlapping triangles where all sides are directly measured. Compare **triangulation** and **traverse.**

trim: The longitudinal axis of a ship not being horizontal on the average, the bow being raised or depressed. Compare **pitch** and **list.**

triplets: A method of determining the elevation correction factor **(ecf)** for gravity data from sets of three readings each. It is assumed that elevation does not correlate with geologic structure. If the difference between the height at a station and the weighted mean of the heights of stations on either side of it is h_i and the difference between the observed gravity reading at the station and the weighted mean at the neighboring stations is g_i (where the weighting is usually taken as inversely proportional to the distance), then the elevation correction factor k is given by:

$$k = -\left(\sum h_i g_i\right) / \sum h_i^2 ,$$

and the probable error ϵ in k is

$$\epsilon = 0.67\left[\left(\sum g_i^2 / \sum h_i^2 - k^2\right) / n\right]^{1/2}.$$

See Siegert, 1942, Determination of the Bouguer Correction Constant: Geophysics, v. 7, p. 29-34.

trisponder: A type of **tellurometer.**

tropospheric correction: A correction to radio-wave propagation for velocity and refraction variations because of meteorological conditions, principally because of variations in atmospheric moisture.

tropospheric scatter: The bending of radio-waves in the atmosphere by scattering instead of by refraction. Responsible for extended-range **shoran** (beyond line of sight).

troubleshoot: To look for the cause of a malfunction.

trough: The lowest part of a waveform between successive peaks.

true: The effect if the medium were homogeneous, as with "true resistivity" or "true IP effect." See **apparent.**

trumpet log: A **microlaterolog** in which the guard electrodes are concentric about the current electrode so that the current flow is concentrated in a tube which gradually flares out.

truncation error: 1. The error resulting from using only a finite number of terms of a series; specifically, the error produced by using only a limited operator length in a convolution, or a finite gate length in a correlation. See also **window.** **2.** The effect of digitizing an analog signal whose corresponding digital value exceeds the maximum value permitted. See **clipping.** Loss of bits of information in digitized data because of truncation of high-order bits or low-order bits also causes errors. **3.** In calculating the total mass from a gravity anomaly, the error resulting from the integration not being carried out to the limit of the anomaly.

T-spread: A seismic spread in which the shotpoint is offset perpendicular to the center of the spread by an appreciable distance. Also called **broadside.** See Figure 171 **(spread).**

TSS: **Time-sharing** system.

tuned voltmeter: A voltmeter containing a band-pass filter.

tuning fork: A U-shaped bar of hard steel, fused quartz, or other elastic material that vibrates at a definite natural frequency when set in motion. Used as a frequency standard.

Turam method: Electromagnetic-survey method employing an energizing source consisting of a long insulated cable grounded at both ends or a large horizontal loop. The cable is often several kilometers long and energized at 100 to 800 hz. Measurements are made of the complex ratio (in-phase to quadrature component) of the voltages induced in two receiving coils about 100 ft apart. Usually the plane of the two loops is horizontal. Many profiles may be made using the same source location.

turkey shoot: A direct comparison of the results of recording with two or more sets of seismic instruments simultaneously under the same field conditions.

turn around: The period of time between submission of data for computer processing and obtaining the results back.

turning point: A point on which a foresight is taken from one instrument station in a line of survey and on which a backsight is taken from the next instrument station; the location of the survey rod. Abbreviated TP.

TV: **Time variant.**

TVG: Time varying gain. See **time variant.**

T-wave: Tertiary wave; seismic waves from earthquakes characterized by travel within the oceans as ordinary sound waves, which are then converted to P-, S-, or surface waves for travel on the continents.

two array: Pole-pole **array.**

two-dimensional: Having no variation in the direction perpendicular to the plane (usually vertical) which includes the line of measurement, as with a plane perpendicular to the axis of an infinitely long feature. "Infinitely long" means so long that the effects of the ends are negligible.

two-dimensional plot: A contour plot of **depth-probe** or sounding data (delay time, apparent resistivity, metal factor, etc.) as a function of position along a line (often plotted below midpoints) and electrode separation or offset. See **pseudosection.**

two's complement: The radix complement form for representing negative binary numbers. It can be found by replacing all ones by zeros and all zeros by ones and then adding one. For example, the decimal number 27 is represented as 0011011 and −27 by 11001 01. When a number is added to its negative all registers are empty. There is only one representation for zero: all zeros. Compare **one's complement.**

T-X curve: **Time-distance curve.**

type curves: Master curves; the computed response of various models. IP, resistivity, and electromagnetic response curves plotted against electrode interval, source-receiver spacing, or frequency are used for interpreting field data where the layering conditions of the models appear to hold. Data are usually compared by using transparent overlays. Abscissa and ordinate are usually normalized so as to be dimensionless. Type-curve derivations often employ the method of images. Models are sometimes given "type" names as in Figure 180 (**three-layer resistivity models**).

ultra-long spaced electric log: A modified long **normal** borehole log mounted on a 5000 ft bridle. Abbreviated ULSEL. The *A* to *M* spacings are 75, 150, 600, and 1000 ft. Differences between the measured resistivities and anticipated resistivities calculated from conventional resistivity logs indicate nearby resistivity anomalies. Used to define the distance to a salt-dome flank.

uncertainty: The precision with which a measurement or value is known. Often implies a 50-50 chance that any one of a series of measurements would fall within a given range. Does not necessarily imply anything about **accuracy,** which is comparison with the true value. See **probable error** and **standard deviation.**

unconformity: A surface of erosion or nondeposition that separates younger strata from older rocks. An unconformity ordinarily is not recognizable in seismic data except where the layers above and below the unconformity are not parallel (angular unconformity).

unijunction transistor: A **transistor** made of *n*-type semiconductive material with a *p*-type alloy region on one side; connections are made to base contacts at either end of the *n*-type material, and also to the *p* region. Primarily used in timing circuits.

union: Disjunction.

unit circle: A circle of unit radius. See **z-plane.**

Universal Transverse Mercator: A standard rectangular map grid. The projection is onto a cylinder tangent to the earth along a central meridian (i.e., with the cylinder axis perpendicular to the earth's axis). See Figure 116 **(map projections)** and **Mercator Projection.** The earth is divided into 60 north-south columns, each 6 degrees of longitude wide. The central meridian is assigned the value of 500,000 meters easting and the equator a northing of 0. Abbreviated UTM.

universe: The complete collection of objects, numbers, functions, etc., with probabilities attached to relevant subcollections; *population.*

update: To correct a positioning system for deviations or drift which has occurred. Satellites are updated by having new position information injected each day. Doppler sonar positions are updated with satellite-fix information.

uphole geophone: A geophone placed a few feet from a shothole to detect **uphole time.**

uphole seis: Uphole geophone.

uphole shooting: The successive detonation of a series of charges at varying depths in a shothole in order to determine the velocities of the near-surface formation, the weathering thickness, and (sometimes) the variations of record quality with shot depth. Used to establish **weathering** thickness.

uphole stack: The combining of seismic records from shots at different shot depth, after time shifting them using uphole-time information so that reflection energy is in register. A method of attenuating ghost energy. Sometimes called (erroneously) vertical stack, which is a different process.

uphole survey: Uphole shooting.

uphole time: Time for the first wave from an explosion to reach the surface at or near the shotpoint. Used in determining weathering or near-surface corrections to seismic data. See Figure 149 **(record).**

upstairs: In the numerator of an equation.

upsweep: Vibroseis sweep in which the frequency increases with time.

upward continuation: Calculation of the potential field at an elevation higher than that at which the field is known. The **continuation** involves the application of Green's theorem and is rigorous if the field is completely known over the lower surface and if no sources are present between the surfaces (which is usually true for gravity and magnetic fields). Upward continuation is used to smooth out near-surface effects and to tie aeromagnetic surveys flown at different heights.

UTM: Universal Transverse Mercator projection.

VA: Variable-area recording. Also abbreviated VAR.

vacuum-tube voltmeter: A voltage-measuring device using electronic circuits to obtain high impedance across the measurement probe so that very little current is drawn. The vacuum tube itself now is often superceded by solid-state circuitry. Abbreviated VTVM.

Vaporchoc: A marine seismic energy source in which a quantity of superheated steam under high pressure is injected into the water. Subsequent condensation of the steam attenuates bubble oscillation. Also called *steam gun.* Trade name of Compagnie Général de Géophysique.

vara: An old Spanish unit of length; about 33 inches.

variable-area: A display method in which the width of a blacked-in area is roughly proportional to the signal strength. See Figure 56 **(display).**

variable-density: A display method wherein the photographic density is proportional to signal amplitude. Abbreviated VD. See Figure 56 **(display).**

variable-density log: A *microseismogram* log, **three-D log,** or *VDL* log.

variance: The variance σ^2 between a set of data X_i and the mean $\bar{X}$ is

$$\sigma^2 = \frac{1}{n}\sum_i (X_i - \bar{X})^2.$$

The variance is the square of the *standard deviation* σ. See **statistical measures.**

variometer: An instrument for measuring small magnetic variations used in magnetotelluric work. Consists essentially of a magnet suspended on a torsion fiber; slight rotations of the magnet deflect a light beam reflected by a small mirror attached to the magnet and the position of the beam is recorded on photographic paper. Used in magnetic observatories to observe diurnal changes in the magnetic field.

VAX: Variable-area record section.

V-band: Radar frequencies between 46 and 56 Ghz; See Figure 68 **(electromagnetic spectrum).**

V-bar: Average velocity: See **velocity.**

VD: Variable-density record section.

VDL log: A **three-D log.** Schlumberger trademark.

vectogram method: A scheme in telluric surveying in which *x-y* recorders rather than strip-chart recorders are used. The technique yields more accurate results and is faster than the triangle or ellipse methods. Exact time ties are not required to establish simultaneity of observations of electrical vectors at a station and at the base station. See Yungul, 1968, Vectogram Method in Telluric Surveying: Geophysics, v. 33, p. 127-131.

vector: 1. A quantity having both a magnitude and a sense of direction. Vectors are usually represented by arrows pointing in the appropriate direction whose length is proportional to the vector's magnitude. Vector quantities may be added by the vector-parallelogram method whereby the tail of the one vector is placed at the head of the other vector; the sum vector then goes from the tail of the second to the head of the first. The negative of a vector is represented by the arrow pointing in the opposite direction; subtraction is performed by adding the negative vector. Vectors are often resolved into component vectors in orthogonal directions, the directions being indicated by the unit vectors **i, j,** and **k.** The *dot product* (or "inner product") of two vectors **A** and **B** is not a vector but a scalar of magnitude **A · B = |A| |B| cos (A,B),** where the cosine is of the angle between their directions. The *cross product* (or "outer product") is a vector perpendicular to both **A** and **B** and of magnitude **A × B = k|A| |B|** sin **(A,B),** where **k** is a unit vector perpendicular to the plane of **A** and **B.** For vectors expressed in orthogonal components, these products are:

$$\text{If } \mathbf{A} = a_1\mathbf{i} + a_2\mathbf{j} + a_3\mathbf{k},$$

$$a_1 = |\mathbf{A}| \cos(\mathbf{A},\mathbf{i}),$$

$$\mathbf{B} = b_1\mathbf{i} + b_2\mathbf{j} + b_3\mathbf{k},$$

$$\mathbf{A} \cdot \mathbf{B} = a_1b_1 + a_2b_2 + a_3b_3, \text{ and}$$

$$\mathbf{A} \times \mathbf{B} = (a_2b_3 - a_3b_2)\mathbf{i} + (a_3b_1 - a_1b_3)\mathbf{j} + (a_1b_2 - a_2b_1)\mathbf{k}.$$

With "vector fields" (i.e., where a vector is associated with every point in space) we define the divergence and curl using the operator del = ∇:

$$\nabla = \mathbf{i}\frac{\partial}{\partial x} + \mathbf{j}\frac{\partial}{\partial y} + \mathbf{k}\frac{\partial}{\partial z};$$

$$\text{div } \mathbf{V} = \nabla \cdot \mathbf{V} = \frac{\partial V_1}{\partial x} + \frac{\partial V_2}{\partial y} + \frac{\partial V_3}{\partial z} \text{ and}$$

$$\text{curl } \mathbf{V} = \nabla \times \mathbf{V} = \mathbf{i}\left(\frac{\partial V_3}{\partial y} - \frac{\partial V_2}{\partial z}\right) + \mathbf{j}\left(\frac{\partial V_1}{\partial z} - \frac{\partial V_3}{\partial x}\right) + \mathbf{k}\left(\frac{\partial V_2}{\partial x} - \frac{\partial V_1}{\partial y}\right).$$

For *rotating vectors,* see **complex notation. 2.** A sequence of values, such as might represent values of a wavelet at successive discrete time intervals. Thus a wavelet which has the values 0.75, 0.25, −0.50 at three successive time intervals and the value zero thereafter might be considered as a 3-term row vector, (0.75, 0.25, −0.50). See **matrix.**

Vela Uniform: A project under the Nuclear Test Detection Office to devise methods of detecting nuclear explosions by the characteristics of the seismic waves which they generate.

velocimeter: 1. An instrument for measuring the velocity of sound in water, used to correct doppler-sonar data for salinity and temperature variations. An acoustic pulse is transmitted over a fixed distance between transducers in the instrument, amplified, and used to generate the next pulse to be transmitted. Thus, the regeneration frequency depends on the traveltime and, hence, on the velocity of the acoustic wave in the water. This technique is sometimes called *sing around.* See Figure 191 **(water velocity). 2.** A device which measures fluid flow; a flowmeter.

velocity: 1. A vector quantity which indicates time rate of change of displacement. 2. Usually refers to the propagation rate of a seismic wave without implying any direction. Velocity is a property of the medium and is not a vector quantity when used in this sense; speed. 3. Seismic velocity is measured (or inferred) from **sonic logs,** normal moveout (see **velocity analysis), well shooting,** and refraction **time-distance curves.** 4. Confusion sometimes arises between instantaneous and apparent velocity. *Instantaneous velocity* refers to the speed of a wavefront in the direction of the energy propagation (perpendicular to the wavefront for isotropic media), whereas *apparent velocity* refers to the apparent speed of a given phase in a particular direction; thus apparent velocity is greater than instantaneous velocity by the secant of the angle between the direction of wave travel and the direction in which the apparent velocity is measured. Apparent velocity is usually measured along the spread direction. 5. *Average velocity* is the ratio of a given depth divided by the seismic traveltime to that depth, often (but not always) assuming straight-raypath travel. If the section is made up of parallel flat layers of velocity v_i and thickness z_i so that the traveltime across each layer is $t_i = z_i/v_i$, then the average velocity is $\sum (v_i t_i)/ \sum t_i$ or $\sum z_i/ \sum (z_i/v_i)$. *Rms velocity* refers to determinations in a horizontally layered section based on normal-moveout measurements using X^2T^2 analysis; velocities so derived are given closely by the relation $[\sum (v_i^2 t_i)/ \sum t_i]^{1/2}$. Rms velocities are typically a few percent larger than corresponding average velocities. These formulas can be modified by the cosine of the dip where the velocity layering is parallel but not horizontal; however, where velocity layering is not parallel, average velocity and rms velocity lose meaning. 6. *Interval velocity* often is calculated from the velocities derived by normal-moveout analysis for separate primary events by the method given by Dix, 1955, Seismic Velocities from Surface Measurements: Geophysics, v. 20, p. 68-86:

$$V_i = \left[\frac{V_n^2 t_n - V_{n-1}^2 t_{n-1}}{t_n - t_{n-1}}\right]^{1/2},$$

where v_n is the rms velocity and t_n is the zero offset arrival time corresponding to the *n*th reflection. Provided the reflectors are parallel, V_i represents the mean velocity between the reflectors which give rise to the events *n*-1 and *n*. If the reflectors are not parallel the calculation is meaningless. 7. *Interval velocity* also is calculated from sonic logs (where it is the average velocity over the log span) and from well surveys (where it is calculated for the intervals between measurements).

velocity analysis: Calculation of **velocity** from measurements of normal move-out. In current usage, generally involves CDP data but includes also **T-ΔT analysis** and **X²-T² analysis.** Most of the analysis schemes assume a normal moveout, measure the coherency at that normal moveout, and vary the normal moveout in order to maximize the coherency. Various procedures are used (sometimes involving crosscorrelations), various methods of measuring the coherency are employed, and various amounts of data (both from a range of locations and from a range of arrival times) are analyzed. Figure 185.

velocity filter: Discrimination on the basis of apparent velocity (or dip moveout). Also called **fan filter.** Coherent arrivals with certain moveouts are attenuated. See also **pie slice** and **LaserScan.** Velocity filtering is different from discrimination based on normal moveout (which implies different average velocity).

velocity focusing: The bending of seismic rays at curved velocity interfaces which act like lenses in optics, focusing or defocusing wavefronts and distorting structure and velocities calculated from normal-moveout measurements. The distortion becomes greater as the curvature of the velocity lens increases and as the depth of the feature being studied beneath the velocity lens increases.

velocity function: Seismic velocity expressed as a mathematical function of depth or time. The most common forms of velocity function are (a) linear with depth and (b) linear with arrival time. The use of any functional form is an approximation because lithology and the other factors involved in velocity do not vary systematically and smoothly. The extrapolation of velocity functions is especially dangerous and may imply impossible velocities. The linear-with-depth velocity function ($V_i = V_o + KZ$) leads to a simple method for migrating data because then both raypaths and wavefronts are arcs of circles; see Figure 186, and Dix, 1952, Seismic Prospecting for Oil: New York, Harper Bros., p. 141-159; also Kaufman, 1953, Velocity Functions in Seismic Prospecting: Geophysics, v. 18, p. 289-297.

velocity inversion: A decrease in velocity with depth. Such a situation can lead to **channel waves** and **hidden layers.**

velocity log: Sonic log.

velocity profile: An arrangement designed to record reflections over a large range of offset distance so that velocity can be determined from the time-distance data for reflection events. Also called **X²-T².** See **expanding spread.**

velocity spectrum: A display of the coherency when various normal moveouts (hence, implying various velocities) are assumed. See Figure 187 and **velocity analysis.**

velocity survey: 1. A series of measurements to determine average velocity as a function of depth, as in **well shooting.** 2. May also refer to running a **sonic log.** 3. Sometimes refers to surface velocity shooting (X² – T² or T – ΔT).

Venn diagram: An illustration of relationships used in logic and in **Boolean algebra;** see Figure 88 **(gate).** If the conditions where *A* is true are denoted by one closed outline and the conditions for *B* by another, then the conditions where both are true, where either is true, where neither is true, and other situations are illustrated by cross hatching.

vernal equinox: The point on the celestial sphere occupied by the sun at the time of the vernal equinox (about

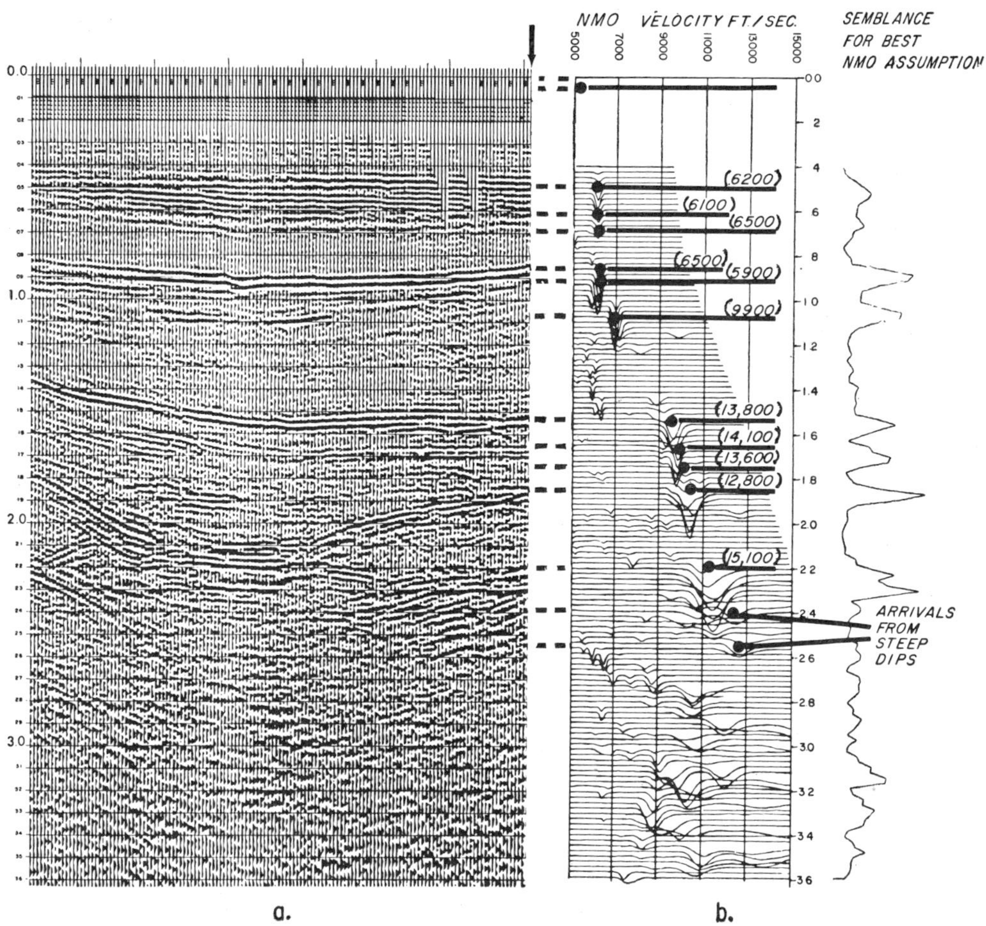

Fig. 185. **Velocity analysis.** (a) Seismic record section. (b) Velocity analysis of data at the right side of the seismic section. The graph shows semblance as a function of the velocity which normal moveout implies. Interval velocities calculated from Dix's formula are shown in parentheses. Velocity picks have been related to specific seismic events. (Courtesy Seiscom.)

March 21), which is the reference point from which right ascension and celestial longitude are calculated. Also called the *first point of Aries* and symbolized by ♈. One of the two points of intersection of the ecliptic and the celestial equator.

vernier: An auxiliary scale used in conjunction with the main scale of a measuring device to obtain a more precise reading. Figure 188.

versine: Versine $\alpha = 1 - \cos \alpha$.

vertical angle: The angle between a direction and the horizontal plane; *altitude.*

vertical electric sounding: See **sounding.** Abbreviated VES.

vertical exaggeration: 1. The use of a vertical scale which is larger than the horizontal scale. Exaggeration makes subtle effects more evident but distorts structural relationships. Seismic time sections involve variable vertical exaggeration because the velocity varies with depth. The picking and interpretation of significant features on record sections is greatly affected by vertical exaggeration. The migration of data is another factor involved in the direct interpretation of record sections. Figure 189. **2.** The ratio of vertical scale to horizontal scale.

vertical intensity: The component of the total-intensity field in the vertical direction.

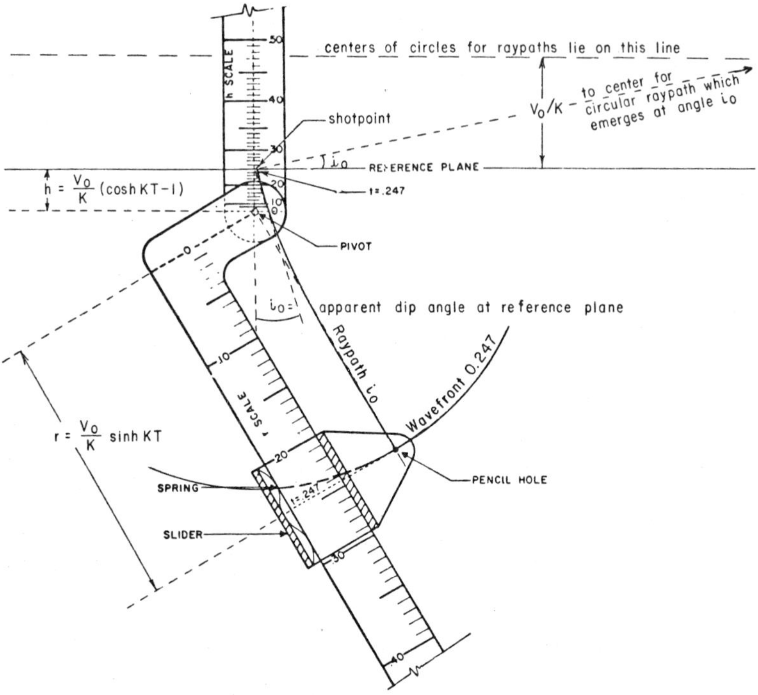

Fig. 186. **Velocity function** relations for linear increase with depth, $V = V_o + KZ$. The plotting machine for this function has the h and r scales calibrated in arrival time *T*. The scales are thus specific for a particular V_o and *K*. (From Musgrave, ed., 1967, Seismic Refraction Prospecting, Tulsa, S.E.G., p.381.)

vertical-loop dip-angle method: An electromagnetic-prospecting method in which the transmitter coil or loop is vertical and the receiver coil is in the plane of the transmitter coil. The null or minimum-coupling orientation is observed for the receiver coil. This orientation will be horizontal except near a conducting body. The angle between the plane of the coil and horizontal *(dip angle)* corresponds to the angle between the major axes of the **ellipse of polarization** and the horizontal.

vertical profile: See **sounding.**

vertical section: **1.** A plot of seismic events directly beneath the point midway between the source and detector locations. Such a section does not represent structural relationships correctly except where reflectors are flat. The vertical scale is usually either **vertical time** or depth (obtained by multiplying vertical time by average velocity). **2.** A plot of projections of seismic data showing where interfaces attributed to seismic events would intersect a vertical plane. Such a section is a correct structural section to the extent the picking, plotting, velocity, and projecting are correct.

vertical stack: **1.** Mixing together of the records of several shots made in nearly the same location without correcting for offset differences. Used especially with surface sources in which the records from several successive weight drops, vibrations, pops, etc., are added together without making static or dynamic corrections to the components before adding. **2.** Sometimes used (incorrectly) to mean **uphole stack.**

vertical time: **1. Uphole time. 2.** The arrival time which would be observed for a given reflection if the travel path had been vertical. The vertical time differs from the observed arrival time if the reflector is dipping so that the reflecting point is not directly under the observation point, if the travel path is longer because geophone and shotpoint are not coincident, or if velocity variations distort raypaths.

Vibroseis: A seismic method in which a vibrator is used as an energy source to generate a wave train of controlled frequencies. A sinusoidal vibration of continuously varying frequency (Figure 190) is applied during a **sweep** period, typically lasting seven seconds or longer. In *upsweeping,* the frequency begins low and

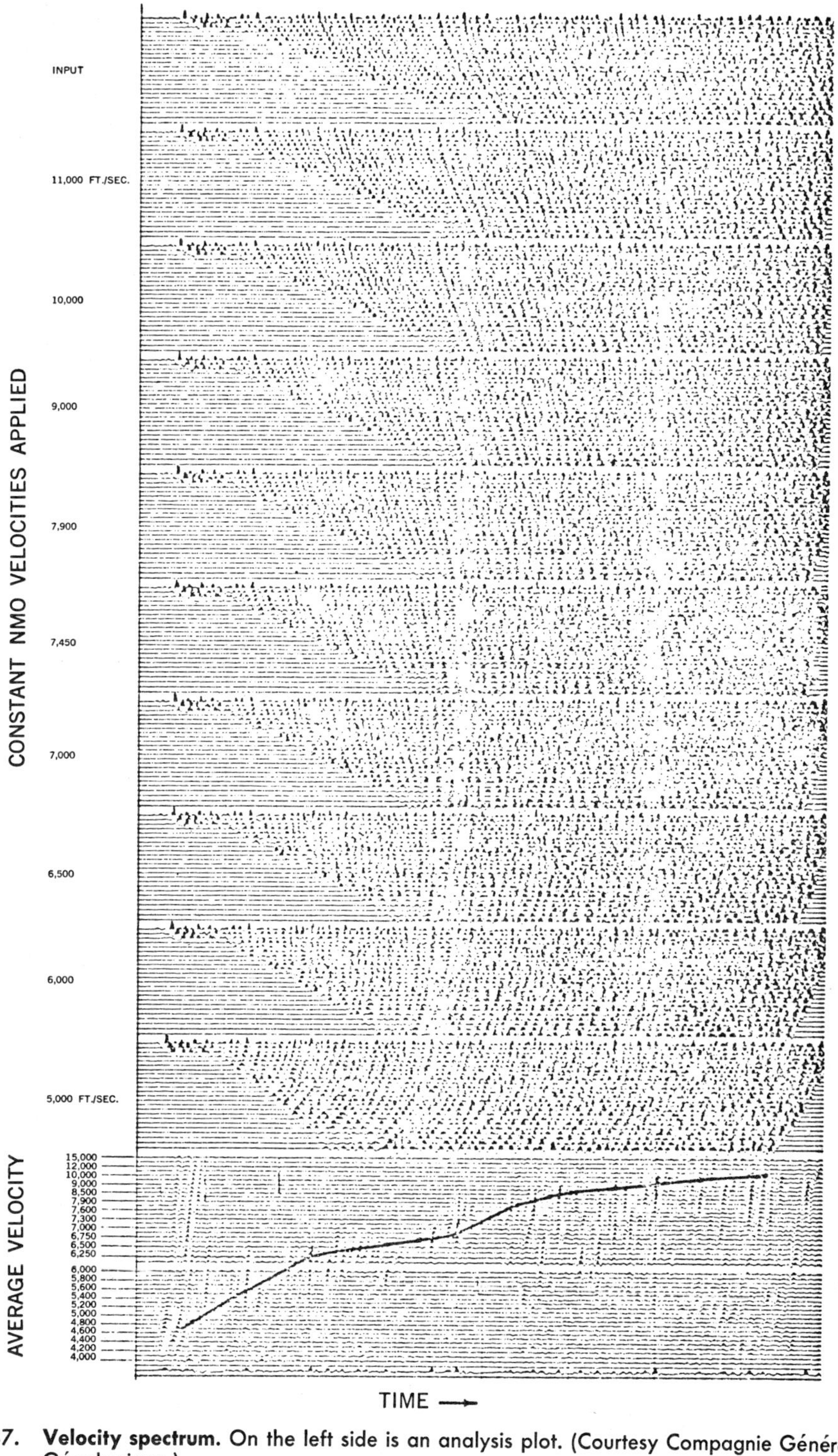

Fig. 187. **Velocity spectrum.** On the left side is an analysis plot. (Courtesy Compagnie Générale de Géophysique.)

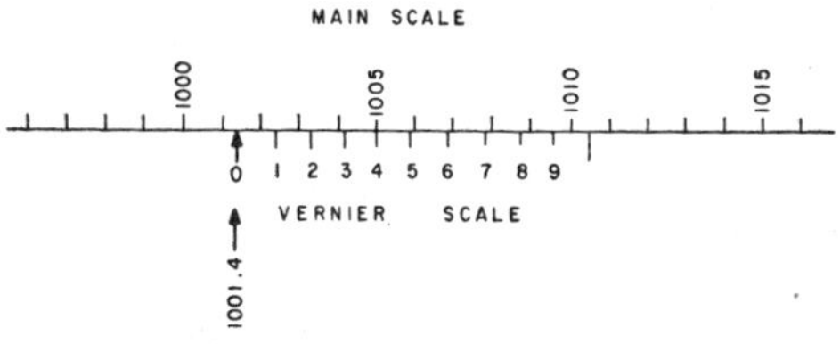

Fig. 188. **Vernier** principle. A Vernier scale to read to $1/n$ contains n subdivisions in the space of $(n-1)$ subdivisions on the main scale. The tenths figure is given by the mark on the Vernier scale which lines up with a mark on the main scale.

increases with time, and in *downsweeping* the highest frequencies occur first. The field record which consists of the superposition of many such long wave trains is correlated with the sweep frequency. The correlated record resembles the conventional seismic record which results from an impulsive source. Continental Oil Company trademark.

viscoelastic: Having a stress-strain relationship which includes terms proportional to both the strain and the rate of strain. Leads to attenuation of seismic waves dependent on the square of the frequency. A viscoelastic solid is also call a **Voigt solid.**

viscometer: A device for measuring **viscosity.**

viscosity: Resistance of a fluid to flow; (stress)/(rate of shear).

viscous magnetization: Remanent magnetization produced by a weak magnetic field over a long period of time. It is generally proportional to the logarithm of the time and parallel to the weak applied field. Viscous magnetization has its origin in the thermal energy which consists, in part, of energy levels high enough to realign the directions of magnetization of even those domains with rather high energy barriers. The weak field of the earth acts to bias the direction of these jumps.

VLF: Very-low frequency. **1.** Radio transmission at frequencies of 3 to 30 khz, used for communication with submerged submarines and long-range radio positioning. **2.** Some electromagnetic prospecting methods use such transmissions as energy sources. **3.** A radiolocation system such as Omega which uses very-low frequency energy.

voice grade: A channel with a frequency range from approximately 300 to 3000 hz.

Voigt solid: A solid in which stress is proportional to both strain and time rate of strain. See **viscoelastic** and compare **elastic solid.**

Voigt waves: Body waves (P-waves) in a Voigt or **viscoelastic solid.**

voltmeter: An electrical instrument used to measure the potential differences between points in a circuit. A voltmeter may respond to average, root-mean-square, or peak-voltage values.

volume magnetization: Magnetic moment per unit volume.

volume reverberation: Water-track mode.

vortex shedding: A mode of fluid flow involved in pulling a cable through the water. See **fairing.**

vote: See **majority vote.**

VTVM: Vacuum-tube voltmeter.

V_2: Subweathering velocity, often determined from first-break refraction arrivals.

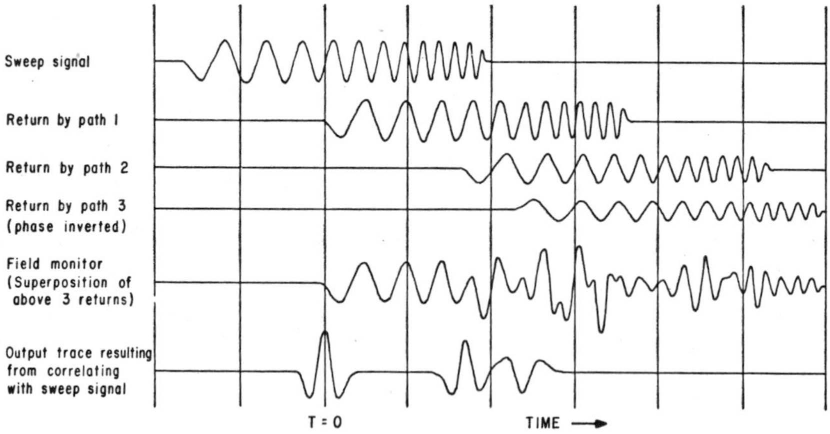

Fig. 190. **Vibroseis** record. (Courtesy Continental Oil Company.)

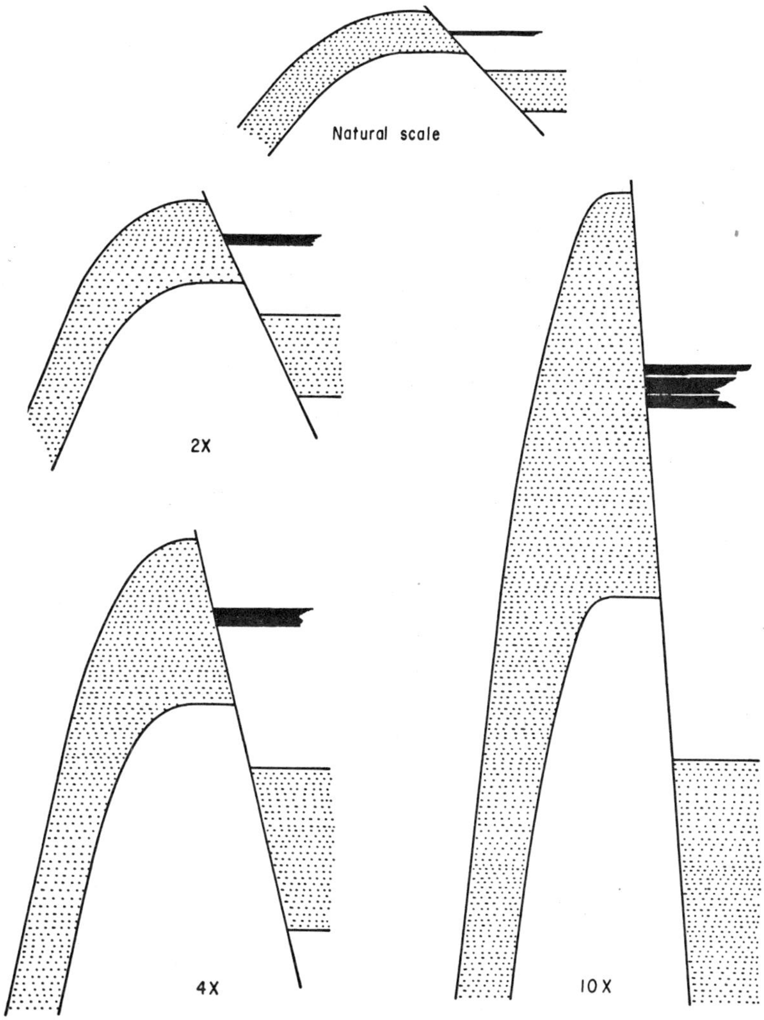

Fig. 189. **Vertical exaggeration** Exaggeration allows one to see both vertical detail and horizontal context but severely distorts bed thickness, structural relationship, fault dip, etc.

walkaway: **Noise test.**

wall resistivity log: A **microresistivity log.**

Warburg impedance: Impedance involved in current transfer at an electrode by a **faradaic path;** it is a measure of the rate of the faradaic ion-diffusion process. Varies inversely as the square root of the frequency.

Warburg region: The steep part of the **resistivity-spectrum** curve near the inflection point where the electrode impedance of a rock is dominated by **faradaic-path** conduction.

washing: Demagnetizing (degaussing) a rock or other material, especially by gradually increasing the demagnetization to remove successively harder portions of the remanent magnetization.

WASSP: **Exploding wire,** a marine energy source. Teledyne trademark.

waterbreak: The arrival of energy which travels in the water directly from the source unit (or shot) to the waterbreak detector. Used to determine the location of the waterbreak detectors in a seismic streamer with respect to the source location, and, hence, the location of the streamer. Filters in the waterbreak circuitry pass components between approximately 500 and 5000 hz, thus avoiding possible confusion with shallow refractions of lower frequency and with background sonic energy of higher frequency.

water cut: The volume fraction of water produced from a well.

water injection: A method of drilling which uses air as the principal fluid for removal of cuttings but with enough water added to make the hole wall firm enough to prevent excessive caving. Used when drilling dry sand.

water saturation: Fraction of the pore volume filled with formation water. Symbol: S_w. See **Archie's formulas.**

water track: An intermittent mode of reflected sonar energy produced by scattering by water particles within the first tens of feet below the transducer. When using **doppler sonar** navigation in water deeper than 400 to 1000 ft, the scattered energy dominates the reflection from a deeper ocean bottom. Accuracy drops by about a factor of four compared with "bottom mode." Water-track positioning also is done with respect to the water which may itself be moving, introducing additional systematic error.

water velocity: The speed of sonic energy (P-waves) in water, which depends mainly on the temperature and salinity:

$$V = 1449 + 4.6T - 0.055T^2 + 0.0003T^3 + (1.39 - 0.012T)(S - 35) + 0.017Z \ ,$$

where V is the velocity in meters per second, T is the temperature in degrees centigrade, S is the salinity in parts per thousand, and Z is the depth below the surface in meters. See Figure 191. Water velocity is involved in marine seismic work, sonar ranging, doppler-sonar, etc. See **velocimeter.**

wave: a. A disturbance which is propagated on the surface of or through the body of a medium, without involving net movement of the material. Waves are often characterized by periodicity (Figure 192). The *wave amplitude* is usually defined as the maximum displacement from the equilibrium or null position. The *rms amplitude* is the square root of the mean of the squares of the displacements which is $\sqrt{2}/2$ times the peak amplitude for sinusoidal waves. The *wave peak* or *crest* is the point at which the displacement is greater (in the positive direction) than at adjacent points. The *wave trough* is the point which is displaced farther than adjacent points in the negative sense. The *wave height* is the difference in displacement between successive peaks and troughs. The *wavelength* (λ) is the distance perpendicular to the wavefront between successive similar points on the wave train. The *wave number* (K) is the reciprocal of the wavelength (sometimes defined as 2π/wavelength). **b.** Seismic waves are sometimes classified as *body waves,* where the wave propagates through the body of the medium, and *surface* or *interface* waves, where the wave propagates along a boundary. Body waves may be either **P-waves** (also called *compressional* or *longitudinal waves)* or **S-waves** (also called *shear* or *transverse waves).* Wave energy which has traveled part of its way as a P-wave and part as an S-wave is called a *converted wave.* Surface waves (or interface waves) may travel by several modes, the most common of which are **Rayleigh waves** (R); other surface waves include **Love waves** (Q), **hydrodynamic waves** (H), **coupled waves** (C), and **Stoneley waves.** A surface wave which travels along the surface of a borehole is the *tube wave.* See also **wave notation.**

wave conductor: **Refractor.**

wave equation: In rectangular coordinates x, y, z,

$$\nabla^2 \Psi = \frac{\partial^2 \Psi}{\partial x^2} + \frac{\partial^2 \Psi}{\partial y^2} + \frac{\partial^2 \Psi}{\partial z^2} = \frac{1}{V^2}\frac{\partial^2 \Psi}{\partial t^2} \ ,$$

where ψ represents wave displacement (pressure, rotation, dilitation, etc.) and V the velocity of the wave. Functions $f(x \pm Vt)$ are solutions to this equation. In spherical coordinates where r is the radius, θ the colatitude, and ϕ the longitude, the wave equation becomes:

$$\frac{1}{V^2}\frac{\partial^2 \Psi}{\partial t^2} = \frac{1}{r^2}\left[\frac{\partial}{\partial r}\left(r^2 \frac{\partial \Psi}{\partial r}\right) + \frac{1}{\sin\theta}\frac{\partial}{\partial \theta}\left(\sin\theta \frac{\partial \Psi}{\partial \Phi}\right) + \frac{1}{\sin^2\theta}\left(\frac{\partial^2 \Psi}{\partial \theta^2}\right)\right].$$

waveform: A plot (usually as a function of time) of voltage, current, seismic displacement, etc., which is involved in wave motion.

wavefront: 1. The surface over which the phase of a traveling wave disturbance is the same. The wavefront moves perpendicular to itself as the disturbance travels in an isotropic medium. **2.** The leading edge of a waveform.

wavefront chart: A plot of horizontal distance versus depth on which wavefronts have been drawn emanat-

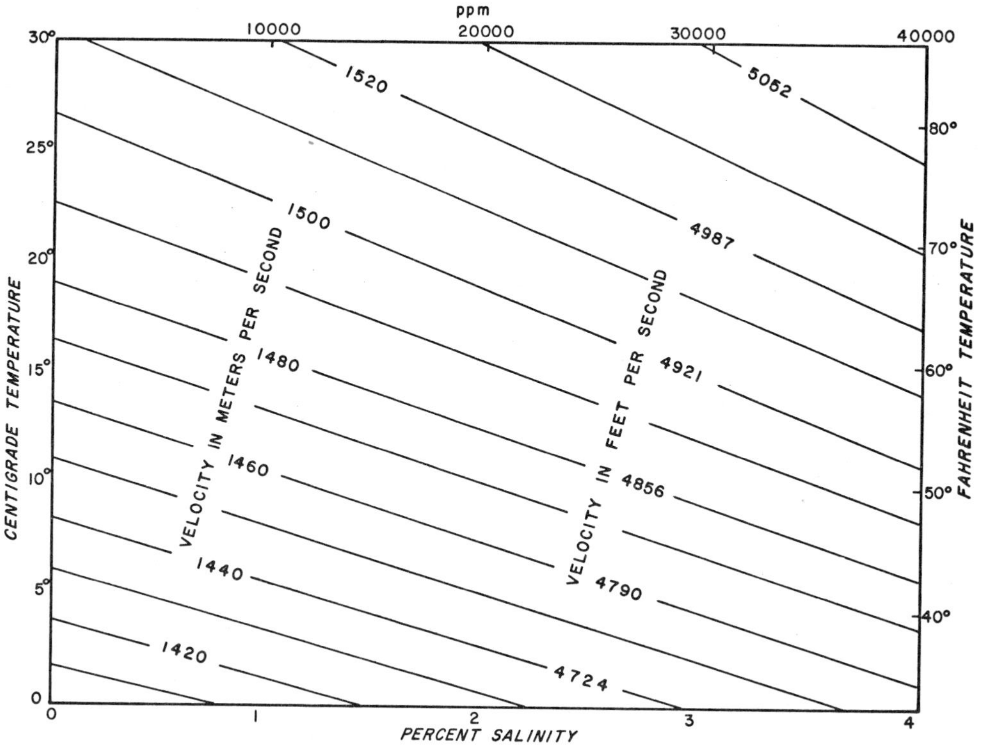

Fig. 191. **Velocity of sound in water.**

ing from a point source. Wavefronts denote the position of a traveling seismic disturbance at successive times and are actually surfaces rather than curves (see Figure 118, **migration**). The shape of the wavefronts depends on the velocity distribution. **Raypaths** corresponding to different moveouts usually are drawn also on such charts; raypaths are perpendicular to wavefronts for isotropic media; see Figure 193a. Asymmetric wavefront charts (Figure 193b) occasionally are drawn where velocity varies laterally but such charts are specific for a particular location because the velocities are necessarily different for an adjacent point.

wavefront healing: Energy diffracts into shadow zones and the shadow effectively disappears when sufficiently removed from the shadow-producing obstacle. This is called "healing" of the wavefronts. Figure 194 shows how the reflection from a reflector containing a hole has nearly completely healed the effect of the hole.

wavefront method: A seismic interpretation method (often graphical) which involves reconstructing the portions of emerging wavefronts from the arrival times at a spread of geophones from a common shot (or the equivalent). A locus of equal traveltime (wavefront) is constructed for the time t by striking circles about each geophone position of radius $(t_i - t)V$, where t_i is the traveltime observed by that geophone and V is the velocity of the upper layer; see Figure 195. The method can be varied to accommodate variation of velocity. Similar wavefronts can be constructed from other shotpoints or from the reversed profile of a refraction interpretation. The solution which locates the reflector or refractor must satisfy the observed arrival times. See Rockwell, 1967, A General Wavefront Method, *in* Seismic Refraction Prospecting, Musgrave, ed., Tulsa, SEG, p. 363-415.

waveguide: A situation which permits **channel waves.**

wave impedance: 1. The ratio of orthogonal components of electric-field to magnetic-field intensity. See **impedance. 2.** The ratio of particle velocity to displacement in a wave.

wavelength: The distance between successive similar points on two adjacent cycles of a monochromatic wave, measured perpendicular to the wavefront. Symbol:

$$\lambda = v/f = 1/k,$$

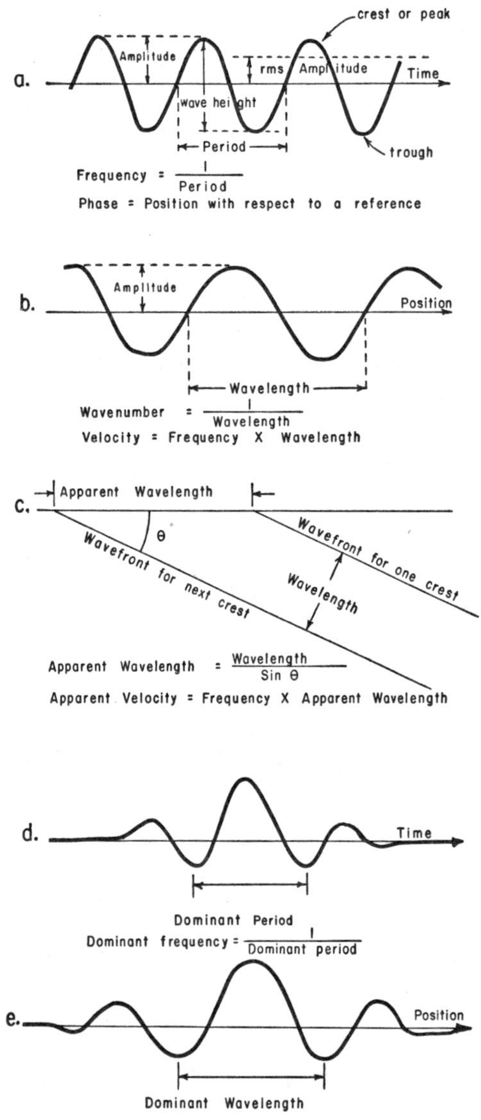

Fig. 192. **Wave** definitions. *For sinusoids*: (a) How displacement at one point varies with time. (b) How wave looks at different places at a given instant. (c) If wavefront approaches at an angle, the apparent wavelength will differ from the true wavelength. *For nonperiodic wave*: (d) Dominant period between principal adjacent troughs (or peaks). (e) Dominant wavelength is similarly measured.

where v = wave velocity, f = frequency, and k = wavenumber. See Figure 192 **(wave)**. *Dominant wavelength* is often used, referring to the dominant frequency component, when dealing with wave trains which contain a spectrum of frequencies.

wavelet: A seismic pulse usually consisting of 1- 1/2 to 2 cycles. A **Ricker wavelet** is a particular type based on the assumption that attenuation is proportional to the square of the frequency.

wave notation: **1.** Arrivals of seismic waves from earthquakes (see **earthquake seismology**) conventionally are identified as to their travel path using letters to designate the nature of the wave over various legs of the path. See Figure 196.

P	=	longitudinal wave outside the core,
S	=	transverse wave,
K	=	longitudinal wave in the core,
I	=	longitudinal wave in the inner core,
G	=	Q = L_Q = Love surface wave,
R	=	L_R = Rayleigh surface wave,
L_g	=	guided wave in the continental crust, which disappears if any of the travel path is through oceanic crust, and
W_2	=	surface wave which has traveled around the earth the long way.

Repetition of P and S indicates reflection at the surface of the earth (as in PP, PS, SS). Reflection at the outer surface of the core is indicated by c PcP, PcS), at the inner surface of the core by KK (e.g., PKKP, PKKS). P′ = PKP. A **Jeffreys-Bullen chart** (Figure 106) shows the normal arrival time for the various types of waves. See also **T-wave, H-wave,** and **C-wave. 2.** Guided electromagnetic waves are classified as TE, TM, or TEM to indicate that the electric or magnetic field or both have transverse components only.

wavenumber: **1.** *Spatial frequency k.* The number of wave cycles per unit distance, the reciprocal of **wavelength** (λ). $k = 1/\lambda$. **2.** Sometimes wavenumber is defined as 2π/wavelength. **3.** See **propagation constant.**

wavenumber filtering: Filtering of certain wavenumbers as performed by spatial sampling and mixing. See Figure 54 **(directivity graph)**.

waveshape kit: A modification of an air gun which bleeds air into the expanding bubble to increase the pressure in the bubble during the collapse, thereby decreasing the sharpness of the collapse to simplify the shape of the seismic waveform which results from firing the airgun.

waveshape stabilization: **Cross-equalization;** shaping frequency spectra so as to make the waveshape for adjacent traces more nearly the same.

wave spreading: See **divergence.**

wave test: **Noise test** or walkaway.

wavetilt: The ratio of the horizontal to the vertical component of the magnetic or electric field. In some cases only the modulus is considered (see **polarization ellipse**). In an electromagnetic method using VLF and higher-frequency radio waves, the apparent resistivity of the earth is calculated from measurements of the quadrature component of the electric wavetilt; for a

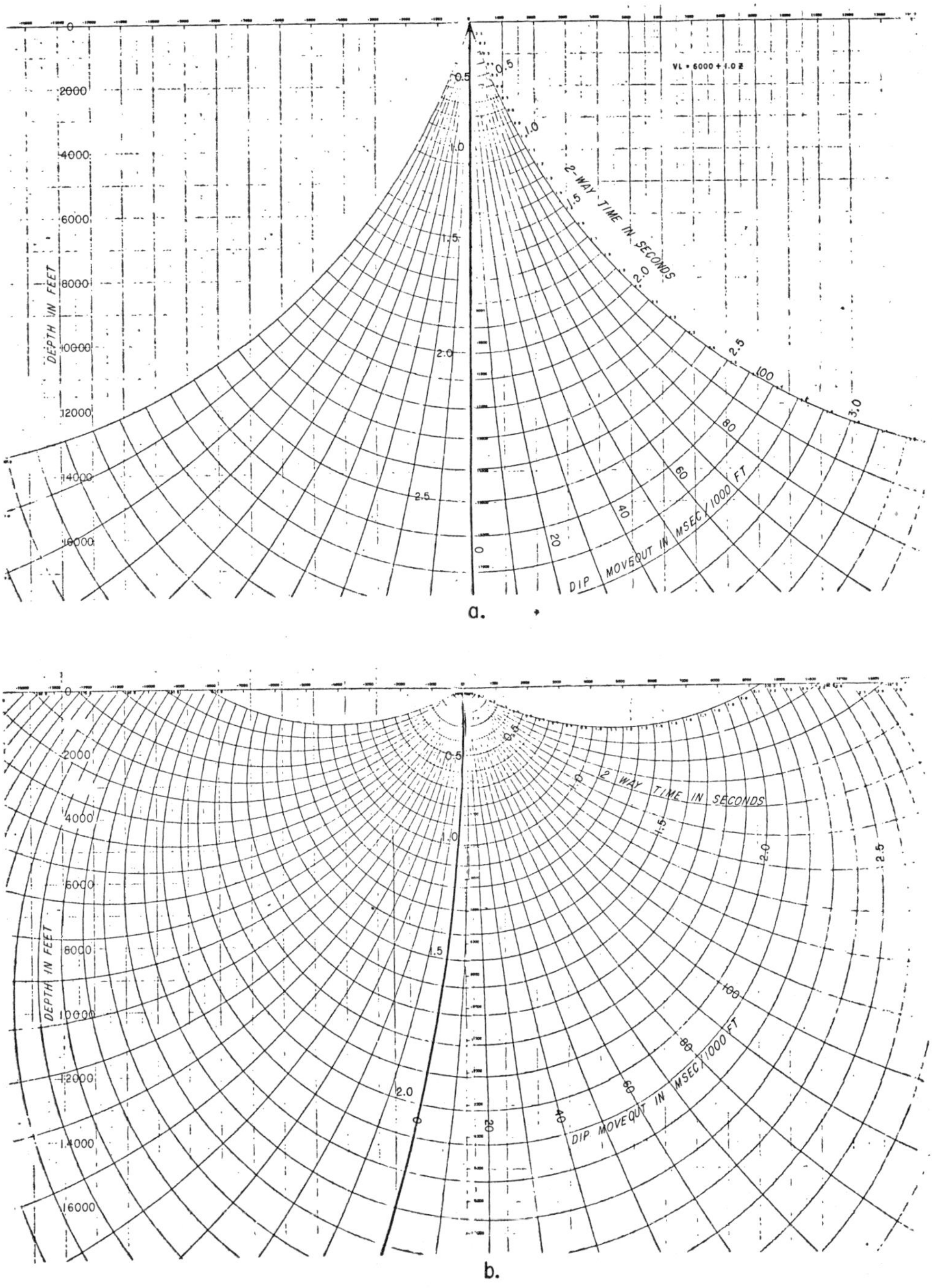

Fig. 193 **Wavefront charts.** (a) Wavefronts for a particular velocity function become farther apart with depth because of the increase of velocity with depth. Raypaths are orthogonal to wavefronts. (b) Wavefronts where velocity increases from right to left as well as with depth. (Courtesy Chevron Oil Company.)

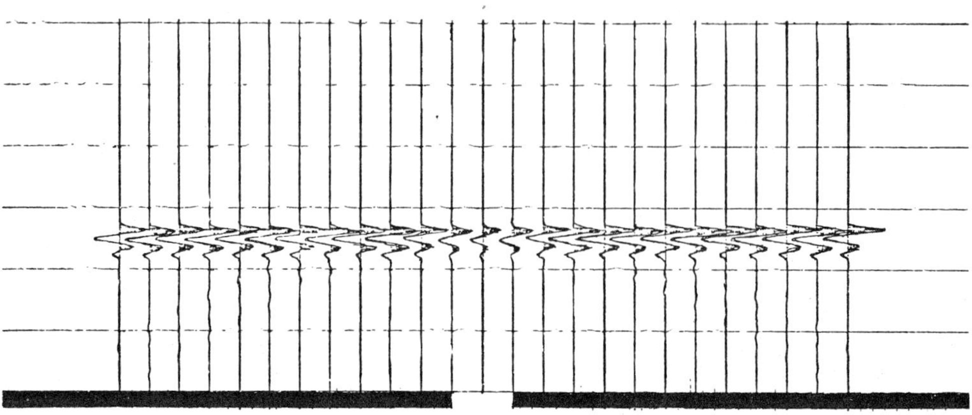

Fig. 194. **Wavefront healing.** A plane wave reflected from a reflector 5000 ft deep containing a hole shows how the diffraction process "heals" the wavefront. (Courtesy Chevron Oil Company.)

horizontally layered earth, the results are equivalent to a magnetotelluric measurement made at the same frequency.

wave train: A wave having several cycles.

wave-vector filtering: Filtering according to wavenumber as performed by spatial sampling. The use of source or geophone arrays involves wavenumber filtering. See Figure 54 **(directivity graph)**.

weathering: The low-velocity layer, a zone of low-velocity material near the earth's surface at the base of which the velocity abruptly increases. The seismic weathering is usually different from the "geologic weathering" and the term LVL (lower velocity layer) is often used. Frequently the base of the weathering is the water table. Sometimes the weathering velocity is gradational, sometimes it is sharply layered. Weathering velocities are typically 1500 to 2500 ft/sec or 500 to 800 m/sec (although weathering velocity may be 500 to 1500 ft/sec for the first few feet) compared to subweathering velocities of 5000 ft/sec (1700 m/sec) or greater. Weathering thickness is calculated from uphole-survey data and from refraction first breaks. See **weathering correction.**

weathering correction: A correction of seismic reflection or refraction times to remove the delay in the weathering or LVL. The simplest correction is based on uphole times from shots in the subweathering layer **(uphole shooting)**. Correction methods based on first-break times include the **ABC method,** the **Blondeau method,** the **summation method,** and the first-break intercept-time method. Automatic **statics** correction programs, based on maximizing the coherence of reflection events, are usually an early stage in digital processing. See also **double-layer weathering.**

weathering map: A map showing the thickness (occasionally velocity) of the weathering or LVL.

weathering shot: A special shot to give weathering data. Where pattern shots or geophone patterns are used or where surface sources are used, the first-break

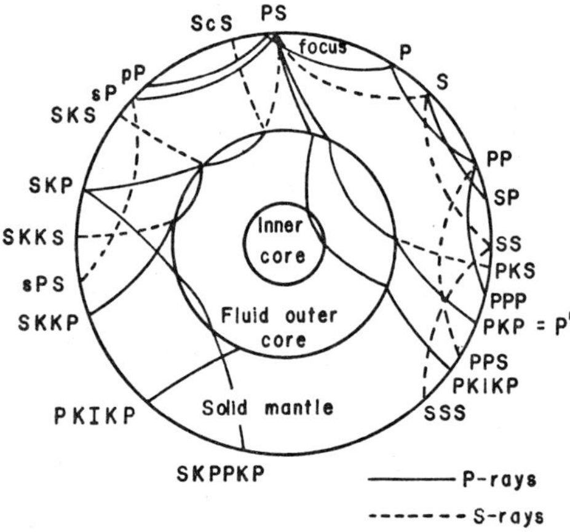

Fig. 196. **Wave modes** for seismic body wave in the earth. (From Bullen, Introduction to the Theory of Seismology, 1947, Cambridge University Press, p.165-168.)

quality may be too poor for good weathering corrections and a separate single-hole weathering shot may be recorded into a single geophone of each group. See also **uphole shooting.**

weber: The unit of magnetic flux in the mks system, being one joule/ampere. The analogous unit in the cgs system is the maxwell = one erg/abampere. 1 weber = 10^8 maxwells.

weight-drop: Use of a dropped weight as a seismic source; **thumper.**

weighted array: Pattern in which geophones or seismic sources are distributed over an area (or along the line)

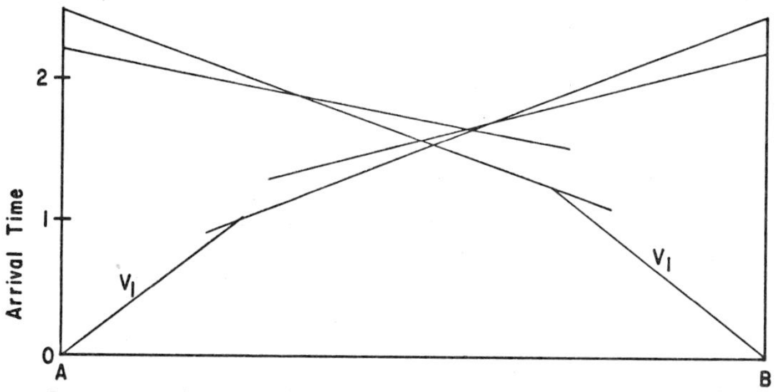

Reversed time distance curves showing both primary and secondary refraction events.

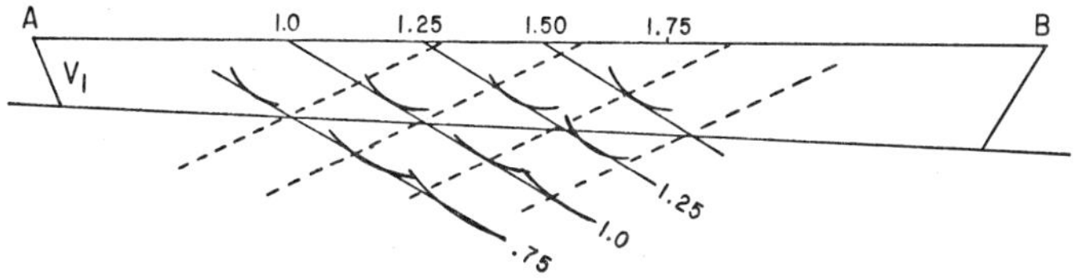

Arcs of radius $(t_n-t)V_1$ about geophone positions (where t_n = arrival time at geophone n) locate wavefront at time t. This is also done for reversed profile for various values of t. Symmetry of critical angles shows refractor dip. Reconciling travel times locates refractor.

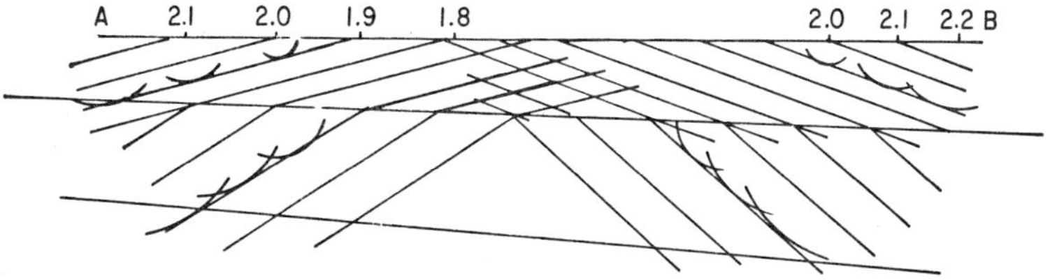

Extension to second refractor; new arcs are drawn about points where wavefronts impinge on first refractor to locate wavefronts in second layer.

Method can be extended for relief on refractors, etc., and also used with reflection events (where wavefronts converge on omage point), with diffractions (where they converge on the diffracting point), etc.

Fig. 195. **Wavefront methods.**

so that the contributions of various parts of the area (or line) are unequal. Achieved by varying the geometric distribution of geophones and/or shotpoints or weight-drops, by varying the outputs of the different geophone elements, or by varying the charge size in different holes. See also **tapered array**.

weighted average: The sum of a set of values x_i multiplied by weighting values, w_i, normalized by the sum of the weights:

$$\sum w_i x_i / \sum w_i.$$

Weiss's theory of magnetism: A ferromagnetic material is made up of small regions or *domains* magnetized to saturation (i.e., spins aligned cooperatively) despite the tendency of thermal agitation to disorient the spins, as in the case in paramagnetism. A weak external field can orient the domains with the field direction, and if strong enough, can align the domains irreversibly and so make a permanent magnet.

well log: A record of one or more physical measurements as a function of depth in a borehole. Also called *borehole log.* Distinction is sometimes made between a log as an entire record which may contain curves showing several measurements and the individual curves themselves, which are also called logs. **1.** *Wireline logs* are recorded by means of sondes carrying sensors which are lowered into the hole by a cable. Examples include logs which record electrical measurements (SP, resistivity, etc.), acoustic measurements (sonic, three-D, etc.), nuclear measurements (natural radioactivity, neutron logs, etc.), and miscellaneous measurements (hole size, temperature, etc.). **2.** Other types of well logs are made of data collected at the surface; examples are core logs, drilling-time logs, etc. **3.** Still other logs show quantities calculated from other measurements; examples are movable-oil plots, synergetic logs, etc. Pickett (1970, Applications for Borehole Geophysics in Exploration: Geophysics, v. 35, p. 81-92) lists types of well-log measurements and applications for well-log measurements, as shown in Table 11.

well shooting: A method of determining the average velocity as a function of depth by lowering a geophone into a hole and recording energy from shots fired from surface shotholes. Often run in addition to a sonic log to supply a reference time at the base of the casing and to check the integrated time, in which case they are called *check shots.* A standard computation form is shown in Figure 197.

well tie: Running a seismic line by a well so that seismic events may be correlated with subsurface information. Well ties are often done incorrectly for failure to consider the migration of the seismic data. Reflections from the vicinity of the well bore at depth may have to be observed some distance from the well head. See Figure 154 (**reverse migration**).

well-velocity surveys: Well shooting.

Wenner electrode array: Electrode arrangement used in resistivity surveying, consisting of four equally

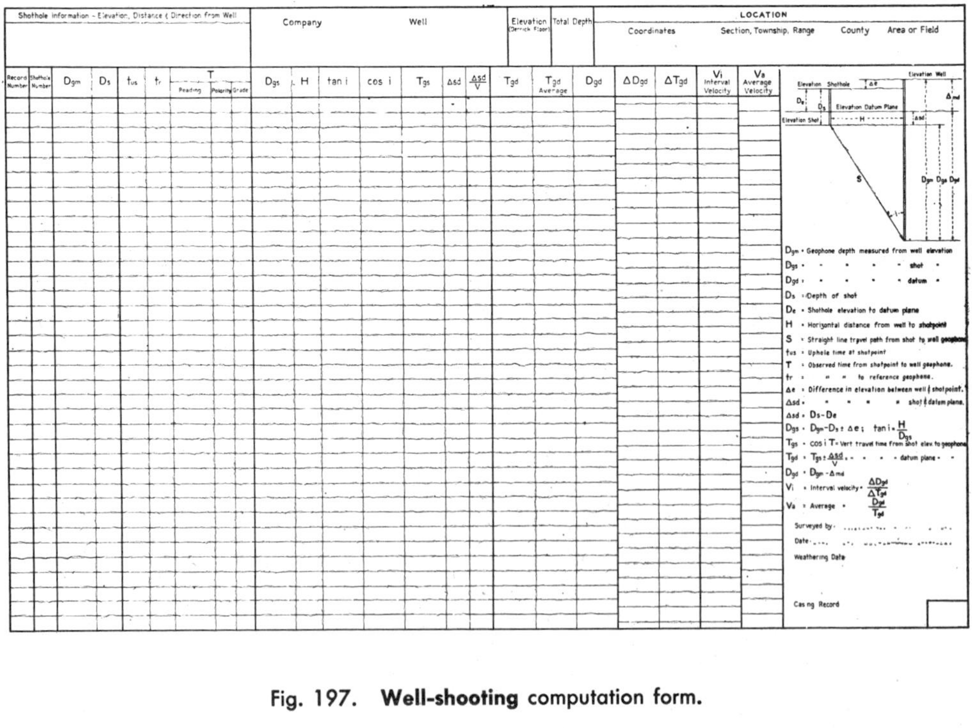

Fig. 197. **Well-shooting** computation form.

Table 11. Types of well-log measurements and applications

a. Cuttings Samples (lithology, "shows," rock types)

b. Cores (lithology, porosity, permeability, grain density, formation factor, saturation exponent, capillary-pressure curves, initial-residual oil saturation, acoustic velocity)

c. Accoustic logs (velocity—Δt, amplitude, waveform displays, variable intensity, cement-bond logs, borehole scanner, shear-wave velocity)

d. Radioactivity logs (gamma ray, neutron, density, chlorine, nuclear magnetism, neutron, lifetime, spectral)

e. SP-Resistivity logs (self potential, electrical, focused resistivity, flushed-zone resistivity, induction, dipmeter)

f. Borehole fluid logs (hydrocarbon component concentrations in drilling fluid)

g. Productivity and fluid testing (wireline formation tests, drillstem tests, production tests)

h. Gravity measurement (borehole gravimeter)

Applications

1. Estimates of hydrocarbon volumes in place.
2. Estimates of recoverable hydrocarbon volume.
3. Rock typing; lithology determination; parameters in log response equations; initial versus residual oil saturation relations.
4. Identification of geologic environments.
5. Water flood feasibility.
6. Location of fluid contacts.
7. Reservoir "quality" maping.
8. Water salinity determination.
9. Determination of reservoir fluid pressure.
10. Fracture detection.
11. Determination of reservoir engineering parameters.
12. Prediction of interzone fluid communication in annulus.
13. Determination of porosity and pore size distribution.
14. Monitoring of fluid movement in reservoir.

spaced collinear electrodes, with the outer two serving as current sources and the inner two as potential-measuring points. See **array.**

westing: See **departure.**

white: Containing all frequencies in equal proportion.

whiten: To adjust the amplitudes of all frequency components within a certain band-pass to the same level. A method of **deconvolving.**

white noise: Random energy containing all frequency components in equal proportions within the bandwidth; phases are random.

white-noise level: The amount of white noise added to data undergoing analysis for an inverse-filter design. An inverse filter tends to build up frequencies in which meaningful data are virtually missing so that noise at such frequencies is magnified: the addition of white noise (or what is equivalent, superimposing an impulsive function or biasing the amplitude-frequency response curve) limits the extent to which this can occur.

wide-angle reflection: Reflection where the angle of incidence is near or greater than the critical angle. Reflection coefficients may have large values near the critical angle so that reflection energy is exceptionally strong. See Figure 39 **(critical distance)** and 201 **(Zoeppritz equations).**

wide-band stack: A stack which does not produce any frequency discrimination. See also **optimum wide-band.**

width: **1.** The width of an anomaly usually is measured at either half amplitude or between inflection points. See **half-width. 2.** The width of a pulse is the width of a **boxcar** with the same peak height and containing the same energy.

Wiener filter: A filter which will optimize (in a least-squares sense) standout of a signal S (which is a function of frequency) in the presence of noise N (also a function of frequency). Each frequency is passed proportional to

$$[S(f)]^2 / \{[S(f)]^2 + [N(f)]^2\} .$$

If the desired output is specified, the Wiener filter will give the output for an actual input which will come closest to the desired output. See **Wiener-Hopf equations.**

Wiener-Hopf equations: **1.** The Wiener-Hopf equation of the first kind, an integral equation in the unknown $f(t)$:

$$\phi_{xz}(\tau) = \int_0^\infty f(t)\, \phi_{xx}\, (\tau - t)\, dt, \qquad \tau \geq 0,$$

This equation is the necessary and sufficient condition for minimizing the mean-square error between a desired output $z(t)$ and the acutal output $y(t)$ which results from passing an input $x(t)$ through a filter with an impulse response $f(t)$. $\phi_{xx}(\tau)$ is the autocorrelation of x and $\phi_{xz}(\tau)$ is the crosscorrelation between z and x. When digital processing is involved, this equation becomes the "normal" set of linear simultaneous equations. **2.** The Wiener-Hopf equation of the second kind which applies to a nonstationary input involves a time-varying filter $f(t,\delta)$ and time-varying correlation functions:

$$\phi_{xz}(t,\tau) = \int_{0}^{\infty} f(t,\delta)\; \phi_{xx}(\delta,\tau)\; d\delta .$$

See **Wiener filter** and Lee, 1960, Statistical Theory of Communication: New York, John Wiley & Sons.

Wiener-Levinson algorithm: See **Levinson algorithm.**

wiggle trace: A graph of amplitude against time, as on a conventional seismic recording with mirror galvanometers. Also called "squiggle" recordings. See Figure 56 **(display).**

wild: Having very large and often unpredictable amplitudes, e.g., a noisy seismic channel at high gain.

wind noise: **1.** Random noise attributed principally to ground unrest caused by the wind moving plants, trees, etc., and shaking their roots. **2.** Seismic background noise (in the absence of a shot) regardless of the source. **3.** Noise voltage induced in a suspended wire of an IP, resistivity, or telluric-current measuring system, caused by oscillations of the wires in the earth's magnetic field.

window: **1.** A portion of a seismic record free from certain disturbances; that is, where certain important noise trains are absent. **2.** The portion of a record chosen for designing operators such as those used for autocorrelation or frequency analysis. Also called *gate.* The values within a window are weighted according to some scheme. Windows in common use are shown in Table 12 and in Figure 198. Note the tradeoff between narrowness of the major lobe and low side-lobe energy. The *equivalent width* of a window is the width of a boxcar with the same peak amplitude which contains the same energy.

window pair: A **window** and its Fourier transform, such as shown in Figure 198.

wind scale: Wind force is often given in terms of a Beaufort number. See Table 1 **(Beaufort number).**

wireline log: A well log recorded by a sonde which is lowered into the hole by a cable.

witness marker: A marked location (such as a blaze on a tree) which helps locate a survey point whose location is known with respect to the witness marker.

word: A group of characters occupying one storage location in a computer. This unit is treated by the computer as an entity.

Worden: A type of gravimeter. Light weight and insulated by a thermos bottle, the meter is very portable. See Figure 91 **(gravimeter).**

wow: Variations in the speed of a magnetic tape or a camera producing irregular timing lines. Often periodic and of low frequency.

WWV: The United States Bureau of Standards radio station which broadcasts time and frequency standards. WWV (Ft. Collins, Colorado) and WWVH (Maui, Hawaii) broadcast continuously on 2.5, 5, 10, 15, 20, and 25 Mhz (the last 2 only by WWV). WWV is off the air for 4 minutes commencing at 45 minutes 15 sec after each hour and WWVH for 4 minutes

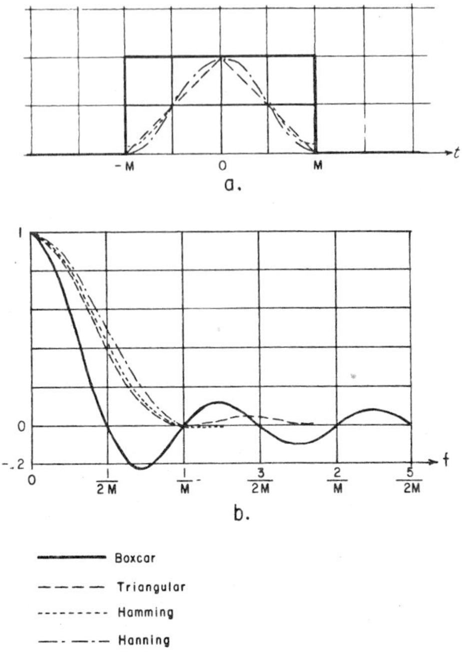

Fig. 198. **Windows** in time and frequency domains. (a) Window shapes. (b) Spectral shapes. The *effective width* of the boxcar is greater than that for the other window shapes and, hence, its central lobe is sharper.

commencing at 15 min 15 sec after each hour. Each second is marked by a signal or tick; the 59th second is omitted. A voice announcement of time is given during the last half of every fifth minute during the hour. At 2235 GMT, for instance, the voice announcement would be: "National Bureau of Standards, WWV, Fort Collins, Colorado; the next tone begins at twenty-two hours thirty-five minutes, Greenwich Mean Time."

Wx: Weathering.

Wyllie relationship: An empirical expression for calculating porosity ϕ from sonic-log transit times:

$$\phi = \frac{\Delta t - \Delta t_{ma}}{\Delta t_f - \Delta t_{ma}} ,$$

where Δt = observed interval transit time, Δt_f = transit time in the pore fluid, and Δt_{ma} = transit time in the rock matrix. This relation works well in clean consolidated formations with uniformly distributed pores. In formations containing vugs, the sonic log may not reflect the secondary porosity; and in unconsolidated formations, this relationship may overestimate porosity. In such cases the formula may be empirically modified to give better values.

Wyrobek method: A refraction-interpretation method

Table 12. Window types

rectangular window (boxcar)	$y = 1$	$\lvert t \rvert \leq M$
	$y = 0$	$\lvert t \rvert > M$
	$y =$	
Triangular window (Bartlett window)	$y = 1 - \lvert t/M \rvert$	$\lvert t \rvert \leq M$
	$y = 0$	$\lvert t \rvert > M$
Hamming window	$y = 0.54 + 0.46 \cos(\pi t/M)$	$\lvert t \rvert \leq M$
	$y = 0$	$\lvert t \rvert > M$
Hanning window	$y = \frac{1}{2} + \frac{1}{2} \cos(\pi t/m)$	$\lvert t \rvert \leq m$
	$y = 0$	$\lvert t \rvert > m$
Daniell window	$y = \mathrm{sinc}(t\pi/2M)$	$\lvert t \rvert \leq M$
	$y = 0$	$\lvert t \rvert > M$

based on applying delay and intercept times to continuous refraction profiling, even where the profiles are not reversed. Consider Figure 199. Total delay times D_i are calculated using an assumed refractor velocity V_R:

$$D_i = t_i - x_i/V_R \ ,$$

where t_i is arrival time and x_i is offset distance. Half-intercept times are also plotted. The departure from parallel of the composite total delay-time curve and the intercept curve is a measure of the error in refractor velocity: in this case it amounts to 40 msec/16,000 ft. Applying this correction allows recalculation of the delay times, which can then be converted into a depth profile if the velocity above the refractor V_o is known.

$$\text{Depth} = Z = \tfrac{1}{2}DF \ ,$$

where the conversion factor $F = V_R \tan i_c = 1/(1/V_o^2 - 1/V_R^2)^{1/2}$ (i_c is the critical angle). See Wyrobek, 1956, Application of Delay and Intercept Times in the Interpretation of Multilayer Refraction Time-Distance Curves: Geophysical Prospecting, v. 4, p. 112-130.

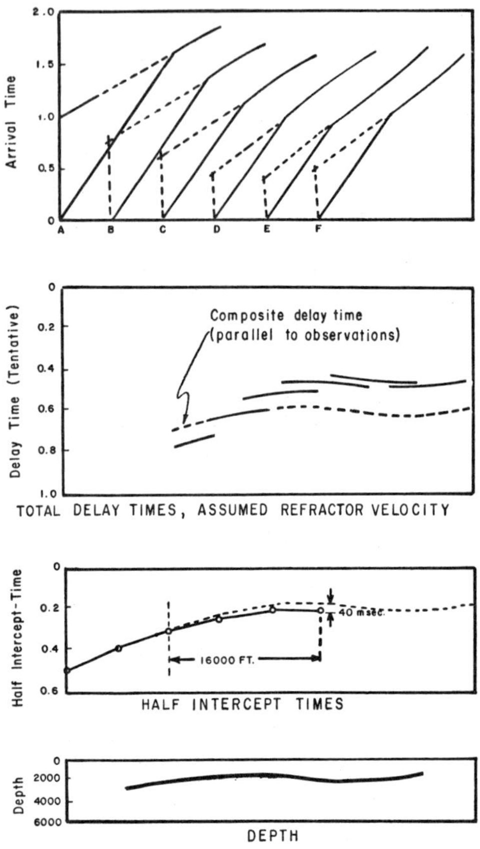

Fig. 199. **Wyrobek refraction method.** (From Wyrobek, Geophysical Prospecting: v. 4, no. 2, p. 121.)

X

x: **1.** The distance from the shotpoint to a particular geophone group; **offset.** **2.** Cross; an *x-spread* is a **cross-spread.**

xo: Subscript used with log terms to indicate values appropriate to the flushed zone adjacent to the borehole.

XR: Extended range; see **extended-range shoran.**

X-spread: Cross-spread.

X^2-T^2 analysis: A method of determining rms velocity *V* and the depth of a reflector *Z* from the arrival time versus offset relationship:

$$V^2T^2 = 4Z^2 + X^2.$$

The square of the horizontal distance or offset (X^2) is plotted (Figure 200) against the square of the reflection time (T^2); the slope gives the inverse of the velocity squared and the depth can be obtained from the intercept. Also used for identifying multiples because multiples tend to show abnormally low velocities for their arrival times. See also **velocity.**

X-Y reader: A device for converting the positions of points on a map or graph to digital coordinates. Coordinatograph.

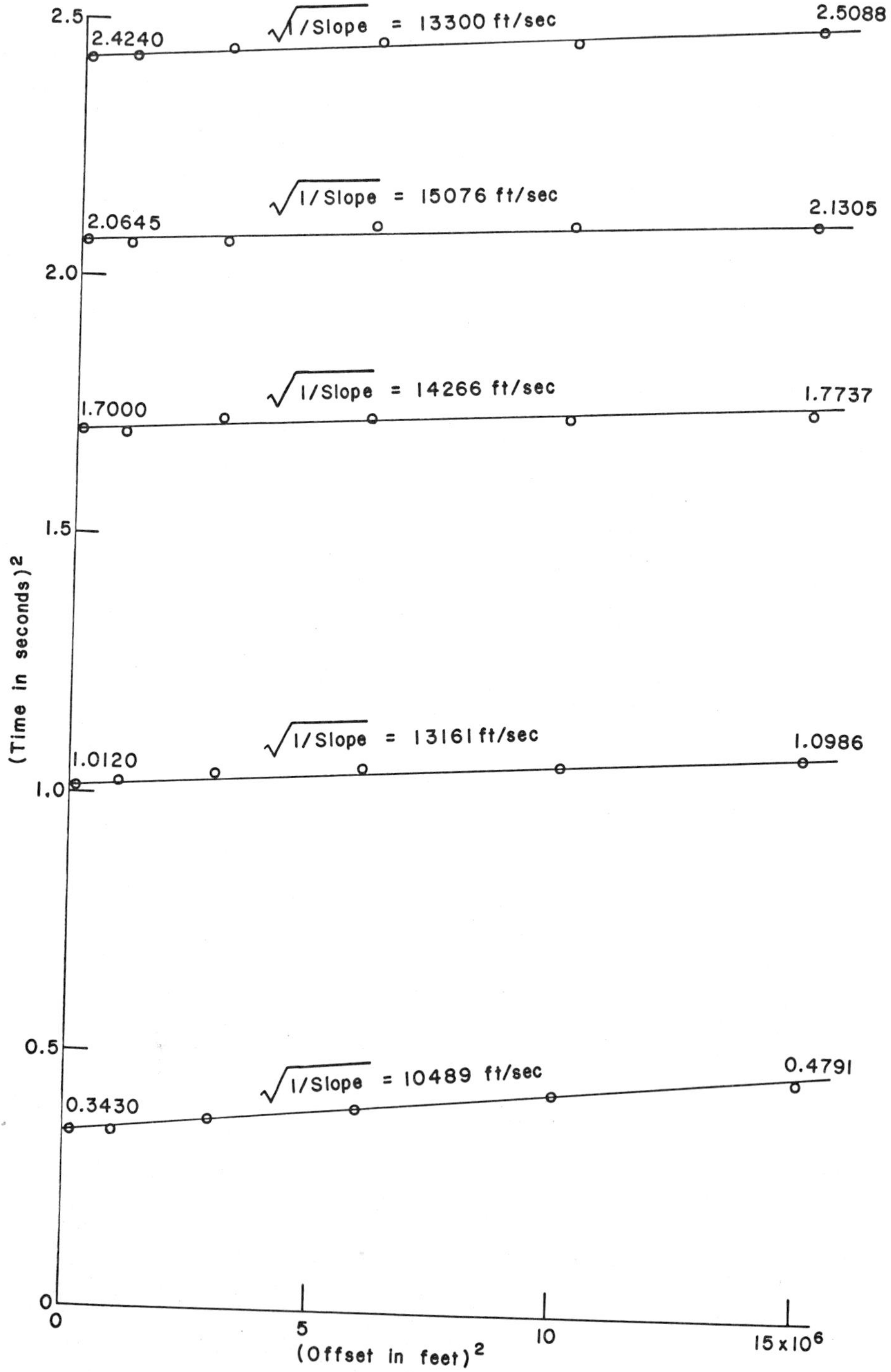

Fig. 200. ***X^2T^2* plot.**

Y

yardstick: Standards of performance for evaluating devices or processes.

yaw: Rotational motion of a ship or aircraft about a vertical axis. A steady heading at an angle to the course (such as to compensate for cross wind or cross sea) is called **crab**. Compare **pitch** and **roll**.

Young's modulus: See **elastic constant**.

yo-yo: 1. A method of marine seismic shooting in which the seismic cable or streamer being towed by the recording boat is alternately released so that it will float freely in the water during a recording, and then reeled in between recordings to catch up with the recording boat which is traveling steadily ahead all the while. **2.** Depth oscillation of a logging sonde because of unequal drag as it is pulled up the hole.

Z/A: The ratio of atomic number to atomic weight and, hence, proportional to the ratio of electron density (as measured by the **density log**) to mass density. "Apparent density" equals true density when $Z/A = 1/2$, which is closely approximated for many minerals (e.g., quartz, calcite, anhydrite, dolomite). The apparent density is larger than the true density when $Z > 1/2$ (e.g., gypsum, oil, 110 percent for water), smaller when $Z < 1/2$ (e.g., salt).

Zener diode: A silicon diode in which the breakdown voltage in the reverse direction (Zener voltage) is used for voltage stabilization or voltage reference.

zenith: A point directly overhead. *Zenith distance* is the angle between zenith and a body.

zero frequency: Alternating-current phenomena extended to zero frequency.

zero-frequency seismology: Study of long-term displacements, strains, and tilts.

zero-lag correlation: The value of an autocorrelation $\phi_{xx}(0)$ or crosscorrelation $\phi_{zx}(0)$ for zero time shift; a measure of the mean power or crosspower.

zero-length spring: A spring whose effective length, as measured from its fixed point of support, is zero when the external forces acting upon it are zero. The stress-strain relationship between its points of attachment projected back to zero length has zero strain. Gravimeters designed with a zero-length spring are linear and have extreme sensitivity combined with stability and reduced sensitivity to leveling.

zero-phase: A filter for which the phase shift is zero for all frequencies. Zero-phase filters are anticipatory and hence are not physically realizable; i.e., some of the energy is pushed forward with respect to the new time reference so that one gets output before the input arrives. If the input to a zero-phase filter is symmetric, then the output will also be symmetric. Zero-phase filtering often can be accomplished by using a linear-phase filter, a mixed-phase filter which shifts component frequencies proportional to their frequency. A linear phase filter has a phase-shift-versus-frequency graph which is linear over the band-pass, and an intercept which is a multiple of 2π, so that each component is delayed equally. The delay so introduced then can be compensated by moving the time reference or time break. Such a filter produces no phase distortion. If the intercept is an odd multiple of π, it will invert a wavelet. See Figure 119 (**minimum-phase**).

zeta potential: Potential drop across the diffuse layer in an electrolyte, which consists of a group of relatively mobile ions at the interface between a solid and a liquid.

Zoeppritz's equations: Equations which express the partition of energy when a plane wave impinges on an acoustic-impedance contrast. In the general case when the incident angle is not zero, four waves are generated: reflected P-wave and S-wave and transmitted P-wave and S-wave. The partition of energy among these is found from the four boundary conditions which require continuity of normal and tangential displacement and stress at the boundary. Assume, for example, a plane P-wave incident on a plane interface using the symbols given in Figure 165 (**Snell's law**). Snell's law states:

$$\frac{\sin\theta_{P1}}{V_{P1}} = \frac{\sin\theta_{S1}}{V_{S1}} = \frac{\sin\theta_{P2}}{V_{P2}} = \frac{\sin\theta_{S2}}{V_{S2}},$$

which defines all angles. Continuity of tangential displacement (x direction) requires that:

$$\sin\theta_{P1} + A\sin\theta_{P1} + B\cos\theta_{S1} = C\sin\theta_{P2} - D\cos\theta_{S2},$$

where we take the P-wave displacement to be positive in the direction of propagation and the S-wave displacement to be positive to the right of the direction of propagation. Continuity of normal displacement (z direction) requires that:

$$\cos\theta_{P1} - A\cos\theta_{P1} + B\sin\theta_{S1} = C\cos\theta_{P2} + D\sin\theta_{S2}.$$

Continuity of normal and tangential stress yields (after some manipulation) the two equations:

$$-\sin 2\theta_{P1} + A\sin 2\theta_{P1} + \frac{V_{P1}}{V_{S1}} B\cos 2\theta_{S1} = \frac{-\rho_2 V_{S2}^2 V_{P1}}{\rho_1 V_{S1}^2 V_{P2}} C\sin 2\theta_{P2} + \frac{\rho_2 V_{S2} V_{P1}}{\rho_1 V_{S1}^2} D\cos 2\theta_{S2}.$$

$$\cos 2\theta_{\theta S1} + A\cos 2\theta_{S1} - \frac{V_{S1}}{V_{P1}} B\sin 2\theta_{S1} = \frac{\rho_2 V_{P2}}{\rho_1 V_{P1}} C\cos 2\theta_{S2} + \frac{\rho_2 V_{S2}}{\rho_1 V_{P1}} D\sin 2\theta_{S2}.$$

Figure 201 shows the variation of amplitude with angle for several sets of parameters. Beyond the critical angles for P- and S-waves, the respective refracted waves vanish. The increase in reflection energy near the critical angle is sometimes referred to as the "wide angle" phenomenon and sometimes exploited in seismic surveying. The same relationships in terms of potentials are called *Knott's equations.*

zone chart: A template for making **terrain corrections** or **isostatic corrections.** The template (Figure 202) is laid over a topographic map with its center at the station being corrected. The difference in mean absolute elevation between each zone and the station's elevation is tabulated without regard for sign (because the correction is always positive for zones either higher or lower than the station elevation) to determine the terrain correction. With a template for isostatic correction the mean elevation above sea level in each zone is used in calculating the correction.

z-plane: A representation of the **z-transform** polynomial associated with a waveform in sampled form.

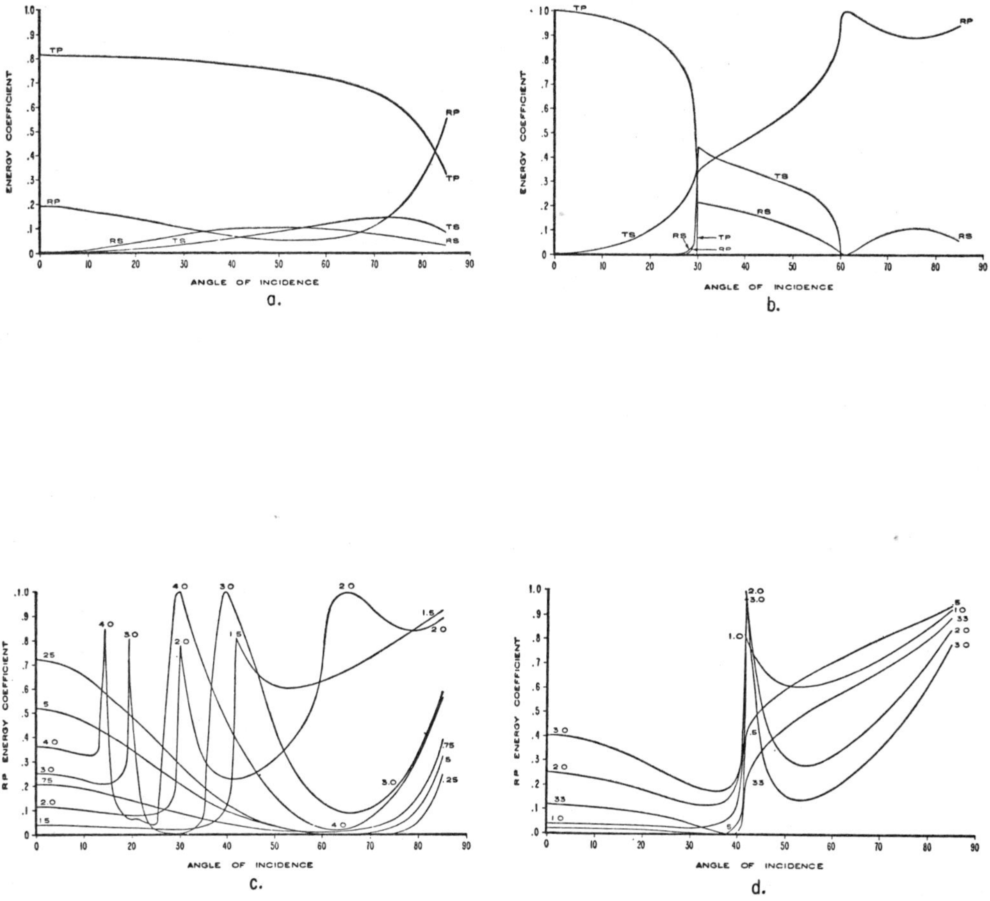

Fig. 201. **Zoeppritz equations** yield the amplitude of the waves which result from oblique incidence on an interface. (a) Energy ratio for reflected P-waves *RP,* reflected S-waves *RS,* transmitted P-waves *TP,* and transmitted S-waves *TS* for a velocity ratio of 0.5, a density ratio of 0.8, a Poisson's ratio of 0.3 in the upper medium and 0.25 in the lower. (b) Similar to (a) except for a velocity ratio of 2.0 and a density ratio of 0.5. (c) Energy ratio for reflected P-waves as a function of P-wave velocity ratio. No density contrast and Poisson's ratios are 0.25. (d) Energy ratio for reflected P-waves as a function of density ratio. P-wave velocity contrast 1.5, Poisson's ratios 0.25. (From Tooley et al, Geophysics: v. 30, no. 4, p.562, 565, and 567.)

z-transform: A method of calculating the effect of various operations on quantized data. If the sample values of a wavelet at successive times are:

$x_t = (x_0, x_1, x_2, x_3, \ldots, x_n)$,

then the *z*-transform of the wavelet is

$$x(z) = x_0 + x_1 z + x_2 z^2 + x_3 z^3 + \ldots + x_n z^n$$

The *z*-transform may be thought of as

$$Z = e^{-i\omega t}.$$

which allows one to relate it to the Fourier transform. The z-transform technique is an easy way of converting (by inspection) from the time domain into a form which can be treated (in some ways) as the frequency domain. Thus the convolution operator can be accomplished by merely multiplying the z-transforms of the waveforms being convolved. The inverse of a filter can be found by finding the reciprocal of the filter's z-transform. Used in digital seismic processing. The z-transform polynomial can be factored and expressed as the product of couplets of the form:

$$f(z) = (z - a)(z - b)(z - c) \cdots (z - n).$$

The values for which this expression vanishes, i.e., $z = a$, $z = b$, etc., (a, b, etc., may be complex) are called the "zeros" or "roots" of the expression. Values greater than unity are said to "lie outside the unit circle;" Figure 203. If all the roots lie outside the unit circle, the function is minimum phase; if all are inside, maximum phase (assuming that in the z-transform higher powers of z represent values later in time.) Values for which an expression becomes infinite [such as r in $1/(z - r)$] are called "Poles" or "singularities." Filters are sometimes designed in the z-plane. See Robinson and Treitel, 1964, Principles of Digital Filtering: Geophysics, v. 29, p. 395-404. Sometimes the opposite convention used, i.e., successive sample values are multiplied by successively higher negative powers of z. Use of this convention reverses the criteria for minimum and maximum phase with respect to the unit circle.

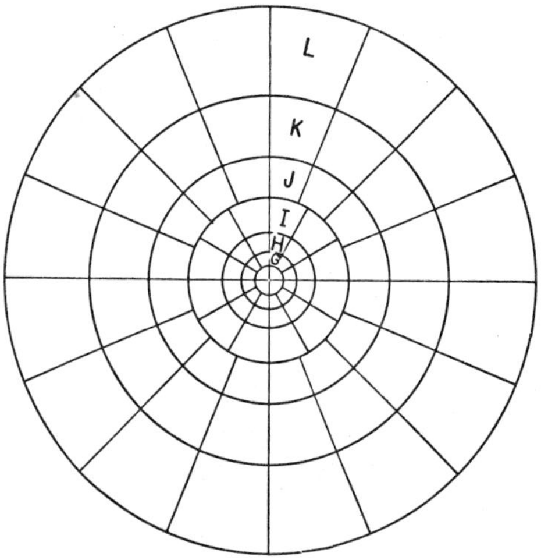

Fig. 202. **Zone chart** for terrain corrections. (From Nettleton, Geophysical Prospecting for Oil, 1940, McGraw-Hill, New York, p.146.)

The wavelet:

10, -2, -1, 1, 1

has the z-transform:

$10 - 2z - z^2 + z^3 + z^4$

which may be factored:

$(2 + j + z)(2 - j + z)(-1 + j + z)(-1 - j + z)$

The roots of this are:

-2-j

-2+j

1-j

1+j

A plot of these roots "in the z-plane":

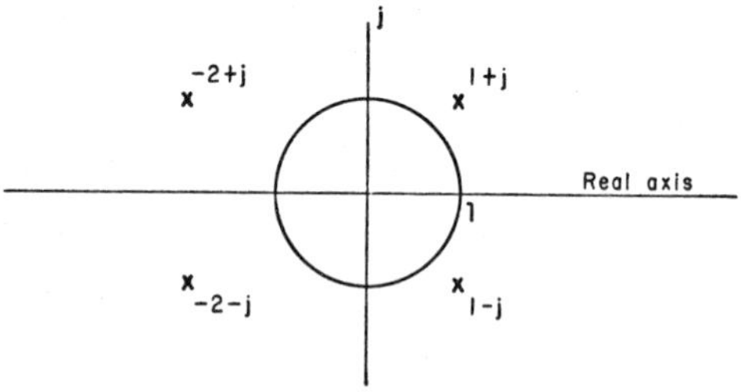

Since all roots lie outside a circle of radius 1, the wavelet is minimum phase.

Fig. 203. **z-plane.**

APPENDIX A
NUMERICAL PREFIXES

Prefixes	Multiple	Symbols
tera	10^{12}	T
giga	10^{9}	G
mega	10^{6}	M
kilo	10^{3}	k
hecto	10^{2}	h
deka	10	
deci	10^{-1}	d
centi	10^{-2}	c
milli	10^{-3}	m
micro	10^{-6}	μ (mu)
nano	10^{-9}	n
pico	10^{-12}	p
femto	10^{-15}	f
atto	10^{-18}	a

APPENDIX B
GREEK ALPHABET

Α	α	alpha	Ι	ι	iota	Ρ	ρ	rho
Β	β	beta	Κ	κ	kappa	Σ	σ	sigma
Γ	γ	gamma	Λ	λ	lambda	Τ	τ	tau
Δ	δ	delta	Μ	μ	mu	Υ	υ	upsilon
Ε	ϵ	epsilon	Ν	ν	nu	Φ	ϕ	phi
Ζ	ζ	zeta	Ξ	ξ	xi	Χ	χ	chi
Η	η	eta	Ο	o	omicron	Ψ	ψ	psi
Θ	θ	theta	Π	π	pi	Ω	ω	omega

APPENDIX C
SYMBOLS USED IN GEOPHYSICAL EXPLORATION

(Symbols used in well logging are in Appendix D)

The choice of symbols to represent quantities is arbitrary and not universally agreed upon. Nevertheless, the following choices are used most frequently:

a	(subscript) apparent
B	magnetic induction (a vector) $= 4\pi\mathbf{I}+\mathbf{H}=\mu\mathbf{H}$
C	capacitance $= Q/V$
E	electric field strength (a vector); Young's modulus
f	frequency
FE	frequency effect $= \Delta\rho/\rho$
G	gravitational field strength (a vector)
H	magnetic field strength (a vector)
i	$(-1)^{1/2}$
I	current; intensity of magnetization (a vector)
IP	induced polarization
j	$(-1)^{1/2}$
J	current density (a vector)
k	wavenumber; bulk modulus
L	inductance; area over decay curve
M	mutual inductance; chargeability
MF	metal factor
P	compressional wave; polarization; dipole moment/volume
PFE	percent frequency effect
Q	charge
R	resistance
S	shear wave; admittance $= 1/Z$
t	time
T	period
V	velocity of a medium; voltage
X	offset distance; reactance
X_C	capacitative reactance $= 1/(2\pi fC)$
X_L	inductive reactance $= 2\pi fL$
Z	depth; impedance
Z^n	time delay of n units

γ	gamma	skewness; gyromagnetic ratio; unit of magnetic field strength
δ	delta	impulse; skin depth; loss tangent
Δ	delta	difference
ϵ	epsilon	capacitivity = permitivity
η	eta	overvoltage
θ_C	theta	critical angle
λ	lambda	wavelength; coefficient of anisotropy; Lame's constant
μ	mu	magnetic permeability = B/H; micro; Lame's shear modulus; damping factor
ρ	rho	density; resistivity
σ	sigma	Poisson's ratio; electrical conductivity; standard deviation
ϕ	phi	porosity; flux; correlation function
χ	chi	magnetic susceptibility
ω	omega	angular frequency = $2\pi f$
∇	del	gradient operator (a vector)
*		convolution operator

APPENDIX D
SYMBOLS USED IN WELL LOGGING

a	activity; air requirement; air; apparent; atmospheric
A	area, areal
b	reciprocal of formation-volume-factor; bank; bubble point; bulk; burned
B	formation volume factor = volume at reservoir conditions/volume at standard conditions; turbulence
c	capillary; contact; conversion; critical; electrochemical; compressibility
cf	flowing casing (pressure)
cs	static casing (pressure)
C	coefficient of back-pressure curve; concentration; conductivity; specific heat; water-drive constant
C_1	methane
C_2	ethane
d	diameter; depletion; dew-point; differential separation; dip; displaced; drainage
D	deliverability; depth; diffusion coefficient; dimensionless; displacement

Appendix D — Well Logging Symbols (continued)

e	influx rate; cumulative influx; effective; external boundary conditions
ext	extrapolated (subscript)
E	efficiency; electromotive force; experimental
f	coefficient of friction; fraction; fugacity; flash separation; fluid; formation; front
F	factor; force; formation resistivity factor $= R_o/R_w$; free; fuel
g	acceleration of gravity; gas
G	gas
h	thickness; hole; hydrocarbon; thermal heat
H	enthalpy
i	injection rate; initial; injected; invaded zone
I	injectivity index; resistivity index $= R_t/R_o$; invasion
j	$(-1)^{1/2}$
J	productivity index
k	absolute permeability; electrokinetic
K	spontaneous electromotive force; equilibrium ratio
L	length; moles of liquid phase; liquid
lim	limiting value
m	mud; fuel; fuel consumption; mass; slope; ratio of initial reservoir free gas volume/initial reservoir oil volume; porosity exponent
ma	solid matrix (subscript)
mc	mud cake (subscript)
mf	mud filtrate (subscript)
max	maximum (subscript)
min	minimum (subscript)
M	mobility ratio; molecular weight; molal
n	net; total moles; back-pressure exponent; saturation exponent
N	dimensionless; initial oil in place
N_2	nitrogen
o	oil (subscript) (except with resistivity); 100 percent water saturated (resistivity)

Appendix D — Well Logging Symbols (continued)

O_2	oxygen
p	pressure; cumulative produced; particle; pore; production time; mean
pc	pseudo-critical
P	pattern
q	charge; flow rate
Q	heat-flow rate
r	radial distance; resistance; reduced; relative; residual
R	recovery; reservoir; gas-oil ratio; resistivity; universal gas constant
Re	Reynolds Number
s	skin effect; skin; segregation; solid; solution; specific; stabilization; surrounding formation; swept region
sb	solution at bubble-point conditions
sc	standard conditions
sw	solution in water
S	saturation
t	time; true; total; gross
tf	flowing tubing (pressure)
ts	static tubing (pressure)
T	temperature
u	flux; flow rate/area; unburned
v	acoustic velocity; specific volume; velocity; vaporization
V	volume; volumetric; moles of vapor phase
w	water; well conditions; mass flow rate
wa	apparent wellbore (subscript)
wf	flowing bottom-hole (pressure); well flowing conditions
wg	wet gas; water in gas cap
ws	static bottom-hole; well static conditions
W	initial water in place; work
x	mole fraction in liquid phase
xo	flushed zone (subscript)
y	mole fraction in vapor phase
z	mole fraction in mixture; compressibility factor = pV/nRT
Z	elevation with respect to datum

Appendix D — Well Logging Symbols (continued)

α	alpha	angle; SP reduction factor
β	beta	thermal cubic expansion coefficient
γ	gamma	specific gravity
δ	delta	displacement ratio
Δ	delta (cap)	difference
ϵ	epsilon	hydraulic diffusivity
η	eta	kinematic viscosity
θ	theta	angle
Θ	theta (cap)	acoustic transit time per unit length
λ	lambda	mobility; wavelength
μ	mu	viscosity
ρ	rho	density; resistivity
σ	sigma	surface tension; interfacial tension; wavenumber
τ	tau	tortuosity
ϕ	phi	porosity
Φ	phi (cap)	potential
Ψ	psi (cap)	stream function

APPENDIX E
MATHEMATICAL SYMBOLS

Bold face	Vectors **V** Matrices **A**
Superscripts	Powers as a^n, $\sin^n x = (\sin x)^n$ (n positive) inverse functions as $\sin^{-n} = \text{arc} \sin x$ order of differentiation as $x^1 = \frac{dx}{dt}$
Subscripts	position in a sequence, set, or matrix as $\begin{Vmatrix} a_{11} & a_{12} \\ a_{11} & a_{22} \end{Vmatrix}$
Parentheses	aggregation as $(a+b)$ argument of a function $f(t)$
Vertical bars	absolute value, modulus, magnitude of vector **V** determinant $\begin{vmatrix} a_{11} & a_{12} \\ a_{21} & a_{22} \end{vmatrix}$ evaluated at $f(x)\|_a = f(a)$

Appendix E — Mathematical Symbols (continued)

Double bars	matrix $\|a_{ij}\| = \begin{Vmatrix} a_{11} & a_{12} \\ a_{11} & a_{21} \end{Vmatrix}$
Brackets	ordered set [x, y, z]
Superscript line	mean
$=$	equal to
$\neq$	not equal to
$\approx$	approximately equal to
$>$	greater than
$<$	smaller than
$\geq$	greater than or equal to
$\leq$	smaller than or equal to
$+$	plus
$-$	minus
$\pm$	plus or minus
$\cup$	union
$\cap$	intersection
$\supset$	contains; implies
$\rightarrow$	approaches; transforms to
/	division as 1/2 = ½
*	convolved with; as superscript: complex conjugate
. . .	and so forth as $a_0, a_1, \ldots, a_n$
!	factorial $3! = 3 \cdot 2 \cdot 1 = 6$
$\int_b^a$	integral from a to b
∞	infinity
Δ	difference; sizeable increment
∇	del
δ	very small increment
∂	partial derivative $\frac{\partial f(x, y)}{\partial x}$

Appendix E — Mathematical Symbols (continued)

Σ	sum $\sum_{i=1}^{3} a_i = a_1 + a_2 + a_3$
Π	product $\prod_{i=1}^{3} a_i = a_1 a_2 a_3$
ϕ	correlation function
σ	standard deviation
ω	angular velocity
abs	absolute
arc	inverse as arc sin $x = \sin^{-1} x$ = angle whose sine is x
arg	argument of
av	average
cis θ	$\cos\theta + i \sin\theta$
d	differential
det	determinant
div	divergence
erf	error function $\operatorname{erf}(x) = \frac{2}{\sqrt{\pi}} \int_0^x e^{-v^2} dv$
exp	exponential function $\exp(x) = e^x$
Im	imaginary part of
lim	limit
ln	natural logarithm
max	maximum
P(E)	probability of E
$P(E \wedge F)$	probability of both E and F
P(E/F)	probability of E given F
Re	real part of
rms	root mean square
sgn	sign of

APPENDIX F
MAP AND ROCK SYMBOLS

A. WELL SYMBOLS

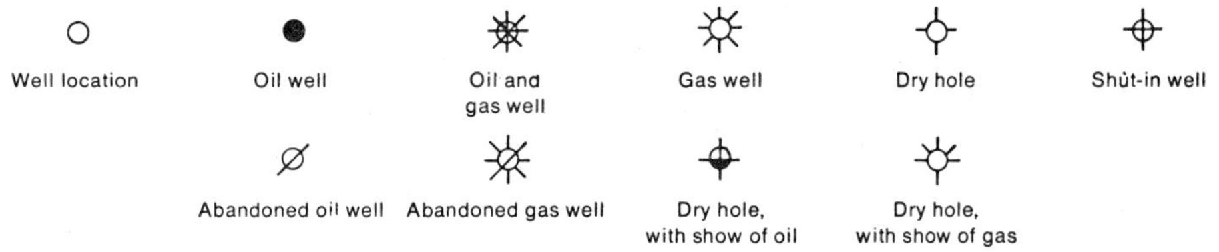

B. STRUCTURE SYMBOLS

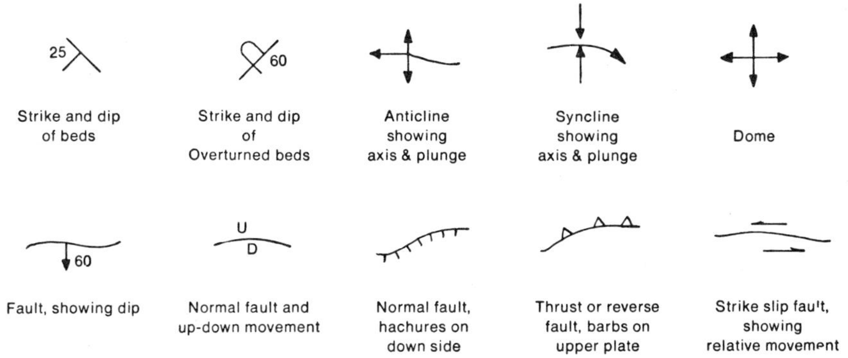

C. ROCK SYMBOLS

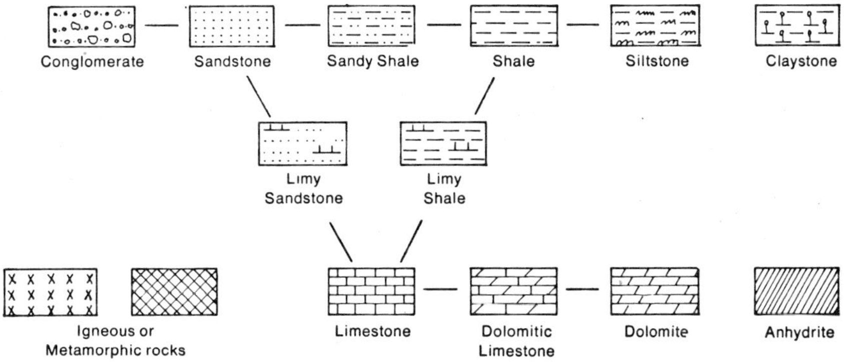

Note: Intermediate rock compositions are represented by combining symbols.

APPENDIX G
SI OR MKS SYSTEM OF UNITS

The meter-kilogram-second-ampere system of units (sometimes called Giorgi system) has become the standard in scientific usage although the older cgs-esu and cgs-emu systems (especially the latter) are still used in much geophysical work. See also Table 6 (magnetic field). The table below attempts to relate these units; although, if one works from the fundamental definitions in the three systems, certain inconsistencies can be found which we are ignoring here.

	mks system	cgs system	
		emu	esu*
length	1 meter	$= 10^2$ centimeter	
mass	1 kilogram	$= 10^3$ gram	
time	1 second	= 1 second	
force	1 newton = 1 kg m/sec^2	$= 10^5$ dyne	
acceleration	1 m/sec^2	$= 10^5$ milligal	
energy (work)	1 joule = 1 newton m	$= 10^7$ erg	
power	1 watt = 1 joule/sec	$= 10^7$ erg/sec	
current	1 ampere	$= 10^{-1}$ abampere	$= 3 \times 10^9$ statampere
charge	1 coulomb = 1 amp/sec	$= 10^{-1}$ abcoulomb	$= 3 \times 10^9$ statcoulomb
electrical potential	1 volt = 1 watt/amp	$= 10^8$ abvolt	= statvolt/300
resistance	1 ohm = 1 watt/amp^2	$= 10^9$ abohm	= statohm/(9×10^{11})
capacitance	1 farad = 1 coulomb/volt	$= 10^{-9}$ abfarad	$= 9 \times 10^{11}$ statfarad
mag. flux density **B**	1 weber/m^2 = 1 nt/amp m	$= 10^4$ gauss	
	= 1 tesla	$= 10^9$ gamma	
magnetic flux	1 weber = 1 joule/amp	$= 10^8$ maxwell	
magnetizing force **H**	1 amp turn/m	$= 4\pi 10^3$ oersted	
inductance	1 henry = 1 joule/amp^2	$= 10^9$ abhenry	

* These units depend upon the speed of light in a vacuum, here taken as 3×10^8 m/sec; it is actually closer to 2.99776×10^8 m/sec.

APPENDIX H
NAMES FOR COMMERCIAL BOREHOLE LOGGING SERVICES
(Principal glossary entry of each type is boldface)

	WELEX	TIDELANDS	BIRDWELL	SCHLUMBERGER	DRESSER-ATLAS
Electrical	1. Electric Log EL			**Electrical Log** ES	Electrolog
Electrical	2. **Induction** Electric **Log** IEL			Induction Electrical Log IES	Induction Electrolog
Electrical	3. Dual Induction Guard Log DIL			Dual Induction Laterolog DIL	Dual Induction Focused Log
Electrical	4. Guard Log GL			Laterolog — 3, LL3 Laterolog — 7, LL7	**Laterolog**
Electrical	5. Contact Log CL			**Microlog** ML	Minilog
Electrical	6. FoRxo Log FRL			**Microlaterolog** MLL Proximity Log PL	Micro-Laterolog Proximity Log
Electrical	7. Resistivity Dip Log			Dipmeter	Diplog
Acoustic	8. Acoustic Velocity Log AVL			**Sonic Log**	Acoustilog
Acoustic	Compensated Acoustic Velocity Log BHC			BHC Sonic Log	BHC Acoustilog
Acoustic	9. **Fracture** Finder **Log** FF			Amplitude Log	Fraclog
Acoustic	10. Micro-Seismogram Log MSG		**Three D Log**	Variable Density Log VDL	Variable Amplitude Density Log
Acoustic	Acoustic Cement Bond Log ACBL			**Cement Bond Log**	
Radioactive	11. **Density Log** DL			Formation Density Log FD	Densilog
Radioactive	Compensated Density Log CDL			Compensated Formation Density Log FDC	Compensated Densilog
Radioactive	12. Gamma Ray-**Neutron Log** GRN			Gamma Ray-Neutron Log GRN	Gamma Ray-Neutron Log
Radioactive	13. **Side Wall Neutron Log** SWN			SNP Neutron Log SNP	Epithermal Sidewall Neutron Log
	14.			Thermal Decay Time Log TDT	**Neutron Lifetime Log** NLL
Gravity	15.	Borehole Gravity Log			
	16. **Formation Tester** FT			Formation Tester FT	Formation Tester
	17. **Sidewall Core**			Sidewall Core	Corgun
	18. **Drill Pipe** Electric **Log** DPL				

APPENDIX I

ABBREVIATIONS FOR GEOPHYSICAL CONTRACTORS

CDP Computer Data Processors
CGG Compagnie Générale de Geophysíque
GAI Geophysical Associates (EG&G)
GMX Gravity Meter Exploration (EG&G)
GSI Geophysical Service Inc.
GUS Globe Universal Services, Inc.
SEI Seismic Engineering
SPC Satellite Positioning Corp.
SSC Seismograph Service Corp.
SSL Seismograph Service Limited
TI Texas Instruments Inc.
ONI Offshore Navigation, Inc.

ABBREVIATIONS FOR ORGANIZATIONS

AAAS American Association for the Advancement of Science
AAPG American Association of Petroleum Geologists, publishers of the *Bulletin of the AAPG*
AEG Association of Engineering Geologists
AGI American Geological Institute
AGU American Geophysical Union
AIME American Institute of Mining, Metallurgical, and Petroleum Engineers
AIP American Institute of Physics
AIPG American Institute of Professional Geologists
AMS American Mathematical Society
API American Petroleum Institute
CSEG Canadian Society of Exploration Geophysicists
EAEG European Association of Exploration Geophysicists
GSA Geological Society of America
IEEE Institute of Electrical and Electronics Engineers, Inc.
KEGS Canadian Exploration Geophysical Society
SEG (1) Society of Exploration Geophysicists, publishers of the journal *Geophysics.* (2) Society of Economic Geologists
SEPM Society of Economic Paleontologists and Mineralogists
SIPES Society of Independent Professional Earth Scientists
SPE Society of Petroleum Engineers of AIME

APPENDIX J
CONVERSION OF UNITS

Each of the fractions below has the value 1 (i.e., numerator and denominator are equal).

CIRCULAR MEASURE

$\frac{0.01745 \text{ radians}}{\text{degree}}$ $\frac{57.30 \text{ degrees}}{\text{radian}}$ $\frac{9.55 \text{ rev/min}}{\text{rad/sec}}$

LINEAR MEASURE

$\frac{0.3048 \text{ m}}{\text{ft}}$ $\frac{3.281 \text{ ft}}{\text{m}}$ $\frac{1.609 \text{ km}}{\text{statute mile}}$ $\frac{0.6214 \text{ statute mile}}{\text{km}}$ $\frac{10^{10} \text{ Angström}}{\text{m}}$

$\frac{9.46 \text{ m}}{10^{-15} \text{ light year}}$ $\frac{10^{6} \text{ micron}}{\text{m}}$ $\frac{66 \text{ ft}}{\text{chain}}$ $\frac{6 \text{ ft}}{\text{fathom}}$ $\frac{1.1516 \text{ statute mile}}{\text{nautical mile}}$

$\frac{1.8532 \text{ km}}{\text{nautical mile}}$ $\frac{60 \text{ nautical miles}}{\text{degree at equator}}$

VELOCITY

$\frac{1.689 \text{ ft/sec}}{\text{knot}}$ $\frac{1.15157 \text{ mi/hour}}{\text{knot}}$ $\frac{0.5148 \text{ m/sec}}{\text{knot}}$ $\frac{1.8532 \text{ km/hr}}{\text{knot}}$

AREA

$\frac{10^{28} \text{ barn}}{\text{m}^2}$ $\frac{640 \text{ acres}}{\text{mile}^2}$ $\frac{2.471 \text{ acres}}{\text{hectare}}$ $\frac{2.590 \text{ km}^2}{\text{mile}^2}$ $\frac{100 \text{ hectares}}{\text{km}^2}$

VOLUME

$\frac{3.785 \text{ liters}}{\text{U.S. gallon}}$ $\frac{4.546 \text{ liters}}{\text{British gallon}}$ $\frac{7.4805 \text{ U.S. gal}}{\text{ft}^3}$ $\frac{7758 \text{ bbl}}{\text{acre ft}}$

MASS

$\frac{2.205 \text{ lb}}{\text{kg}}$ $\frac{0.4536 \text{ kg}}{\text{lb}}$ $\frac{1.120 \text{ short ton}}{\text{long ton}}$ $\frac{1.102 \text{ short ton}}{\text{metric ton}}$ $\frac{0.9842 \text{ long ton}}{\text{metric ton}}$

PRESSURE

$\frac{1.0133 \text{ nt/m}^2}{10^{-5} \text{ atmosphere}}$ $\frac{1 \text{ bar}}{10^{5} \text{ newtons/m}^2}$ $\frac{29.92 \text{ in of Hg}}{\text{atmosphere}}$ $\frac{14.22 \text{ lb/in}^2}{10^{4} \text{ kg/m}^2}$

$\frac{0.06895 \text{ bar}}{\text{lb/in}^2}$ $\frac{703.07 \text{ kg/m}^2}{\text{lb/in}^2}$

WORK

$\frac{1055 \text{ joules}}{\text{BTU}}$ $\frac{4186 \text{ joules}}{\text{kilocalorie}}$ $\frac{3600 \text{ joules}}{\text{watt hr}}$ $\frac{1.6020 \text{ joule}}{10^{19} \text{ electron volt}}$ $\frac{0.2930 \text{ watt hr}}{\text{BTU}}$

POWER

$\frac{745.7 \text{ watts}}{\text{hp}}$ $\frac{0.001341 \text{ hp}}{\text{watt}}$

APPENDIX K
NUMERICAL CONSTANTS

π		$= 3.1415927$
e		$= 2.71828$
vel. light	c	$= 2.99776 \times 10^{8}$ m/sec
gravitational constant		$= 6.670 \times 10^{-11}$ newton m²/kg²
Planck constant	h	$= 6.626 \times 10^{-34}$ joule sec
Boltzmann constant	k	$= 1.3805 \times 10^{-23}$ joule/deg
Avogadro's number		$= 6.0226 \times 10^{23}$/mole
gas constant	R	$= 8.314$ joules/mole deg
volume of gas at STP		$= 22.414$ liters/mole
absolute zero	0°K	$= -273.15$°C
Faraday		$= 96494$ coulombs/gm equivalent
electronic charge		$= 1.6021 \times 10^{-19}$ coulombs
rest mass of electron		$= 9.109 \times 10^{-31}$ kg
proton mass		$= 1.672 \times 10^{-27}$ kg
standard gravity		$= 9.81274$ m/sec^2 $= 32.1937$ ft/sec^2
equatorial radius		$= 6{,}378{,}388$ m $= 3963.34$ mi
polar radius		$= 6{,}356{,}912$ m $= 3949.99$ mi
mass of the earth		$= 5.983 \times 10^{24}$ kg
mean density of the earth		$= 5.522$ g/cm^3
solar day		$= 86400$ sec
sidereal day		$= 86164$ sec

APPENDIX L
GEOLOGIC TIME SCALE

	Era	Period (system)	Epoch (series)	Age (stage)	Beginning of epoch Millions of years
Phanerozoic	Cenozoic	Quaternary	Recent		0.005
			Pleistocene		2.5
		Tertiary	Pliocene	Upper	
				Lower	7
			Miocene	Upper	
				Middle	
				Lower	26
			Oligocene	Upper	
				Middle	
				Lower	38
			Eocene	Jackson	
				Claiborne	
				Wilcox	54
			Paleocene	Midway	65
	Mesozoic	Cretaceous	Late (Upper)	Maestrichtian	
				Campanian	
				Santonian	
				Coniacian	
				Turonian	
				Cenomanian	
			Early (Lower)	Albian	
				Aptian	
				Barremian	
				Hauterivian	
				Valanginian	
				Berriasian	136
		Jurassic	Late (Upper)	Purbeckian	
				Portlandian	
				Kimeridgian	
				Oxfordian	
			Middle	Caliovian	
				Bathonian	
				Bajocian	
			Early (Lower)	Toarcian	
				Pliensbachian	
				Sinemurian	
				Hettangian	190

Appendix L — Geologic Time Scale (continued)

	Triassic	Late (Upper)	Rhaetian	
			Norian	
			Carnian	
		Middle	Ladinian	
			Anisian	
		Late (Lower)	Scythian	225
Paleozoic	Permian	Ochoa		
		Guadalupe		
		Leonard		
		Wolfcamp		280
	Carboniferous	Pennsylvanian*	Virgil	
			Missouri	
			Des Moines	
			Atoka	
			Morrow	325
		Mississippian*	Chester	
			Meramec	
			Osage	
			Kinderhook	345
	Devonian	Bradfordian	Famennian	
		Chautauquan		
		Senecan	Frasnian	
		Erian	Givetian	
			Eifelian	
		Ulsterian	Coblenzian	
			Gedinnian	395
	Silurian	Cayugan		
		Niagaran		
		Albion		430
	Ordovician	Cincinnatian	Ashgillian*	
			Caradocian*	
		Champlainian	Llandellian*	
			Skiddavian*	
		Canadian	Tremadocian*	500
	Cambrian	Late (Upper)		
		Middle		
		Early (lower)		570
Precambrian	Proterozoic			2000
	Archeozoic			4600

*Pennsylvanian and Mississippian are periods rather than epochs. Their subdivisions shown are epochs.
The Ordovician subdivisions are the British names of epochs.
The terms in parentheses refer to the time-rock scale (as compared with the time scale).